THE COMBAT SOLDIER

The Changing Character of War Programme is an interdisciplinary research group located at the University of Oxford.

The Combat Soldier

*Infantry Tactics and Cohesion in the
Twentieth and Twenty-First Centuries*

ANTHONY KING

UNIVERSITY PRESS

Great Clarendon Street, Oxford, OX2 6DP,
United Kingdom

Oxford University Press is a department of the University of Oxford.
It furthers the University's objective of excellence in research, scholarship,
and education by publishing worldwide. Oxford is a registered trade mark of
Oxford University Press in the UK and in certain other countries

© Anthony King 2013

The moral rights of the author have been asserted

First published 2013
First published in paperback 2019

Impression: 2

All rights reserved. No part of this publication may be reproduced, stored in
a retrieval system, or transmitted, in any form or by any means, without the
prior permission in writing of Oxford University Press, or as expressly permitted
by law, by licence or under terms agreed with the appropriate reprographics
rights organization. Enquiries concerning reproduction outside the scope of the
above should be sent to the Rights Department, Oxford University Press, at the
address above

You must not circulate this work in any other form
and you must impose this same condition on any acquirer

Published in the United States of America by Oxford University Press
198 Madison Avenue, New York, NY 10016, United States of America

British Library Cataloguing in Publication Data
Data available

Library of Congress Cataloging in Publication Data
Data available

ISBN 978-0-19-965884-8 (Hbk.)
ISBN 978-0-19-884377-1 (Pbk.)

Printed and bound by
CPI Group (UK) Ltd, Croydon, CR0 4YY

Links to third party websites are provided by Oxford in good faith and
for information only. Oxford disclaims any responsibility for the materials
contained in any third party website referenced in this work.

Contents

List of Figures — vi
Preface — viii

1. The Elementary Forms of the Military Life — 1
2. Cohesion — 24
3. The Marshall Effect — 40
4. Combat Motivation — 62
5. Mass Tactics — 98
6. Modern Tactics — 129
7. The Persistence of Mass — 164
8. Battle Drills — 208
9. Training — 266
10. Professionalism — 338
11. The Female Soldier — 376
12. The Professional Society — 419

Notes — 446
Bibliography — 503
Index — 521

List of Figures

3.1.	Makin Islands	47
6.1a.	Trench to trench attack. Platoon in the first wave	141
6.1b.	Trench to trench attack. Platoon in the second wave	142
6.2.	Attack in open warfare	143
6.3a.	Plate I (a platoon advancing to attack)	144
6.3b.	Plate II (a platoon attacking a strong point)	145
8.1.	Phase line battle drill	229
8.2a.	Demonstrating the 90 degree pivot	245
8.2b.	He pivots with the feet	246
8.2c.	Assuming the combat position	247
8.3.	Canadian soldiers practise the combat glide	248
8.4a.	Stacking, Royal Marines CQB course	252
8.4b.	Marines stack in the compound	253
8.5a.	The five-step entry sequence	255
8.5b.	No. 3 enters and clears in depth	255
8.5c.	No. 4 enters and team establishes dominant position	256
8.5d.	Two-man dominant position	256
8.6.	A room clearance demonstration by directing staff	257
8.7.	Body armour to body armour drill	259
9.1.	Infantry officer course ROC drill	285
9.2.	ROC drill	286
9.3.	British army ROC drill	287
9.4.	French army ROC drill	292
9.5.	The infantry immersion trainer, Camp Pendleton	306
9.6.	Range 220, 29 Palms, MCAGCC	308
9.7a.	The compound, Commando Training Centre	309
9.7b.	The compound, Commando Training Centre	310
9.7c.	The compound, Commando Training Centre	311
9.8.	Stanford Training Area	312
9.9.	Shin Kallay, Stanford Training Area	312

List of Figures

9.10a.	CENZUB, Jeoffrecourt. The modern town	314
9.10b.	CENZUB, Jeoffrecourt. The old town	315
9.10c.	CENZUB, Jeoffrecourt. The suburbs	316
9.10d.	CENZUB, Jeoffrecourt. Attack on new town	317
9.11a.	Groningen urban combat facility	318
9.11b.	C-Can Village	318
9.12a.	Royal Marines practise pivoting	322
9.12b.	Pivoting	322
9.13a.	Canadian soldiers practise the pivot	323
9.13b.	Pivot	323
9.14.	Jäger Regiment training at Bonnland	335

Preface

There are few obvious advantages to being a sociologist in Britain today. The status of the discipline in the academy and wider society has never been as high as its intellectual influence has perhaps deserved and, after a high point in the 1950s and 1960s, its political influence in the United Kingdom has now also declined, notwithstanding Anthony Giddens's relationship with New Labour. With the ending of the post-war settlement and especially after the end of the Cold War, sociologists in the UK seem to have struggled to fire the imagination of students in a way their predecessors did up to the 1970s. Yet, for all these ills, the sociologist still enjoys one often overlooked but extraordinary privilege. While they might complain about their status or the condition of the contemporary university, sociologists are accorded the rare benefit of being allowed and, indeed, required by their profession to engage with individuals and groups, with whom, in reality, they have no proper business to interact and whom, as mere citizens or even as scholars in other disciplines, they would never have the opportunity of meeting. These acquaintances may be initiated for purely intellectual reasons in the first instance but the experience of participating in other life-worlds and alternative cultures can often become among the most enriching and edifying of any sociologist's life, even if the insights gained from the field find no place in formal academic journals or monographs. In my previous work, I have enjoyed the hospitality of First Peoples in British Columbia and football fans in Manchester and been lucky enough to participate in events of rare intensity with them. During the research for this book, I have once again been privileged to have been invited into a world to which I was a stranger; to be surprised, shocked, and delighted by innumerable unexpected and fascinating experiences, from the deserts of California and Afghanistan to the woods of Hammelburg.

Yet, in some ways, this research has been more poignant than any of my previous engagements. Samuel Johnson famously claimed 'that every man thinks meanly of himself for not having been a soldier'. Perhaps. Yet, during this research, it was difficult to envy the soldiers whose duty it was to patrol daily out from their bases in Afghanistan. I felt no desire to experience the deprivations and dangers endured by combat soldiers and, rather than inadequacy, I felt relieved that, unlike previous generations from the First World War to Vietnam, I had not been conscripted into forced military service by the state. Accordingly, whatever an individual might think about the strategic wisdom and morality of these campaigns in Iraq or Afghanistan or even the very legitimacy of the armed forces themselves, it would be difficult not to feel humble in the presence of the young men and women who have served on the front line in these theatres, especially when it is remembered how young most of them have been. Although senior non-commissioned officers and commissioned field officers approach maturity, the vast majority of soldiers currently in Afghanistan were born in the late 1980s, some in the early 1990s; they are half my age—or less. Yet, despite their youth, they have experienced a level of hardship, trauma, and, above all, personal loss which is unimaginable to me and probably to most of us, spared the ordeal of combat. It is hard not to be moved by the depth of their experiences. I am grateful for the time which these soldiers—male and female—gave me and their interest in the project. Perhaps their stories might reinvigorate the interest of their civilian peers in sociology as a discipline, convincing them that it has an enduring, even unique, relevance to the world-historical processes which are currently under way.

There are too many individual servicemen and women to thank personally. Indeed, in many cases, individual soldiers and Marines whom I watched or with whom I spoke were unknown to me, even though they may have provided me with important insights. In other cases, I cannot thank certain individuals publicly here since it would undermine the anonymity which I have sought to protect in the text. I have tried to thank them individually through personal communications; I hope they know who they are and that their absence from this list does not in any way reflect ingratitude on my behalf. I would like to thank the following individuals and organizations for their assistance without which this project would not have been possible. While my own perspective will not accord at all points with the views of these contributors, I hope that the analysis forwarded here is recognizable to them and that, perhaps, the work is useful in illuminating their own experiences. I am particularly grateful to: Neil Brown and A Company 5 Scots for their extraordinary Hibernian hospitality especially in Paris and to Eugene Berger who assisted with translations in France. In Kandahar, I am grateful to Nick Carter and Dickie Davies, the Commander and Chief of Staff Regional Command South 2009–10, respectively, for allowing me to take up a position in their headquarters, and to Ewen McLay, J. D. Stevens, Cassie Dunlap, Randy Brumit, Lee McCarthy, Guy Harrison, Keith Adams, and Debi Lomax, who made that headquarters a good deal less depressing than in reality I, as a civilian, found it. I have only fond memories of the Prism Cell's shack and its recurrently entertaining (but eventually removed) trip hazard; I will never look at an exercise ball in quite the same way again. I am especially indebted to Ewen McLay's continued support and advice.

The Royal Marines have always been very supportive of my research and it is not an exaggeration to say that without their help both in Britain and the United States, much of the research for this project would have never happened. I am particularly indebted to Ged Salzano and the Commando Training Centre Royal Marines and, above all, the Platoon

Weapons Branch there. Al Livingstone kindly gave me permission to watch training which was more than facilitated by Russ Coles, Richard 'Chippy' Thornton, Nick Olive, Mick O'Donnell, and Al Weldon; these Marines formed the core of a highly experienced training team; the novelty of what they do is evident in the text. I learnt much from them and thank each of them personally. The staff at the Royal Marines' Junior Command Course in 2009 were extremely helpful, especially Phil Robinson for facilitating my visits. I am grateful to Royal Marines Liaison Officers in the USA, without whose efforts visits to the United States Marines Corps' Infantry Officer Course would not have happened: Martin Price and Matthew Jones. Individually, Tom Dingwall, Ross Preston, Richard Cantrill, Andy Watkins, and Kate Nesbitt gave me much time. Robert Thompson, an old friend, and the OPTAG Team at Camp Bastion Helmand and in the UK provided essential support and assistance; I cannot name each member of staff individually but would like to acknowledge their help here. I am indebted to Edward Short and Sean Leach.

In Germany, the Infantry School at Hammelburg and its staff were generous with their time and I am particularly grateful to the assistance of Alexander-Nils Simon. In France, Colonel Hubert Legrand and Lieutenant Colonel Jan de Kermenguy at CENZUB were very kind in allowing me to visit what is clearly the premier urban training site in Europe and demonstrates French vision and commitment. That trip was enabled by Paul Newton, in his capacity as Commander Force Development and Training, Sebastian and Emma Miller, and Paul Corden. To observe the US Marines Corps is a privilege accorded to few foreign civilians and I am thankful to Colonel Julian Alford, Major Carlos Barela, Captains Cummings, Antolini, and Perkins of the United States Marines Corps' Basic School and Infantry Officer Course for allowing me to visit and for looking after me at the extraordinary Range 220 in 29 Palms. Involving memorable dawn walks and evening runs in Joshua Tree National Park and concluding with a couple of surfs at San Clemente, it was one of the most interesting and enjoyable weeks of research I have ever conducted. In Canada, I am grateful to the Gregg Centre, University of New Brunswick, and the Combat Training Centre at Gagetown and especially to Lieutenant Colonel Gallinger, Major Andrew Hartson, Captain Giselle Holland, Marc Milner, Valerie Gallant, and, above all, Captains Greg Grant and Chris Anderson (for whose patient driving I am very indebted). I am also grateful to Irena Goldenburg for her generous provision of materials for Chapter 11.

As Randall Collins described in his work on intellectual production, scholarship is only an apparently isolated activity. In fact, any programme of research necessarily involves a conversation with a multiplicity of scholars both living and dead. Scholarship is a communal activity too. Much of the material which I have studied for this project was unusual terrain for a sociologist but I have found my conversations with even long-dead scholars enlightening. This book traverses a number of different fields including military history, international relations, security studies, military anthropology, and sociology. Each of these fields is vibrant and I have found my interactions with colleagues in them enriching, entertaining, and instructive. I have benefited immeasurably from my discussions and indeed friendships with a variety of fellow scholars, and this work is dependent upon their interventions and support. This work is intended as a continuation of many of those conversations and could not have been possible without the help and advice of numerous individuals. Specifically, the research was, in fact, initiated in a debate in 2007 in *Armed Forces & Society* with Guy Siebold on the question of cohesion in the military. In my response to his intervention, I first articulated the central thesis of this book in a short footnote,[1] when I expressed an only half-formed thought which began to occur to me as I began to research more deeply into the question of cohesion. This entire work might be seen as an elaboration of that footnote and, therefore, as an extended response to Guy Siebold himself. I am indebted to him for stimulating these ideas. There are many, many other scholars without whom this work could not have been completed. I am particularly grateful to Hew Strachan, Antulio Enchevarria, Ingo Trauschweizer, Richard Carrier, Martin Thomas, Garth Pratten, Victoria Basham, Rachel Woodward, Andrew Godefroy, Paul Higate, Anthony Forster, Paul Cornish, Huw Davies, Patrick Porter, Tim Bird, John Hockey, Philip Langer, David Segal, Eyal Ben-Ari, Karl Yden, Alexander Watson, Jonathon Fennell, Jonathon Boff, Pat Shields and *Armed Forces & Society*, the Inter-University Seminar on Armed Forces and Society, Rene Moelker, Tibor Tresch, Manon Andres, and the European Research Group on the Military and Society; all of whom provided essential support at certain points in the project. Theo Farrell and Rob Foley provided perceptive guidance on the book proposal submitted to Oxford and, thereby, improved the book. I am also very grateful to Mattias Varul, a close and trusted colleague here at Exeter, who helped me with a number of awkward German translations as well as many other less practical but no less important forms of assistance. Since the start of my research on the armed forces, Christopher Dandeker has consistently provided wise and much appreciated counsel. I have benefited from the help of Jeffrey Alexander and Philip Smith (and their Centre for Cultural Sociology at Yale University where I presented an early version of this work); this project cannot be defined as part of their 'strong programme' but I do see it as compatible with their advocacy of cultural sociology. I was delighted to be able to discuss the issue of combat with the Vietnam veteran and now celebrated author Karl Marlantes.

His interest in the project and comments on my thesis on the basis of his own genuine experiences were deeply appreciated.

There are five scholars for whom a special mention has to be made. I have worked with Bruce Coleman for over a decade at Exeter and have enjoyed our many conversations about cricket and other things. He read Chapters 3 to 7 and provided detailed and apposite comments on them; I thank him personally for his efforts and am confident the chapters are better for his contribution, though I accept that there are wider political questions which he raised that I have failed to answer here. Approximately a decade ago, Randall Collins visited Exeter to give a talk on his recently published work *The Sociology of Philosophies*. It was a pleasure to meet him then and to retain his generous support since that time. I visited the University of Pennsylvania in November 2009 where I presented the initial thesis of this project and when Randall was able to give me a guided 'Mafia' tour of the Italian area of the city, pointing out venues in which various hits had been made; at one point we seemed to be improbably confused for a Mafia don and his lieutenant by a local citizen bemused by two obvious outsiders wandering about his neighbourhood. As with the Yale strong programme, *The Combat Soldier* is not a straightforward example of Randall Collins's interaction ritual chain theory. Yet, with a Durkheimian influence evident throughout the text, the connections with Randall Collins's work and my indebtedness to both Randall personally and his scholarship are clear. Randall kindly read and commented on Chapters 8–10 and provided me with detailed and highly pertinent observations. My responses to his questions are likely to have been inadequate but his observations forced me to improve the text and I hope will provide a more advanced starting point for future discussions. I also record my great appreciation to Jeremy Black, another close colleague, near neighbour, and military historian, from whom I have always learnt a great deal. He read the entire manuscript and was not only hugely supportive of the work but crucially gave me honest critical advice on it. Tim Edmunds, a long-time friend and colleague, who had previously assisted me with *The Transformation of Europe's Armed Forces*, read and commented on the entire manuscript. I remain indebted not only to his careful critical reading and the identification of tensions and deficiencies in the argument but also to his warm enthusiasm for the project. Authors are often the worst judges of their work and, with a project which came to take such a specialized interest in the minutiae of infantry tactics, he was able to assure me that I had not become an obsessive pedant narrowly focused on a topic of interest to no one but myself but that the issue of combat was of import to a much wider audience. That constituted some of the most important support and guidance I received. The same is true of Andrew Dorman, another long-standing friend and colleague from whose work I have learnt a great deal. His advice and support throughout the project has been very important to me and must be fully acknowledged.

The project could not have been completed without institutional support. The research was supported by a Nuffield Small Grant ('The Profession of Violence: infantry tactics in the twenty-first century', SG 04403) and by an ESRC grant ('Combat, Cohesion and Gender' ES/J006645/1); I am deeply grateful to both institutions and I hope that my work shows that numerous small grants to a diversity of projects may be a productive way of funding research. At Exeter, I am grateful to Hannah Pike, Rosamund Davies, Paul Woolnough, and Debra Myhill whose advice and assistance on my applications was invaluable and without whose help I certainly would not have received the ESRC award. The very final stages of the book were completed at All Souls College Oxford, where I currently hold a visiting fellowship. The support of the college and Hew Strachan and his Changing Character of War programme, of whose series this book is a part, has been very important to me.

I am grateful to the support of a number of libraries. First and foremost, I am very grateful to Conrad Crane's Army. Not only is the archive held in AHEC impressive but, somewhat disconcertingly for a British academic, the staff at this library seem to regard it as their duty to assist visiting scholars and to make every endeavour to help them; even more surprisingly, they seemed even to relish their role. It is a rare thing today: a library for readers. I am grateful to one of the staff (whose name I never knew) for kindly driving me to a local mall to purchase a digital camera; it did not work—mainly, as I now recognize, due to my incompetence—but the gesture was not forgotten. I am also indebted to the library at the Joint Services Command and Staff College and its always accessible staff, especially Rhys Thomas, Alison Pratt, and Olga Wronecki who were all exceptionally helpful, to the point of sending me scanned material by email. Research was also conducted at the British Library and the National Archive whose holdings proved essential. The Bundesarchiv-Militär-archiv in Freiburg im Breisgau was also very helpful and provided some useful supporting material. I am very grateful to the American Military History Centre, both for its public provision of the official US Army history series, but also for its permission to reproduce the map of the Makin Islands (a scan of which they sent to me) without cost. The MOD's Defence Intellectual Property Rights at Abbey Wood gave me permission to reproduce a series of images from British army First World War doctrine for Chapter 6; I am grateful to them and particularly to Tim Allsop for their assistance. I would also like to add my gratitude to the team at OUP: Dominic Byatt, Lizzie Suffling, Karen Villahermosa, Jackie Pritchard, and Edwin Pritchard.

The private, domestic sphere has become an increasing focus of public concern and fascination in contemporary culture and it is typical for families to be acknowledged in ways which would have been inconceivable and unnecessary in the past. I prefer privacy. Nevertheless, some obvious acknowledgements have to be made to a small group of people who are neither soldiers nor scholars but without whom this project could never have happened. While the actual writing of this book has been less difficult, despite its length, than some of my previous works, the research for it has involved numerous long trips away from Exeter. I am grateful to my family for accepting those absences and especially to my wife, Cathy King, who has endured the pressures it has placed upon her in terms of raising our children Sam, Megan, and Madeleine, even while establishing herself as a successful printmaker.

1

The Elementary Forms of the Military Life

THE CULTURAL REPRESENTATION OF WAR

On 19 June 1879, General William Tecumseh Sherman recalled his experiences of war to the graduating class of the Michigan Military Academy: 'I've been where you are now and I know just how you feel. It's entirely natural that there should beat in the breast of every one of you a hope and desire that some day you can use the skill you have acquired here. Suppress it! You don't know the horrible aspects of war. I've been through two wars and I know. I've seen cities and homes in ashes. I've seen thousands of men lying on the ground, their dead faces looking up at the skies. I tell you, war is Hell!'[1] In declaiming against war at the end of the nineteenth century, Sherman unwittingly created a leitmotif which would become dominant in popular perceptions of industrial warfare in the twentieth century. Unbounded by the limits of human strength and endurance, mechanized war constituted a uniquely traumatic experience for soldiers, placing an almost insupportable physical and emotional load upon their senses. The noise of the guns, the speed and violence of projectiles, the simultaneous and multiple movement of soldiers, tanks, planes, and vehicles overloaded the capacities of human comprehension to inflict catastrophic physical and psychological harm. Not only had the battlefield become more deadly but, where combatants even up to the middle of the nineteenth century had once operated in close proximity to each other, the modern battlefield had also become an anonymous arena dominated by blind chance; a soldier might be killed randomly at any point by a shell or bullet fired by unseen and distant opponents who did not even know where their shots had fallen.[2] Mutilation and death were ubiquitous, instantaneous, and apparently agentless.

The idea that war was indeed a hell began to crystallize as a central reaction in western culture to the First World War and was articulated by European writers, poets, artists, and composers in new and often shocking pieces. In one of the most important and troubling memoirs to emerge from the Great War, Ernst Jünger described the chaos of battle as a senseless 'inferno' propelling defenceless and powerless soldiers to their deaths.[3] In order to communicate this sense of transcendent horror, Ernst Jünger's accounts of the First World War are punctuated by graphic descriptions of the injuries inflicted on the human body by modern weaponry. A scene following a shell-strike in the midst of an advance is illustrative: 'The rolling motion of the dark mass [of injured and dying soldiers] in the bottom of the smoking and glowing cauldron [the shell crater], like a hellish vision, for an instant tore open the extreme abyss of terror.'[4] Jünger recalled a similar scene from a trench following another artillery strike: 'A figure stripped to the waist, with ripped-open back leaned against the parapet. Another, with a triangular flap hanging off the back of his skull, emitted short, high-pitched screams. This was the home of the great god of Pain, and for the first time I look through a devilish chink into the depths of his realm.'[5] Significantly, wounds did not seem to be inflicted by human action but by an unseen satanic agent. Confronted daily by these appalling scenes, Jünger memorably summarized the front line as 'the red-hot chambers of dread';[6] like Sherman, Jünger seems to have found himself in hell, although at the end of the novel, he affirms its existential and political purpose. Erich Maria Remarque's work *All Quiet on the Western Front* is replete with such imagery and it too represents the Great War as hell, without any prospect of redemption.

French memoirs from the First World War traverse similar terrain. Jules Romain was equally dismayed by his experiences of combat where normal standards of decency had been inverted. The entire front line was like an open sewer in which human excreta mixed with the rotting corpses of a nearby graveyard to produce a repellent odour; 'the sewage was still fresh and living, fresh too the graveyard and, in a sense, living too.'[7] Some bodies were never properly interred.

At one point in the trench a hand stuck out, a think sheath of black and sticky flesh—the colour of black flies—that barely hid the bone beneath projecting from a torn coat-sleeve. The hand had been there during the battalion's last stay in the sector, but then it had been an ordinary hand and of quite a different colour: whitish, like a dead and drooping flower.[8]

Other French writers also demonstrated a similarly morbid interest in recording what industrial weapons could do to the human body. Mutilation is a recurrent theme of Barbusse's narrative: 'They [the dead] are pressed against one another, each making a different gesture of death with his arms or his legs. Some exhibit half-mouldy faces, their skin rusted or yellow with black spots. Several have faces that have turned completely black, tarred, their lips huge and swollen; Negro heads blown up like balloons. Between two bodies, belonging to either one or the other, is a severed hand with a mass of filaments emerging from the wrist. Others are shapeless, fouled by larvae with vague pieces of equipment or fragments of bone.'[9]

British writers were no less disgusted by the war and its grotesque effects. Edmund Blunden recorded seeing 'a lance corporal reduced to gobbets of blackening flesh, the earth-wall spotted with blood, with flesh, an eye under the duckboard, the pulpy bone',[10] as a result of shell fire. Wyn Griffiths's account of the First World War climaxes in his gruesome description of the corpses in Mametz Wood in 1916 during the Battle of the Somme: 'Limbs and mutilated trunks and here and there a detached head, forming splashes of red against the green leaves, and, as an advertisement of the horror of our way of life and death, and of our crucifixion of youth, one tree held in its branches a leg, with its torn flesh hanging down over a spray of leaves.'[11] The juxtaposition of natural vegetation and mechanized destruction highlighted the obscenity of industrial warfare for Wyn Griffiths. Even the phlegmatic and disciplinarian medical officer of the Second Battalion, the Royal Welch Fusiliers, Captain John Dunn, was plainly troubled by some of the worst scenes he encountered. In his chronicle of the battalion, he describes (drawing on contributions from Siegfried Sassoon) the line the Fusiliers held in March 1917 near the Somme as a 'dreadful place', a 'monstrous region of death and disaster' which might have 'enriched' Dante, Milton, or Blake; 'the World War had got our insignificant little unit in its mouth; we were there to be munched, maimed and liberated.'[12] Explicitly recalling one of Dante's circles of Hell, Dunn recalled: 'I can remember looking down, as I blundered and gasped my way along, and seeing a mask-like face floating on the surface of the flooded trench. The face had detached itself from its skull.'[13] Elsewhere, 'I can remember two mud-clotted hands protruding from the wet ashen soil like the roots of a tree turned upside down.'[14] J. R. R. Tolkien's description of the landscape of Mordor in *The Lord of the Rings* was drawn directly from his experiences of the Western Front and echoed Dunn's experiences closely: 'As the light grew a little he saw to his surprise that what from a distance had seemed wide and featureless flats were in fact all broken and tumbled. Indeed the whole surface of the plains of Gorgoroth was pocked with great holes, as if, while it was still a waste of soft mud, it had been smitten with a shower of bolts and huge slingstones. The largest of these holes were rimmed with ridges of broken rock, and broad fissures ran out from them in all directions.'[15] Tolkien, like John Dunn, seems to have seen ghostly faces floating under the water in trenches or craters and the image appears in passages as his heroes, Frodo and Sam, cross the Dead Marshes, the site of a previous battle between the forces of good and evil:

Hurrying forward again, Sam tripped, catching his foot in some old root or tussock. He fell and came heavily on his hands, which sank deep into sticky ooze, so that his face was brought close to the surface of the dark mere. There was a faint hiss, a noisome smell went up, the lights flickered and danced and swirled. For a moment the water below him looked like some window, glazed with grimy glass, through which he was peering. Wrenching his hands out of the bog, he sprang back with a cry. 'There are dead things, dead faces in the water', he said with horror.[16]

For all these writers, industrial warfare with its mechanical desecration of the human body was barbarous. The body was ripped apart by a demonic power.

Yet, literary accounts of the First World War do not merely point to the obscenity of mutilation. Commentators interpreted this obliteration of the physical body as a symbol of the spiritual destruction of the human community. Indeed, in one of the most interesting passages of his book, Jünger contemplates this collapse of community at length. As the Germans retreated to the Siegfried Stellung in 1917, they were ordered to destroy every village, placing booby-traps wherever they could. Jünger found the scenes of vandalism disturbing. Indeed, he was not alone in this. Crown Prince Rupprecht opposed the order strongly on the grounds that it was damaging to the morale and discipline of his army.[17]

The scenes were reminiscent of a madhouse, and the effect of them was similar: half funny, half repellent. They were also, we could see right away, bad for the men's morale and honour. Here, for the first time, I witnessed wanton destruction that I was later to see to excess; this is something that is unhealthily bound up with the economic thinking of our age, but it does more harm than good to the destroyer, and dishonours the soldier.[18]

Here, it is possible to see that, while Jünger was himself never a member of the Nazi party, his atavistic desire for a noble brotherhood in arms had resonances with the extreme right. More importantly, he specifically relates the barbarity of mechanized warfare to the economic philosophy of the nineteenth and early twentieth centuries with its concept of rational individualism. Industrial warfare is a product of industrial capitalism and reflects its social and historical origins. For Jünger, the First World War was dominated by a utilitarian principle in which strategic ends were achieved in the most efficient manner, corrupting the bonds between soldiers and defiling their human dignity.

Jünger saw the implication of this strategic logic for the human community clearly. In one of the last actions in which Jünger was involved as part of the final German offensives of the war in March 1918, Jünger describes the intensity of his feelings; 'As we advanced we were in the grip of a berserk rage. The overwhelming desire to kill lent wings to my stride. Rage squeezed bitter tears from my eyes. The immense desire to destroy that overhung the battlefield precipitated a red mist in our brains. We called out sobbing and stammering fragments of sentences to one another, and an impartial observer might have concluded we were all ecstatically happy.'[19] The description is significant. Although part of a regimental advance, Jünger describes the collapse of social relations in the German infantry. Jünger and his comrades were driven on by individual rage. Although they could not communicate with each other, they stampeded forward. Their actions were not so much collective as the mere coincidence of individualized, bestial instincts. He recalls this battle earlier in his work: 'It wasn't until much later that I experienced the direct coming together, the climax of battle in the form of waves of attackers on an open field, which for decisive, murderous moments would break into the chaos and vacuity of the battlefield.'[20] Paradoxically, despite Jünger's numerous discussions of esprit de corps and comradeship, at this decisive moment in 1918 as the First World War reached its climax and the true reality of war was revealed, and all sense of community disappeared: 'Every man ran forward for himself,'[21] so that 'in the midst of these masses that had risen up, one was still alone'.[22] The German troops were no longer fully human; they were propelled by innate drives for survival and revenge. For Jünger, industrial war had dissolved the human community into a brutal herd of individuals. In this, the war was, for all these writers, as Sherman predicted, a hell.

While the Western Front retains a special place in European imagination, much of the literature about the Second World War conveys a similar message about war. Guy Sajer's memoir of his service as an infantryman in the Gross Deutschland Division on the Eastern Front recurrently describes the conflict as horrific in a manner which is directly consistent with the work of both Jünger and Remarque. Following a withdrawal and an intense Russian bombardment during the Battle of Belgorod, Sajer describes how 'we were unable to speak': 'Abandoned by a God in whom many of us believed, we lay prostrate and dazed in our demi-tomb.'[23] Later a fellow soldier is amazed that Sajer has escaped 'that inferno'.[24] Interestingly, while the United States' short participation at the end of the First World War, perhaps, ensured that popular and literary responses were muted in the 1920s, reactions to the war were very pronounced in American writing after the Second World War. Drawing on his experiences in the Pacific, Norman Mailer too emphasizes the incomprehensibility of combat illustrated by the random and callous mutilation of the human body:

The Japanese had been dead for a week and they had swollen to the dimensions of obese men with enormous legs and bellies, and buttocks which split their clothing. They had turned green and purple and the maggots festered in their wounds and covered their feet. Each maggot was about a half inch long and it looked like a slug except that it was the colour of a fish's belly.[25]

One of his characters notably summarizes the status to which the human has been reduced by war: 'Goddam carrion, that's all we are, men, goddam carrion.'[26] James Jones's work on the Pacific campaign echoes Mailer's account. Perhaps the greatest piece of literature to emerge from the Second World War, Joseph Heller's *Catch-22*, also revolves around the theme of the senselessness of war. Although Heller's novel is a far more encompassing satire of western society, he too draws upon the motif of bodily mutilation to illustrate war's special horrors. The central scene of the novel which is gradually revealed in the course of the narrative to explain Yossarian's breakdown involves a grotesque evisceration. Returning from a bombing mission, Yossarian is called to the back of his plane to tend to a young gunner, Snowden, who is wounded. Although disturbed by the desperate pleadings of Snowden that he is 'so cold', Yossarian gains confidence as he tends to a serious but not mortal wound on the airman's leg.[27] However, as he inspects the injured gunner more closely he notices a 'strangely coloured stain seeping through the coveralls just above the armhole of Snowden's flak suit'.[28]

Yossarian ripped open the snaps of Snowden's flak suit and heard himself scream wildly as Snowden's insides slithered down to the floor in a soggy pile and just kept dripping out... He forced himself to look again. Here was God's plenty, all right, he thought bitterly as he stared—liver, lungs, kidneys, ribs, stomach and bits of the stewed tomatoes Snowden had eaten that day for lunch.[29]

The motif of bodily mutilation serves the same purpose in *Catch-22* as it does in the works of Jünger, Wyn Griffiths, Dunn, Romain, and Barbusse. It points to the mindless destruction of the human community by industrialized warfare. Specifically, repeating Mailer's own observation, Snowden's death demonstrated the debasement of human life: 'Man was matter, that was Snowden's secret.'[30]

For Heller, this desecration of the human body was an inevitable result of the irrationality of war which is famously encapsulated in the eponymous concept of Catch-22. The Catch is explained to Yossarian as he recovered from the experience of watching Snowden die by reference to a fellow pilot, Orr, who was regarded as obviously mad primarily because he flew missions without complaint. In order to stop flying, Orr merely had to tell his commanders that he was no longer sane enough to fly missions and they would automatically relieve him.

There was only one catch and that was Catch-22, which specified that a concern for one's own safety in the face of dangers that were real and immediate was the process of a rational mind. Orr was crazy and could be grounded. All he had to do was ask; and as soon as he did, he would no longer be crazy and would have to fly more missions. Orr would be crazy to fly more missions and sane if he didn't, but if he was sane he had to fly them. If he flew them he was crazy and didn't have to; but if he didn't want to he was sane and had to. Yossarian was moved very deeply by the absolute simplicity of this clause of Catch-22 and let out a respectful whistle.[31]

Catch-22 is a condensation of the meaninglessness of war which is manifested multiply in the book with officers who can only be seen, when they are not in, chocolate bars made out of parachute silk, open and shut cases without any charges, and senior officers' orders which had to be obeyed even if it meant disobeying orders. Snowden is disembowelled by this madness.

At this point, there is a surprising connection between the work of Jünger and Heller, despite their radically differing political perspectives. Both writers claim that by destroying the possibility of community, industrial war necessarily generates a hell of individualism. Accordingly, in response to the death of Snowden, Yossarian refuses to leave hospital and to fly again: 'From now on I'm thinking only of me.' Major Danby replied indulgently with a superior smile: 'But, Yossarian, suppose everyone felt that way.' 'Then, I'd certainly be a damned fool to feel any other way, wouldn't I?'[32] For Heller, individualism is not just a product of the irrationality of war but offers the only possible redemption from it. The character Orr (with the pun on his name perhaps representing the alternative to the irrational world of military capitalism for Heller) disappears on one of his missions over Italy. It is assumed that he is dead, drowned in the Mediterranean. However, at the very end of the novel, reports emerge that he has been discovered in Sweden, having rowed the entire distance in his plane's small inflatable life-raft, a feat he had planned for months.[33] The novel ends with Yossarian running out of hospital to imitate Orr's escape attempt.

'I'll keep on my toes every minute'
  'You'll have to jump.'
  'I'll jump.'
  'Jump!' Major Danby cried.
  Yossarian jumped. Nately's whore was hiding just outside the door. The knife came down, missing him by inches, and he took off.[34]

The tone, content, and politics of Heller's book are totally distinct from Jünger's work on the First World War. Yet, the description of Yossarian's dash from the hospital and Jünger's account of the German infantry assault in March 1918 share fundamental similarities. In the face of the horror of industrial war, the human community dissolves into herds of isolated individuals propelled by a desire for salvation or revenge. In the twentieth century, war began to be represented as a duality; individual soldiers, wrenched from their social bonds, were thrown into a hellish abyss of industrialized combat with its red-hot chambers of dread.[35]

TOWARDS A SOCIOLOGY OF COMBAT

The motif of war as hell is powerful and persuasive and it would be certainly difficult not to be disturbed by the imagery which emerged from the First and Second World Wars. Nevertheless, while these artistic representations are impressive and they have been taken up in popular imagination as definitive depictions of these conflicts, it may not be necessary to accept them as entirely accurate, still less comprehensive, accounts in sociological terms. While combat on the modern battlefield was undoubtedly terrifying and twentieth-century representations of industrial warfare communicate this horror effectively, the prose of Jünger and Heller does not need to be taken as definitive accounts of the reality of combat itself. Indeed, in his famous work on the Great War, Paul Fussell plausibly suggests that the literary representation of the trenches should not be primarily read as factually accurate descriptions of the Western Front. On the contrary, these works utilize a narrow repertoire of 'recognition scenes',[36] some of which are sometimes at least partly fictional, not primarily to represent that conflict as it actually was but to create an imagined recollection of it. Indeed, Fussell noted the anomaly that 'such a myth-ridden world could take shape in the midst of a war representing the triumph of modern industrialism, materialism and mechanism'.[37] In many cases these works were explicitly inspired by political or critical motivations; they are not an attempt to represent war dispassionately. Specifically, while wanton and senseless destruction was a necessary feature of industrial warfare and modern combat was certainly profoundly chaotic and terrifying, it may be inaccurate to suggest that warfare was an a-social and indeed antisocial reality, as these representations have often tended to do. The theme of social nihilism might better be seen as a reaction of writers confronting social conditions after their respective wars than as truly fundamental to the experience of war itself. They seem to have projected post-bellum concerns about social disintegration onto their memories of combat itself. Indeed, in the case of both Jünger and Heller, their works are explicitly designed as critiques of the post-war social orders in which they lived; in combat, both saw the corrosiveness of utilitarian rationality.[38] By contrast, one of the most troubling aspects of modern war may be that its horrors might best be understood not as the negation of community but as the manifestation of specific kinds of social solidarities and interactional dynamics. The domain of Jünger's God of Pain may, in fact, be just another sphere of social activity, whose undoubted inhumanity does not deny that it is finally a collective human event; war too for all its barbarity may paradoxically be a social enterprise. Once combat is recognized as a social activity and as a domain of a strange but nevertheless real human community, it is possible to go beyond the standard motifs of twentieth-century literature: the blind fury of war, the destruction of the human body, the elimination of the human community, and the isolation of the individual. It is possible to understand combat itself, its obscenity notwithstanding, as a form of social interaction and as a sphere of collective activity. While terrible, it might be treated as simply another, albeit uniquely extreme, domain of human practice with its own recurring patterns. As a form of social activity, even the 'red hot chambers of dread' might then be analysed systematically in order to identify the social dynamics between soldiers and their enemies. In this way, it might be possible to begin to understand the activities of soldiers on the field of battle and, indeed, to comprehend—but not diminish—the very horror which they face. It may be possible to put a meaning on the apparent senselessness of war.

Indeed, social scientists have recurrently sought to do this. Sociologists, for instance, have had a long-standing interest in combat. Tony Ashworth's work on trench warfare constitutes only one of the most well-known and widely cited examples of this attempt to apply sociological analysis to the front line. He rejected the concept that soldiers were forced merely to submit to war and the commands of their generals. On the contrary, 'the aim [of Ashworth's book] is not to show how the decisions of a few generals affected thousands of soldiers, but, rather, how the decisions of thousands of soldiers affected a few generals'.[39] For Ashworth, the ordinary soldier exercised genuine agency over the prosecution of the war, which agency should be the focus of sustained sociological attention. Informed by this populist approach, Ashworth explored the way in which 'resentment was also translated into subtle, collective action, which thwarted the high command trench war strategy'.[40] Against the common image of war as a senseless vortex into which soldiers were thrown, Ashworth explicitly insists that 'the fact is, soldiers strove with success for control over their environment and thereby radically changed the nature of their war experience'.[41] Specifically, and against the vision of trench warfare as terrifying, Ashworth explores how Allied and German soldiers on the Western Front periodically created live and let live systems to their mutual benefit.

Thus, throughout the war, opposing troops, against the orders of their generals, would contrive to create or sustain quiet sectors. In effect, an elaborate system of communication was instituted by which troops signalled their mutual intentions to each other, while warning of the consequences of any breach of the informal truce. There were numerous, in many cases surprising, methods by which live and let live systems were sustained. Most basically, soldiers would simply experiment by not shooting to see if the enemy would respond in kind: 'He doesn't fire if you lie doggo.'[42] Alternatively, there were more ritualized methods of assessing the intentions of opponents. Troops on either side of the lines would call to each other; 'We are Saxons, you Anglo-Saxons, don't shoot', or by written messages held up above the trenches. Perhaps surprisingly, smells were also used. Basil Liddell Hart noted that the smell of bacon would initiate temporary truces on the lines; both sides recognized that the other was breakfasting and would avoid firing in order not to have their own meal disturbed. Ashworth notes, however, that 'verbal truces' were 'not pervasive nor continuous'.[43] The lines had to be very close for such an agreement to endure. Moreover, visible truces of this kind were 'easily repressed by high command'.[44] Typically, therefore, troops on either side of the lines employed military methods to signal their intention. They would fire their weapons as commanded but they would deliberately fire them high, wide, or short. The troops engaged in 'ritualized aggression' which opponents became sophisticated at distinguishing from effective fire.[45] Ashworth emphasizes that passivity was not unconditional. On the contrary, accords were enforced by the threat of a return to hostilities should opponents unilaterally break the live and let live system. A side which broke an informal truce would be punished by retaliatory firing and this retaliatory fire was itself part of the signalling system in the trenches for sustaining relations. Ashworth's analysis is, at a certain level, politically motivated. It aims to undermine the legitimacy and power of political and military leaders who prosecute war in which normal people have to fight and suffer. It is perhaps because of this populist orientation which downplays the fact that many soldiers fought willingly that Hew Strachan has called it 'overdrawn'.[46]

Nevertheless, while Ashworth predominantly concentrates on soldiers who resist the war, he is fully aware that not all soldiers wanted to avoid fighting. Thus, he repeatedly emphasizes not only the orders of high command in breaking live and let systems, but also the growing British commitment to the offensive from 1916, the development of specialized tactics,[47] and the voluntary actions of elite units. Indeed, the refinement of tactics and the offensive orientation of elite units were typically interrelated. Ashworth correctly identifies the Royal Welch Fusiliers, in which John Dunn (medical officer), Siegfried Sassoon, Robert Graves, and Edmund Blunden all served, as an elite regiment who were both at the forefront of raiding tactics and of sustaining intense hostilities. He notes: 'it was still a point of honour to dominate no-man's land.'[48] Ashworth does not make combat any less brutal but he serves the crucial role of puncturing the presumption that war is an external, quasi-natural force which exceeds the human capacity to understand and control it. On the contrary, Ashworth demonstrates that the war was a social phenomenon prosecuted by groups of belligerents whose intra- and interrelations generate complex geographies of violence. War may indeed be a hell but it is a hell of thoroughly human and collective making.

Ashworth was primarily interested in how the common man resisted the Great War. However, his sociology of combat refusal evidently suggests that it would be entirely possible for sociologists to study active combat participation. Here, the sociologist is interested not in how unwilling soldiers avoid fighting but how troops participate in military violence; the social scientist analyses the dynamics of combat itself. There is a long and, indeed, eminent literature on precisely this topic which can be traced immediately back to the Second World War, including the famous article by Janowitz and Shils, which will be discussed later. Yet sociologists, today, remain intrigued by the problem of combat and military violence. Indeed, in his most recent work on violence, Randall Collins has performed the important service of attempting to bring sociology to the domain of combat itself. Collins dismisses standard myths about violence. Signally, he rejects the notion that violence is easy for humans who have a natural facility for it or that fighting is quickly contagious. On the contrary, Collins maintains that fighting is very difficult for humans, primarily because they are biologically programmed to be social. Moreover, even when violence does occur, it follows the interactional dynamics which are observable in all other spheres of human action. Social relations and especially group membership structure violent interaction and, consequently, although violent acts take a multiplicity of forms, the interactional dynamics of violence recur. Violence is not the random outcome of deviant or overly aggressive individuals, according to Collins. Rather it is a patterned, situational product; it is generated through the process of interactional dynamics. Here Collins takes Ashworth's study forward. Where Ashworth describes humans stopping war through social means, Collins explores the way in which participants prosecute violence collectively. Collins's work extends far beyond warfare to domestic and ritualistic youth violence but combat is a major focus. In particular, Collins notices that combat, in contrast to cultural and especially filmic representations of it, is typically characterized by collective bluster and incompetence. Skilled performances and individual heroism are rare. Drawing on the well-known example of warfare among the Yanomama of Brazil, who engaged in regular low-level confrontations with other tribes, Collins describes how groups of fighters stage displays of aggression, testing the resolve of their own group and that of the opponent, before committing themselves to the fight decisively. Indeed, when a belligerent group commit themselves fully, their opponents often flee before them. Warfare is rarely the struggle of equals; typically it involves relatively long-range assaults followed by the fragmentation of one side which then suffers disproportionate casualties, if they are unable to flee quickly enough.[49]

One of the most powerful analyses in the book discusses the concept of forward panic. When they panic, humans typically freeze or run away. However, in pressurized situations, panic can take a different direction. It can become an uncontrolled and apparently boundless form of aggression. Instead of fleeing, individuals protect themselves, by assaulting the threat which terrifies them.[50] This phenomenon has been widely observed in warfare; Ardant Du Picq discussed 'flight to the front' in his work on combat in the late nineteenth century. Collins builds upon these observations to show how forward panic arises in an identifiable context. Typically in war it occurs when troops have been under intense pressure from their opponents for a long period; 'a forward panic starts with tension and fear in a conflict situation. This is the normal condition of violent conflict, but here the tension is prolonged and built up.'[51] Suddenly, 'there is the shift from relatively passive—waiting, holding back until one is in a position to bring the conflict to a head—to be fully active. When the opportunity finally arrives, the tension/fear come out in an emotional rush.'[52] Collins uses Philip Caputo's description of his experiences in Vietnam as a Marine lieutenant to illustrate the process: 'Then it happened. The platoon exploded. It was a collective emotional detonation of men who had been pushed to the extremity of endurance. I lost control of them and even myself.'[53] Forward panics are overwhelmingly associated with atrocities and massacre. While planned and controlled mass killings do take place in war, many are the accidental, situational result of forward panics. Collins cites the example of a US airborne platoon in Normandy that went on a frenzy of killing which included the massacre not only of German soldiers but of the livestock in their position.[54] 'A variant, common in ancient and medieval warfare, was for a massacre of inhabitants to take place at the end of a siege.'[55] Collins cites the examples of Cromwell at Drogheda and Alexander at Thebes, but the destruction of Badajoz by Wellington's forces during the Peninsula War was also manifestly explicable as a forward panic. More recently, the massacre of thirteen civilians by British paratroopers in Derry on 'Bloody Sunday' on 30 January 1972 or the killing of twenty-four Iraqis in Haditha on 19 November 2005 by United States Marines, following an IED attack in which a Marine corporal was killed, seem to display all the characteristics of forward panics.

For Collins, even the most outrageous and apparently gratuitous atrocity is comprehensible in sociological terms, then, as a form of collective action. Even at their most brutal, humans are still social animals and the possibility of extreme violence arises in distinctive social contexts. Collins's work is important. It dissolves the easy presumption that war in even its most terrible forms is a force majeure beyond human agency. It is not. Even the largest-scale industrial war is a social reality impelled by complex, interrelated social dynamics at every level. War is ultimately a produce of inter- and intra-group dynamics and combat is, therefore, utterly human inhumanity. Significantly, Collins's descriptions of the forward panic facilitate a direct reinterpretation of Jünger's description of his final assault. Although Jünger implied that he and his comrades were driven forward by some bestial individual instinct, the assault becomes, in the light of Collins's work, a forward panic. Jünger and his unit had been under pressure for weeks and, indeed, years. In March 1918, they were at last being given the opportunity to assault, immediately before and during which they took heavy casualties. They seem to have been driven forward by a force which was irreducible and perhaps difficult for each individual to articulate; the moral force of this social group. So united were they by their collective fury that the soldiers barely needed to communicate with each other; grunts and half sentences were enough to sustain solidarity and each soldier was imbued with the collective moral power which the troops had mutually generated among themselves. The British soldiers who received this assault may have experienced it as senseless inhumanity. In fact, they were the unfortunate brunt of the intense and reciprocal dynamics of inter-group violence.[56]

COHESION AND INFANTRY TACTICS

Ashworth and Collins explore the interactions between belligerents in order to demonstrate the way in which war remains a collective enterprise. The aim here is to extend their work. However, instead of focusing on the hostile interactions between combatants, this book focuses on the internal dynamics within combat groups. It examines the question of cohesion among combat soldiers to analyse the different ways in which soldiers have generated and sustained their unity as a social group even in the intense environment of combat. It is widely recognized both within the armed forces themselves and in scholarly analysis that group cohesion is essential among combat soldiers, for without that sense of solidarity and mutual obligation, individual soldiers are likely to save themselves or to run away. It is a striking and extraordinary fact that, despite the evident attractions of desertion, soldiers have often preferred to fight and die together. Social scientists and historians have rightly been fascinated by this commitment of individual soldiers to their comrades and to their group, and their interest is of long standing. The question of cohesion was central to the writings of both Machiavelli and Clausewitz on war. Machiavelli disparaged mercenary armies totally, arguing in the *Art of War* that since professional soldiers profited from war, it was in their interests to perpetuate it whether it was in the interests of the city state which hired them or not. Indeed, he gives a number of examples, including that of Francesco Sforza, where mercenaries actively undermined or overthrew the polity which employed them.[57] In *The Prince*, his rejection was even more forthright and speaks directly to the question of cohesion:

Mercenaries and auxiliaries are useless and dangerous. If a prince bases the defence of his state on mercenaries, he will never achieve stability or security. For mercenaries are disunited, thirsty for power, undisciplined, and disloyal; they are brave among their friends and cowards before the enemy; they have no fear of God, *they do not keep faith with their fellow men*; they avoid defeat just so long as they avoid battle; in peacetime you are despoiled by them, and in wartime by the enemy. The reason for this is that there is no loyalty or inducement to keep them on the field apart from the little they are paid, and this is not enough to make them want to die for you.[58]

Instead of mercenaries, Machiavelli advocated the citizen army, like that of Rome, Sparta, or the Swiss, because he believed that the republican sentiments of its soldiers would motivate them to fight for each other and for their state, as opposed to mercenaries who fought only for their own individual remuneration.[59] Significantly, Machiavelli noted that, in order to rid 'themselves and their soldiers of any cause for fear or need for exertion', mercenaries avoided the role of infantry, preferring to operate as cavalry; 'instead of fighting to death in their scrimmages they took prisoners.'[60] Although Carl von Clausewitz's *On War* is very substantially about generalship, one of the most famous passages in the book deals with the question of cohesion. Indeed, he lyrically suggests that only 'an army that maintains its cohesion under the most murderous fire; that cannot be shaken by imaginary fears and resists well-founded ones with all its might; that, proud of its victories, will not lose the strength to obey orders and its respect and trust for its officers even in defeat' could be said to be 'imbued with true military spirit'.[61] Without the possession of an army willing to endure the rigours of campaigning and the terrors of combat, no general, however brilliant a strategist he might be, could succeed. The question of cohesion has remained a central question in military scholarship from that time and is paramount in the work of scholars such as Janowitz and Shils, Stouffer et al., Roger Little, Charles Moskos, and a host of other commentators across the disciplines. In the following chapter and in the rest of the book, some of the most important examples of this literature on cohesion will be discussed. Indeed, the question of cohesion is, in fact, very evident even in those cultural representations discussed earlier in this chapter. Even against the motif of the destruction of community in warfare, numerous memoirs—Jünger's among them—recorded a level of solidarity and unity among combat soldiers which was rarely experienced again away from the front. This book examines how soldiers generate and sustain this distinctive cohesiveness in combat which has so fascinated and bewildered scholars and observers of war.

Of course, the exploration of cohesion on the battlefield has much wider sociological relevance, which in fact explains the interest which many social scientists have shown in the question of cohesion. The problem of how groups cohere has been a fundamental problem for sociology since the origins of the discipline. The great American sociologist of the twentieth century, Talcott Parsons, regarded the question of human cooperation as the fundamental philosophical question for sociology:[62] 'when social thought became secularized about the seventeenth century its central problem was that of the basis of order in society'.[63] For Parsons, Thomas Hobbes was the first philosopher to recognize 'the problem of order';[64] 'This problem, in the sense in which Hobbes posed it, constitutes the most fundamental empirical difficulty of utilitarian thought.'[65] Hobbes's solution to the problem was famously authoritarian—he raised up the leviathan of the sovereign over the people—and, sometimes, Parsons's problem of order has been similarly interpreted merely as a concern with authority and conformity. Yet, Parsons's concept of the problem of order was in fact deeply philosophical. His problem was not how a society remained stable but how human beings were able to engage in collective action at all. He was interested in the remarkable fact of cooperation itself and, therefore, the very emergence and maintenance of social groups. Without the ability to cooperate with each other and to address collective ends, humans could never form or maintain social groups. The problem of order, as he called it, defined the discipline for him. He famously claimed that what he called individualist theories or 'utilitarianism', of which Hobbes was a prime example, could not account satisfactorily for the manifest fact of collective action. By contrast, sociology, recognizing the importance of 'common values' which were collectively understood by actors, was able to solve the problem of order. For Parsons, the central concern of sociology was, therefore, the question of collective action or group formation and, indeed, he believed that sociology justified itself as an academic discipline insofar as it was able to resolve this problem. Precisely because combat is among the most brutal human experiences and the human group is placed under the most intense pressure to perform extraordinary and terrible acts, the battlefield is a potentially fruitful environment for the sociologist. It may be an arena in which the possibility and problems of collective action are highlighted with particular clarity. Combat is certainly a highly distinctive, even unique, environment and yet its very extremity may recommend it as a focus of sociological investigation. In particular, the processes of group formation, maintenance, and disintegration which are universal features of social life may be displayed on the battlefield in an especially vivid way. The battlefield may represent an accentuation of the general patterns of group formation and social interaction rather than, as twentieth-century writers and artists sometimes suggested, an annihilation of community.

Clearly, the question of combat and group formation is a potentially vast, almost limitless, topic. Every war, every kind of warfare, every kind of soldier (from general to private, logistician to infanteer), and every kind of military unit from the army of hundreds of thousands of troops to the section or squad of ten soldiers might validly be chosen as the focus of investigation. This book does not claim to provide a universal account of cohesion in combat. On the contrary, it deliberately focuses on the smallest autonomous military unit in the modern army: the platoon, a group of approximately thirty to forty soldiers. There are evident sociological advantages to focusing on this small group. In his most mature work *The Elementary Forms of the Religious Life*, Émile Durkheim sought to identify the universal features of religion in order to develop a general theory of group formation and solidarity.[66] He adopted a useful method. Although he was interested in the universal question of religion, he deliberately sought to analyse the simplest forms of worship in order to isolate the fundamental features of beliefs and practices which are held in common by even the most complex and developed religions. 'These are the permanent elements which constitute that which is permanent and human in religion; they form all the objective contents of the idea which is expressed when one speaks of *religion* in general. How is it possible to pick them out?'[67] Durkheim answered his own question: 'Surely it is not by observing the complex religions which appear in the course of history.'[68] As religions develop historically, they have become syncretic, diverse, and multiple with different rituals, rites, groups, individuals, and disputed theologies so that many are a 'confused mass of many cults'; 'in these conditions, it is difficult to see what is common to all'.[69] 'Things are quite different in lower societies. The slighter development of individuality, the small extension of the group, the homogeneity of external circumstances, all contribute to reducing the differences and variations to a minimum.'[70] Durkheim concludes: 'Primitive civilizations offer privileged cases, then, because they are simple cases.'[71] He continues: 'But primitive religions do not merely aid us in disengaging the constituent elements of religion; they also have the great advantage that they facilitate the explanation of it. Since the facts are simpler, the relations between them are more apparent.'[72] Not only are the fundamental elements of all religions particularly clear in primitive worship but, because of its small size and its social and theological homogeneity, it is easier to explain the development and role of these religious elements. There were other advantages to tribal religion which emerge later in Durkheim's work. The religious ceremonies which Durkheim described were extreme social events which involved acts of sometimes violent transgression, unobscured by abstract theology. They vividly demonstrated the potency of social interaction which, he believed, lay at the heart of all religious worship. The small scale and simple format of the tribal ceremonies illuminated fundamental processes not only of religion but also of social interaction and group formation which Durkheim believed to be universal with particular, even unique, clarity.

Indeed, combat has sometimes been seen as a religious experience. At the beginning of *Storm of Steel*, Jünger described the exuberance which he and his fellow soldiers felt at the prospect of war: 'Surely the war was about to supply us with what we wanted; the great, overwhelming, the hallowed experience.'[73] It was to be, for him, a sacred rite of passage into manhood. In his recent memoir of his experiences as a United States Marine in Vietnam, Karl Marlantes has explicitly drawn a parallel between religion and combat. Religion involves an intense and abnormal communal experience directed at the transcendental. Perhaps surprisingly, combat includes all these features which are typically associated with religion. Combat involves an intense and communal encounter with death: 'Many will argue that there is nothing remotely spiritual in combat. Consider this. Mystical or religious experiences have four common components: constant awareness of one's own inevitable death, total focus on the present moment, the valuing of others people's lives above one's own, and being part of a larger religious community such at the Sangha, ummah or church.'[74] It is a religious experience for soldiers, even though, as Marlantes claims, western armies have consistently failed to prepare their soldiers for the spiritual challenges of war. This book consciously seeks to imitate Durkheim's method; it takes the elementary military form—the infantry platoon—in order to explore the specific question of group formation and maintenance in combat and more general theoretical concerns about the possibility of social solidarity per se.

The platoon now consists of some thirty to forty infantry soldiers but the term 'platoon' is of long-standing military usage going back to at least the seventeenth century, where the term was used to refer to a small rank of musketeers. A platoon of musketeers was typically organized to fire their volleys together and, consequently, at that point, the word platoon was also a verb referring to the act of firing by this small group. The platoon remained as an organizational element within the military establishment from that time onwards. However, the platoon became an independent and critical tactical unit only during the First World War when the effects of industrial firepower were realized. Before the First World War, infantry tactics prioritized the company of approximately 100 soldiers who assaulted in lines. Although infantry battalions had already begun to break down company assault lines into platoons in the light of costly experiences on the Western Front, by 1917, all the major combatants on the Western Front had formally reorganized the platoon in the light of the recognition of its new independence on the battlefield. As Captain Laffargue's 1915 work *The Attack in Trench Warfare* demonstrates, the French army were rapid in recognizing the need for platoon tactics. *Instruction sur le combat offensif des petits unités*, published in 1916 and revised in 1918, and the 1916 *Manuel du chef de section d'infanterie* officially established the French platoon as central to combat, and institutionalized new fire and movement tactics. Indeed, the *Manuel* notably described the platoon (*le section*) as 'the elementary cell [*cellule*] of the battalion'.[75] The Germans captured and copied Laffargue's suggestions and subsequently published their own doctrine, *Ausbildungsvorschrift für Fusstruppen im Kriege*, replacing the company line which had featured in pre-war doctrine with platoon and section tactics in January 1917. At almost exactly the same time, in February 1917, the British army issued the *Instructions for Training of Infantry Platoons for Offensive Action* which reorganized the platoon, once comprised of thirty riflemen, into four sections each of nine soldiers. The Canadians, Australians, and Americans all adopted this format. This was a decisive moment in modern western infantry tactics. Crucially, then, in this period between 1915 and 1917, the platoon crystallized as an independent tactical unit on the battlefield; 'the platoon is therefore, the tactical unit on which all infantry tactics are built'.[76] The platoon was divided into the three or four specialist sections or squads consisting of approximately eight to twelve soldiers, armed with different weapons and able to mutually support each other in the assault. The combatants of the First World War typically developed sections of riflemen, bombers, and light machine-gunners supported by a light mortar team. As an independent tactical unit, the platoon with its sections was, then, an invention of modern warfare and especially of the First World War. Despite the advances in military technology, the small-group infantry tactics of the platoon with its sections has remained central to the combat experience. The platoon, structured in a recognizably compatible way, remains central to the organization, training, and operations of western forces currently fighting in Afghanistan approximately 100 years later. The platoons fighting in Afghanistan are still organized into three or four specialist, mutually supporting, sections. The platoon represents the smallest combat group whose evolution from its origins in the First World War to the present can be traced more or less directly. The platoon has a historical coherence, therefore. For the past century, it has constituted the simplest and smallest military unit capable of independent action in combat.

The function of the platoon further recommends it as an elementary form for sociological investigation. Since the First World War, the infantry platoon has been assigned to engage in extreme violence at close range; the platoon has normally engaged the enemy at a maximum range of some 300 metres, but face-to-face engagements at 10 metres or less have been commonplace. Like Durkheim's aboriginal clans, the infantry may have a methodological advantage over other military formations therefore. Unlike the vast manoeuvres of mechanized divisions, the movements and actions of the infantry platoon or section can be easily plotted and observed in time and space. In battle, the members of a platoon are immediately contiguous and in constant interaction with each other. Indeed, they can often see each other or verbally communicate with each other through shouting. Their actions are defined within a very circumscribed sphere of time and space and can be easily observed. In the academic literature, cohesion is typically understood to persist in groups all the members of which engage in regular face-to-face relations so that all individuals know each other personally: 'Past studies on the American Army reveal that, in conflict, the unit of cohesion tends to be the squad.'[77] Citing the work of Siebold and Kelly,[78] who have argued that the platoon of forty to fifty soldiers is the optimal size for measuring cohesion, and Marlowe, that 'only teams, squads, platoons and companies possess cohesion',[79] MacCoun confirms that cohesion refers to small groups: 'group cohesion is inversely related to group size.'[80] This study focuses on the platoon of about thirty to forty soldiers in order to examine a group which is typically taken as the prime site of cohesion. Certainly, there is clear evidence that dense solidarities exist at a higher level than this in the armed forces, in battalions, brigades, and divisions, but in these cases, while compatible, the solidarity takes on a different form because while all the soldiers may share a special relationship to each other, they do not all know each other personally. At this higher level, military forces can demonstrate 'esprit de corps', a sense of organizational unity, rather than face-to-face cohesion. Cohesion in this study refers to the special solidarity which necessarily pertains in a platoon because all of its members are immediately known to each other and their actions are interdependent, mutually supporting, and reciprocal. The platoon can serve, therefore, as an elementary form of combat from which much wider claims about the nature of social solidarity might be made. Precisely because of the simplicity of small-group tactics, the dynamics of group violence and solidarity are demonstrated with particular clarity by the infantry platoon. Yet, the actions of small numbers of soldiers on the battlefield may provide an insight into much wider processes of collective action.

The size of the platoon and the relative simplicity of its practices especially recommends it as a focus for the study of combat performance, but there are further reasons for selecting the platoon as the focus of analysis. The infantry platoon has a further methodological advantage. Since that time, there have been startling developments in military technology. Indeed, in many cases, military technology today is unrecognizable from the equipment and arms of that conflict; armed forces now regularly employ precision-guided munitions, advanced surveillance technology, and digital communications. Even in the case of planes and tanks, which were developed during the First World War, the capabilities of the current generation of airframes and vehicles are of a different order from those of the Great War. Given this technological divide, historical comparison can become problematic because the provision of new technology has so changed the practice of warfare. However, the 'elementary' method employed here seeks to obviate the problem of technology and, thereby, to sustain historical and cross-national comparison. Specifically, by focusing on the infantry platoon, as the elementary military form, this analysis seeks to examine the military unit which has changed least since the First World War not only in structure but also in weaponry. While arms and equipment have self-evidently improved, there has been no revolution in platoon weaponry; in the First World War, the infantry platoon was armed with rifles, bayonets, pistols, portable light machine-guns, grenades, rifle-grenades, and mortars. The platoon in Afghanistan is equipped with a compatible suite of weapons. Current models of each of these weapons are undoubtedly more capable than 1917 issues but they are broadly equivalent and serve the same function. The M203 underslung grenade launcher is the modern equivalent of the rifle-grenade; the Minimi 5.56 machine-gun a more effective version of the Lewis gun. Communications have certainly improved dramatically in the armed forces with the dissemination of personal radios to every soldier. However, at the platoon level, verbal commands remain vital because of the noise of battle and the exhortatory collective effect which shouting has in the confusion of battle. While digital communications have transformed higher-level operations, western soldiers clearing compounds in Afghanistan or buildings in Iraq still shout instructions at each other, therefore, as their predecessors did in the trenches on the Western Front. The result is that while evident developments in tactics are observable, and will be discussed at length later, the tactics employed by western forces in Afghanistan are recognizably similar to those of their forebears on the Western Front.

This book seeks to explore how the infantry platoon maintains its cohesion in combat. However, although the identification of this elementary form of the military life as the focus of study narrows the investigation substantially, a further refinement of the topic of research is still necessary. Although the infantry platoon began to be institutionalized into the infantry at the beginning of the twentieth century, it might be possible to investigate small-group infantry tactics in almost any historical era. This book examines infantry tactics from the First World War to the present era; it examines the infantry platoon over the last century, from 1914 to the present. The First World War is selected as a starting point for this analysis because this conflict is widely regarded as the first truly industrial war. In his work *Military Power* Stephen Biddle identifies the First World War as the decisive modern conflict and other scholars would concur with him. It is certainly true that the wars of the late nineteenth and early twentieth centuries, the American Civil War, the Austro-Prussian War, the Franco-Prussian War, the Boer War, and the Russo-Japanese War, all displayed aspects of industrial warfare. The effect of firepower began to be recognized and heavy artillery, capable of indirect fire, and the machine-gun began to be deployed onto the battlefield in these conflicts; combatants sometimes began to camouflage, conceal, and cover themselves as a result. Nevertheless, while these conflicts displayed elements of industrialism from which the most perceptive commentators could infer the future of warfare,[81] mechanization was underdeveloped. The potential of indirect artillery fire or the machine-gun was not fully realized in any of these conflicts and, as a result, armies still practised close-order infantry tactics advancing *en masse*. Moreover, it was only during the First World War that aviation and, later, tanks began to be used. There were precedents and it would be unwise to presume its utter uniqueness despite the horror it induced in European consciousness, but it seems uncontroversial to claim that the First World War constituted the first genuinely industrial war. As such, it seems to be an appropriate starting point for the study of modern infantry tactics, especially since that conflict saw a radical revision in infantry organization and performance. From the First World War, this study follows Peter Kindsvatter's analysis of the American soldier,[82] to examine the major conflicts of the twentieth century in which western citizen armies were involved: the Second World War, the Korean War, and the Vietnam War. The work concludes with an investigation of recent and current campaigns in Iraq and Afghanistan fought by today's professional forces.

The selection of these conflicts is partly justified by the fact that they were the largest wars of the last 100 years but also because this work focuses on western armies and specifically on the largest and most important western military powers of the last century: Australia, Canada, German, France, Italy, the United Kingdom, and the United States of America. The two world wars, Korea, and Vietnam were indisputably the most significant conflicts in which the selected nations were involved over the last 100 years and must, therefore, be central to the analysis. Clearly, by focusing exclusively on the major western powers, the findings of this study are perforce limited. Indeed, Jeremy Black has highlighted the often unwitting ethnocentricism of military historians who assume the existence of a 'Western Way of Warfare',[83] ignoring developments in other parts of the world or presuming that western military practices are simply replicated elsewhere.[84] He has rightly argued for the need for a genuinely global history of war and warfare which recognizes the great differences which have historically pertained between western ways of warfare and those practised in other parts of the world. Nevertheless, despite Black's salutary observations, there may be an advantage in focusing on the selected western forces and the major conflicts in which they have been involved for this study.[85]

Durkheim was interested not only in demonstrating the centrality of solidarity to human existence. He wanted to show that different patterns of solidarity would generate different kinds of practice. To this end, a central element of this research is historical. It seeks to explore the potentially different ways in which cohesion was generated in the mass, citizen armies of the twentieth century and in today's all-volunteer professional forces. The citizen army refers to those armies, typical in the twentieth century, in which the male population as a whole was obliged to serve or volunteered for service, often to pre-empt inevitable conscription. By the first decade of the twenty-first century, however, only Germany, of the major western military powers, retained conscription and, by the summer of 2011, even it had abolished national service. As western forces have professionalized, they have simultaneously been engaged in ever more intense operations, especially in southern Afghanistan. These new operations represent the largest and longest deployment of military force by western nations since Vietnam and they have demanded tactical transformation from the armed forces. Any book on the development of infantry tactics since the First World War has, therefore, to examine the current conflict in Afghanistan and the performance of the infantry in this war. Since only the six selected western forces have been involved as major combatants from the First World War to Afghanistan, they have been selected as the focus of this study. America and Australia fought in all four of these wars; Britain, Canada, and France in three (the two world wars and Korea), and Italy and Germany in the First and Second World Wars. All the infantries examined here fought on one or other of the fronts in the First World War, in one or more of the subsequent international conflicts of the twentieth century, then, and are now deployed in Afghanistan. The United States Army and Marine Corps have been, of course, heavily committed to Afghanistan since 2001 especially in the east and south of the country. British forces too have been involved from the earliest operations and since 2006 have been involved in an intense war in Helmand where Danish forces are also operating, while the Canadians were deployed in Kandahar until their withdrawal in August 2010. German forces have been operating in an increasingly unstable Kunduz since 2003 while the French, initially deployed in Kabul, have been committed to increasingly dangerous operations in the east. The Italians have taken responsibility for the west of the country which still represents a challenging environment. As a result, there is an almost century-long line of continuity from the Great War to Afghanistan for these armies. Focusing on the major western powers, this study seeks to analyse how the platoon has maintained its cohesion from the First World War to present operations in Afghanistan by exploring the development of western platoon tactics over this century. It is fully accepted that some of the ways in which cohesion has been generated by western platoons from the First World War to Afghanistan, typically under more or less liberal and democratic regimes (with the exception of Fascist Italy and Nazi Germany), may be highly culturally and historically distinctive. Cohesion in non-western armies or under authoritarian regimes like Imperial Japan, Soviet Russia, or communist China might have displayed a quite different character. However, since western forces have enjoyed a near monopoly of military innovation throughout the modern period, are still regarded as exemplars of military practice, and consequently wield extraordinary influence over the armed forces of all other nations, including China, the examination of cohesion in the western platoon may have much wider relevance.[86] Indeed, it is my intention that the analysis, while avowedly based on the West, will illuminate processes which are currently in evidence around the globe.

To this end, the book contrasts the performance of the citizen armies of the major powers in the four biggest wars of the twentieth century with the professional forces of those powers in current or recent operations in Iraq and Afghanistan. Specifically, the book explores the thesis that the ways in which a professional platoon generates cohesion on the battlefield are, despite the common presumption of continuity, quite different from its citizen forebears. The solidarity exhibited by citizen soldiers and professionals is quite different, it may be claimed, giving rise to quite different patterns of practice in combat. This book aims to investigate this divide between professional cohesion and the solidarity of citizen soldiers, especially conscripts. In this way, the book aims to trace the ways in which cohesion typical of the mass citizen army has been superseded by a new kind of solidarity among the all-professional volunteer forces which are now the (nearly universal) norm among western militaries.

This book is a contribution to the sociology of the armed forces, therefore. Yet, it is not intended to be merely a piece of military scholarship. On the contrary, the armed forces are chosen because, like Durkheim's aboriginal clans, they may represent a simple but perspicuous example which may illustrate contemporary social transformations much more widely. The armed forces are distinct from civilian society and yet, due to institutional and cultural overlaps, the military is rarely absolutely separate from civil society; typically they share many legal, organizational, and cultural presumptions and practices. The armed forces have a unique function but they may be seen as broadly isomorphic with the other major institutions in the public sector. Indeed, the armed forces often consciously imitate or are imitated by civilian organizations. The book explores changing patterns of cohesion in the infantry platoon in order to shed light on wider social transformation, then, and specifically the development of new patterns of social solidarity in an increasingly globalized world. Specifically, the implication of this book is that this transformation in the social solidarity of the elementary combat unit may usefully reflect wider shifts in interactional dynamics in western culture today. The way in which professional soldiers on the front line unite themselves may have resonances with the way in which civilian communities and groups sustain social solidarity among their members. It is possible that members of western cultures are not friends or colleagues of one another in the way in which their parents and, especially, grandparents were in the twentieth century. This study may illustrate in one small but important sphere that, in the face of globalization, the very basis of social solidarity in western society has imperceptibly but inexorably changed. Community in the twenty-first century may be a very different form of association—neither superior nor inferior—from that which existed in the twentieth century.

2

Cohesion

THE STANDARD DEFINITION OF COHESION

This book addresses the question of how cohesion has been generated and sustained in the western infantry platoon from the First World War to present-day operations in Afghanistan. Although cohesion is a commonly used term, its usage has been diverse and sometimes contentious and, consequently, for the purposes of this study, some discussion of academic debates about the concept of cohesion is required in order to develop a definition of sufficient precision for the analysis of the platoon. The issue of cohesion certainly pre-dates the famous article on cohesion and disintegration in the Wehrmacht by Morris Janowitz and Edward Shils, first published in 1948, but this piece might be taken as the seminal moment for contemporary debates about cohesion in the armed forces. In that article, Janowitz and Shils laid out a definition of cohesion which has influenced practically all subsequent discussions of the issue. Like many others during the Second World War, Janowitz and Shils were struck by the 'extraordinary tenacity of the German Army' which continued to fight well even when defeat was inevitable.[1] The Wehrmacht constituted an interesting historical case which seemed to defy normal expectations of human behaviour, and Janowitz and Shils sought to identify the special social factors which generated this level of performance. Janowitz and Shils were not unaware of the specifically military skills of Wehrmacht soldiers, a hard core of whom were politically motivated 'enthusiasts for military life',[2] strict discipline,[3] hierarchy, or the personal dedication of soldiers to Adolf Hitler.[4] However, for Janowitz and Shils, the performance of the Wehrmacht was ultimately explicable only by reference to the cohesiveness of its primary groups: 'it is the main hypothesis of this paper, however, that the unity of the German Army was in fact sustained only to a very slight extent by the National Socialist political convictions of its members, and that more important in the motivation of the determined resistance of the German soldier was the steady satisfaction of certain *primary* personality demands afforded by the social organization of the army.'[5] Primary groups, consisting of small groups of soldiers of up to perhaps thirty individuals, were held together by bonds of comradeship produced by 'spatial proximity, the capacity for intimate communications, the provision of paternal protectiveness by NCOs and junior officers, and the gratification of certain personality needs, e.g., manliness, by the military organization and its activities'.[6] Janowitz and Shils identified the interpersonal relations within the primary groups as critical to the fighting effectiveness of the Wehrmacht. As long as primary groups fulfilled the 'major primary needs' of the individual soldier, it possessed a 'leadership with which he could identify himself',[7] and he was able to give 'affection to and received affection from the other members of his squad and platoon',[8] the German soldier would be willing to perform in combat. The German soldier would continue to fight 'as long as he felt himself to be a member of his primary group and therefore bound by the expectations and demands to its other members'.[9] Notwithstanding Janowitz and Shils's discussion of the Wehrmacht as a military organization, the bonds between the soldiers were primarily interpersonal.[10] Janowitz and Shils fully recognized the distinctive solidarity (or cohesiveness) of the primary group but these bonds were explanatorily significant for them because they provided motivation for Wehrmacht soldiers. Cohesion, for Janowitz and Shils, refers to the special kind of combat motivation which arises only in interpersonal relations, then. It is very important to recognize this distinction between solidarity and motivation because it has led to much confusion in subsequent debates. The point is that while the special kind of motivation, of which Janowitz and Shils write, arises only in small groups and is indivisible from them, their concept of cohesion refers to the motivation of soldiers, not to the actual cohesiveness of the primary group in combat. In the first instance, it refers to their desire to fight, not the ability of these groups to act in concert in combat.

The concept of primary group cohesion as a form of motivation, first advocated by Janowitz and Shils, has had a very significant influence on academic debates about cohesion since the end of the Second World War. Many scholars have similarly sought to show how soldiers have been motivated to fight as a result of a collective sense of comradeship. In her work on the Falklands/Malvinas War in 1982, Kinzer Stewart forwards a similar definition of cohesion as motivation. Thus, citing Sam Stouffer et al.'s work, she notes that research in military psychology recurrently reaffirms the interrelations of small-group ties and military performance.[11] She notes:

> Herein lies the crux of military cohesion. Disparate men from varied socioeconomic backgrounds, of different ethnic origins and levels of education are expected to become not just a collective of individuals but a unit in which an individual will sacrifice his life and die in order to preserve the group. Because of well-developed friendship or camaraderie, men will fight individually as part of a unit to defend the group as a unit... Friendship ties men to each other.[12]

For Kinzer Stewart, friendship is a—perhaps, the—fundamental means of motivating the soldier. She, like Janowitz and Shils, assumes motivation to be coterminous with cohesion.

In a well-known study, Darryl Henderson defines cohesion, which he regards as critical to military success, in a similar way. Henderson was concerned that the professional US Army of the 1980s faced a crisis of cohesion and sought to offer remedies to this crisis. Operating with an economic model, in which 'you can pay a person enough to be a good soldier',[13] the US Army had forgotten the importance of generating cohesion with the result that, apart from elite or geographically isolated units, the US Army was incapable of maintaining small-group integrity. Soldiers were motivated by money not by the obligation to fight for their comrades. Henderson contrasted the creeping individualism of the US Army with the North Vietnamese and Israeli armies which were 'almost textbook examples of how to create and maintain a cohesive army'.[14] While the US Army may have trained its professional soldiers adequately, it failed to generate bonds of mutual obligation within its small groups. Soldiers were trained to fight but, without the bonds of comradeship, were not motivated to do so. He suggests that 'the creation of a cohesive unit is best accomplished upon its initial formation', requiring re-socialization processes and rites of passage.[15] Henderson notes the importance of training which 'is demanding', leaving little time 'available for other activities',[16] but, he notes, 'it must be emphasized that the creation of a cohesive unit is equally important in teaching skills to the soldier'.[17] Cohesion is the feeling of solidarity and comradeship, fostered in training but relying on interpersonal interaction: 'Ideally, both [personal bonding and military training] occur simultaneously.'[18] For Henderson, 'cohesive units will significantly benefit from barracks and mess halls designed to increase the frequency and duration of unit members' association'.[19] In order to foster cohesion, Henderson recommends that 'clubs, athletics and social events should be organized to promote unit participation'.[20] Henderson explicitly understands cohesion as a special form of motivation, then. As with Janowitz and Shils, such interpersonal motivation requires dense face-to-face bonds, and this bonding is important because it motivates the individual soldier to fight.

Leonard Wong's definition of cohesion adopts a similar position. For instance, in their commissioned work on the Iraq invasion, Leonard Wong et al.[21] conducted forty (group) interviews with US soldiers and Marines who participated in the operation. Wong et al. restate the importance of 'a cause' to the US combatants,[22] but crucially they emphasize the centrality of cohesion to battlefield performance; American soldiers reported that their prime combat motivation was that they were 'fighting for my buddies'.[23] Wong et al. are explicit that by cohesion, they refer to the special motivation which arises in interpersonal relations: 'The soldiers were talking about social cohesion—the emotional bonds between soldiers.'[24] Or again: 'this desire to contribute to the unit mission comes not from a commitment to the mission, but a social compact with the members of the primary group.'[25] Crucially, Wong et al. claim that this cohesion is a powerful motivation because it is a bond which extends beyond professional competence: 'This is not simply trusting in the competence, training or commitment to the mission of another soldier, but trusting in someone they regarded as closer than a friend who was motivated to look out for their welfare.'[26] Wong et al. note that soldiers would support each other emotionally, in one case offering hugs to each other.[27] The obligation to assist and protect immediate comrades, who were intimate associates often equated with family members, was rated as the single most important factor in explaining why US soldiers fought in 2003. For Wong, cohesion explicitly refers to the personal bonds between soldiers which motivate them to fight; 'social cohesion remains a key component of combat motivation.'[28] Significantly, 'cohesion is not just developed in training' but 'in the long, often mundane, periods of time spent neither in training or actual combat', when soldiers talked extensively with each other about their personal lives.[29] Here informal exchanges affirm and intensify the bonds between them. For all these scholars then, cohesion refers to a special sense of comradeship which motivates soldiers to fight.

There are numerous other scholars who affirm this perspective. In their work on cohesiveness and command styles among tank crews, Tziner and Vardi defined group cohesiveness as mutual liking, confirming that the 'concept of group cohesiveness in self-selected groups focuses attention on interpersonal relations',[30] even though they eventually suggest that cohesion defined in this manner was not critical to performance. More recently, James Griffith has explored the issue of cohesion in the armed forces utilizing a similar definition. Thus, in a work on the different rates of cohesion among Unit Replacement and Individual Replacement Units, Griffith draws on the classic literature to define cohesion in terms of interpersonal attraction, warmth, and satisfying personal relations.[31] His data suggest that cohesion (as a sense of collective closeness) is consistently higher among Unit Replacement soldiers (who have been trained and assigned together) than among Individual Replacement soldiers (who are assigned individually). Unlike Individual Replacement soldiers, Unit Replacement soldiers have had time to develop close personal bonds to improve both affective and instrumental functions of cohesion. Because of this feeling of solidarity, soldiers not only worked together more effectively but provided better social support for each other.[32] In short, cohesion motivates.

Guy Siebold has usefully summarized the general contours of debates about cohesion in the armed forces since the 1980s to propose that a 'Standard Model' of cohesion has been detectable in this literature. This model focuses on the generation of primary groups and their incorporation into a network of vertical and horizontal relations which together constitutes and sustains the armed forces as a whole. For Siebold, formal military training is central to the creation of primary groups and the maintenance of these all-important institutional links up the chain of command and across the service hierarchies. Nevertheless, in his model, Siebold stresses that 'the locus of bonding [cohesion] is in the relationship, not in the actions or interactions between the service member and the group, organization, or institution, although such actions or interactions are influenced by and feedback into the relationship'.[33] For Siebold, cohesion is a sentiment which emerges specifically out of interpersonal relationships. Its relevance to the armed forces is that cohesion has a special power to motivate soldiers in the extremis of combat. In short, echoing the definition used by Janowitz and Shils, social cohesion refers, for Siebold, to a personal bond between soldiers which motivates them. It is the collective sentiment of fellowship which encourages them to contribute to their unit and conform to its standards. In a very useful critical article, David Segal and Meyer Kestnbaum have ironically described this standard concept of cohesion,[34] referring to the special motivation which arises out of interpersonal relations as 'pure cohesion', which they see as the conventional approach to the problem of solidarity and combat performance in military scholarship.

In defining cohesion as a form of motivation, scholarship on the armed forces parallels the extensive research conducted on cohesion in social psychology in the 1950s and 1960s and, again, from the 1980s to the present. In 1950, Festinger, Schachter, and Back published a study of married veterans' housing communities, 'Westgate' and 'Westgate West', at the Massachusetts Institute of Technology.[35] The work has been regarded as the first serious attempt to formalize a theory of group cohesiveness in social psychology.[36] Festinger et al. famously defined cohesiveness as 'the total field of forces which act on members to remain in the group'.[37] Decisively, for Festinger et al., the 'total field of forces' referred to the attractiveness of the group to its members and the ability of the group to mediate goals for individuals; that is, the ability of the group to deliver goods which the individual could not attain alone. However, although the 'mediation of goals' was an important element in stimulating cohesion, as they conducted their research, cohesiveness in Westgate and Westgate West became more exclusively identified with the attraction of the group and its members for the individual; and specifically with friendship. Indeed, in a subsequent study, Leon Festinger made the point explicit. He reiterated that the 'cohesiveness of a group is here defined as the resultant of all the forces acting on the members to remain in the group. These forces may depend upon the attractiveness or unattractiveness of either the prestige of the group, members in the group or the activities in which the group engages.'[38] The field of forces referred to the pressure which group members felt from their peers to conform as a result of their interpersonal bonds with each other.

Festinger et al. demonstrated that participation in a new tenancy organization was dependent on friendship ties within the two housing areas and, very suggestively, showed how these relationships were based on nothing more than contingent geographical proximity. Friendships developed among those individuals who lived in the same courts or buildings of the two projects and whose front doors were near to each other; 'the most striking item was the dependence of friendship formation on the mere physical arrangement of the houses'.[39] This was probably Festinger et al.'s most intriguing finding, but they were primarily concerned with trying to establish how the social bonds between the occupants of Westgate and Westgate West motivated them to participate (or not) in the tenancy organization. For Festinger and his colleagues, cohesiveness referred to the attractiveness of the group for members. Cohesion was, then, the sentiment of mutual attraction and sense of collective obligation which encouraged individuals to join, participate in, and sustain the group of which they were a member. Cohesiveness referred in short to the motivation of individuals to act in a way which was consistent with their friendships, respectively encouraging them to join or refuse to join the housing associations in Westgate and Westgate West.

The concept of cohesion as motivation arising from interpersonal attraction has remained central to social psychological investigations since that time. In their review of group cohesiveness, Lott and Lott confirmed the equivalence between cohesion and attraction. They defined cohesiveness as 'that group property which is inferred from the number and strength of mutual positive attributions among the members of a group'.[40] From this definition, they deduce that 'there is a good reason to assume that interpersonal attraction, liking, or positive attitudes among group members, is central to the cohesiveness of small groups' and, citing Bonner, that 'without at least a minimal attraction of members to each other a group cannot exist at all'.[41] For Lott and Lott, the cohesiveness of a group refers to the mutual affection of members for one another (and the group generally). However, this 'interpersonal liking' is critical to Lott and Lott precisely because these affective ties motivate the individual to behave in ways in which they would not, independently of their friendships. This position has been very broadly accepted in social psychology. Thus, Shaw describes cohesiveness as the extent to which a group 'hangs together'. He maintains there are at least three different meanings to this 'hanging-together'; attraction, morale (motivation), and the coordination of individual efforts.[42] Confirming Festinger's definition, he concludes that 'most persons who use the term, however, agree that it [cohesion] refers to the degree to which members are motivated to remain in their group'.[43] Cohesion refers to a form of motivation, then. Zander confirms the point precisely; 'the cohesiveness of an established group is the strength of the member's desire to remain members. Cohesiveness increases as individuals become more attracted to the group.'[44] The more motivated its members become, the more they like one another, the stronger the group becomes. Ultimately, cohesion refers to a collective sense of fellowship which motivates individuals to want to conform to group standards and contribute to group goals. Not implausibly, social psychologists suggest that the more individuals like each other, the greater will be their motivation to act in ways of which their friends approve. In the social psychology literature as in military scholarship, cohesion refers to the special motivation which arises out of interpersonal relations.

Perhaps ironically, even among some of the critics of Festinger's approach, the concept that cohesion refers to motivation, arising out of group membership, is evident. Thus, in 1952, Gross and Martin forwarded some plausible criticisms of Festinger's concept of cohesion and methodology for analysing it. Yet, in rejecting cohesion as the attractiveness of the group for its members, they too confirmed that cohesion was to be understood as a collective sentiment. For them, cohesiveness referred to 'sticking togetherness'; it was a sense of solidarity, although they argued against Festinger et al. that it could be best measured by assessing not how attractive the group was to its members but how resistant the group was to disruptive forces.[45] Gross and Martin may be methodologically correct that it is more fruitful to examine the collective ability of a group to withstand fragmentation rather than to ask what individuals think of each other, but their concept of cohesiveness is equally based in shared affect; it too relates cohesiveness to the affection of group members for each other. Decisively, they too are interested in motivation. In his critique of the literature on cohesion, Hogg has usefully summarized the way social psychologists have generally understood cohesion. It is, according to Hogg, captured by other expressions like solidarity, comradeship, team spirit, unity, one-ness, or we-ness.[46] In each case, social psychologists are interested in the special kinds of motivation which arise from interpersonal bonds.

RE-DEFINING COHESION

The definition of cohesion as interpersonal motivation has obvious validity. There is little doubt that 'cohesion', as social scientists and social psychologists have defined it, has played a role in combat performance. Scholars who employ the term, from Janowitz and Shils onwards, point to a large body of evidence in which soldiers themselves record that in combat itself, their personal commitment to their 'buddy' was decisive to their actions. Nevertheless, while it would be inappropriate to dismiss the significance of personal friendships in motivating soldiers, it is dubious whether, for the purposes of analysis of the infantry platoon, cohesion is optimally defined purely as 'interpersonal attraction' and, more specifically, analysed only in terms of motivation. On the contrary, while in no way rejecting the concept of cohesion as 'interpersonal attraction' or dismissing the research done utilizing this definition, in this specific context, there are significant disadvantages to understanding cohesion solely as motivation.

The central thesis of Janowitz and Shils's work is that the cohesiveness of the primary group is fundamental to combat performance because it generates a unique motivation for soldiers. This relationship between interpersonal attraction and motivation has been central to the work on cohesion in social psychology. The presumption throughout all this literature is that high levels of interpersonal attraction generate good performances because it motivates the individual members to fight. The presumption that there is a connection between cohesion, defined as motivation, and performance is comprehensible, predictable, and, in many cases, it surely pertains. Nevertheless, this connection between cohesion (as interpersonal motivation) and performance is not always automatic. Famously, Janis, for instance, noted that high levels of social cohesion can generate the phenomenon of 'groupthink', where social solidarity is prioritized above performance. Here the cohesiveness of the group actively undermines the attainment of goals and, indeed, the more bonded a group is, the more in danger of groupthink it becomes. Cohesion, ironically, motivates group members not to perform. Janis entertainingly records the case of a group dedicated to helping its members give up smoking but which actually encouraged individuals to continue their habit as they could continue to attend meetings only as long as they were smokers; as a result, 'pressures towards uniformity subvert the fundamental purpose of group meetings'.[47] Other studies have shown a similar correlation between very high levels of group cohesion and poor productivity. Indeed, high levels of cohesion can 'result in excessive socializing that interferes with group performance'.[48] Individuals are motivated into deviant activity.

The problem of deviant motivation has been widely noted in the armed forces. In a creative piece, Leora Rosen et al.[49] noted that the excessive (hypermasculine) cohesiveness of military units may impede task performance. Donna Winslow has provided a vivid ethnographic account of how exactly this kind of hypermasculine cohesion undermined the military effectiveness of the Canadian Airborne in Somalia in 1993 when members of the Regiment tortured and murdered a local. Winslow rightly identifies failings of command, training, and recruitment as well as alcohol abuse as factors. However, she also maintains that the distinctive cohesiveness based on an 'exaggerated masculine ethic' developed by the Canadian paratroopers provided a social context in which violence and torture might occur.[50] For Winslow, informal rituals were a critical part of the creation of this extreme fraternity. Paratroopers 'hazed' initiates into the regiment and regularly participated in liminal practices in order to assert their shared masculinity. Urine and faeces would feature as a central part of these games in which 'the

Airborne would do things which are often associated with homosexuality'. Parties would involve 'soldiers dancing erotically, men in drag and mock sodomizations'.[51] Soldiers who were unable 'to meld into Airborne group identity were excluded'.[52] The intense interpersonal solidarity of paratroopers could not be doubted and it generated very high levels of motivation so that paratroopers were willing to do things which most other people would have refused. However, in this case, airborne cohesiveness produced a deviant and criminal performance.

Indeed, many scholars have noted the potential problem of deviant cohesion where small groups become so internally bonded that they ignore or subvert their obligation to the army in which they serve.[53] During the collapse of the French army in the First World War, the cohesiveness of infantry units actively aided the mutiny.[54] Australian troops in that conflict were also regarded as some of the most cohesive and yet they were also most prone to mutiny precisely because the bonds between the 'Diggers' were so strong.[55] Fuller notes that in March 1918, while only one in a thousand British soldiers was incarcerated for military offences, almost nine Australians in a thousand were imprisoned at the same time;[56] 'it is easy to see how it [the bush legend and the concept of mateship] could have been corrosive of strict discipline'.[57] An obvious and more recent example of this problem of excessive cohesion of an inappropriate type was the conduct of many black troops in Vietnam especially after 1968. As a result of their disaffection with the war and the influence of the civil rights movement at home, black soldiers in Vietnam began to form themselves into tightly knit primary groups. These groups were, in their own way, as cohesive as any in the US armed forces at the time but they were dedicated not to prosecution of the war and to following orders but to refusal, mutiny, and ultimately to the murder of superiors.

These cases confirm the point that while interpersonal solidarity may be central to the motivation of soldiers, its relationship to the combat performance it is invoked to explain is not direct. Interpersonal bonds can undermine combat performance just as well as encourage it. Since interpersonal motivation can at best be only one factor in explaining combat performance, it cannot be the sole focus of attention in this analysis and, ultimately, cohesion cannot be defined purely in terms of comradely bonds. A more extensive definition of cohesion is required which recognizes the role of interpersonal bonds and the special motivation they inspire but which is not limited to them.

There is a further problem with the concept of cohesion as interpersonal attraction. If cohesion were vital to group performance and individual motivation, then it would have to be present in any successful form of collective activity. Soldiers who did not like each other would never be motivated to fight. Yet, in fact, while the connection between interpersonal attraction and performance is highly likely and probably desirable, it is not essential. There is no necessary connection between successful performance and high levels of interpersonal solidarity.[58] The literature on sports teams is particularly interesting here. For instance, in their study of 1,200 basketball players,[59] Martens and Peterson surveyed high, medium, and low performing teams and could find no direct correlation between measures of cohesiveness monitoring interpersonal relations and subsequent performance, although they also stressed a good deal of ambiguity about their own and others' findings.[60] However, Landers and Lüschen's research on bowling teams[61] revealed that interpersonal attraction and friendship measures of cohesiveness were significantly negatively correlated with performance.[62] The data is ambiguous but it seems that interpersonal attraction was not decisive and sometimes irrelevant in each of these cases to group performance.

Similar claims have been made in military scholarship. Charles Moskos was one of the first to suggest that the presumption of a causal connection between interpersonal bonding and combat performance was questionable. 'Rather than viewing soldiers' primary groups as some kind of semi-mystical bond of comradeship, they can be better understood as pragmatic and situational responses.'[63] Against the sentimental view of cohesion as deep personal attraction, he has emphasized the temporary and contingent nature of the solidarity of the primary group in relation to his work on Vietnam. For Moskos, primary group cohesion was a function of necessity; 'my observations in Vietnam, however, indicate that the concept of primary groups has certain limitations in explaining combat behavior and motivation even beyond that suggested by Little. At least in Vietnam, the instrumental and self-serving aspects of primary relations in combat units must be more fully appreciated.'[64] In the face of danger, it was rational for the individual members of combat units to unite in an act of collective self-protection; 'primary group ties are best viewed as mandatory necessities arising from immediate life and death exigencies'. Moreover, 'one can view primary group processes in combat situations as a kind of rudimentary social contract which is entered into because of the advantages to individual self-interest'.[65] 'If the individual soldier is realistically to improve his survival chances, he must necessarily develop and take part in primary-group relations.'[66] For Moskos, soldiers in Vietnam were rational individuals who cooperated with each other for their own calculated benefit; a soldier 'gets such support largely to the degree that he reciprocates'.[67] Consequently, despite close but pragmatic personal bonds in the field, 'in most cases, nothing more is heard from a soldier after he leaves the unit. Once a soldier's personal situation undergoes a dramatic change—going home—he makes little or no effort to keep in contact with his old squad. Perhaps even more revealing, those still in the combat area seldom attempt to initiate mail contact with a former squad member. The rupture of communication is mutual despite protestations of lifelong friendship during the shared combat period.'[68] For Moskos, comradeship is a product of rational calculation alone. In order to survive individually, soldiers need to be comradely. Moskos may overstate the case in terms of individual rationality because in many cases soldiers sacrifice themselves quite deliberately for their comrades as a result of these bonds of interdependence; such sacrifice could not be explained if soldiers were mere rational calculators. Moreover, the subsequent loss of contact between soldiers may have been particularly specific to Vietnam because of the traumatic circumstances of that war; there is considerable evidence of the maintenance of connections between Second World War veterans. However, Moskos's point is important for it emphasizes that even the supposedly very deepest bonds of comradeship are in fact often situationally produced and are not necessarily so much the result of interpersonal attraction but rather more immediate military necessity: above all, shared danger. Others have confirmed the point. As Lott and Lott noted, the sociometric preferences of bomber crew members were reliably predictable from a knowledge of their location or function on the plane and the same was true of the Norwegian navy.[69] Military personnel who are functionally co-dependent learn to like one another, irrespective of their previous dispositions. Soldiers have to be motivated to fight but that motivation does not necessarily come from the bonds of interpersonal attraction; it can come from mere functional interdependence.

Moskos's comments about cohesion in Vietnam begin to unpick the presumed connection between interpersonal affection and combat performance more widely. More recently, Elizabeth Kier and Robert McCoun have sought to disrupt this causal link between friendship and performance radically in order to criticize the US armed forces' discrimination against male and female homosexuals. To this end, they draw on the distinction between task and social cohesion which social psychologists first began to draw in the 1980s.[70] While social cohesion 'refers to the nature and quality of the emotional bonds of friendship, caring and closeness among group members',[71] 'task cohesion refers to the shared commitment among members to achieving a goal that requires the collective efforts of the group'.[72] For Elizabeth Kier, social cohesion refers to 'interpersonal attraction' while task cohesion refers to 'instrumental bonding' and 'a shared commitment among members to achieving a goal'.[73] The point for Kier and MacCoun is that social cohesion (based, of course, on social discrimination by groups whose members like and are like one another) is not necessary for performance; task cohesion is entirely adequate. Accordingly, Robert MacCoun notes the examples of the fractious yet very successful Oakland As of 1973–5 and the New York Yankees between 1977 and 1978. He also cites Aronson's study of black and white coal-miners in Virginia who lived in segregation on the surface but worked in completely integrated teams under the ground. On this basis, MacCoun and Kier conclude that there is no necessary relationship between social cohesion as interpersonal attraction and subsequent performance. On the contrary, 'the sense of group cohesion based on "teamwork" has little to do with whether members enjoy one another's company, share an emotional bond or feel part of some "brotherhood of soldiers"'.[74] Crucially, Kier prioritizes teamwork—or task cohesion—over the motivation which arises from interpersonal bonds. Kier naturally regards teamwork as vital in the armed forces but it is not dependent on interpersonal attraction. She claims, on the basis of a diversity of research, that ten factors are typically vital for the generation of teamwork. Significantly, the most important for her is 'interaction'. Simply by interacting with each other, teams can begin to form, whatever the prior social backgrounds and experiences of the original participants. Indeed, echoing Festinger et al.'s research on Westgate, she records that 'merely being placed in a group—however random or arbitrary—creates positive attitudes towards other group members'.[75] With this displacement of interpersonal relations, and this is Kier's central purpose, it becomes entirely possible to conceive of a military force comprised of quite heterogeneous groups and individuals, all of whom, merely through the process of interaction, begin to unite into an effective body. Robert McCoun has forwarded a similar argument. He too questions whether the notion of cohesion, presuming social homogeneity, can be correlated to combat effectiveness in any simple way. At best, he admits that 'there appears to be a modest positive relationship between cohesion and performance'. However, citing a number of psychological studies, he argues that 'only task cohesion was independently associated with performance; social cohesion and group pride did not correlate with performance'.[76] He concludes: 'it is task cohesion, not social cohesion or group pride, that drives group performance.'[77] The point is well established in civilian society where Ivan Steiner has claimed that 'work groups sometimes persist in the face of adversity even though members have little affection for one another'.[78]

It is not absurd to define cohesion as a collective sentiment of interpersonal affection from which a distinctive kind of motivation arises. Janowitz and Shils, Henderson and Wong—and the wider social psychology literature—are not wrong. On the contrary, it is clear from the historical record that such a sense of commitment has been important to many soldiers. However, defining cohesion exclusively as interpersonal motivation produces two anomalies; there are recorded cases of cohesive primary groups that have spectacularly failed to perform in combat. Alternatively, there are other cases when soldiers have performed well in combat, yet lacked the interpersonal affection which the scholarship on cohesion would suggest was a prerequisite. These anomalies suggest the concept of cohesion needs to be revised for this study of the infantry platoon on the battlefield. Kier's work is useful here. For her, the social scientist should be concerned not merely with how interpersonal bonds motivate soldiers but with how teamwork itself is generated by a number of factors including those personal relations. She is interested in collective performance. Similarly, this study does not ignore interpersonal relations or the special motivations which arise from them but it is specifically concerned with exploring the teamwork of the infantry platoon on the battlefield. Consequently, this study explores the way in which small groups of infantry soldiers are able to coordinate their actions and to cooperate in combat. Combat performance, the phenomenon of collective action by the infantry platoon in battle, is the *explanandum* here; this study seeks to identify the various factors which generate this collective performance. It is interested in elucidating the conditions which have historically made collective performance in combat possible. In *The Structure of Social Theory*, Talcott Parsons famously adopted a Kantian 'transcendental' method in order to address the problem of order. Given that social order exists, he sought to identify the 'transcendental' conditions which made that order possible just as Kant had sought to identify the structure of pure reason which made coherent human experience possible.[79] This study follows the Parsonian—or indeed Kantian—technique of establishing the 'transcendental' conditions which have made combat performance possible for the infantry platoon in combat. It will become evident in the text that interpersonal bonds and the special motivation they engender have been historically very significant but they are not the only sociological factors which have influenced combat performance. On the contrary, this book aims to elucidate the different ways in which combat performance has been generated in the western infantry platoon by the citizen armies of the twentieth century and professional forces of the twenty-first. As the previous discussion of cohesion in the military and social psychology literature demonstrated, the term has typically been understood as a special form of motivation. For the sake of clarity, a terminological revision is suggested here. Since this study explicitly focuses on the phenomenon of collective performance which involves cohesive cooperation between soldiers, cohesion is no longer taken to refer to motivation or to interpersonal bonds but exclusively and specifically to the successful coordination of actions on the battlefield. Cohesion refers to the ability of the soldiers in an infantry platoon to act together and to achieve their mission in the face of the enemy; cohesion is demonstrated when soldiers are able to shoot, move, and seek cover together or in mutually supportive ways. In short, in this study and against conventional accounts, cohesion refers to collective combat performance itself.[80]

This may seem to be a radical, even unwarranted, revision. Yet, there is a clear precedent, even in the classic literature which seems to equate cohesion with interpersonal motivation. Although Janowitz and Shil's 1948 article eventually became a meditation on the importance of interpersonal motivation, it was originally animated by the question of the 'extraordinary' combat performance of the Wehrmacht.[81] Janowitz and Shil's original problematic was that, in combat, German units were more effective than Allied ones. The cohesiveness of the Wehrmacht in combat referred not to the affection of individual soldiers for each other but to the ability of its unit to bring greater and more effective firepower to bear on the Allies, even though its soldiers were under intense pressure and often in a vastly disadvantageous position. German units were willing and able to coordinate their activities and to cooperate with each other—to fight together—to a higher level than their opponents. The act of fighting together—the combat performance itself—constituted their cohesiveness, to which interpersonal bonds between soldiers contributed, as a motivating factor. Janowitz and Shils initially defined cohesion as military performance. Unfortunately, they were quickly diverted into a discussion of combat motivation, arising, in their view, primarily out of interpersonal bonds, but their original concept of cohesion specifically referred to combat performance, as the title of the piece demonstrated.

Indeed, their definition of cohesion as collective combat performance accorded with sociological thinking at the time. It has been noted in the previous chapter that in the *Structure of Social Action*, Talcott Parsons was concerned with resolving the 'problem of order'. For Parsons, order referred to the cohesion of social groups; it referred to the ability of social actors to engage in collective activities together. Thus, Parsons's work begins with a long and technical discussion of the 'unit act', the basic atom of human practice, consisting of 'an actor', 'an end', 'a situation', and a 'normative orientation'.[82] The problem of order addresses the question of how all the unit acts of multiple social actors are to be organized in a coherent way so that they are oriented to common ends. Order—or cohesion—emerges when these acts are coordinated. Parsons was very well aware of the importance of motivation and, specifically, of shared values. Indeed, he regarded the 'normative orientation' of actors towards 'common values' as fundamental. However, while 'normative orientation' was critical in generating order, it was not order itself. Order referred explicitly to the fact of successful cooperation itself; to collective action. To extrapolate from Parsons, it might be claimed that cohesion refers most accurately to collective action itself and, specifically, to successful collective performance, not to the sentiments which encourage that performance. On this account, a group is cohesive insofar as it is successfully able to act together, and cohesion refers to its collective performances. Certainly, as the literature recurrently demonstrates, successful performance typically strengthens the affection group members have for each other (which facilitates improved performance), but cohesion is most accurately understood as referring to the collective performance itself.

Indeed, social psychologists have themselves been troubled by the implications of the interpersonal definition of cohesion. Hogg claims that 'the overriding reason why group cohesiveness, as a distinct group-level concept, has made little headway in social psychology is perhaps a metatheoretical one'.[83] The problem is that social psychologists have tended to develop 'theories of human social behaviour that explain group phenomenon in terms of interpersonal interaction or the properties of individuals';[84] 'group level theories readily tend to dissolve into theories of interpersonal processes'.[85] For Hogg, precisely because cohesion became associated with individual attraction rather than group performance, social psychologists were unable to generate any interesting research findings. They ignored the phenomenon of collective action itself and preferred to examine what individuals privately thought of each other and their group. In the light of these difficulties, it seems plausible to suggest that cohesion is best understood as the successful coordination of the activities of group members to the same goal, utilizing compatible methods. Cohesion refers ultimately to successful collective performance, then. Indeed, in this study of infantry tactics, cohesion will refer purely to the phenomenon of combat performance. Accordingly, in this study, platoons which are defined as cohesive are those which are able to perform in combat effectively while ineffective platoons on the battlefield are deemed *ipso facto* to demonstrate low cohesion.

It might be useful to refine the precise area of interest in this study yet further. In discussing the combat performance of infantry platoons and sections, this study necessarily focuses on the way in which these groups generate accurate and effective fire on the battlefield with their personal weapons. This may seem an overly simple and brutal definition but on the industrial battlefield, as Stephen Biddle has shown,[86] success is ultimately determined by the application of combat power, a central component of which is firepower. Biddle's examination is conducted at the level of the army but his point seems to remain valid for the platoon. At the level of minor tactics, accurate firepower is the key to combat performance. There is a great deal more to infantry tactics than merely firing a weapon, of course. However, all the other skills of infantry work, camouflage, map-reading, cover, observation, communication, personal administration in the field, manoeuvre, and coordination, are ultimately but ways of maximizing the eventual effect of the firepower at the disposal of the platoon. It is possible that the infantry never fires its weapons on an operation. This is widely regarded as the acme of military skill and is much emphasized on today's counter-insurgency operations, but the infantry only enjoys the privilege of not firing because it has positioned itself in a way where it overmatches its opponent so thoroughly that resistance is plainly futile. In this case, the infantry has gained a position of such superiority that its potential firepower acts as a deterrent. Yet, this observable situation when fire is threatened but not used in no way undermines the claim about the centrality of firepower to the infantry platoon. For the platoon, combat effectiveness relies on the application of effective firepower.

This book examines the question of how the soldiers in a platoon together generate firepower which is essential to its performance and, indeed, survival on the field of battle. A further qualification about focus is also required. Although defensive action is discussed throughout the book, defence is not the decisive or most difficult form of performance of the battlefield. In order to prevail on the field of battle, infantry doctrine insists that the infantry must attack. Indeed, contemporary US Army and Marine doctrine defines two truths of infantry action: '(1) In combat, Infantrymen who are moving are attacking. (2) Infantrymen who are not attacking are preparing to attack.'[87] For the armed forces, there are good reasons for prioritizing the offensive role of the infantry. There are many cases where effective defence has led to a victory for the infantry; the Battle of Waterloo or the Battle of Gettysburg would be obvious examples. However, that victory is possible only by moving over to the offensive. Indeed, military doctrine recurrently emphasizes that defence can only be the prelude to the offensive. In order to prevail, the infantry has to attack, specifically putting itself in a position where its firepower neutralizes the enemy threat. Any resistance from the enemy, in the form of firing, is immediately overwhelmed by effective counter-fire. Accordingly, this study is primarily interested in platoon attacks. Specifically, it explores how infantry platoons and their sections or squads coordinate their movements to maximize the effect of their firepower as they seek to engage and destroy their enemies and to seize ground. In this way, this book examines the possibility of teamwork in one of the most extreme forms of human activity: war. Consequently and although focused on the details of military action throughout, this study aims to investigate the collective action problem, as one of the central issues in sociology, in its most simple but also brutal forms. Through the analysis of combat, the study ultimately aims to contribute to debates about collective action much more widely. It is interested in the question of how human cooperation is possible at all. Specifically, the study explores how cohesion or combat performance, often assumed to be unchanging and universal across wars, may have changed in the course of the last century, as armies have moved away from the citizen towards the all-volunteer professional model. This book explores the different ways in which the platoon has fought on the industrial battlefield from the First World War to the present. It seeks to examine the way in which the infantry platoon has performed in combat. In no way does it dismiss the classic studies of 'pure cohesion'. The motivations of soldiers and their potential affection for each other are always recognized in this study. However, these sentiments are always placed in a wider institutional context. This study seeks to identify a multiplicity of factors including comradeship, political motivation, doctrine, tactics, and training, which may explain the performance of the platoon in battle from the trenches of the Western Front to the patrol bases of Afghanistan.

3

The Marshall Effect

This book investigates the combat performance of the infantry platoon over the past century and is specifically interested in exploring the different ways in which cohesion (as performance) was generated in the mass citizen armies of the twentieth century in comparison with today's professional forces. In order to initiate this analysis and to assess this thesis, it is essential that some general account of the performance of the citizen infantry in the twentieth century is established. Although extensive qualification and refinement will be made later on (in Chapter 6), some broadly sustainable assessment of the infantry platoon from the First World War to Vietnam is required. Clearly, great care needs to be taken here because the political, geographic, climactic, and military conditions which pertained in each of the major conflicts that are the focus of this study were quite different. Nevertheless, despite the requirement for caution, some overarching sense of how the citizen infantry performed from the First World War to Vietnam has to be established so that the comparison with today's professionals can be drawn. It is not easy to develop a general picture of the performance of the citizen infantry from 1914 to 1973, not least because many scholars understandably seek to analyse specific conflicts.

Yet, a number of studies begin to draw comparisons across conflicts to develop more general historical theories about combat and cohesion. The work of Charles Moskos, Peter Kindsvatter, and Stephen Wesbrook would be obvious examples here and, indeed, their work is drawn on heavily below. However, although over the last two decades he has been the focus of intense critique and sometimes outright rejection, the work of S. L. A. Marshall represents an obvious and indeed indispensible starting point for this project, notwithstanding the scepticism which now attends his writing. Samuel Lyman Attwood Marshall was commissioned into the American army at the end of the First World War and became an official army historian during the Second World War with the rank of colonel, recording the campaigns in both the Pacific and European theatres, where he was eventually appointed Chief Historian. He went on to write numerous accounts of the battles in Korea and Vietnam which he followed as a historian and journalist (significantly aided by his close association with the army). Marshall, consequently, personally witnessed three of the four wars examined here, writing extensively on major battles in each, and while he did not serve at the front in the First World War, he was a member of the US Army in 1918; he experienced the First World War at least indirectly. Consequently, despite the recent reaction against his work, it would be difficult to deny that Marshall enjoyed a privileged position as a commentator on twentieth-century warfare.

There is a further reason for beginning with Marshall. Marshall's work specifically and indeed exclusively sought to address the reality of combat at the level of the infantry soldier. At the beginning of one of this books on Vietnam, *Ambush*, Marshall eloquently reiterates his Tolstoyan understanding of combat; 'most battles are more like a schoolyard in a rough neighbourhood at recess time than a clash of football giants in the Rose Bowl. They are messy, inorganic and uncoordinated.'[1] A battle is not merely the execution of a general's concept; rather its reality is in the grubby actions of small groups of men, often confused and frightened. Accordingly, Marshall's understanding of the realities of warfare led him to focus almost entirely on the actions of precisely these small groups, the squad, the platoon, and occasionally the company described and discussed in extraordinary detail. Even his detractors recognize Marshall's often compelling attempts to communicate the confusion and trauma of battle. Although deeply critical of his methods, the historian Roger Spiller fully recognized the validity of Marshall's approach: 'The axiom upon which so much of his reputation has been built overshadows his real contribution. Marshall's insistence that modern warfare is best understood through the medium of those who must actually do the fighting stands a challenge to the disembodied, mechanistic approaches that all too often are the mainstay of military theorists and historians alike.'[2] Spiller concludes: 'Forty years later, as the quest for universal laws of combat continues unabated, Marshall is still right.'[3] Moreover, of direct relevance to this study, Marshall was concerned specifically with combat performance. As a US Army officer, and subsequently as a historian, Marshall wanted to document the conduct of the US infantry soldier in combat in order to improve the army's combat performance. It may be possible to reject his findings, as some scholars have, but this historical pedigree suggests that he represents at least a fruitful starting point for understanding cohesion and the combat performance of the infantry platoon in the twentieth century. After all, it was the US infantry platoon to which he dedicated over thirty years of his professional life and about which he collated more material than almost any other historian.

In his famous work on US combat soldiers in the Second World War, *Men against Fire*, S. L. A. Marshall claimed to have made a surprising and controversial discovery; less than one in four US infantry soldiers actually fired their rifle at the enemy: 'In an average experienced infantry company in an average stern day's action, the number engaging with any and all weapons was approximately 15 per cent of total strength. In the most aggressive infantry companies, the figure rarely rose about 25 per cent of total strength from the opening to the close of action.'[4] Marshall's findings were all the more striking since he did not require a soldier to have killed or wounded an enemy to be recorded as an active firer nor did recorded firers have to fire an inordinate number of rounds before they were dubbed active. Marshall counted any soldier that had fired one or two rounds as active.[5]

Marshall's figures were politically unpalatable because, against the rhetoric of mass mobilization during the Second World War in which America's citizen army fought nobly for the protection of liberal democracy, he demonstrated that in fact America's conscript army was not the outstanding organization which many took it to be. Of course, he also caused deep concern to the American military. Tactical success on the battlefield is, in the age of gunpowder, won through firepower. The army which is most able to bring effective fire to bear on its enemy will triumph. As Marshall himself stated, 'What we need in battle is more and better fire.'[6] Yet Marshall recorded an army whose soldiers seemed to be unwilling to defend themselves. Since the publication of *Men against Fire*, S. L. A. Marshall's work—and especially his notorious one-in-four claim—has become a consistent reference point for scholarly debates about the question of cohesion. It remains a useful starting point in considering the question of the combat performance of the citizen army.

For Marshall, the explanation of the poor performance of American riflemen lay partly in the morals of western culture. Marshall described non-firing soldiers as 'conscience objectors' rendered impotent by their social conditioning; they had internalized the taboo against killing so thoroughly that even on the battlefield they were reluctant to kill their enemy.[7] However, although Marshall regarded morality as a significant problem for the western soldier, his fundamental explanation of non-firing was sociological; it was a situational product of the peculiar environment of the twentieth-century battlefield. Marshall noted that there was a decisive difference between the training environment and the battlefield. In the course of training, soldiers were continually in contact with friendly and 'enemy' units. In training, the battlefield was a populated place. Moreover, soldiers were motivated by the desire to be noticed by their superior officers. The actual battlefield was a quite different space; it was strangely empty. At the front, despite being populated by hundreds of thousands of troops, there was rarely anything to be seen; either enemy or friendly forces. On coming under fire, soldiers immediately went to ground. Consequently, their own field of vision was restricted to inches before their faces while their colleagues had similarly concealed themselves from fire and were almost always hidden from view. The enemy meanwhile was camouflaged and concealed often in trenches or bunkers. The modern battlefield had become an empty space and, paradoxically, despite the increasingly large number of troops involved in ever greater encounters, the experience for the soldier was of isolation and loneliness.[8] In moral isolation, it was natural and understandable why riflemen froze and were unable to fire their weapons. Alone, they felt weak and exposed to the attention of the enemy. Marshall noted the higher performance of those crews working machine-guns, bazookas, or mortars which he specifically related to the social context of their action. These soldiers were no longer isolated; they were part of a team, and the mutual support and pressure which crew members offered each other encouraged more active performance. Marshall, although not a sociologist, relied primarily on sociological—not psychological—explanations of poor firing rates. The riflemen in the mass conscript armies of the Second World War performed poorly because, given the gap between their inadequate training and the reality of the battlefield, they were disconcerted by the isolation which they experienced and, as part of a mass, did not feel the weight of collective responsibility upon them. They could shirk without fear of social repercussions because they knew that they were invisible.

Marshall's work is deservedly famous and has become the focal point of much of the discussion of cohesion in the armed forces. However, his findings have become deeply controversial. Indeed, several critics have rejected Marshall's work outright, maintaining that he grievously underestimated firing rates.[9] Some of the criticisms of Marshall's work can be dismissed fairly easily. General James Gavin, for instance, rejected Marshall's claims outright as 'absolutely false': 'All of our infantry fired their weapons. I know because I was there and took part.'[10] James Gavin was a brigadier commanding the 507th and 508th Parachute Infantry Regiments on the night of D-Day. General Gavin's demonstrativeness is perhaps excessive and his 'nearly visceral reaction to Marshall's probings' might be partly explained by Marshall's treatment of his division, the 82nd Airborne, in *Night Drop*, Marshall's account of the US airborne landings on D-Day.[11] While the performance of the 101st Airborne is consistently praised through the book, Marshall notes the errors made by the 82 Airborne's commanders and especially General Ridgway, around Sainte-Mere-Église.[12] A number of other soldiers who fought in the Second World War had a similar reaction to Marshall's claim about firing rates. Harold Leinbaugh and John Campbell, who served with the 333rd Infantry Regiment, similarly dismissed the claims on the basis of their own personal experiences of combat.[13] They too saw his claim as a personal affront and their refutations of it are based on mere assertion.

However, a number of scholars, especially Roger Spiller and John Whiteclay Chambers II, have made more serious methodological criticisms of Marshall based on substantial evidence. Roger Spiller's critique, specifically commissioned by Leinbaugh and Campbell to support their personal case, is particularly important here. Spiller examined Marshall's notes and interviewed John Westover, Marshall's assistant, about the post-combat interviews. Westover never recalled Marshall ever talking about firing ratios in their private correspondence and his notes 'show no signs of statistical compilations that would have been necessary to deduce a ratio as precise as Marshall reported'.[14] Moreover, Spiller notes that while Marshall initially claimed to have interviewed 400 companies, by 1952, he claimed to have interviewed 603. On Spiller's calculations, Marshall could not have interviewed 400 companies (each taking two to three days) by 1946 when he began writing *Men against Fire*.[15] Indeed, Spiller could not find any evidence that Marshall had conducted any company-level interviews anywhere; 'the systematic collection of data' was, according to Spiller, 'an invention'.[16] Spiller did not deny that Marshall interviewed many soldiers but his findings from those interviews, and specifically the famous statistical claim that only one in four riflemen ever fired their weapon, were fabricated. For Spiller, non-firing could be explained by relation to the ground and to circumstances.[17] Sometimes US riflemen could not fire because they could not see the enemy or would hit their own comrades but, at other times, it was deemed inappropriate to fire. Spiller cites Leinbaugh and Campbell's work *The Men of Company K*. In that work, Leinbaugh and Campbell describe how riflemen and machine-gunners would deliberately not open fire against the Germans but use artillery instead. By discharging their weapons, they would give away their positions, exposing themselves to furious German counter-fire: 'Marshall should have known that there are times in combat when one should not fire his weapons.'[18] Crucially for Spiller, Marshall himself seemed to admit his own mistake. In 1952, Marshall observed: 'I came out of the war with too much of a conviction that our basic difficulty was in the development of fire.'[19] For Spiller, this was a decisive moment when 'Marshall had very nearly admitted in a moment of weakness that his interpretation, fixed in its scientific pose, was too frail a vessel to bear the weight of explaining a soldier's conduct in combat'.[20]

Building upon Spiller, Whiteclay Chambers II interviewed Frank Brennan, who accompanied Marshall in Korea, in 2003. On the question of rates of fire, Brennan could recall only that Marshall 'brought up the question but that he did not put it in regard to the amount of firing. That came up incidentally rather than as a result of a specific question from him.'[21] Brennan also confirmed that during the interviews, Marshall took only minimal notes (insufficient for the generation of precise statistics).[22] On the basis of this interview, Whiteclay Chambers concludes: 'But without further corroboration, the source of Marshall's contentions about shockingly low rates of fire at least in some US Army Divisions in World War II appears to have been based at best on chance rather than on scientific sampling, and at worst on sheer speculation.'[23] This methodological inadequacy is accentuated by the fact that Marshall does in his private and professional life seem to have been prone to exaggeration and self-promotion. As Spiller points out, he gave the impression that he had commanded troops in the trenches in the First World War. In fact, he had been commissioned after the Armistice and had served only in a depot. He also exaggerated the amount of support which the army gave him in terms of staff and transport in order, it seems, to inflate his importance.[24] These criticisms undermine Marshall's claim about the poor performance of the US citizen soldier. According to these accounts, there were no significant shortcomings in the conduct of the infantry and Marshall's claims are ill founded and ultimately scurrilous.[25] It is plainly critical to establish which account of combat performance is nearer the truth if any general understanding of the citizen infantry is to be generated.

The methodological charges—and their empirical implications about combat performance—are serious but some defence might be made of Marshall which might affirm his account of the poor performance of the citizen infantry in the Second World War. Marshall seems to have exaggerated the number of companies he interviewed and probably the systemic nature of his research on firing rates. However, two decisive interviews occurred which seem to have been the basis of his claim about firing rates: his interviews with the 3rd Battalion from the 165th Infantry on the Makin Islands (which occurred immediately after combat) and an interview with Lieutenant Colonel Robert Coles, the decorated battalion commander of the 3rd Battalion, 502nd Parachute Infantry Regiment in Normandy.

Indeed, the Makin Islands fight, 20 to 23 November 1943, constitutes perhaps the crucial piece of evidence in Marshall's work for empirical and methodological reasons. On the Makin Islands, US defensive positions were directly assaulted by Japanese troops. In this situation, it might have been thought that American soldiers had to defend themselves or be killed by their assailants. However, even on the Makin Islands, only thirty-six men of the 3rd Battalion, 165th Infantry Regiment fired at the enemy.[26] Marshall records how the dead Japanese troops were found in front of those few American positions occupied by willing fighters. Since the battalion was under intense fire (suggesting it might be overrun) and it was apparently in the immediate interests of survival that every man use his weapon, it seems an ideal case when firing rates would be expected to be very high. Yet, only a small minority of the battalion actually fired their weapon and Marshall was rightly struck by the anomaly. Marshall noted that 'thereafter the trail of this same question [about firing rates] was followed through many companies with varying degrees of battle experience, in the Pacific and in Europe'.[27] On Makin, Marshall also developed the post-combat collective interview technique; he landed on the island soon after the initial assault and, as part of the official history, was able to conduct extensive interviews with the 165th Infantry Regiment as they waited for their next mission. Marshall emphasized the importance of his experiences on Makin, claiming that the material which he subsequently gathered about the Kwajalein Atoll campaign between 31 January 1944 and 5 February 1944 'really developed out of the Makin operation'.[28] On Kwajalein, he was able to spend four days 'reconstituting minute by minute'.[29] The

Makin Islands were then the prime evidential basis for Marshall's programmatic claim about firing rates.[30]

The importance of the Makin Islands fight on the night of 22–3 November, the so-called 'Fight on Saki Night', is evident in *Men against Fire* which includes a brief description of the action. However, in order to clarify the evidence on which Marshall generated his one in four firing ratio, it is useful to consider the action in more detail. The official US Army's history of the Second World War, to which Marshall contributed, is instructive here.[31] Makin is a small archipelago of atolls which is part of the Gilbert Islands. The main feature of the islands is a long thin strip of land some 5 miles in length called Butaritari Island. On 20 November, 1st and 2nd Battalions/165th Regiment landed on red and blue beaches on the west of this island and fought their way east. 3/165 was held in reserve and on 22 November moved through 2/165 which had fought its way halfway along the island to just beyond the wharf and the eastern tank barrier (a major feature).[32] The third battalion advanced during the day. According to the official history, 'the day's activity had been easy' with the battalion tasked to 'mop up a now thoroughly disorganized enemy trapped on the extreme northeastern tip of Butaritari'.[33] At nightfall, 3/165th was positioned by some ponds near the end of the island. I and K Companies faced east in company parameters separated by the ponds, I to the north and K to the south, the heavy machine-guns of M company interspersed between the positions. Their positions were poorly dug, consisting of shallow foxholes and barricades with coconut logs.[34] One lieutenant, Robert Wilson, emphasized that 'many of us had the idea there were no Japs left'.[35] As mentioned in Marshall's personal account, the Japanese assault was initiated by the use of 'baby cries' followed by 'intermittent fire fights' and 'uncoordinated small unit, sometimes, individual fights'.[36] The attacks were not particularly serious but 3/165 were compromised by their poor preparations. Consequently, 'the brunt of the attack fell on a few machine guns and heavy weapons positions that were covering the right and left of the line' and 'to the crews of these weapons the attack naturally appeared formidable indeed'.[37] The result was significant: 'although from three to four hundred men of the battalion were under Japanese mortar, machine gun, rifle and grenade fire from time to time, the enemy onslaught broke and disintegrated around relatively few positions held down by the heavy weapons and machine guns on the front':[38] M Company's machine-guns. It was here that Marshall's 36 firers were located and from which he generated the one in four figure.

The night of 22–3 November provided an excellent example for Marshall which can be illustrated by elaborating from these descriptions. The battalion consisted primarily of riflemen who were poorly informed of Japanese troop numbers and had constructed weak positions for themselves. Surprised by the assaults and lacking adequate cover from extensive direct and indirect fire, these riflemen remained passive. As Marshall described in *Men against Fire*, isolated and frightened, they went to ground. The official history provides some defence of 3/165's riflemen. Because of the direction of the Japanese attack, it records that 'those who were only slightly to the rear of the guns were in a position of uneasy onlookers, bound by the character of the defence to take relatively little hand in the repulse given to the enemy'.[39] The official history excuses their passivity on the grounds of the location, though given the small front which the battalion defended (approximately 300 yards) and the possibility of riflemen moving to support their comrades, it is perhaps surprising they did not try to intervene. There was no suggestion that the attack was not serious enough to merit their participation. Indeed, in contrast to Marshall's stated figure of one in four firers, the firing ratio among the 3/165 on this night was closer to 1 in 10 soldiers actually firing their weapon (see Figure 3.1).

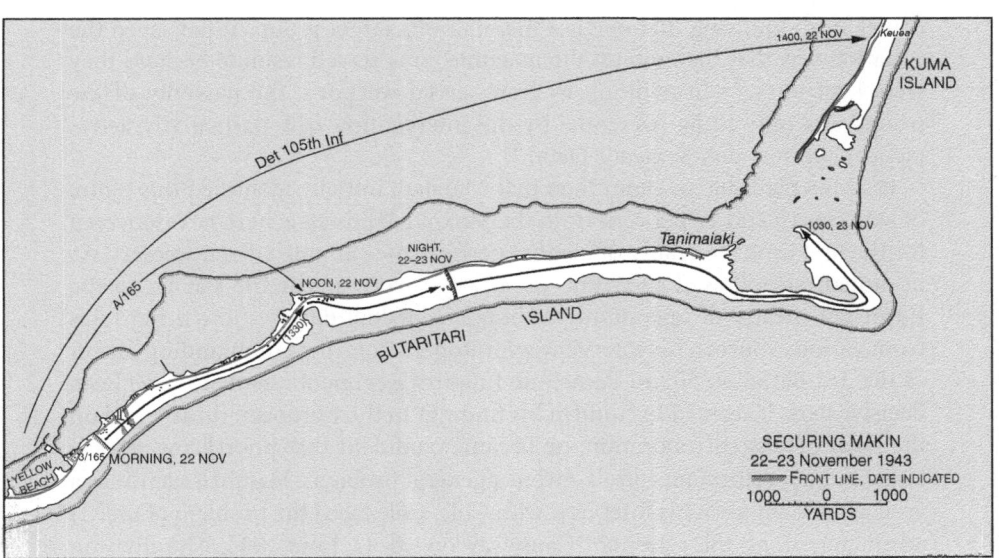

Fig 3.1. The Makin Islands, Butaritari. The 3rd Battalion's fight on the night of 22–3 November took place at the marked defensive line; the small ponds, around which the companies were positioned (I to the north, K to the south, with M interspersed between them), are just visible.

Instructively, the official history concludes with a condemnation of the 165th Regiment's performance on Makin. 'The seizure of Makin at first appears to have been cheap'; 218 casualties of whom 58 were killed in action and a further eight died of their wounds.[40] 'In view of the tremendous superiority of American forces to those of the enemy and the comparatively weak state of Japanese defences, the ratio of American combat casualties to those of Japanese combat troops was remarkably high.' That is, 'about two to three': for every three Japanese fighters killed, 'two Americans were killed or wounded'.[41] This performance was not only serious in itself but compared very poorly with the United States Marine Corps, which was about to assault Tarawa in one of the most celebrated and hard-fought battles of the campaign. Indeed, the next chapter of the official history, which stands as a rebuke to the Makin operation, describes US Army involvement in Tarawa. It seems plausible to suggest that not only did Marshall begin to recognize the problem of poor infantry performances and low firing rates as a result of his detailed investigation of the Makin Islands operation and its implications for operational effectiveness, but that this performance level undermined the reputation of the army especially in relation to its rivals, the US Marine Corps. Marshall's account of the subsequent assault on Kwajalein only confirmed Marshall's concern. Once again, the lacklustre performance of the Makin Islands was repeated: 'The men of the companies lost that strength which comes from knowing that one is a member of a larger group. What saved the situation was that the men on the machine guns stayed resolute because they were kept busy.'[42] In addition to crew-served weapons, the passivity of the troops was only to be overcome by the intervention of a particularly active tactical commander, Sergeant Deini.[43]

It seems plausible to claim then that Marshall initially identified this figure of between 15 and 25 per cent from the Makin Islands case; here he discovered for the first time the reality of combat performance among citizen soldiers. As he noted, he followed the 'trail' of this discovery across the Pacific to the European theatre of operations. Although Marshall deduces low firing rates from various sources, his interview with Robert Cole, the commanding officer of the 3rd Battalion 502nd Parachute Infantry Regiment, seems to have been decisive here; it seemed to confirm his findings in the European theatre and on the basis of this corroboration, he became confident that poor firing rates—and combat performance itself—were a general problem. Marshall identifies a crucial incident from his interview with Cole. Cole faced the problem of inertia advancing along the Carentan Causeway on 10–11 June 1944. The division had been tasked to seize Carentan and the 502nd had been ordered to secure the causeway to the north-west of the town to allow the 506th Regiment to take the town.[44] On the night of 10 June, the battalion came under fire in a defile near the causeway and Cole was forced to run up and down the column ordering his men to fire without success: 'I walked up and down the line yelling, 'God damn it! Start shooting!' But it did little good. They fired only while I watched them or while some other officer stood over them'.[45] They remained reluctant: 'I found no way to make them continue fire. *Not one man in twenty-five* voluntarily used his weapon'.[46] His men's reluctance to fire forced Cole to mount a now celebrated bayonet charge the following morning, for which Cole himself was eventually awarded the Congressional Medal of Honour (see Chapter 5). Cole's confession about the failure of his men to fire was extraordinary and it was understandable why Marshall was so struck by it. Cole was the commander of an elite airborne battalion whose performance throughout the war was exemplary. Indeed, for its participation in the Normandy campaign, the 502nd Regiment received a Presidential Unit Citation. Nevertheless, its own commander confessed that in one of its most important and celebrated actions, only a minority of his soldiers were willing to fire their weapon. Indeed, participation in the charge was lacklustre. Lacking artillery support, Cole 'was like a quarterback picking a play without thinking'.[47] Cole yelled to one of his majors: 'We're going to order smoke from the artillery and then make a bayonet charge on the house'.[48] Yet, Cole did not inform his men that the signal for the charge was his whistle with the result that only 21 men followed him, 'a picket squad'.[49] Whether feigned or not, other soldiers pleaded ignorance: 'I heard someone yell something about a f—— whistle'.[50] His men made a credible but not wholly convincing excuse for their non-participation. Although it is impossible to be certain, it does not seem coincidental that it was Cole who specifically identified a one in four firing rate or less; 'not more than one in twenty-five men' fired.

From the numerous other interviews which he conducted and his own extensive study of combat, Marshall convinced himself that this low firing ratio among infantry soldiers was approximately accurate. His tendency towards self-promotion may have encouraged him to be more demonstrative about his evidence than he was strictly entitled to be but his figure was based on some significant, if narrow evidence; it was more than guesswork and certainly much better than speculation as Whiteclay Chambers has argued. The complaint of Spiller and Frederic Smoler that Marshall was trying to pass off the 15 to 25 per cent figure as rigorous statistical evidence seems to be less compelling, especially since Marshall himself always qualified the figures with phrases like 'on average', 'approximately', or 'rarely more than'. *Men against Fire* communicates very clearly that firing rates were low, much lower than typically presumed, but Marshall only ever put the firing rates within a definable but broad band rather than forwarding a precise statistical calculation. Indeed, Marshall regarded the figure as illustrative rather than demonstrative. His collaborator, Dr Hugh Cole, reported that he had warned Marshall not to use such a precise figure in the strongest terms, to which Marshall had replied: 'You don't understand. I'm just making a point.'[51] It is minimally unsympathetic to criticize Marshall, an army historian concerned with recording and improving combat performance, as though he were a statistician whose prime concern was numerical accuracy.

Indeed, reading any one of Marshall's combat histories, the issue is not Marshall's cavalier treatment of the truth. The problem for the reader is being overwhelmed with excessive detail. Marshall focuses so closely on the immediate actions of small groups of men that it is sometimes very difficult for the reader to reconstruct the general contours of the battle. *Night Drop* is one of the clearer examples of Marshall's work but the combat descriptions in both his major pieces on the Korean War, *Pork Chop Hill* and the *River and the Gauntlet*, as well as his work on the Vietnam War, *Bird*, *Ambush*, and *Battles in the Monsoon*, are so intricate that they are difficult to follow. By being dragged into the bewildering detail of the fire-fight, the narrative loses sight of the whole and so does the reader. This is a significant complaint about Marshall's work, but his ability to generate these detailed histories should dissuade critics from easily assuming that Marshall merely invented his data. On the contrary, he used his technique of the post-combat interview recurrently in order to reveal the realities of the schoolyard fight from the Second World War to Vietnam. The care Marshall displayed in compiling his accounts may not constitute the 400 company group interviews which Marshall claimed he conducted in the Second World War but neither does it suggest a cavalier or cynical approach to evidence.

This is not to say that Marshall's broad assertions about low rates of fire were always totally accurate. Certainly, there were cases when firing rates were much higher than 25 per cent. In exemplifying his concept of forward panic, Randall Collins cites Marshall's own account of some US paratroops under Lieutenant Millsap in Normandy soon after D-Day who 'had been under heavy German fire for three days'. Having charged the enemy, they slaughtered 'every German in sight' but then 'ran into the barns of the French farmhouses where they killed the hogs, cows, and sheep'.[52] Marshall describes the episode as a 'terrible orgy' when the soldiers 'had all become victims of mass hypnosis'.[53] Indeed, on the Makin Islands on 21 November, the morning after the amphibious landing, an engineer falsely identified the presence of a significant Japanese force which precipitated a wave of hysterical shooting which commanders could stop only by direct orders to individual soldiers to cease fire.[54] Having survived the stress of an amphibious assault in which they expected to be opposed, these US troops overreacted to the slightest suggestion that they were under threat. Other examples of 'forward panic' could also be identified in the Second World War. One of the most famous, which was likely to have been known to Marshall, was the taking of Nijmegen Bridge by 82nd Airborne on 20 September 1944. Having failed to seize both ends of the bridge in the initial assault, 3rd Battalion 504th Parachute Infantry Regiment was ordered to paddle in canvas boats across the Waal in broad daylight and in plain view of established German defences, including mortar and machine-gun positions.[55] The battalion suffered very high casualties during its crossing—only 11 of the 26 boats survived—but on reaching the east bank of the Rhine, the soldiers assaulted the German positions with ferocity. As one participant observed:

Many times I have seen troops who are driven to fever pitch—troops who, for a brief interval of combat, are lifted out of themselves—fanatics rendered crazy by rage and the lust for killing—men who forget temporarily the meaning of fear. However, I have never witnessed this human metamorphosis so acutely displayed as on this day. The men were beside themselves. They continued to cross that field in spite of all the Krauts could do, cursing savagely, their guns spitting fire.[56]

Firing rates among these soldiers were total and Marshall himself recognized them. Consequently, Marshall certainly overstated the case when he maintained that the low firing rate 'varied little from situation to situation'.[57] There were documented cases during the Second World War when a mass army produced total or nearly total firing rates among their citizen soldiers.[58]

As a general critique of the performance of the typically poorly trained and inadequately prepared US citizen soldier, Marshall's work seems to be valid, therefore. Moreover, on the basis of his observations of the US soldier, it is possible to extrapolate his findings across to the infantry of other mass citizen armies of the twentieth century. Certainly, a number of commentators have concurred with his general point. David Lee broadly confirms Marshall's findings arguing that the 'ratio of fire identified by Marshall has a valid feel to it' and that 'no doubt' many did not fire.[59] Although sometimes critical of elements of his work, other analysts have concurred broadly with Marshall's figures.[60] David Grossman draws heavily upon Marshall's subordinate psychological claim that, fundamentally, humans are conditioned not to kill, finding it difficult to overcome their socialization. However, although Grossman's interests are primarily psychological, he provides much evidence to support Marshall's claim for the poor performance of soldiers in battle. In order to avoid killing, Grossman claims that soldiers, especially in untrained forces, resort to posturing and shirking. Most of the soldiers on the battlefield did very little and major casualties were inflicted only '*after* one side or the other had won the battle'.[61] Typically, soldiers would fire high or not fire at all. They would 'provide support for those who are willing to shoot',[62] but sometimes they would imitate the actions of firing without actually shooting their weapon. Randall Collins has argued forcefully in favour of Marshall's broad figure. Using photographic evidence, Collins suggests that a 13 to 18 per cent firing ratio was typical reaching 46 to 50 per cent if the photographs are interpreted at their most generous and 'at least one person in the photo is firing'.[63]

Perhaps most significantly, Marshall's observation was widely supported by other officers at the time. General Lucian Truscott, commander of the US 3rd Division in Italy, made a similar observation, suggesting that only a third of the troops in his formation were effective:

I was surprised to find what a relatively small proportion of individual weapons available in any unit were ever employed in action. One would think that every unit engaged in combat would employ every possible weapon, for battles are won by destroying or threatening to destroy the enemy. But this had not been so in the actions in which our infantry and armoured battalions had been engaged. Our investigation revealed that the battle had been waged with a series of small unit actions and that each of them had been won by a fraction of the unit with the assistance of supporting arms—artillery, naval gunfire, machine guns, bombers and tanks.[64]

He also noted the tendency of his troops to panic: 'It had never occurred to me that naval gunfire passing over the heads of an infantry battalion could cause such panic that the battalion would take to its heels and disperse so that it required most of two days to collect the stragglers.'[65] The willingness of soldiers to surrender also dismayed him: 'I was astounded by the relatively large number of American soldiers who surrendered to our French opponents when they still possessed means of continuing to fight or who could have withdrawn to continue the fight elsewhere.'[66] He was not alone in his scepticism about the citizen soldier. General Patton also thought that the fundamental problem with his infantry soldiers was their reluctance to fire. He emphasized that on taking cover 'he must not hit the dirt and stay supine. He must shoot fast at the enemy.'[67] Yet, he regretted to note: 'It is a sad commentary on our troops that frequently we get the report that such and such a unit is pinned down under fire, and later the same unit comes back.'[68] Having failed to return fire, the unit retreated. Patton instituted a practice of 'marching fire' in the 3rd Army in order to overcome the problem.[69]

These were not isolated views. There were deep concerns in the higher echelons of the US Army generally. On 3 December 1943, Brigadier General John Lentz of the Army Ground Forces wrote to Brigadier General Marcus Bell of 81st Division to discuss central deficiencies of training: 'Combat firing . . . is our major weakness. It is the one phase which I am discouraged by . . . Officers with years of background and peacetime safety concerns simply will not cut loose with realistic combat firing as a general thing. There are so damn many flags and umpires and control they no more resemble a battlefield than a kindergarten.'[70] Lentz was talking about training but it was clear that his concerns about poor firing in training reflected inadequate performances and especially firing rates in combat in which the US Army had at that point been involved for two years.

In the United States forces, there was substantial evidence for poor performance. In his history of the 29th Infantry Division, Joseph Balkosi frequently records poor levels of tactics and participation in firing, even though the army emphasized the importance of attaining fire superiority in doctrine and in training.[71] In Normandy, the US infantry found that it was very difficult to establish fire superiority over the German MG 42s which had a very high rate of fire and around which German tactics were built. Consequently, opening fire on a German position typically encouraged the unwanted and accurate attention of intense machine-gun fire, especially since American gunpowder was neither smokeless nor flashless. Lieutenant Colonel Cooper, of 110th Field Artillery, explicitly recognized the problem of non-firing: 'You really couldn't blame the men for holding their fire under these conditions.'[72] Yet, while individually rational, this institutionalized reluctance to fire ensured that American infantry conceded fire superiority to their enemies. Indeed, General George Patton explicitly rejected such accounts: 'Remember that the life of the infantry squad depends on its capacity to fire. It must fire.'[73]

The performance of the 29th Infantry Division does not seem to have been unusual. Harold Leinbaugh and John Campbell, ex-soldiers from the 84th Division, have been among Marshall's most vocal critics; yet these military critics ironically provide some of the most compelling evidence for the 'Marshall effect'. As we have noted, they rejected outright Marshall's non-participation figures, on the basis that they witnessed combat, which he only reported inaccurately upon. However, in their memoir of their experiences in a rifle company in the 84th Division, the problem of poor performance and non-participation is a recurrent theme, even though the work is explicitly written as a festschrift for the fighting men of the company. For instance, the company commander, Captain George Giezl, echoing Marshall, emphasized the sense of isolation on the modern battlefield; 'The terrible thing about that experience is the utter and complete aloneness. There isn't anything else—just you. Someone could be just two feet from you, but there's nothing there.'[74] Although the company acquitted itself as well as any, this isolation produced a pattern of performance which Marshall recognized. While the company performed well in one of their first assaults on Chateau Leerodt near Geilenkirchen, firing and moving as they attacked, with high levels of participation, this action involved no German resistance, as the authors themselves admit.[75] Typically, when German resistance was organized, the company would rely on artillery or tanks or a small minority of soldiers would have to initiate, lead, and, indeed, prosecute any action in combat while the rest of the company was passive. For instance, on 28 February 1945, the company was pinned down by German fire as it assaulted the town of Hardt across some fields and began to take casualties. Eventually, a tank provided crucial support, knocking out a machine-gun. At that moment, Lieutenant Masters, taking individual initiative, led the final charge, personally killing machine-gunners who had been firing on the company. Sergeant George Pope recalled: 'We were all pinned down, and I see guys turning their heads. I felt like doing that myself. It was flat as a floor. There wasn't a blade of grass you could hide under. I'm yelling, "Shoot, you sons of bitches!" That was a tough time . . . Talk about Masters, then you saw a guy with some cool head, man.'[76] Pope's testimony usefully illustrates Marshall's own analysis. While Masters's platoon lay passive, desperately seeking to avoid German attention, and even its sergeant could only call for fire while failing to shoot himself, the platoon commander jumped up and rushed the position, firing at the enemy gunners. Emphasizing the level of non-firing elsewhere in their work, one of the company notes that 'maybe our own weapons training wasn't really that good',[77] while another recalls that he received a direct fire order (which Marshall emphasized as crucial to the stimulation of fire) only very rarely in combat.[78] Leinbaugh and Campbell may well be correct to dismiss Marshall's blanket assertion that firing rates were always less than one in four but their own descriptions of combat in which they themselves participated suggest that Company K often failed to fire their weapons at the appropriate moment. Their work includes frequent admissions of non-firing which demonstrate precisely the point which Marshall was trying to make. Generally, the performance of infantry platoons in the Second World War was poor.

Indeed, the US Army was very concerned about the quality of its infantry throughout the Second World War and these worries were recorded at length in the army's official assessment of its performance published in 1948.[79] Specifically, the 'quality of manpower in the ground arms, when mobilization was nearly complete in the latter part of 1943, compared unfavourably with other elements of the Army'.[80] One of the problems here was the selection process. As recruits presented themselves, their aptitude was assessed by the Army General Classification Test (AGCT) which divided them into five categories from I to V: 40 per cent of the recruits went to Ground Combat Arms (infantry, cavalry, artillery, armour) but of those only 34 per cent were graded highest in aptitude and intelligence (Grade I and II) while 44 per cent were graded lowest (Grade IV and V). The army introduced a policy where it tried to place recruits into specialisms which matched civilian employment: longshoremen were assigned to the Quartermaster Corps, miners to the engineers or infantry, boilermakers, bricklayers, and riveters to the engineers.[81] 'The problem of technical training in the Army was thereby simplified, but the problem of tactical and combat training was rendered more difficult ... The loss of the type of men who acquired skills in civilian life left the ground arms with a sub-average portion of available manpower.'[82] Skilled workers tended to receive higher grades on the AGCT and their skills disqualified them from the combat arms and above all the infantry. The result was that the dregs of the army's recruits were overrepresented in the combat arms to the consternation of combat commanders:

Field commanders of 1942 protested repeatedly to Headquarters, Army Ground Forces, that they were receiving men of too low a mental quality to be trained. They said it was dangerous to entrust lethal weapons to men in AGCT Class V, and wasteful to develop elaborate and expensive equipment and then place it in the hands of men incapable of using it properly.[83]

A letter to Lieutenant General Lesley McNair, Commandant General of the Army Ground Forces (AGF), and therefore responsible for the recruitment, assignment, and initial training of troops, from Lieutenant General Ben Lear, Commander of the Second Army, was condemnatory about army policy especially in terms of combat leaders: 'We are scratching the bottom of the barrel now for officer candidates. We are decidedly short of the right material for non-commissioned officer leaders. We will pay for this dearly in battle.'[84]

Army Ground Forces observers with the 5th Army noted in the middle of 1943 in Italy that 'squad leaders and patrol leaders with initiative were scarce' and that 'the assignment of Grade V intelligence men to the infantry is murder'.[85] A surgeon attached to XIV Corps during the New Georgia Campaign demonstrated a connection between poor leadership and breakdown. The failure of combat leaders under fire had induced panic among the men, 'with a needless sacrifice of manpower'.[86] Indeed, General McNair agreed with field commanders entirely, although he resisted some of their suggestions to amend the situation. 'He came increasingly to believe that American soldiers were sustaining avoidable casualties and perhaps taking longer than necessary to win the war, because men assigned to the ground combat units did not represent a fair cross-section of the nation's manpower.'[87] In this he came very close to Marshall's position. Indeed, he generated more rigorous data which sought to show 'the shortage of combat leadership statistically' during the recent North Africa campaign.[88] While the Army Ground Forces' combat units suffered battle combat ratios of 2.6 times those of the Air Corps, only 5.2 per cent of the men in the combat units were rated as Class I in the AGCT, 26.5 per cent in Class II, as opposed to 9.3 per cent Class I and 38 per cent Class II in the Air Corps.[89] It was axiomatic that strong leadership was in almost all cases found among men in the upper classes of the AGCT.[90] On the basis of these figures, McNair sent his findings to the army's Chief of Staff on 17 December 1943: 'this headquarters has pointed out that certain procedures in distributing manpower discriminate against ground forces. The enclosed charts show the cumulative effect of such measures. While the situation is viewed as unfortunate, it is realized that it is now too late for effective remedial action.'[91]

Marshall focused exclusively on the US Army and there is substantial evidence to support his findings in relation to the US Army in the Second World War, but it is possible to find evidence of the 'Marshall effect' in other armies at the time.[92] Indeed, there are many other examples of poor performance among the citizen armies of all the combatant nations from the First World War to Vietnam, which will be further discussed in Chapter 7. British officers in the Second World War certainly noted something similar to Marshall. For instance, during the Second World War, Major Lionel Wigram became a well-known trainer in the British army. He played an important role in the creation of the infantry battle school at Chelwood Gate of which he was appointed the Chief Instructor.[93] Chelwood Gate was recognized as one of the most innovative and significant training establishments in Britain at this time. As chief instructor, Wigram published his own manual on infantry tactics which he circulated to those who attended the school. He introduced new battle drills and training innovations at Chelwood Gate which attenders on his courses were expected to disseminate to their battalions on return to their units.[94] Wigram was well placed to judge the quality of British infantry. Yet, when he was eventually deployed to Sicily on operations as a commander, he was dismayed to find that 'only around a quarter of the men in a typical platoon were reliable in battle'.[95] In August 1943, he recorded that: 'every platoon can be analysed as follows: six gutful men who will go anywhere and do anything, twelve "sheep" who will follow a short distance behind if they are well led and from four to six ineffectual men.'[96] He identified that this lack of willingness to participate in combat—to fire—had to be overcome if any degree of sophistication in infantry minor tactics was to be achieved.[97] Wigram's firing rates and their centrality to military performance accord almost exactly with Marshall's argument.

Sidney Jary served as a platoon commander in D Company, 4th Battalion Somerset Light Infantry from July 1944 to the end of the war, fighting across France and into Germany. In his memoirs, he emphasizes many of the qualities of his own soldiers and the British infantry in general. However, even Jary, an advocate of Britain troops and intensely loyal to his own men, recorded a highly differential rate of performance among his own platoon: 'Unlike characters in novels and films, most men react nervously to real battle conditions.'[98] Jary estimated that only a third of his platoon could be trusted to be active in combat; 'in an infantry platoon you will find a small band of brothers—about fifteen to twenty per cent of the platoon—who create an atmosphere that get things done and wins battles.'[99] Jary does not specify which soldiers but it does not seem implausible to suggest, in line with Marshall, that his most reliable soldiers were those manning the heavier weapons and, in particular, the platoon's Bren guns. Thus, isolated on the battlefield, soldiers went to ground and shirked. Indeed, many soldiers were explicit about their reasons for not firing. For instance, Rifleman Joseph Belzar of the Royal Sussex Regiment was fighting near Perugia on 19 June 1944. He was involved in an assault during which his battalion were engaging the enemy. However, in an unfavourably exposed location, he declined the opportunity to fire: 'I realized that, being on the forward slope of the hill, I was exposed to view and firing from such a position would not be to my advantage—to put it mildly—I realized that discretion was the better part of valour.'[100] Belzar, lying prone, opted for passivity, allowing his comrades to prosecute the assault. It is not difficult to assume that the decision which Belzar made in this situation was recurrently and consistently made by British soldiers in the firing line during the Second World War.

In Chapter 7, more comprehensive and conclusive evidence for the poor performance of the citizen infantry in the Second World War will be considered; British and American material will be supported by evidence from the other major western combatants, Canada, France, Germany, Italy, and Australia, to suggest that Marshall's insights about the US infantry during this conflict are broadly applicable. However, in order to sustain the central thesis about the performance of the citizen army, Marshall's observations about the Second World War have to be valid for all the major conflicts of the twentieth century. A general pattern of poor infantry performance has to be evident in those conflicts too. It is possible to provide some indications which may at least suggest that the Marshall effect was a general problem even if at this point it is impossible to prove the claim. Thus, it seems likely that the low level of participation which Marshall noted in the Second World War may also have been a feature of the Great War. Robert Graves observed a wide differential between infantry divisions: 'About a third of the troops forming the British Expeditionary Force were dependable on all occasions: those always called on for important tasks. About a third were variable: divisions that contained one or two weak battalions but could usually be trusted. The remainder were more or less untrustworthy.'[101] He regarded the 'Second, Seventh, Twenty-ninth, Guards, First Canadian' Divisions as 'top-notch', noting that 'English southern county regiments varied from very good to very bad'.[102] Although Graves is concerned purely with divisions rather than platoons, it is noticeable that the idea of a third of the force being effective is articulated by him, as it was by Marshall.

It is perhaps more difficult to extrapolate Marshall's claims about combat performance to Korea and especially Vietnam. He asserted that, partly as a result of his interventions, American army firing rates improved dramatically after the Second World War. In Korea, he asserted that firing rates reached 55 per cent, in Vietnam, 'not infrequently 100 per cent'.[103] Jordan has claimed that, in fact, any improvement in the performance of the US infantry was due to the increase in the number of Browning Automatic Rifles in the infantry squad rather than any improvement in the performance of the rifleman himself as a result of training or Marshall's recommendations.[104] Effectively, according to Jordan, the US Army obviated the problem of firing rates by reducing the number of riflemen and assigning more men to squad weapons which, as Marshall perceptively recognized, the operators would be more willing to fire. In fact, despite this innovation, there is little evidence to support the general claim about the performance of the infantry in Korea where 'bugging out' was all but endemic.

The same phenomenon may be observable in Vietnam. Russell Glenn has produced one of the most comprehensive considerations of infantry performance in Vietnam, taking Marshall's total firing rate as his starting point.[105] Glenn's argument is based on a survey questionnaire of 258 soldiers from the

1st Cavalry Division.[106] This division was one of the best and most famous divisions in Vietnam, representing the new concept of airmobility for the US army and deploying to Vietnam in 1965 after intensive training. It was involved in the first major offensive operations of the war in November 1965 in the Ia Drang Valley and subsequently participated in most of the major fighting including the battle for Hue during the Tet Offensive in February 1968. It was an appropriate formation on which to conduct research since good performance and high firing rates were most likely to be found in this division. Glenn's findings are instructive. On the basis of his data, Glenn rejects Marshall's 100 per cent figure to record that firing rates were typically about 80 per cent; respondents estimated that about 80 per cent of the soldiers fired their weapon at the enemy at some stage during their tour.[107] However, his account records numerous examples of very poor performance. Signally, Glenn notes that there were very significant training deficiencies which compromised the performance of the troops in Vietnam. Some elements of 1st Cavalry were among the most highly trained soldiers which the USA had ever deployed but the division quickly had to draw on less trained recruits.[108] These training deficiencies worsened as the conflict continued so that by 1969, troops were being deployed into theatre with inadequate preparation. Indeed, some members of the division claimed that bad weapons training cost the USA more casualties than the Vietcong.[109] General Abrams wanted to introduce a four-week training programme in Panama to prepare troops for jungle fighting but there were inadequate time and resources to implement such a scheme.[110] By 1970, the division's performance was substantially affected by this inadequate training. Moreover, various of Glenn's respondents recorded precisely the passivity which Marshall observed on Second World War conscripts. One sergeant reported the shock of newly arrived soldiers, who were totally overawed by combat. He describes having to pick these inexperienced soldiers up physically in fire-fights in order to get them to move, although he claims that they were firing.[111] They had frozen, as Marshall predicted. Like Marshall, Glenn also emphasized that crew-served weapons provided most of the fire because the best soldiers were put on these weapons, they felt responsible for their comrades, and because of the collective effort their use demanded.[112]

Marshall's three major works on Vietnam confirm the point. They would all suggest that his figure of almost 100 per cent firers, and, more significantly, the high levels of combat participation which that figure illustrated, cannot be sustained. Marshall's work *Bird* records the overrunning and eventual eviction of an NVA regiment from a fire base near Pleiku in 1967. The assault was eventually halted by a staff sergeant, in a desperate measure, firing two 'beehive rounds';[113] beehive rounds are fragmentation munition which disperse thousands of flechettes of steel in a wide area. In the event, the beehive rounds destroyed the NVA attack while inflicting few US casualties precisely because most of the defending US force were cowering under cover. The effectiveness of the beehive round only demonstrates Marshall's wider point about the defence of Bird; most of the defending US forces contributed very little. The defence was, in effect, a reprise of the action on the Makin Islands, conducted by a very small number of troops who became actively involved in the fighting while the majority remained passive. In his other major works on Vietnam, Marshall consistently records reluctant infantry performance. On numerous occasions, he reports the shock of riflemen in combat which reduces them to passivity. In the face of a Viet Cong or NVA surprise attack, soldiers typically froze and were unable to respond with fire. There is significant additional evidence to support the point. Despite their elite status, the 173rd Airborne Brigade illustrated the problem at the Battle of Dak To in June 1967 when they assaulted a number of hills in order to find and destroy enemy forces. On 10 July on Hill 830, in response to heavy enemy fire, Bravo Company 4th Battalion, 503rd Parachute Infantry Regiment 'went to ground, unable to fire back' and one private fled, dropping his weapon.[114] 'Only a few of Bravo's paratroopers were firing back; Severson (the company commander) screamed at the others to shoot . . . Slowly, as the new combatants responded to their training and survival instincts, they began returning the enemy's fire.'[115] Other officers had to persuade their men to fire, many of whom 'were shocked into inactivity'.[116] There is little evidence to suggest that the citizen infantry in Vietnam were significantly different from those in previous conflicts.

Indeed, Glenn himself suggests that the performance of US troops in Vietnam was not unique. Although his research addresses Marshall's assertions about total firing rates in Vietnam, his real target is Marshall's claim that only one in four US soldiers in the Second World War used their weapon. Glenn wants to use Vietnam to invalidate Marshall's imputations about the citizen soldier in the 1940s.[117] Glenn's point is that since the infantrymen in Vietnam were mainly conscripts as they were in the Second World War, identifying firing rates in Vietnam would retrospectively provide a more realistic and palatable account of the performance of US troops in the Second World War. Glenn's evidence suggests continuity across conflicts but of a different kind than he might approve. Rather than refuting the Marshall effect, Glenn's evidence, notwithstanding that perhaps 80 per cent of soldiers in an elite division may have used their weapon at one point in their tour, implies the general inadequacy of citizen infantry and the low level of cohesion typically demonstrated by it. In the Second World War as in Vietnam, poorly trained infantry were disoriented by combat, especially initially, so that they were unable to execute their roles effectively. Indeed, the point might be extended to the First World War and Korea. In this, Glenn's argument, while intended as a refutation of Marshall's position, is consonant with it. Specifically, it affirms the generally low level of cohesion displayed by US citizen infantry throughout the wars of the twentieth century. Marshall's work is certainly not definitive in terms of non-US forces but it at least allows a broad observation about the citizen infantry to be forwarded at this point. As we shall see at the end of Chapter 7, it is possible to extend Marshall's point and to maintain that it was not only US citizen soldiers who collectively performed poorly in combat. At this point, it may be possible to suggest that low levels of cohesion were, in fact, the norm in the citizen platoon and were regularly evident in the case of the mass British, French, Italian, Canadian, and Australian armies in the twentieth century and even the much vaunted German armies of the First and Second World Wars. This proposal cannot be more than a suggestion at this point but it may be sufficient to initiate a deeper investigation into the problem of cohesion. It may provide at least a plausible working framework in which the evidence can be assessed.

Roger Spiller was surely correct to highlight the shortcomings of Marshall's methods and to caution against the self-promotional certainty with which Marshall made his claims. He certainly did not prove that only 15 to 25 per cent of US soldiers in the Second World War ever fired their weapons. However, Marshall's broadly sustainable if imprecise estimate about firing indicated something much more important. Developing a line of military analysis which extended back to at least Ardant Du Picq, Marshall recognized the organizational difficulties which the modern battlefield posed for the military. The close physical proximity of combatants to each other which had been a typical feature of warfighting up to that time had been displaced by mechanized weapons. The mass armies of the twentieth century struggled to find a solution to the problem of collective action which their industrial weaponry had set them; collective combat performance remained a challenge for citizen infantries. Despite the care with which some of his material needs to be treated, Marshall provided a broadly sustainable account of the underperformance of the citizen soldier, unpalatable though it was to soldiers who fought in these campaigns, to politicians, and, indeed, to many civilians. This distaste may be particularly pronounced in relation to the so-called 'Great Generation' which fought the Second World War. They have now become so enmeshed in national mythologies and contemporary nostalgia that dispassionate analysis of their performance has become a potentially emotive issue. There is a further problem. The performance of the citizen soldiers of this generation and the generations which fought the First World War, Korea, and Vietnam is seen through the prism of contemporary military operations and specifically through the lens of professional infantry performances. It is very easy to presume, especially since the citizen soldiers of the Great Generation fought in a major war of national survival, that they replicated the performance levels of today's professionals. History is elided with the present to produce a distorted reading of the past. It is important to avoid such a conflation and to assess the conduct of the citizen soldier as accurately as possible. Generally, as a result of the special conditions which confronted the citizen soldier in battle, his poor preparation for the reality of combat, the rifleman, as Marshall claimed, demonstrated low levels of cohesion and performed poorly as a collective body. Ultimately, in the major conflicts of the twentieth century, the infantry was unable to overcome the problem of dispersal and to engage in the complex tactics which industrial warfare demanded. They struggled, as Marshall had argued, to overcome the force of inertia among individual soldiers and to stimulate collective action. Marshall's firing rates may not be strictly always accurate in themselves but they point presciently to a severe social problem which modern armies had generated for themselves. They struggled to encourage their platoons to perform.

4

Combat Motivation

The mass armies of the twentieth century faced a predicament on the industrial battlefield of which many generals and military commentators were only too acutely aware. The devastating effect of mechanized fire forced their troops to disperse and take cover, but once they had done so it was very difficult to encourage them to participate in combat again. The survival of individual soldiers implied the emasculation of the army as an effective force. Marshall himself emphasized the importance of comradeship as a means of overcoming mass inertia on the modern battlefield: 'I hold it to be one of the simplest truths of war that the thing which enables an infantry soldier to keep going with his weapons is the near presence or the presumed presence of a comrade. The warmth which derives from human companionship is as essential to his employment of the arms with which he fights as is the finger with which he pulls a trigger.'[1] In a moment of sociological insight, he insisted that the soldier 'is sustained by his fellows primarily and by his weapons secondarily'.[2] Soldiers require 'some feeling of spiritual unity'.[3] In this way, Marshall not only begins to suggest a sociological answer to the problem of cohesion but he points to some of the strategies which mass armies from the First World War to Vietnam historically implemented in order to encourage their soldiers. Armies actively sought to accentuate the bonds of comradely affection, interdependence, and mutual obligation between soldiers which might motivate them to participate in combat, even at great personal risk. Armies—and the wider society—sought to unite their troops around a common concept of masculinity so that the notion of manliness became the standard of appropriate performance.

At the same time, armies appealed to the patriotism or national identity of their soldiers. Twentieth-century armies sought to unite their troops around common forms of social identity and civic obligation; they sought to indoctrinate their soldiers with a sense of national mission in order to motivate them politically. As will become clear, it is, in fact, very difficult empirically to separate masculinity from nationality as a basis of cohesion. In reality, the appeal to manhood was often simultaneously an appeal to nationality. In his work on nationalism and sexuality, George Mosse maintained that there was a conceptual isomorphism between masculinity and nationalism.[4] Both were conceived as unified and discrete entities which had to defend themselves from external threat. It is unnecessary in this context to go as far as Mosse but the internal connection between masculinity and nationality in the armies of the twentieth century emerges clearly from the historical record. An upright and respectable man was not only a rational, self-disciplined citizen who controlled himself but he was also, therefore, capable of defending the borders of his nation. Men proved their masculinity by protecting their nation which they were expected to defend precisely because they were men. The soldiers in the twentieth-century army earned the respect of the comrades and their nation but, in return for that honour, they had to be prepared to risk their lives in combat. Masculinity and nationalism, comradeship and political motivation, were, therefore, always intimately associated in the citizen army. Indeed, they were often integral to each other. They were recurrently drawn upon by political leaders, commanders, and soldiers themselves to motivate themselves and each other in combat and to overcome the Marshall effect. They were central to the generation of cohesion among the infantry in battle. However, although often mobilized simultaneously, for the purposes of clarity, the question of masculinity and nationalism or political motivation will be discussed separately in the following analysis.

MASCULINITY

The tactical performance of the Wehrmacht has been regularly applauded by soldiers and military scholars alike. Both Trevor Depuy and Martin van Creveld have emphasized the superior performance of the German army, maintaining that in all theatres, the German forces were approximately 50 per cent more combat effective than any other western force of equivalent size.[5] As already noted, Janowitz and Shils famously explained the high level of cohesion among the Wehrmacht by reference to the existence of strong 'primary groups' in the Wehrmacht. Janowitz and Shils recognized the importance of tactical doctrine, training, and formal discipline to the Wehrmacht but prioritized the comradely bonds between soldiers which fulfilled their masculine personality needs.[6] For Janowitz and Shils, the primary group was united around a common concept of masculinity. The men in it were unified by and related to each other by reference to their manhood; as equals, leaders, and followers. Indeed, the primary group serviced their fundamental psychological needs as men. It provided them with the recognition of other men for the successful performance of activities which they all agreed were definitive of masculinity.

In his work on the Freikorps, Klaus Theweleit highlights the centrality which concepts of masculinity played in the self-definition of these ex-soldiers who would go on to form the core of the SS units in the Second World War. However, since these soldiers were initially veterans of the First World War, Theweleit simultaneously illustrates the importance of collective concepts of masculinity to Ernst Jünger's generation of combat soldiers. Theweleit's work begins intriguingly with an examination of the letters which the men of the Freikorps sent home from the front line during the Second World War. The letters were ostensibly addressed to the wives of the soldiers but, in every case, the partners of these soldiers and indeed the entire domestic sphere were effaced from these notes. Instead these soldiers described themselves, Germany, and the front. They consciously washed themselves of potentially emasculating associations with the feminine world of their homes. From this starting point, Theweleit explores the masculinity of this Nazi vanguard. These soldiers regarded their masculinity as fragile and requiring constant vigilance and effort to maintain. In particular, the self-identity of the soldiers of the Freikorps was threatened by internal sexual drives and the external threats; particularly, women and communists. These threats were conceived as pulp, slime, and floods which could be extirpated only through parades, drill, and, finally, the gun.[7] The soldier male's self was created through its incorporation into social institutions and, above all, the army. Through its rigid institutionalization, the threat of self-dissolution was suppressed. Drawing on the letters of ex-Freikorps soldiers serving during the Second World War, Theweleit concludes the absolute priority of the national mission over family commitments. While wives and children disappear from view, the dedication to the national mission is paramount: 'His male self-esteem is dependent on the status of Germany, *not* on his actual relations with a woman.'[8] The masculinity of the Freikorps lacked the fraternal intimacy which is evident in Janowitz and Shils's analysis of the Wehrmacht. Nevertheless, Theweleit's work demonstrates the important role which masculinity played in unifying elite German soldiers. Indeed, General Gunther Blumentritt, who was Chief of Staff of the western higher command (OKB) during the Normandy campaign and was tried at Nuremburg, claimed that it was precisely this over-masculinized identity which had been at the root of German militarism in the Second World War: 'the majority of officers had little time to devote to their families and in which to occupy themselves with other than military matters.'[9] Through their collective incorporation into the army, soldiers mutually sustained each other's manhood from the threat of dissipation. Theweleit's soldiers bonded as men against the defiling mass of the world. Afraid of being swamped by that mire and motivated by the need to prove themselves to their fellow males, these soldiers committed themselves to a crusade against their enemies.

At the same time, German doctrine emphasized, in a perhaps more prosaic way, the importance of fraternal comradeship to military duty. In 1940, a reserve officer, Major Dr Reibert, published a widely circulated work, *Dienstunterricht* (Service Instruction), which was an edited collection drawn from a number of important doctrine publications. The *Dienstunterricht* began by declaring: 'The Service Instruction is an important means to achieve a mental and spiritual feeling between superior and subordinate.'[10] Following a long discussion of military honour, faith, and virtue, to which all German soldiers should dedicate themselves, Reibert went on to specify the nature of this spiritual communion between soldiers: 'Comradeship is the bond, which binds all soldiers closely. They feel with conviction that one will not leave another in need and danger. "All for one and one for all!" Comradeship is not bound by origins, age, education, rank or anything else.'[11] German infantry doctrine of the time (on which Dr Reibert drew) affirmed the point. The famous 1922 version of *Ausbildungsvorschrift für die Infanterie* (AVI) (Training Rules for Infantry) described the section as the smallest tactical unit and emphasized its distinctive cohesiveness: 'Its members are bound in true comradeship through life and death.'[12] *Ausbildungsvorschrift für die Infanterie* of 1936 affirmed the point. Closely echoing the paternalistic role which Janowitz and Shils claimed that the junior commander played in the Wehrmacht, the 1942 version of *Ausbildungsvorschrift für die Infanterie* described the platoon commander's duties in significant terms: 'He should know every man in his platoon by age, profession, personality and domestic situation. He ought to build up a judgement about the fundamental character, bodily and spiritual capabilities of the individuals, about his position in the group of comrades and in particularly his conduct in combat. The platoon commander is the friend and adviser of his squads [note the use of *Mannschaften* here not *Gruppen* possibly to emphasize the masculine bond]. He should through encouragement and enthusiasm seek to promote joy in service and enterprise as well as to strengthen the feeling of fellowship.'[13] It was clear that Reibert conceived of this comradeship as a male brotherhood.

Doctrinal statements, especially characterized by this level of rhetoric, can fail to reflect the realities at the level of the combat soldier. On the front line, soldiers might be motivated by entirely different considerations and the basis of their solidarity might be quite different. However, during the war itself, the evidence shows that the overwhelming majority of German combat soldiers were united by a common concept of manhood. They actively drew on this concept to sustain themselves. Lieutenant Colonel Dr Henry Dicks, a prominent psychologist drafted into the British army, produced a long research memorandum for the Directorate of Army Psychiatry on the psychology of the German soldier in February 1944 on the basis of archival research and interviews with prisoners of war, including Rudolf Hess, over a fourteen-month period. Although ponderous, he concluded that a heightened concept of masculinity (which he claimed was the result of over-indulgence in childhood and a compensation for a weak ego) was central to the self- and collective definition of the German soldier: 'It is not stretching the imagination to see in the German devotion to authoritarian rule the derivative of father-rule or patriarchy, with its emphasis on the virtues of manliness, by which they mean "hardness". The Germans have persuaded themselves and the world that they are the most virile and martially-minded people on this earth.'[14] Dicks noted the low status of women in Germany and recorded how German prisoners of war not only held up their hands 'in horror when asked if they would like daughters' but also doubted 'the capacity of their wives to bring up their sons in their absence'.[15] 'It is characteristic of this "manliness" cult that it identifies femaleness with inferiority and instils a sense of guilt about any qualities connected in some way with the tender relations between little boys and their mothers. This perfectly natural relation becomes one of the forbidden needs, condemned by the German equivalents of such expressions as "milk-sop", "Mummy's boy", "cry-baby".'[16] The ideal of masculinity was essential to soldiers' own self-esteem and the esteem in which they were held by their peers. This close masculine interdependence is demonstrated in personal accounts of the war. For instance, in his memoir of the Eastern Front, Guy Sajer recorded the importance of these personal masculine associations. During one battle, one of Sajer's closest comrades confessed his emotional dependence on him when Sajer threatened to give up and accept death, since it made no difference whether he was killed now or later; death was inevitable, 'what difference does it make?' His friend cries out: 'The difference is that I need to see your face from time to time.'[17] While they had lost all loyalty to the Wehrmacht and commitment to their war, their personal bond to each other and to their officer endured. Dinter elaborates upon Janowitz and Shils's argument to emphasize the importance of masculine small-group solidarity to the effectiveness of fighting forces. He examines the German army, specifically at Stalingrad. For Dinter, masculinity was particularly important for leadership. For instance, Dinter records a case where a group of German soldiers on the Eastern Front refused to fight any more as a result of a lack of food. However, carefully handled by their highly respected commander, Captain Münch, the rebellious troops agreed to follow orders so long as they were under his command.[18] They were willing to follow a leader whom they saw as an ideal man and who therefore demanded similarly honourable conduct from themselves.

The performance of Germany's primary groups in the Second World War may have been exceptional and, certainly, as Theweleit emphasizes, the masculinity of those former Freikorps SS soldiers was extreme—even pathological. Nevertheless, masculinity was an important resource for Allied soldiers in both wars. Tragically, in the First World War, Kitchener's volunteer armies of the First and Second One Hundred Thousand which would suffer terribly on the first day of the Somme actively sought to exploit the civilian social relations between male volunteers. Pals battalions, consisting of friends or workmates, were raised from local communities:[19] 'Some men volunteered as a result of pressure from their employees or social superiors.' Alternatively, 'another class of recruits was made up of men whose chief reason for joining the army was that all their friends were going'.[20] Of course, these friends were all men and the bond between them was explicitly seen in masculine terms. Accordingly, those individuals who ignored this moral pressure were subject to ridicule and dishonour; reluctant warriors could expect to receive a white feather to denote their cowardice. The concept of masculine honour was explicitly employed to encourage performance. The friendships between male soldiers were regarded as a key means of generating cohesion at the front.

In the Second World War, masculinity was no less important for British soldiers. Dinter shows the Allied soldiers were willing to fight out of the fear of disgrace in front of their male comrades.[21] In the British army, this comradeship assumed a rather more homely friendship than the stern Teutonic fraternities which Theweleit and Reibert describe but it was nevertheless an important reference point. Thus, in May 1940, the British army's 30 Brigade fought a stubborn rearguard action at Calais against advancing German troops in order to allow for the evacuation of the troops, even though the members of the brigade knew that no escape was possible for them. Their commander, Brigadier Claude Nicholson, explained their performance in terms of personal bonds: 'The regiment had always fought well, and we were with friends.' Kellett adds: 'Most of the personnel of the regular battalions had been together for years.'[22] Instructively, in emphasizing friendship, the brigadier was in line with British doctrine: for instance, *Infantry Training* (1944), one of the most important and useful documents to be produced by the British army during the Second World War. It includes extensive material on sophisticated infantry tactics but, nevertheless, it too stresses the importance of existing social ties to military performance. When describing the structure of sections, it states that these units are organized into two or (more unusually) three groups, one or two rifle groups and one Bren gun group.[23] The manual recommends that 'Groups are formed from friends as far as possible, in order that friends keep together and fight together.'[24]

The appeal to masculinity was consistently prominent in the United States Army throughout the twentieth century. Indeed, Sam Stouffer in his survey of US soldiers in the Second World War concluded that 'combat posed a challenge for a man to prove himself to himself and others'.[25] Masculinity was a key motivating factor used to encourage solidarity on the line and 'the man who lived up to the code of the combat soldier had proved his manhood'.[26] Conducted during the Korean conflict, Roger Little's research provides a useful insight into the enduring importance of masculinity to the cohesion of the citizen army. Little explores the 'buddy system' within the US infantry platoon in order to establish the mechanisms by which cohesion was established. Little demonstrates that everyone in the platoon was a soldier's buddy to whom each man was self-consciously bound through a 'network of interpersonal linkages' in a round of reciprocity and mutual support;[27] 'thus buddies also constituted a status group'.[28] Buddies shared everything, including decisively the dangers of combat;[29] 'You don't always have a chance to do a favour in combat, but if you share everything, you can be pretty sure that your buddy will remember it if you need help.'[30] Interestingly, Little observed that within the community of buddies, each soldier had a special attachment to one other man; his personal buddy. This friend was often not part of his formal section or sub-unit but they would choose to spend time together in the 'chow line', at night, and during work squads when the combat formation was broken up.[31] These personal buddy relationships affirmed and strengthened the wider buddy system but they were also a potential threat; should these relations become too obvious, other soldiers in the platoon would become jealous or annoyed that two attached individuals were not contributing sufficiently to the platoon as a whole: 'You've got to make every man in the squad your buddy...But I never show one man is my buddy because a lot of guys may think that I'm a buddy.'[32]

Based on emotional need, the buddy system enforced conformity through the exercise of moral judgement. Soldiers were ridiculed, shunned, and finally excluded for being 'duds' or 'heroes'. Duds were selfish individuals unwilling or incapable of contributing to the platoon, while heroes risked their comrades' safety through self-serving displays of bravado. Yet, both suffered similar social sanctions. Significantly, while traditional concepts of 'being a man' were central to the buddy system, excessive displays of aggression and independence were disparaged because they led to selfish heroism.[33] Masculinity was not irrelevant to the buddy system; on the contrary, it remained at the heart of the system of interpersonal solidarity. Being a man in the buddy system was behaving in a buddy-like way, helping comrades, performing common duties, and sharing danger.

In the First and Second World Wars, the Australian army perhaps utilized the most self-consciously institutionalized appeal to masculine comradeship to encourage battlefield performance. As an important predominantly white member of the British Empire, Australia committed itself to these conflicts from the outset and, relative to its size, made significant military contributions to the Western Front and to the Gallipoli campaigns in the First World War and the North African, Italian, and Far Eastern campaigns in the Second. The Australian contribution was all the more striking since only the Borneo and New Guinea campaigns represented the defence of Australian national territory, yet numerous Australian soldiers fought and died thousands of miles from their homeland in the First and Second World Wars. While Australia at this time identified closely with the British Empire and indeed with the monarchy, Australians contrasted their society with the rigid class hierarchy of Britain. Australians saw themselves as part of a more egalitarian and open social order and this concept had a major influence on the Australian army which fought in the two world wars. While the British army was derided as hierarchical and disciplinarian, Australian soldiers were bound by relations of 'mateship'. They understood themselves to be part of a fraternity accepting the dangers of the front together and fighting for each other, not for orders or their officers. This was seen to reflect the more open frontier society from which they supposedly came, even though most soldiers were recruited from urban areas.

It is important to avoid the idealization of the institution of mateship in the Australian army. Garth Pratten's work on the battalion commanders in the Second World War clearly demonstrates that the concept of mateship has become mythologized so that it does not so much reflect the realities of the Australian army as provide an emotive resource for collective national memories in the latter part of the twentieth century and early decades of this one.[34] Many battalion commanders in the Second World War were harsh martinets whose bonds with their men could not be described as in any way comradely; they were ferocious disciplinarians just like their British counterparts with similarly middle-class and professional backgrounds (in stark contrast to many of their men).[35] It seems likely that the same may have been true in the First World War. Nevertheless, while perhaps exaggerated later, mateship was an important point of reference for Australian soldiers and consciously understood by them to oblige soldiers to perform in combat.[36] Mateship remained an important resource for soldiers into the Second World War and, indeed, even in Vietnam it remained an important aspect of the collective identity of the Australian soldier. Mateship bound Australian soldiers as men in primary groups obligating them to perform in combat.[37]

Primary groups unified themselves as bands of men on the front, but the wider society also utilized a similar concept of manhood to enjoin appropriate behaviour from these male groups. Moreover, as Dinter also noted, it was in this performance of the sanctioned masculine role that the family became important. Interestingly, and in contrast to Theweleit's Freikorps, Dinter emphasizes that while women had to be excluded from the comradeship of the front,[38] wives and families played an important role in sustaining these fraternities. They provided emotional support for the soldiers but, perhaps just as significantly, they exerted significant moral pressure on them. Effectively, soldiers on the front line were compelled to perform well—and to act as good primary group members—out of fear that the shame of cowardice would redound upon their families at home. Soldiers' wives and families mobilized a concept of masculinity which delineated honourable and dishonourable forms of military service for their husbands and sons.

Soldiers might never have genuinely believed they were fighting for democracy or their country but they did understand themselves to be defending their wives and families; their failure to live up to the male ideal in theatre would dishonour those people at home they invested with the most importance. This was, in fact, a common strategy by governments and the armies in the First and Second World Wars. For instance, during the First World War, the British War Office, recognizing the power of feminine influence, petitioned women to ensure that their young men fulfil their national duty in a series of public announcements; 'To the young women of London. Is your best boy wearing khaki? If not don't you think he should be? . . . If your young man neglects his duty to his King and Country, the time may come when he will *neglect you.*'[39] Roger Little noted the use of similar moral leverage during the Korean War. Little noted that 'events which reinforced the soldier's relations to some meaningful element of the larger society—especially his family—correspondingly strengthened his relation to his organization'.[40] The family were actively employed to encourage honourable performance from US soldiers. The US Army recognized that the letters were vital in maintaining this connection and pressure: 'The conception of his role as the citizen of a community or a member of a family was influenced by the letters written to him by persons whose evaluations of him were very important.'[41] Moreover, the army were well aware of the way in which decorations and insignia earned while on operations could enhance the soldier's masculine standing in his community on his return from duty. While disparaging the status markers when on the line, since they represented the chain of command and potentially promoted the individual over his comrades as a despised hero, soldiers actively participated in this process once they were recalled from duty and returned as individual servicemen back home: 'When the time for rotation approached and he anticipated the reaction of his family and civilian friends, he became increasingly aware of the symbols that he had previously deprecated with his peers.'[42] This does not undercut Little's argument that soldiers generally rejected the concept of heroism. Rather, it reveals that the reputation of soldiers as honourable men in their home communities was an enduring influence upon the citizen soldiers of the twentieth century even on the front line.

The question of cohesion began to animate American sociologists in the 1970s precisely because the morale and discipline of the US Army had collapsed so spectacularly in the final years of Vietnam. One of the most important contributors here was Charles Moskos with his discussion of latent ideology and racial conflict which will be considered more fully below. As noted in Chapter 2, Moskos was sceptical about how much weight could be put upon the concept of the primary group. Moskos disparaged the view that primary groups represent 'some kind of semi-mystical bond of comradeship'.[43] However, he maintained that soldiers committed themselves to their fellow platoon members whose cooperation they sought to enjoin and reciprocate by comradely activity. Appropriate conduct in this context which would engender reciprocity was defined as manly. Indeed, Moskos notes that the armed forces themselves mobilized a concept of masculinity throughout the Vietnam War to encourage US males to join particular services: 'The Marine Corps builds men' and 'Join the army and feel like a man' were examples of this appropriation of masculinity. Moskos emphasizes the role which 'manly honour' played in attracting men to and sustaining them in combat units. Thus, airborne units in Vietnam disparaged non-airborne personnel as 'legs' while the virility of rear-echelon troops was constantly disparaged by all forward troops; they were 'REMFs', rear-echelon motherfuckers, i.e. they were not proper men. For Moskos, 'the observations made on the ethic of masculinity among the *Wehrmacht* during World War II seem equally appropriate to American soldiers in Vietnam'.[44] Male soldiers, united around a common concept of masculinity, were prepared to risk themselves for their fellow soldiers on this basis. Their reputation as men in the eyes of their immediate peers was actually in the end more important to them than a rational preservation of their own lives.

Robert Eisenhart provides a compelling, even shocking, account of how the US Marines—and by implication the army—actually built men for the Vietnam War. He illustrates how the concept of masculinity became central to the collective self-definition of US Marines. He was himself drafted into the US Marines for Vietnam and went through boot camp. He records a degrading system of brutalization in which violence was routine and every transgression was punished with physical assault. 'The conscripts were constantly accused of being inadequate as men; they were women, girls or homosexuals.' 'You can't hack it, you goddamned faggot' was a favoured form of abuse.[45] In one imaginative piece of sadistic punishment, Eisenhart and his fellow recruits were forced to place their penises in the breeches of their weapon as a censure for ineffectiveness.[46] The punishment was very painful but it also physically demonstrated the centrality of the concept of masculinity to soldiering at that time. The recruits' weapons were just an extension of their masculine personalities and poor military performance was a dereliction of their manhood; it emasculated them. At one point during training, Eisenhart choked a fellow recruit during unarmed combat which led, not to rebuke as might be expected, but to approbation; 'the drill instructor gleefully reaffirmed my masculinity in front of the platoon saying I was a lot more of a man than he had previously imagined.'[47] The article reaches its climax with a discussion of Private Green, an 'effeminate' recruit who 'could not keep up'. He became the object of hatred not only for the training staff but for the recruits themselves who were punished for his misdemeanours.[48] He represented precisely the weakness and femininity, the 'exorcism' of which was one of the main purposes of basic training.[49] Green was eventually severely beaten by Eisenhart and his fellow recruits. Eisenhart concluded that in the US Marines during Vietnam masculinity was equated with the performance of military activity and that sexual identity was linked to military function.[50]

Eisenhart's experiences were, perhaps, extreme; it is clear that he feels that he was personally damaged by his experiences in basic training and in Vietnam. However, responding to Vietnam, other sociologists also emphasize the critical role of masculinity in uniting the US conscript forces. Between 1974 and 1980, Cockerham, Aran, and Arkin and Dobrofsky produced significant pieces in prominent sociology journals which addressed the issue of cohesion. Arkin and Dobrofsky's piece most implicitly identified masculinity as central to solidarity in the armed forces. The process of military training aims to turn 'the boy into a man'.[51] Closely echoing Eisenhart's work, Arkin notes that basic training was a rite of masculinity in which all those habits and traits which were interpreted as feminine were expunged from the recruits.[52] To this end, the symbolic connection between the rifle and the penis was regularly emphasized, both represented as the tools of 'pleasure and fun'.[53] In order to become true men (and to be able to use their penises properly), recruits had to master their weapons. Recruits were constantly accused of being 'faggots' while femininity was recurrently denigrated and objectified in an attempt to 'strengthen the intended masculine imagery'.[54] Both Cockerham and Aran employ parachuting training as a focus for their investigation of cohesion. Cockerham's analysis is initially consistent with the work of Arkin, Eisenhart, and, insofar as he discusses masculinity, Moskos. Cockerham claims that 'the very act of parachuting' 'tends to promote feelings of self-worth, accomplishment and personal mastery. It also tends to foster group bonds and social cohesion.'[55] Cockerham concludes that airborne training is a 'status passage': it is a 'test of manhood'.[56]

While Cockerham ultimately puts forward an alternative explanation of cohesion, Aran's analysis of parachuting more obviously foregrounds the role of masculinity. In that piece, Aran describes, in Durkheimian fashion, how military parachuting constituted a functionally useful moment of anomic individuality which assisted in reinforcing the social solidarity among paratroops.[57] Crucially for Aran, the jump experience resembles 'rites of passage'.[58] Paratroopers are part of a highly cohesive social group but, as they launch themselves from the plane and in the brief moments as they hang beneath their parachutes, they are free of all social ties. They become individuals. For Aran, this affirmation of the individual has the ultimate effect of uniting paratroopers even more closely once they have landed. Anomic individuality promotes mechanical solidarity because airborne units are filled only with individuals who have proved their bravery and self-discipline; 'parachuting is a sort of test' which 'helps the organization distinguish between the action seekers and those who are not'.[59] Arkin does not explicitly discuss masculinity but his notion of the strong individual forged by parachuting who is an action seeker is plainly gendered; he is male.

There is ample evidence that in the citizen army of the twentieth century itself, masculinity was recurrently and regularly drawn upon to motivate troops in the line. Male citizen soldiers understood themselves and sought to justify and accept their sacrifices by reference to their manhood. Masculinity was not then simply an abstract idea; it was an active collective reference point for civil society, armies, and soldiers themselves by means of which they engendered tactical cohesion. The moral obligation of one buddy to another was identified and used as a means of overcoming the problem of the modern battlefield. Bound together by close personal bonds, soldiers might be encouraged to fight rather than shirk or run away. Since their status as men—both in the eyes of their comrades and at home—was central to the generation of this sense of obligation, masculinity as a regular and self-conscious means of collective self-definition was a central means by which the mass armies of the twentieth century overcame the Marshall effect. Citizen soldiers were encouraged to fight because they were men, and a soldier who failed to perform in combat had voluntarily relinquished his masculine status. The appeal to masculinity and to peer-group reputation would seem to be a frail basis on which to build the combat power of an army. Yet, the evidence demonstrates that overwhelmingly the threat of emasculation and the loss of social credibility was often sufficient to encourage groups of citizen soldiers to fight even if, in many cases, it meant their deaths.

POLITICAL MOTIVATION

Masculinity was an important means of unifying soldiers in the twentieth century but there were obvious limitations to the kind of cohesiveness which it could generate, which were noted in Chapter 2. As Stephen Wesbrook has noted, while primary groups might be essential to combat performance and to motivating citizen soldiers normatively, these bonds might just as easily undermine the army: 'The most serious problem in relying on primary group cohesion to prevent disintegration is the well-documented fact that bonds between members of primary groups can develop into a basis of opposition to the larger organization.'[60] In this case primary groups manifest 'resistance to organisation goals'. The problem was discussed in Chapter 2. Consequently, Westbrook suggests that the prevention of military disintegration requires not only primary groups, whose members are united at a personal level, but an overarching commitment to the political goals of the mission and to the military-political hierarchy. As Wesbrook observes, 'the basic problem is that the soldier must not only respond to the demands of his peers while fighting but also the demands of the nation and the military organization to fight'.[61] Some political ideology is, in short, necessary to bind primary groups to organizational goals. Indeed, Wesbrook appositely notes that while professional armies can sometimes rely on unit affiliation to ensure allegiance to higher organizational goals, the fostering of regimental and divisional loyalties which would link the citizen soldier automatically to larger organizational interests is 'often impractical for the mass army' which has insufficient time to generate these expansive bonds of military solidarity.[62] Here, in the mass army, the masculinity which binds the primary group together reasonably effectively requires some form of nationalist consciousness to align primary groups with overarching military and strategic objectives. Wesbrook concludes: 'The prevention of military disintegration in modern military organizations first depends on the soldiers' moral involvement with their primary groups and either their unit or their national socio-political system, but ideally both.'[63] Citizen soldiers need political motivation in order to fight effectively. They need to feel collectively that their sacrifices promote the shared interests of their nation (and are recognized by citizens at home as doing so). This motivation is, of course, heightened when their enemies are seen to represent a genuine threat to their community, its territories, or its way of life.

The role of political motivation in the form of patriotism was very obvious in the First World War, where fervently patriotic nationalism was used to stimulate the troops. In his study of Kitchener's first army, Simkins tries to show how other factors, such as economic insecurity or boredom, contributed in certain cases to the recruitment boom in 1914 and 1915. However, overwhelmingly, his study demonstrates that even if not all British men were energized with missionary zeal, 'the bulk of Kitchener's volunteers were motivated by a sense of duty and obligation'.[64] Illustrating this sense of national duty, Kitchener's famous recruiting poster addressed the conscience of each individual man with the demand that 'Your King and Country Needs You'. Commentators at the time noted that 'Lord Kitchener is confident that this appeal will be responded to by all those who have the safety of our Empire at heart'.[65] It was not just politicians or field marshals who understood the First World War in terms of the defence of the Empire. There was 'widespread indignation at Germany's action'; 'our country was one hundred per cent right and Germany one hundred per cent wrong... We had been taught to worship God one day a week but to worship Country and Empire seven days a week.'[66] Letters from soldiers before the Battle of the Somme affirm a strong sense of patriotism. 2nd Lieutenant Jack Engall of the London Regiment wrote: 'The day has almost dawned when I shall really do my little bit for the cause of civilization.' He concluded his letter: 'I am quite prepared to go; and I could not wish for a finer death; and my dear Mother and Dad, will know that I died doing my duty to my God, my country, my King.'[67] Another lieutenant wrote that, despite 'our babe may grow up without my knowing her or her knowing me', 'I pray God I may do my duty'.[68]

Interestingly, the British army saw sport as a critical means of realizing a national identity and pride in the action of small groups: 'In all the national games men of British race submit with enthusiasm to training and discipline for the sake of the side; they have an inborn instinct which makes them naturally work for the side and play the game.'[69] The connection between sport and nationalism was manifested in a number of different ways in the First World War in the British army. Perhaps most obviously, it was illustrated by the popularity of Henry Bolt's poem *Vitai Lampada* which celebrates stoic British sacrifice. Even in the face of certain death in battle, the officer in the poem is able to recall his schoolboy sporting experiences, challenging his enemy to 'Play up and play the game'. On the first day of the Somme, this connection between sport, nationalism, and teamwork was demonstrated with perhaps most pathos by Captain Nevill of the East Surrey Regiment. Commanding an assault company in the 18th Division, he conceived the idea of procuring a football for each of his platoons which would be kicked into no man's land at H-hour and followed by them; the winner being the first platoon who got its ball and themselves into the German lines. On one of the footballs, Nevill wrote, 'Great European Cup-Tie Final, E Surreys v Bavarians, Kick off at Zero'. The footballs were meant to distract his men from dwelling excessively on the potential dangers which they faced,[70] but Nevill was also explicitly seeking to encourage cohesion among his men by appeals to British national identity; the football was a collective reference point communicating a sense of imperial pride and destiny. On 1 July, Nevill's footballs soared into no man's land followed by the East Surrey Regiment. In the end, the 18th Division assault of which the East Surreys were part was completely successful, though the Surreys themselves took heavy casualties, including Nevill himself, who was killed.

Perhaps in a somewhat wiser form, nationalism continued to play an important role for the British army in the Second World War. Indeed, although very mild in comparison with Reibert's rhetoric, British doctrine too recognized that a concept of national mission was essential to competent military performance: 'Throughout all training, therefore, while skill is being acquired, the necessity for fostering morale must always be borne in mind so that the two qualities may grow together. The moral qualities to be developed by the soldier include self-control, self-respect, *patriotism*, loyalty, *pride of race* and a high sense of honour.'[71] Patriotic nationalism and imperial pride was certainly less strident in the 1940s but it was nevertheless very evident among British soldiers. Indeed, the army sought to encourage officers to foster this sense of moral obligation among their troops: 'To develop morale in the people as a whole is an essential part of the policy of the nation; to foster moral qualities in his force is the imperative duty of every commander, because without them full development of fighting power is not possible.'[72] It was not simply generals who saw the moral component—the sense of collective national mission which demanded honourable self-sacrifice—as essential to victory. There is manifest evidence that soldiers on the front line, who had to make this sacrifice and who were typically more sceptical of abstract notions of national mission, also understood themselves in these terms. In her analysis of letters sent from the front during the Second World War, Margaretta Jolly maintains that 'these [are the] kinds of letters in which we see the "private man" in the soldier, defined by feelings not action and by the individual woman he loves, not the army he serves'.[73] Jolly is certainly right to explore this subterranean world of intimacy obscured by the workings of a vast state-military machinery during the war. Interestingly, however, while the letters which she has examined avoid the virulent ideology of Theweleit's archive, the sense of national purpose is palpable, even alongside the expressions of private sentiment. Thus, Jolly cites a 1941 posthumous letter from one soldier, Ivor Gwyn Williams. The soldier notes that 'my last thoughts were of my family and you' and he wishes that his sweetheart would have 'the happiest married life' (having found a new man). He also offers his thoughts to his sweetheart's 'Mother, Father and all your friends and relations'.[74] Decisively, however, Ivor Williams acknowledged his role as a soldier fighting for his country: 'I have died in fighting for our and others countries.'[75] Even British conscripts articulated a developed sense of political purpose.

A similar nationalism can be observed in the Canadian army in both the First and Second World Wars. In the First World War, the first contingent of volunteers consisted of British-born Canadians who were eager to serve King and Empire, as they understood it.[76] In English-speaking Canada, young men not in uniform were regularly accosted in the street by women who would pin white feathers on their lapels.[77] Indeed, this patriotism could sometimes find expression on the front. In order to motivate troops about to counter-attack at Ypres in April 1915 following a German gas attack, a padre exhorted the men by assuring them that 'It's a great day for Canada, boys', although he may have been greeted with some scepticism by soldiers about to go over the top.[78] In the Second World War, a sense of Canadian nationalism and imperial obligation continued to motivate Anglo-Canadians to volunteer or to be willing to be conscripted. The nationalist motivation of Anglo-Canadians becomes most evident when it is contrasted with the francophone community. In stark contrast, Québécois Canadians were generally unwilling to serve in either conflict: 'Years of propaganda about the "Anglos"' and 'absence of recent links to Europe' ensured that French Canadians, with an ambiguous if not hostile relation to English-speaking Canadians and to Britain, did not feel it their patriotic duty to defend either the Empire or France.[79] As a result of this patriotism, Canada was able to raise a corps in the First World War and a full army in the Second primarily from the anglophone community, which participated in the major campaigns of those conflicts on the Western Front, North Africa, Italy, and Normandy, taking heavy casualties in each theatre.

Defending its own national territory from over-aggression, patriotism was important to the French army during the First World War, although it was, as

Hew Strachan has shown, a more contested and less enthusiastic affiliation than is often recognized.[80] France may not have universally celebrated the approaching war, as subsequent accounts often presumed, but there was as total a consensus as was ever likely in any liberal society that the war was justified, had to be fought, and that the male population were willing to fight it. Significantly, as troops began to mobilize, patriotism became more evident; 'As units marched off to the strains of the Marseillaise and *Chant du Depart*, tricolours waving over their heads and flowers falling at their feet, public and private feelings fused.'[81] These public displays gave substance to the *union sacrée* and united the nation and its army into a sustained effort.[82] In the Second World War, Gallic patriotism was not sufficient to prevent the humiliating capitulation of the French army in 1940. Nevertheless, it was explicitly identified as an important motivational resource by the French army throughout the inter-war period and the Second World War itself. This is particularly clear in the second part of the Ministry of War's *Règlement de l'infanterie* of 1940 which was republished in 1944. According to these regulations and despite the mechanization of warfare: 'The infantry is charged with the principal mission in combat.'[83] Not only is this mission the most important but it is the most difficult: 'The infantry has the hardest task, but also the most important in combat: the hardest because it leads the fight right up to hand to hand combat in order to conquer and retain ground: the most important because it is this, by definition, which ensures success.'[84] The *Règlement* invokes the existential realities of combat for the infantry soldier:

Of all combatants, the infantry soldier is the one who is subjected to the greatest degree to the sufferings of war. The fatigue of long marches and hard work, sleep deprivation, hunger, thirst, bad weather, heat as well as cold, the pain of injuries threaten to overwhelm him [*viennent tour à tour l'accabler*]. He holds onto the objective steadfastly against all the enemy's power of destruction; he not only lives through a constant struggle against death; he also constantly witnesses the most horrible aspects of the battlefield, every day he sees the death of his leaders and dearest comrades around him.[85]

Echoing Clausewitz's famous comments on military virtue, the ability of the French infantryman to bear his privations was a badge of honour since 'it is by his suffering that he wins victory'.[86] Significantly, a special type of man was required for this extraordinary role: 'The infantry soldier must therefore be a man of particularly strong body and spirit. The task of the infantry is above all glorious. The good citizen should consider it an honour to serve in its ranks.'[87] For the French army, the duty of soldiering could be executed only by those men who were not only physically and morally sound, but who were inspired by the highest sense of national duty and patriotism. Their reward for suffering was glory; the recognition and gratitude of the French people and their fellow citizens. Despite the experience of the First World War and its recognition of the primacy of fire on the battlefield, the *Règlement* nevertheless still regarded combat as 'in the final instance a moral struggle',[88] and identified patriotism as the basis of all morale:

> The moral force of the soldier has as its basis his faith in the grandeur and the destiny of the *Patrie*, his conviction that he is fighting for a just cause, his confidence in his leaders and his comrades and a feeling of his own worth as a combatant. Patriotism is born in the family, it is developed in the school and encouraged in the barracks. It is the love of France, not only for its beauty and riches of its landscape, its intellectual and moral heritage and the glory of its history, but also for its radiance in the world.[89]

The point was confirmed in the first part of the *Règlement* which dealt purely with the instruction of the infantry, but also emphasized the importance of moral training: 'To encourage (*exalter*) patriotism and the spirit of sacrifice, to inspire confidence, to make the necessity of discipline understood, to develop the sentiment of duty [*du devoir*] and of camaraderie, that is the object of the moral education of the soldier.'[90] In a work of instruction which prescribed the ways in which the infantry was to be organized, these passages are surprisingly lyrical. They are deliberately designed not merely to communicate the importance of patriotic duty but to invoke nationalist sentiment and pride itself. Indeed, emphasizing the significance with which patriotism was invested, in its report to the ministry, which was published as a foreword to the work, the military commission which composed the *Règlement* emphasized the centrality of nationalism to the performance of the infantry, repeating verbatim the requirement to have only good citizens in its ranks.[91] In this way, the centrality of nationalism as the motivating factor for France's citizen army as it approached the Second World War is unignorable.

Despite the catastrophes of 1940, nationalism remained an important source of motivation for those Free French forces which continued to fight for the liberation of France and, indeed, for the Vichy regime which saw itself as loyalist. Indeed, when de Gaulle became openly complicit in British attempts to destroy what remained of Vichy's colonial forces with the failed assault on Dakar in September 1940, he was portrayed as a traitor and fratricidal murderer by the Vichy regime.[92] Despite these accusations, not only did Charles de Gaulle have a clear nationalist agenda in the commitment of French troops (along with a suspicion of British motives) but republican patriotism was central to the definition of de Gaulle's Free French troops themselves. General Leclerc's famous Deuxième Division Blindée was an extraordinary division which united 'arch-Catholics of *la vieille France*, Communists, monarchists, socialists, republicans and even some Spanish anarchists' in the common pursuit of the liberation of France.[93] Their arrival at Utah Beach on 1 August was 'intensely emotional' since most had not seen their country for four years and some 'scooped up handfuls of sand on Utah beach to preserve in jars'.[94] Many civilians in Normandy volunteered to join the division in a wave of nationalist euphoria following liberation. The subsequent performance of Leclerc's division in the Battle of Normandy and liberation of Europe demonstrated the deep commitment of his troops to national goals.

Charles Moskos has emphasized the role which political ideology has played for US conscripts throughout the twentieth century. Although US combat troops may have abjured excessive displays of patriotism and rejected the cruder attempts at ideological indoctrination outright, they were still motivated by a common sense of national purpose. In particular, they viewed themselves as defending their families and friends and a general American way of life. For Moskos, the 'salient' values of this latent ideology were materialism and a belief in the 'worthwhileness of US society'.[95] US soldiers wanted to protect the standard and style of living which they and their families enjoyed. In fact, especially in the early years of that campaign, there is evidence that many US soldiers did feel highly politically motivated. They understood their participation in patriotic terms and regarded the mission as strategically important in defending their United States. For Moskos some commitment to worthwhile national goals is essential to the performance of citizen soldiers. Significantly, Moskos suggestively claimed that latent political ideology did not simply motivate the individual at a psychological level but became a collective resource for small groups of American soldiers: 'individual behaviour and small-group processes occurring in combat squads operate within a widespread attitudinal context of underlying value commitments.'[96] Like appeals to masculinity, general commitments to national goals were drawn upon in the micro-interactions within primary groups and collectively used by soldiers to comprehend their sacrifices and to impose expectations on each other. Nationalism was not a substitute for the primary group but defined, justified, and augmented the mutual obligations between the members of the platoon. Anticipating Wesbrook, Charles Moskos suggests that 'combat motivation arises out of linkages between individual self-concern, primary group processes and the shared beliefs of the soldiers'.[97] Moskos usefully concluded: 'The fact that the American soldier has a general aversion to overt ideological symbols and patriotic appeals should not obscure latent ideological factors which impinge upon the soldier's willingness to exert himself under dangerous conditions.'[98]

Consequently, the Vietnam War represents an almost ideal case study for Charles Moskos because it was a war which a citizen army was sent to fight with insufficient political motivation. The linkages between primary group solidarity and political motivation collapsed during that campaign. Moskos notes that the Vietnam War can be divided into three periods. Between 1965 and 1967, there was high cohesion and morale. In the following two years up to 1969, the army's performance was mixed, followed by a breakdown in discipline between 1970 and 1973. Moskos seeks to explain this collapse of US morale. For Moskos, the American rotation system which deployed soldiers on an individual basis for one year corroded primary groups; small-group cohesion could not be generated because individual soldiers were necessarily oriented to their own Days Remaining of Service (DEROS)—and surviving to see it—not to the mission or to their comrades.[99] After nine months, soldiers were not combat effective. In addition, the army was plagued by alcohol and drug abuse which further undermined unity and the performance of military duty. However, while Moskos emphasizes that the US combat soldiers disparage overt nationalism, he regards the absence of any semblance of a latent ideology to legitimize the campaign, to unite the soldiers, and to justify their suffering as fundamental to the collapse of the US Army. Precisely because the Vietnam War was opposed by a large and very vocal element of US society, soldiers in Vietnam became increasingly disillusioned with the campaign; they could not see the purpose of their sacrifices nor the violence which they were being ordered to perpetrate. Unlike their forebears in the Second World War, who suffered minimally equal deprivations and periodically felt alienated from civilians who could not understand the reality of combat, soldiers in Vietnam could not connect their activities to the preservation of any worthwhile values in American society; not only did civilians not comprehend them but a significant proportion actively opposed them. There was no national mission to which soldiers could appeal in order to make sense of the dangers which they faced. Consequently, in the absence of even a latent ideology, the American army collapsed. In fact, according to Peter Kindsvatter, there was a forgotten crisis of morale in the Korean War due to the soldiers' uncertainty about its political value, and he suggests that in this way Korea was much closer to Vietnam than is often recognized: 'The Korean War GI not only resented what he believed was an insufficient effort by the ROK [Republic of Korean] Army, but he also came to resent the lack of support from the home front. American soldiers never displayed the antiwar sentiments that they did during the Vietnam War, but neither did they exhibit much enthusiasm for the "forgotten" war.'[100] Due to the more favourable military situation for most of that war, the US Army, despite the evident tendency of troops to panic and desert, was able to avoid collapse.

It is understandable why Vietnam has attained a position of such importance for American scholars and the American psyche more widely. It is a deeply interesting case. Yet, in terms of the citizen armies of the twentieth century, there are numerous other examples when a lack of political motivation undermined military effectiveness in a manner consonant with Vietnam.

The woeful performance of the Italian army, especially in the Second World War, has been the subject of extensive scholarly interest and is a very useful example of the role of political ideology in generating cohesion. John Sadkovich has sought to restore the reputation of the Italian army against nationalist stereotypes,[101] and he has plausibly argued that their performance in the western desert was far better and much more important than subsequent historiography—and German propaganda—has allowed. In particular, after 1941, the Italian army and especially the mechanized Ariete Division fought well.[102] Sadkovich's point is well taken. However, while the Italian performance in the western desert improved from 1941 to its eventual defeat in March 1943, even Sadkovich cannot obscure the deep problems which attended the Italian army in this war. It suffered debacles in the Alps, the Balkans, Greece, and Africa in 1939 and 1940 and was further humiliated in 1943 when it surrendered to the Allies. It was only effective therefore when it fought alongside the Germans. Both Brian Sullivan and John Gooch have recognized the shortcomings of the Italian army, whose performance Sullivan has described as 'disastrous'.[103] Scholars have ascribed Italian military failures to poor strategic leadership by Mussolini, uneasy relations between the political and military leadership, limited material and economic resources, poor military command, and pitiful training.[104] However, the lack of political motivation has also been identified as an important factor in explaining the poverty of Italian military performance. Indeed, in a now old but widely read debate on Italian military efficiency, this factor was identified as central. Thus, Stanislav Andrevski has suggested that the explanation for consistently poor Italian military performances is sociological. Specifically, they reside in the peculiar political structure of the Italian polity which consists of a weak central state, fragmented regionalism, and, uniquely, the power of the Papacy. Thus, Stanislav sees the Catholic Church as critical since it has opposed 'the creation of a unified Italy':[105] 'to appreciate the importance of this factor we must bear in mind how universally important was the backing of religion in instilling into people loyalty to the state and especially the readiness to die in battle.'[106] For Andrevski, the strength of Catholicism in Italy undermined adherence to nationalism and a political mission. It was directly responsible for poor military performance since soldiers were not motivated to fight. Andrevksi's thesis is interesting, though uncorroborated. However, while the other scholars in this debate, perhaps sceptical of the Papacy thesis, do not explicitly engage with the argument, they broadly concur with Andrevski. Thus, John Gooch records a number of political and military failings but concludes by identifying a lack of nationalism as central to Italian defeats. Authorities tried to generate a sense of national unity and patriotic duty (Italianita) through a system of conscription which mixed recruits from different regions. Rather than generating a sense of national consciousness, it merely enhanced the ill feelings between the regions, however; 'the attempt to amalgamate Italians through conscription may well have had the effect of increasing rather than diminishing the distance between north and south.'[107] Gooch concludes:

> In a country where national political consciousness, and a national political culture, ranked far behind local and familial bonds, the conscription system and the leadership and example of junior officers failed to bridge the gap between the individual and his patriotic duty. Among the many factors which help to shape a nation's performance in war consciousness of national identity and national purpose must be one of the more influential.[108]

Christie Davies emphasizes the point, arguing that Italian campanilism has been consistently more important than national bonds: 'the prolonged disunity and powerlessness of Italy led to a situation in which the only strong loyalties were local ones—to a city or community and not to the nation.'[109] Davies eschews the stereotype of Italian pacificism or cowardice; she perceptively notes the great Italian proclivity for small-scale violence in the form of crime or the vendetta: 'the incapacity of leader and follower alike is simply a reflection of the general Italian inability to be loyal to anything beyond an immediate network of personal relationships.'[110] Accordingly, 'Italian military operations indeed seem only ever to be effective where they are limited to the operation of very small groups of men (as with the crews of the midget submarines that penetrated the defences of Gibraltar in the Second World War) or where a group of locally recruited men are defending their home territory (as in the case of the Alpini).'[111] Davies summarizes: 'The Italian army could not appeal to national sentiment for this did not exist and it dared not appeal to more local loyalties since these were not securely attached to the new Italian polity.'[112]

None of the scholars of the Italian defence policy would suggest a monocausal explanation of the disasters of the Second World War. However, clearly, the lack of patriotic unity and national mission has been widely identified as an important factor in explaining the poverty of the Italian army's performance. Nevertheless, this leaves scholars with a potential anomaly. While the Italian army's defence of northern Italy was deeply problematic in the First World War and it suffered a major catastrophe at Caporetto in 1917, the army demonstrated extraordinary resilience throughout that campaign and, with Allied assistance, was ultimately successful against a formidable enemy who enjoyed geographic advantages on the battlefields of the Isonzo Valley. The liberal use of punishment was not insignificant here (and will be discussed in Chapter 10) but the ability of the commander, General Cadorna, to institute such a repressive system in the First World War, in contrast to the paradoxical leniency of the Fascist regime, demonstrates an underlying difference between the two campaigns which may highlight the importance of nationalism to the citizen army. While Italians in the First World War were no more patriotic than their successors in the Second, the political circumstances of the Great War were potentially more favourable to instilling an at least nascent sense of national unity. In the First World War, of course, the Italian army was explicitly defending national territory against the predations of the Austro-Hungarian Empire while simultaneously seeking to liberate irredentist movements in the Balkans and Tyrol regions. Maltreated and poorly commanded as they were, Italian soldiers were fighting for a concrete political cause; the land which was pledged to them at the London Treaty in April 1915. It was noticeable that Mussolini himself renounced his socialism to become a pro-war nationalist during the First World War,[113] and attacking Italian soldiers cried either 'Savoy' or 'Trento and Trieste' as they charged enemy trenches.[114] This patriotism was effusively expressed in the florid nationalist poetry of the First World War whose celebration of national sacrifice was starkly at odds with the British, German, and French poetry of the Western Front.[115] Plainly, as Mark Thompson emphasizes, many illiterate and poor Italian peasants drafted into the ranks had little idea why they were fighting. Nevertheless, patriotic sentiments and a national purpose animated many Italian soldiers and commanders in the First World War. It seems likely that it was precisely because this war was seen as a conflict of national defence and self-assertion that General Cadorna was permitted to use such extreme disciplinary measures. This was quite in contrast to Mussolini and the Second World War. His entry into that conflict was late and entirely opportunistic in June 1940, when he believed that German victory was inevitable and that Italy could benefit politically from an alliance of temporary military convenience. Political elites and the population were never as supportive of the war as they had been in the First World War. Indeed, Lucia Ceva notes that Italians had a negative attitude to the Germans in contrast to their affection for Americans and a degree even of Anglophilia.[116] They were never committed, in the way in which their forebears had been committed to the struggle in the Dolomites and on the Isonzo, to a war which did not initially threaten Italy and which was fought, for the most part, in distant parts of the Mediterranean and Africa. The lack of political motivation of the Italian army, especially in the Second World War, seems to highlight the importance of patriotism and nationalism to the citizen army.

British troops were generally reliable precisely because they were imbued with a sense of patriotic duty which the Italians singularly lacked in the Second World War. It is all the more striking, then, what happened when the political motivation of British troops failed and they no longer understood themselves in terms of patriotic duty. The performance of the 7th Armoured and 51st Highland Infantry Divisions in Normandy are among the most useful and indeed notorious examples of the importance of political motivation to British troops. 7th Armoured Division was one of the most successful British units in the North African campaign. As a result of its exploits in Libya, it became known as the 'Desert Rats', taking a red jerboa as its divisional symbol. It was subsequently transferred to Britain to prepare for the Normandy campaign, where it was held back until after D-Day by Montgomery, on the not unreasonable presumption that it would constitute his elite shock force once the bridgehead had been established. He was to be severely disappointed. The division had become, in the view of Major General Verney who took over when the original commander was sacked in August 1944, 'swollen-headed' as a result of its successes in the desert,[117] and thought that additional training in preparation for Operation Overlord was unnecessary. Partly as a result of this arrogance, there were serious morale problems in both formations, as one staff officer noted: 'The 3rd Royal Tank Regiment were virtually mutinous just before D-Day. They painted the walls of their barracks in Aldershot with such slogans as "No Second Front".'[118] A young Lieutenant Edwin Bramall, who would go on to become Chief of the General Staff, on joining his infantry regiment in 4th Brigade of 7th Armoured Division, was dismayed to find that the battalion 'had shot their bolt'.[119] Major General Verney concurred:

A great many of the officers and men were war-weary by the time they got home and unfortunately there was no large-scale system of relief; in fact, very few left the Division, and it was owing to this that they made a very poor showing in Normandy. Looking back it is quite easy to see that there ought to have been considerable changes; whole units, particularly in the Infantry Brigade, should have been exchanged for fresh troops who had spent the last few years training in England.[120]

Horrocks similarly noted that: 'They begin to feel it is time they had a rest and someone else did the fighting.'[121] 7th Armoured Division's morale did not improve once deployed and was reflected in their lacklustre performance on every mission they conducted. Eventually, Montgomery sacked the divisional commander, Major General W. Erskine.

The famous 51st Highland Division which had distinguished itself in North Africa suffered a similar fate and, indeed, Major General Verney diagnosed their problems together: 'Both these divisions did badly from the moment they landed in Normandy. They greatly deserved the criticism they received.'[122] Montgomery reported its failings to General Alan Brooke, the Chief of the Imperial General Staff: 'Regret to report it is considered opinion of Crocker, Dempsey and myself that 51st Division is at present not—NOT—battleworthy. It does not fight with determination and has failed in every operation it has been given to do. It cannot fight the Germans successfully.'[123] Montgomery and other commanders identified the failings of divisional commanders as critical and this does seem to have been a factor; 50 Northumberland Division fought well despite having served in North Africa, and the performance of the much mauled 43 Division was notable primarily because of its ferocious commander. However, given that the troops themselves expressed reluctance to fight and dissatisfaction that they were being risked again, it seems plausible to suggest that a lack of political motivation was a central issue. The absence of political motivation was somewhat peculiar in the case of the 51st and 7th Divisions since these soldiers, like most British citizens, supported the war wholeheartedly, as their excellent service in Africa showed. Yet, their patriotism was not sufficiently deep to compel them to fight again, before others who had not yet played their part. Their sense of patriotic duty was highly conditional, therefore, and as such it did not provide adequate motivation in the intense and very difficult fighting which they were to experience in Normandy. Once again, political ideology and a sense of national mission which the soldier feels obligated to heed seems to have been essential to the performance—and non-performance—of the citizen soldier.

In the First World War, there were a number of interesting examples when conscripted troops did not associate themselves with the national mission and their reluctance both affected their combat performance and concerned military and political leaders. Alexander Watson has recently identified one of the most interesting examples of this lack of patriotic or ethnic commitment to the war in his work on the Polish forces in the German armies of the First World War, 850,000 of whom fought for the Kaiser in that conflict. Although Poles were recruited, trained, equipped, and dressed in exactly the same way as their German peers, their performance in battle often demonstrated a marked difference. Specifically, there were a number of significant cases of Polish desertions with forces raised from Posen, a heavily Catholic and nationalist area, being most prone to it. In February 1915, large numbers of Poles crossed over to the French lines from V Army Corps and V Reserve Corps.[124] Suggesting Polish nationalism and a corresponding disaffection with German nationalist objections was central to the collapse of Polish units, problems were particularly acute when the proportion of Poles in any unit was high and 'especially in such units, in which the percentage of Poles surpassed that of the Germans'.[125] Watson's point is that in units where there were strong primary groups of Poles, low combat motivations, disaffection, and desertion were most pronounced precisely because Polish rather than German nationalist consciousness was particularly developed in these groups. In the light of the poor performance of Polish troops, especially from Posen, in the early years of the war, the German high command sought to integrate Polish soldiers into German units in small numbers, so that they always represented a small minority who could be supervised closely.[126] Harsher disciplinary measures were also introduced. In this way, commanders hoped to discourage the emergence of deviant groups of Polish soldiers and to incorporate individual soldiers into an ethnic group which was committed to German goals.[127] Polish troops, especially from Posen, were probably never converted into enthusiastic soldiers but, overwhelmingly, this ethnic integration was successful. Watson concludes: 'The conduct of Prussia's Polish-speaking soldiers during the First

World War indicates, as other recent studies have also argued, that patriotic commitment was of great importance in generating effective performance by conscript armies on the twentieth century battlefield.'[128]

Political motivation was manifestly important to citizen soldiers in the twentieth century. Typically, these motivations were informed by nationalist and patriotic ideas and sentiments. However, in many cases, this political ideology was inflected with a more radical ethnic and, indeed, racist dimension. Soldiers understood themselves to be fighting not simply as citizens for the political goals of their state but actively mobilized themselves as ethnic or racial groups. These ethno-political motivations became a powerful if sometimes unstated means by which combat soldiers generated cohesion, forming themselves into ethnically or racially distinct groups. Omar Bartov's work on the Wehrmacht represents one of the most obvious and important attempts to expose this ethno-political dimension to combat performance. Omar Bartov highlights the importance of political motivations to soldiers in order to rectify what he sees as fundamental misconceptions in the work of Janowitz and Shils. In the latter's work, the cohesion of the primary groups and the extraordinary endurance of the Wehrmacht is attributed to the personal bonds between male soldiers. In this, the Wehrmacht demonstrated simply a highly developed form of cohesion which was displayed all but universally by armies in the twentieth century; soldiers united around concepts of manhood. Although it is difficult to deny the importance of masculinity to the primary groups of the Wehrmacht and other armies, there have been a number of criticisms of Janowitz and Shils's work. David Segal and Meyer Kestnbaum note that both Janowitz and Shils were attached to the intelligence sections of SHAEF's Psychological Warfare Unit and conducted the interrogations ultimately as a part of a de-Nazification programme. Consequently, their data base was information collected through interrogation of German prisoners of war by American military personnel. This was not necessarily the most objective or reliable source of social scientific data, 'particularly since German soldiers could cite primary group bonding rather than Nazi party loyalty as justification for their fighting'.[129] In a more recent piece, Simon Wessely has made precisely the same point. 'It is based partly on POW interrogations carried out during the war and it is plausible that captured soldiers would be more likely to emphasize the role of local social factors and their own military professionalism, than of National Socialist ideology.'[130]

Notwithstanding the methodological critique, Bartov raises a much more profound interpretative objection to the theory of the primary group. He sees the dignification of the Wehrmacht primary group as false and potentially politically dangerous.[131] Decisively, it ignores the political context in which the Wehrmacht operated and which ultimately determined the kind of organization it was and warfare which it prosecuted; 'while "primary group" theory "depoliticized" the Wehrmacht, one consequence of writing the history of the Third Reich "from below" was to create an impression of a "depoliticized" civilian society, most of whose members presumably considered the "normality" of daily life as far more important that the "abnormality" of Nazi ideology and actions.'[132] For Bartov, political ideology—and specifically an extreme nationalist racism—was fundamental to the performance of Germany's armed forces during the Second World War. Indeed, Bartov denies that the primary groups, to which Janowitz and Shils attach so much importance, could have been vital to the performance of the German army on the Eastern Front. There, attrition rates were so high that primary groups could not continue to exist, since so many of their members were killed. The very large numbers of replacements had insufficient time to form new primary groups. Thus, Bartov reports how in the 12th Infantry, there was inadequate time for primary groups to form since 'many of them [replacements] hardly spent more than a few days in its ranks before they were killed or sent back to the rear in hospital trains'.[133] In fact, while replacements were killed very quickly, veterans (who formed the core of the primary groups) suffered a slower rate of attrition. Bartov underestimates the endurance of the primary group and, indeed, he revised his original position and accepted the importance of primary groups in the east 'as long as the rate of casualties permitted' it.[134] Nevertheless, his work provides an important insight into one of the ways in which loyalty to the primary group was engendered. Crucially, the German army and the divisions themselves actively sought to indoctrinate their soldiers with Nazi political ideology in order to intensify primary group loyalty. The cult of the Führer was particularly important here where soldiers were collectively mobilized around a concrete image of a heroic, Aryan leader. Loyalty to the Führer and, therefore, the fatherland was a collective reference for the German soldiers. Closely associated with this myth of the Führer was a highly politicized view of the war. The war in the east was regarded by the normal soldier not as a conventional conflict but as a war of annihilation against a racially inferior Judaeo-Bolshevik enemy. 'There were many such enthusiastic young men among Wehrmacht's soldiers and junior officers, powerfully imbued both with racist sentiments and with the notion that war was a climax of human existence.'[135] Their extreme nationalist belief enjoined the highest level of solidarity among serving German soldiers on the Eastern Front, even to the point of being willing to commit mass atrocities in the name of Germany.

Bartov's emphasis on the role which nationalist ideology played upon the German soldier is supported by the statements of numerous 'normal' German soldiers. In his analysis of the German combat soldier, the Landser, Fritz, like other commentators, points to the rigorous training of its recruits in explaining the performance of the Wehrmacht.[136] However, he notes the Wehrmacht 'attempted to strengthen the morale and stiffen the resolve of its forces in a number of ways. One was to stress camaraderie and peer pressure by grouping friends together.'[137] Crucially, however, Fritz identifies ideological indoctrination as a critical mechanism for inducing solidarity. 'The average *Landser* was encouraged in this racist hatred by the *Wehrmacht* high command.'[138] The concept of a war against an Asiatic horde or Jewish Bolshevik threat was actively employed to motivate German soldiers. Indeed, 'determined efforts were made by the *Wehrmacht* to influence the Landsers through both written propaganda, such as front newspapers, and spoken propaganda, initially from "education officers" and, later, National Socialist Leadership Officers.'[139] It might be argued that this political indoctrination was ineffective and, as Little suggested, had little impact at the front line. Yet, Landsers themselves seemed to draw upon these political resources to sustain themselves on the front. Indeed, some Landsers found the Führer's speeches more uplifting than a letter from home.[140] Significantly, even if the more virulent elements of Nazi ideology were not a constant resource for unifying primary groups, the less contentious but nevertheless heavily politicized concept of a *Volksgemeinschaft* was evident among the Wehrmacht. A *Volksgemeinschaft* referred to an ideal ethno-national community, which overcame all class divisions, in pursuit of common goals.[141] The soldiers of the Wehrmacht not only understood themselves to be fighting for their homeland and 'for everything they knew and cherished',[142] but because they necessarily created a *Frontsgemeinschaft* in order to survive the front, they were understood to be the representatives and ideal of that ethno-national community. The very concept of a community purified through conflict and in pursuit of its historic destiny actively motivated German soldiers in the Second World War.[143] As one soldier recalled: 'My generation was brought up to believe that no sacrifice was too great for the [*Volksgemeinschaft*].' Indeed, even after the war, soldiers clung to the 'tried and tested nationalism of the community'.[144]

There is clear evidence in German doctrine of precisely the processes which Fritz described. Indeed, the official political indoctrination of the German army began very soon after the First World War. Thus, in his technical description of the development of tactics during that conflict, published in 1922, Lieutenant General William Balck concludes his discussion ominously:

The army at home knifed the undefeated field army in the back, like Hagen of old did to the unconquerable valiant hero, Siegfried... The German Spirit will continue to live! When later on a new aurora of history embellishes our days, then may our people remember also our heroes of the World War. Work, suffer and fight like they did, and thus Germany's future will be secure. And then, also, the blood of our heroes who fell in the belief of Germany's victory will not have been spilled in vain![145]

Indeed, Wolfram Wette has explicitly sought to demolish the legend of a 'clean' Wehrmacht. He draws on much of the existing literature including Bartov and Fritz but pertinently explores the way in which not only was the

Reichswehr of the 1920s bitter about Versailles but this bitterness was inflected with an anti-Semitism which produced a culture that had close elective affinities with Nazism. Although Wette acknowledges that a small number of Reichswehr generals, such as Walther Reinhardt, rejected anti-Semitism, insisting that Jews who had fought in the First World War 'must be recognized and honoured', the Reichswehr was fundamentally anti-Semitic. Excluded from the veterans' association, *Stahlhelm*, formed in 1919, Jews had to form their own veterans' association after the First World War, and although formally allowed into the officer corps of the Reichswehr, the right-wing culture of the force ensured, in fact, that they were excluded. Indeed, many senior Reichswehr commanders openly dismissed the black, red, and gold flag of the Weimar Republic as a 'Judenfahne' (a Jews' flag),[146] and, affirming a virulent radicalism, Reichswehr officers were involved in the murder of a number of communist and Jewish leaders, including Rosa Luxemburg and Karl Liebknechts.[147] The institutionalization of the Aryan Paragraphs into the Reichswehr's constitution in 1934 represented no major cultural or organizational rupture. On the contrary, according to Wette, 'the basic ideas of national socialist racist teachings already found increasing usage in the Reichswehr by 1934'.[148]

By the Second World War, the myth of the *Dolchstoss* had been fully incorporated into Wehrmacht infantry doctrine itself. Thus, Dr Reiberts's *Deinstunterricht*, which was designed to be read by young men of draftable age and by serving soldiers, and included detailed instructions on such military banalities as how to make a bed, keep a locker, and pack a rucksack (including a spare horseshoe), featured a long chapter which melodramatically invokes a mythical history of the Third Reich. In an elaborate exercise in historical imagination, entitled 'The Patriotic Part', the Third Reich is traced back to the origins of Germany in the Bronze Age. Reibert asserted an enduring geo-political problem for the German people which, he claimed, united their history: 'The history of our people is for the most part from the beginning determined by the geographic situation of its living space.'[149] The Germans have always been surrounded by enemies across open borders. However, 'a look into the past shows that Germany has been unconquerable, when the German people were united'.[150] Reibert forwards a teleological narrative in which the Teutoburger forest, Charlemagne, Ludwig the German, Henry the Fowler, Otto the Great, Frederick I, Frederick II, the Second Reich of the Prussian State, the First World War, and the *Dolchstoss* all lead inevitably to the foundation of the Nazis' Third Reich.[151] The ideological basis of this state forms a central part of the narrative. Reibert notes that the Greeks and Romans praised 'German tribal purity and courage',[152] and that Hitler's regime is the ultimate expression of this purity; in 1919, as a result of his experiences of the war, Hitler conceived 'the creation of a single German people and the foundation of a strong, clean nation-state'.[153] This project could be implemented once the Nazi revolution succeeded 'in an ordered and disciplined way' and Hitler took power in 1933,[154] which has subsequently led to war with France and 'England' due to the insidious activities of the Poles stirring up ethnic terror against the German people in Poland.[155] Reibert concluded his narrative by explicitly tying Hitler's ethno-nationalist policy to the duties of the German people and to the individual soldier.

There is no doubt that Germany will be victorious in this struggle; since Germany could never be defeated as long as it was united. That the German people are united today, is proved not only by the million person cry on the establishment of Greater German Reich, 'One People—one Reich—one Leader!' but also in the first instance by the love and respect for the saviour of our Fatherland which every German shows: 'The Führer is Germany' and 'Germany is the Führer'.[156]

National duty was embodied in the physical person of the Führer himself to whom every soldier had to take an oath: 'I swear by God this holy oath, that I will be unconditionally obedient to the Führer of the German Reich and People, Adolf Hitler, the Commander in Chief of the Wehrmacht, and am ready as a brave soldier or I forgo my life for this oath.'[157] It is perhaps not insignificant that the frontispiece of Reibert's volume featured a large photograph of Adolf Hitler staring imperiously into the middle distance.[158] In Adelbert Holl's account of his experiences as an infantry officer in Stalingrad, political ideology and a personal dedication to Hitler played a prominent role in motivating him and his comrades. Once they were trapped in the Kessel, Holl reports that 'a slogan made the rounds: "Hold out, the Führer will get you out".'[159] Moreover, even when their fate was sealed and Russian troops were pouring over the Volga, Holl observed: 'That was not welcome news for us. Nevertheless, we were soldiers who had to carry out our orders. We had to fulfil our duties and be faithful to our oath for Führer, Volk und Vaterland!'[160]

Jürgen Förster has emphasized the importance of political ideology to the Wehrmacht. Field Marshal Werner von Blombert's notorious directive of 16 April 1934 identified the educational goals of the Wehrmacht as being 'not only the thoroughly trained soldier and master of his weapon, but also the man who is aware of his race (*Volkstum*) and his duties towards the state',[161] one of the results of which was, at von Blombert's instigation, the oath to the Führer. All commanders-in-chief agreed that officers in the Wehrmacht had to be National Socialist in spirit, and to engender these political sentiments among the rank and file extensive efforts were made to disseminate daily newspapers, broadcasts, films, sports, and Wehrmacht pamphlets (like Dr Reibert's).[162] The members of the Wehrmacht were not simply united as fraternal comrades, therefore. On the contrary, they were actively imbued with a sense of national destiny, which engendered heightened levels of solidarity among them. Through thorough instruction and the presence of insignia and symbols, Nazi ideology became a common resource for uniting them in the most desperate of circumstances.[163] Nazi ideology was certainly not the only factor in explaining the performance of the Wehrmacht in the Second World War.[164] As Thomas Schulte has noted, it represents poor scholarship to assert that '*every* German soldier was in some measure a war criminal' and that the Wehrmacht could be seen merely as an expression and tool of Nazi ideology.[165] Yet, to deny its extraordinary collective hold over the soldiers of the Wehrmacht ignores the historical evidence. It was minimally a factor in the generation of social cohesion in the Wehrmacht both as an organization and in the primary groups themselves.[166]

It would be easy to dismiss the ethno-nationalist mobilization of the German army during the Second World War as an aberration. Its ideological fanaticism was certainly a product of extraordinary historical circumstances. Yet, although the virulent racism of Nazism was, of course, absent from the armies of western democracies in the twentieth century, citizen soldiers were much more deeply influenced by ethno-racial ideas than is often admitted. The ethno-political ideology of the Allied powers seems very tame in comparison with Nazi supremacist ideology. Yet, there was a consistent but often overlooked ethnic and racial dimension to Allied political ideology. In both the First and Second World Wars, there was explicit discrimination on the basis of ethnic and racial identity in both the US and British armed forces. Indeed, although, as we have seen, primary groups in western forces may have utilized notions of manhood dedicated to the defence of the national, presumptions of white ethnicity were central to group formation and cohesion. Primary groups were always assumed to consist of white men. It is at this point that there are uncomfortable historical parallels between the ethno-politics of the Wehrmacht and that of the Allied forces.

The United States Army and Marine Corps were particularly obvious examples of this racism. In both the First and Second World Wars, although black Americans were formally equal to white Americans, the War Department imposed a colour bar upon the force; blacks were segregated from white forces. Even in the Second World War, Jim Crow laws were de facto in place and, although they were meant to be 'separate and equal', the reality was that they were consistently discriminated against.[167] Indeed in his major study of US troops in the Second World War, *The American Soldier*, Sam Stouffer was well aware of the problem of discrimination; 'The stereotype of the shiftless, ambitionless plantation hand did not, as this chapter has already suggested, characterize the Negro soldier, although examples of the stereotype were not too hard to find.'[168] On the contrary, 'a concept of the average Negro as a happy, dull, indifferent creature who was quite contented with his status in the social system as a whole and the military segment of that social system, found little support in the study'. A third of black soldiers were from the North and were as well educated as Southern whites,[169] although Stouffer also averred that 'they were Americans in spirit.... though less likely than whites to feel the war was their war'.[170] The segregation of black soldiers proved to be very damaging for black troops. Black troops in both world wars fought well when commanded adequately and furnished with the appropriate level of training, equipment, and fire support; in the First World War, the black 93rd Divisions fought well under French command,[171] while the black 332nd Fighter Group (the Tuskegee Airmen) was decorated in the Second World War.[172] Despite their potential, the discrimination which black soldiers faced had the tendency of affirming prejudices about them. Thus, deliberately fewer black soldiers were recruited than whites and were assigned to menial support duties for the most part.[173] The level of ethnic discrimination in the US Army in the 1940s is demonstrated by the response of senior commanders to the proposal of the Secretary of War's civilian aid, Judge William Hastie, for desegregation of the armed forces in 1941. Hastie had long been a well-known advocate of desegregation and black equality. In rejecting Hastie, Lieutenant Colonel James Boyer usefully highlighted the self-understandings of most white US soldiers:

> It is the purpose of Judge Hastie and his backers to advance the colored people as a race at the expense of the Army... There may be some super-tolerant people that would join a Negro outfit but their numbers would be few. Other whites that would join a Negro outfit would be of the same class of whites that would live in a Negro community.[174]

Responsible for recruiting and training, General Lesley McNair, Commanding General Army Ground Forces, was not quite so obviously discriminatory. He regarded the problem of the inclusion of blacks into the US army not as an issue primarily of ethnicity but of intelligence: 'I am unalterably opposed to the incorporation of negroes in small units with white soldiers. Inevitably such action would weaken the unit, since it would introduce men of comparatively low intelligence. We have a sufficiency of such men among white soldiers.'[175] The AGCT scores among black soldiers were indeed low in comparison with white units, with 84 per cent of black soldiers being graded in category IV and V as opposed to 33 per cent of white soldiers.[176] However, Macnair elided the intelligence and race of black Americans and conveniently ignored the poor training and poor leadership (provided by typically substandard white officers) of black formations.[177] In point of fact, although education was a problem among the blacks of the Southern states in this era, there were many capable black Americans. Indeed, in his important report following the war on the performance of black soldiers, Ulysses Lee demonstrated that in many cases capable black soldiers and officers were deliberately sidelined because they were seen as a challenge to white officers; 'Those Negroes who exhibit the manliness, self-reliance and self-respect which are "sine qua non" in white units are humiliated and discouraged.'[178] Indeed, the situation was so pernicious that Lee observed: 'I was astounded by the willingness of the white officers who preceded us to place their own lives in a hazardous position in order to have tractable Negroes around them.'[179]

Most sociologists and historians have understandably explored the nefarious effect which discrimination had on black American soldiers and on black divisions. However, it is important to recognize the impact which this racial ideology had on the white soldiers who were favoured by it. By discriminating against blacks, the army implicitly elevated the status of the white soldier and invested his skin colour with increased importance. Just as black soldiers were not allowed to fight because of their skin colour, white soldiers were expected to fulfil their duty precisely because of their race. Whiteness became a symbol of distinction which was used to encourage higher levels of performance and cohesion. Indeed, animosity and violence, in some cases leading to murder, was typical between white and black units. This violence was significant in that it demonstrated that for white units, race was an important issue. They actively sought to enforce their racial superiority, especially over access to desirable resources such as bars, clubs, and women. In Alexandria and Fayetteville in 1941 and 1942, there were serious clashes between black soldiers and the Military Police and, in April 1941, Private Felix Hall was lynched at Fort Benning.[180] White violence enforced segregation and subordination on black troops, demonstrating the importance of white ethnicity to their own self-conception. On the line, where blacks were notable by their absence and where troops were almost all white, this ethnic ideology was not as immediately relevant as it was for the indoctrinated Wehrmacht. Nevertheless, the fact that blacks had already been excluded from primary groups does not mean that a shared skin colour was not implicitly important in generating social solidarity on the front line. US soldiers may not have understood the conflict as part of an ethno-national mission as their German opponents often did, but their front-line solidarity was based on ethnic presumptions.

Although the American army was integrated during the Korean War and there are many examples of successful multi-racial units in Vietnam, very serious problems were also evident. Many black soldiers certainly believed that they were fighting in a racist army for a discriminatory white society. They were convinced that white solidarity was central to the prosecution of the war by the US Army and by officers and soldiers themselves who conspired to exploit black soldiers for their own racial benefit. By the early 1970s, black soldiers in Vietnam, especially in the rear areas, had become nearly mutinous to such an extent that Moskos suggested that 'black soldiers may find they owe higher fealty to the black community than to the United States Army'.[181] Black soldiers in Vietnam, inspired by the new black consciousness embodied by Malcolm X and the Black Panthers, developed their own subculture within the US Army, voluntarily segregating themselves from other US soldiers, especially after the assassination of Martin Luther King in 1969. They developed a symbolic code to communicate their solidarity, including special handshakes (the dap), slave bracelets, and their own patois (in which they called each other 'brother-me' and 'nationtie').[182] Black soldiers increasingly challenged official army authority, 'talking it out' with their officers and, in extreme cases, engaging in 'fragging': intentionally wounding or killing superior officers deemed to be endangering the lives of (black) subordinates. In the last years of the Vietnam War, there was an epidemic of 300 to 400 officially recorded fraggings, though, as Moskos notes, the real figure was 'undoubtedly several times that number'.[183] Significantly, less than 20 per cent of these fraggings involved solitary individuals engaged in a personal vendetta.[184] In most cases, they were acts of collective mutiny typically executed by groups of black soldiers against white superiors.

The racism among US forces, a product of the institution of slavery in America, was not replicated in the British army whose Empire tended to encourage a more inclusive attitude. Nevertheless, while far less extreme, at least comparable patterns of discrimination were often evident in the British army. There, no formal segregation rule existed but since, during the early part of the twentieth century, Britain was almost exclusively white, the issue of race was not relevant. British regiments generally drew on local county or regional affiliations which presumed a white Christian constituency. The exception for the British forces was in the Commonwealth divisions and here a very similar policy was enacted to the US Army. Raised from the colonies, and especially from British India, the Commonwealth divisions fought alongside British divisions, principally in the Far Eastern theatre in the Second World War, but were separate from them. In his recent work on African soldiers in the Second World War, David Killingray asserts that 'race was a determining element in the command and order of British African colonial forces'.[185] The British army recruited half a million officers from Africa, 120,000 of whom served in Burma, but these troops were segregated from white units. Black African soldiers were organized into regiments like the King's African Rifles which were overwhelmingly commanded by white officers and non-commissioned officers.[186] Racism was far more virulent in the South African army and, indeed, in Burma, General Giffard refused to employ South African officers because 'of their peculiar ideas of the treatment of natives'.[187] By contrast, black soldiers from South Africa were shocked by the equal treatment they received from British soldiers who 'shake hands with us', 'talk to us', and 'sit next to us in cinemas'.[188] Nevertheless, the suspicions which attended the American perceptions of the performance of black soldiers were evident in the British army. When recruiting troops in Africa, the British identified particular ethnic groups as 'martial races', such as the Hausa, which produced the best soldiers. Yet, criticisms of African soldiers were commonplace. In Burma, General Hugh Stockwell, commander of the 82nd Division, reprimanded his white officers whose indiscreet comments about the quality of their black soldiers were having a negative effect on morale, though Stockwell himself admitted that their performance was suboptimal.[189] He accepted that the African 'has not a fighting history, and as a rule therefore battle does not come naturally'.[190] Reflecting their low esteem, the pay and conditions of service were systematically worse for African soldiers throughout the war in comparison with their white British peers. Rations, access to alcohol, leave, and demobilization were particularly emotive issues. When the discrepancy of treatment became clear to African soldiers, it frequently led to unrest.[191] In 1945, Gold Coast troops in the 37th General Hospital in Accra rejected 'native food' and demanded 'white man's food', and in Sierra Leone 150 soldiers refused to parade because, in contrast to white British soldiers, they had no boots.[192] Indeed, one British brigadier noted that: 'In my view, we were lucky to have escaped with a few flare-ups instead of a more general revolt.'[193]

In the Indian army, there were similar tensions. In the 1920s and 1930s, the Indian army was commanded by white British officers.[194] However, in the light of the Japanese threat, the army was formally expanded from April 1940, which necessarily included a process of 'Indianization'; not only were the number of Indian recruits increased but Indian officers began to be commissioned.[195] While the degree of racism varied, white British officers demonstrated regular hostility towards native Indian officers with whom they had to share a mess.[196] Indian officers were not allowed to dance with the wives of British officers and there were cases of racial abuse.[197] Perhaps more seriously, there were significant concerns about whether Indian officers should have the right to punish white British soldiers under their command. This situation was finally resolved in June 1943, when an official communiqué confirmed that Indian officers did have the powers of punishment.[198] Nevertheless, incidents continued. Captain Nair, posted to the 14th Army Headquarters as a staff officer, was returned to his unit because the staff, expecting an Irishman, were dismayed when an Indian officer arrived.[199] The Commander-in-Chief India, General Auchinleck, signalled General Slim about the incident and both agreed that this type of incident could not be tolerated.[200] Although Lieutenant Colonel Guatum Sharma claimed that 'discrimination was practiced on a liberal scale' in his work on the Indian army and regarded racism as endemic in the Indian army, Daniel Marston has claimed that it was practised by only a small minority of British officers in the 1940s.[201] It was nevertheless a possibility and the fact that significant efforts had to be taken to eliminate it suggests that white British soldiers were highly sensitive to the question of race and ethnicity even if they were not personally discriminatory. Barkawi, like Marston, notes the importance of formal military training to 'revivify bonds of solidarity' between soldiers of 'vastly different backgrounds' on the basis of 'general human capacities'.[202] Yet colour remained at least a surface issue.[203]

The French army of both the First and Second World Wars drew heavily on its colonial subjects and its attitude towards them echoed those of the British. While the virulent racism evident in the US Army might have been avoided sometimes, a racially informed paternalism formed the French view of black French soldiers, who were seen as 'colonial children'. Thus, in the First World War, the French regarded some French West Africans as innately warlike, which ultimately led to their profligate expenditure as shock troops in frontal assaults.[204] In the period before the Second World War, some French generals noted that colonial troops could perform well but there was concern about their reliability, which in fact echoed wider worries about the preparedness of the French army as a whole.[205] Indeed, French generals did not believe that the Senegalese would be able to withstand the shock of modern warfare with its tanks and planes. Consequently, the colonial forces were seen purely in terms of a 'back-up' army.[206] Like the British imperial troops, French colonial soldiers were seen as predisposed to show low levels of leadership and initiative and they were accordingly paid much less than their white peers.[207] Nevertheless, the colonial troops from West Africa fought extremely bravely in the defence of France in May 1940 and, indeed, shocked the Germans with their pugnaciousness.[208] Indeed, as a result of the casualties inflicted by African troops and, doubtless, recommended by the racist beliefs of the Germans, a number of massacres were perpetrated against them by the Wehrmacht. Nevertheless, despite their performance and their losses, French high command, which had, in fact, been centrally responsible for the catastrophe, still harboured anxieties about the reliability and discipline of their colonial troops.[209] That anxiety could be interpreted as direct discrimination. The celebrated intellectual Franz Fanon, originally from Martinique, fought for the Free French in the Second World War, was wounded, and decorated. However, his 'painful discovery that a black man was not treated as an equal in the French army' was substantially responsible for his political radicalization after the war and his eventual involvement in the Front de Libération Nationale in Algeria in the 1950s.[210] In the First and Second World Wars, French African troops played an important role. Nevertheless, the treatment of these troops reveals that there was a presumption that the duty and honour of defending France lay, in the first instance, with white Frenchmen born and raised in the metropole. Albeit obliquely, whiteness seems to have been a motivating factor for French troops.

In the Allied armies, ethnicity and race were perhaps not weighted with the intensity which they were in the Wehrmacht; even the most extreme white racist in the US Army did not propose the elimination of black Americans as a sub-species. However, that it played a role in fostering solidarity can be demonstrated by the reactions when the exclusivity of white soldiers was threatened. Whiteness was a non-issue for most of the time for American, British, and French soldiers, only because the exclusion of other ethnic groups from their primary groups was always guaranteed. Although it is an unpalatable observation which tarnishes comfortable assumptions about why and how the Allies fought the Second World War, ethnicity played a subordinate, unacknowledged, but not irrelevant role in engendering cohesion even in the armies of the liberal democracies in the twentieth century.

The industrial battlefield presented the armies of the twentieth century with a dilemma; in order to preserve their troops, they had to disperse and seek cover from fire but, in so doing, commanders risked their armies becoming totally ineffective especially since citizen soldiers typically lacked the training to be able to engage in complex tactical manoeuvres demanded by modern combat. Armies wrestled with the problem of how to overcome the so-called Marshall effect. In the next two chapters, the main tactical techniques adopted by the citizen army in order to overcome the inertia which Marshall observed as a structural feature of the modern battlefield will be discussed. However, one of the central means by which armies recurrently sought to overcome the Marshall effect was not ultimately military at all. The civil society, the army, and troops themselves resorted to appeals to masculine honour, nationalism, ethnicity, and patriotic duty in order to encourage participation on the field of battle. Ultimately, quite arbitrary cultural criteria which aimed to shame soldiers into fighting out of fear of being ridiculed and excluded by their immediate peers and wider social communities were central to enjoining combat performance on the part of the citizen soldier. In Britain, one of the most poignant examples of the power of these arbitrary social criteria to influence the individual and to generate mutual obligations among troops was the first day of the Somme, as the fate of Captain Nevill shows. On that day, 20,000 British soldiers died, inspired by nationalist ideas and obliged, by their concept of masculine honour, to go over the top.

5

Mass Tactics

THE BAYONET

Comradeship and political motivation were essential to engendering cohesion among citizen soldiers. However, even with the most developed motivation, citizen soldiers had to be trained and ordered to conduct particular kinds of manoeuvres on the battlefield. They could not be allowed merely to express their commitment in any way they individually chose. They had to be unified into a coherent military body. The question was how to channel masculine and patriotic sentiments into effective tactical practices. From the end of the nineteenth century, as strategists tried to come to terms with the mechanized battlefield dominated by rifles, machine-guns, and indirect artillery, there were substantial debates about the problem of dispersion and inertia. Military commanders and theorists recognized that close order tactics, in which shirking was difficult because of the physical proximity of troops to each other and the surveillance of officers, had become problematic. However, in many cases, close order skirmish lines continued to be recommended for assault because although losses would be heavy, it would maintain the momentum of the assault.[1] Fritz Honig and Jakob Meckel, Emory Upton and Ardant du Picq all advocated close order tactics in order to sustain the assault.[2] Furthermore, the size of armies, the level of their training, and the lack of communication systems seemed to rule out more complex tactical manoeuvres which might offset the problems posed by firepower. In particular, the bayonet charge was favoured as the decisive and most effective form of infantry tactics during this period. The bayonet features prominently in Joanna Bourke's critique of militarism. For Bourke, the bayonet was an ideological means of infusing aggression into the soldiers in order to make them more willing ultimately to kill or be killed: as one Vietnam veteran noted, it was 'a charm to ward off fear'.[3] There is evidence to support Bourke's interpretation but she also misses the important organizational benefits of the bayonet which were all too clear to Du Picq and many of his military peers. Unlike more elaborate tactical manoeuvres, the bayonet charge is extremely simple and requires minimal training. Crucially, it physically unites potentially poorly trained troops into a mass, thereby inspiring the assailants. In the bayonet charge, the charging mass provided mutual moral support for its members, discouraging wavering individuals whose cowardice is visible to their comrades. Du Picq fully recognized the moral effect of this loss of mutual observation on troops: 'The bewildered men, even the officers, have no longer the eyes of their comrades or of their commanders upon them, sustaining them. Self-esteem [i.e. recognition by one's peers] no longer impels them.'[4] The bayonet charge solved this problem since individuals were always under the observation of their comrades. At the same time, the bayonet charge demoralized defenders not primarily because they feared the wounds the bayonet inflicted (bullets and grenades caused worse injuries) but because a force charging with bayonets demonstrates its unshakable resolve; they would close and kill. In the light of this display of collective determination, Du Picq plausibly claimed that no force ever withstood a bayonet charge.[5]

Reflecting Du Picq's advocacy, the bayonet charge was a central part of doctrine and training both before and during the First World War.[6] Indeed, First World War infantry tactics have often been defined by the bayonet charge—and taken as a sign of its futility—and, indeed, there is significant evidence to demonstrate its centrality to military thinking at that time. For instance, before the war there was a dispute within the British officer corps as to whether the bayonet was obsolete. In published debates in Britain before the Great War, Major McMahon proposed a minority view that firepower had made the bayonet charge obsolete, but Brigadier General Kiggell, who would later become Haig's Chief of Staff in the British General Headquarters during the First World War, insisted that:

After the Boer War the general opinion was that the result of the battle would for the future depend on fire-arms alone, and that the sword and the bayonet were played out. But this idea is erroneous and was proved to be so in the late war in Manchuria. Everyone admits that. Victory is won actually by the bayonet, or by fear of it.[7]

The Russo-Japanese War became an important resource for the British army in these discussions about infantry tactics, as it did for all European militaries. For British military observers, Japanese tactics during this conflict represented a political and military ideal. Although their interpretation of Japanese culture was more a reflection of their anxieties about their own society than an accurate account of Nippon culture, commentators were impressed by the 'human bullet' tactics which the Japanese employed during their assault on Port Arthur. Massed Japanese infantry successfully charged Russian positions with bayonets. Not only did this appear to be a successful tactic in itself but the British observers admired the social preconditions which facilitated the use of bayonet charges. To mount such a charge presumed a high state of morale which in itself demonstrated 'the impact of nationalist sentiments and social cohesion, generated partly by mass education'.[8] In a speech to the Royal

United Services Institute in April 1910, Sir Alexander Bannerman identified Japanese patriotism and monarchism as fundamental to the cohesion which Japanese troops demonstrated at Port Arthur. Bannerman and others compared the nobility of Japanese nationalism with the contemporaneous dangers of British liberalism with its prioritization of the individual over the collective good.[9] Prioritizing the highest morale, it matched precisely the sense of national mission which military commanders, political leaders, and commentators idealized. The bayonet charge was then not merely a tactic but the belligerent expression of a democratic nation.

There were potentially very different tactical lessons which the British army could have taken from their experiences at the turn of the twentieth century. If they had invested their own troubled campaign against the Boers with greater significance than the Russo-Japanese War, then, they might have seen the importance of firepower, concealment, and dispersion to the infantry assault. However, the concept of mass tactics predominated and it is important to recognize this preference was not a matter of stupidity. The bayonet charge had evident organizational advantages over other forms of more complex skirmishing tactics.

Western infantry doctrine before the First World War clarifies the widespread preference for the bayonet charge as a tactic on the industrial battlefield among armies at this time. The British army's pre-First World War doctrine, for instance, both demonstrates that the bayonet was given priority as a weapon for the infantry and explains the rationale behind the advocacy of the bayonet charge as the principal tactic of the infantry battalion. The 1914 version of *Infantry Training* instructed the infantry to advance quickly to within 'effective ranges'. When enemy fire prevented further advance, the infantry was to form itself into a firing line in order to attain fire superiority: 'as the enemy's fire is gradually subdued, further progress will be made by bounds from place to place, the movement gathering renewed force at each pause until the enemy can be assaulted with the bayonet.'[10] Finally, 'the commander who decides to assault will order the *charge* to be sounded, the call will at once be taken up by all buglers, and all neighbouring units will join the charge as quickly as possible. During the delivery of the assault the men will cheer, bugles be sounded, and pipes played.'[11] Although *Infantry Training* stressed the importance of fire superiority in setting the conditions for the bayonet charge, the bayonet and the bayonet charge was not merely one tactical procedure among others. It was regarded as the decisive element of an infantry attack in British military culture. Indeed, the point was officially recorded in pre-First World War doctrine when the 1909 Field Service Regulations replaced the phrase 'the decision is obtained by superiority of fire'[12] to record that 'the climax of the infantry attack is the assault, which is made possible by superiority of fire'.[13]

The bayonet and the bayonet charge became even more prominent in British doctrine and practice as the army expanded itself to include Kitchener's volunteers precisely because the new volunteer had so little training and could not hope to perform more complex tactical manoeuvres. Following infantry doctrine during First World War assaults, the distances between men in each line (3 to 5 yards), generally consisting of a full company led by an officer, and between waves (200 yards) was strictly enforced to create the correct density for the bayonet attack. As Britain's first major offensive of the war, the Battle of the Somme provides a useful insight into the emphasis which was placed on a combination of high morale and the bayonet. In British military history, the first day of the Somme has been typically employed as an example of the bayonet charge (and its futility). Peter Hart has used numerous memoirs to emphasize the dense linearity of the assault on 1 July: 'Line after line advanced and disappeared in the clouds of smoke.'[14] There is now some controversy about these claims. Prior and Wilson explicitly reject Hart's account, claiming that this was part of the myth of the First World War perpetuated by Basil Liddell Hart, John Buchan, and, indeed, James Edmonds, the official British army historian.[15] They stress the variation in tactics on the Somme; many units did not advance 'shoulder to shoulder'.[16] Different divisions and the different brigades within the divisions used a number of alternative assault techniques. Of the eighty battalions that went over the top from these Corps, fifty-three crept into no man's land close to the German wire and then rushed the German line, ten others rushed from their parapet. Twelve other battalions walked at a steady pace.[17]

Nevertheless, Prior and Wilson do not dispute that even those troops that rushed the German lines from the wire were self-consciously engaged in a bayonet charge. General Haig did not trust the volunteers of Kitchener's First Army to do more than advance in dense lines, bayonets determinedly fixed, and he himself embraced the pre-war view that 'victory almost always went to the side with the highest moral, discipline and offensive spirit'.[18] Although, as Paddy Griffith and Gary Sheffield have stressed, British tactics did improve during the conflict, morale, enthusiasm, and shock remained uppermost in the minds of commanders in their understanding of infantry assault for most of the conflict. Indeed, the British army employed Major (later Lieutenant Colonel) Ronald Campbell to tour the BEF giving brutal instruction in bayonet technique before the Somme offensive. Siegfried Sassoon himself witnessed the 'homicidal eloquence' of Campbell as he prepared with the Royal Welch Fusiliers for battle;[19] with 'an ultra-vindictive attitude' Campbell exhorted them to 'put on the killing face' when using the bayonet. Liddell Hart also witnessed these lectures and was appalled not only by their barbarity but also by their tactical crudity. Liddell Hart noted that while 'their [Campbell and his team] inflammatory efforts were not taken very seriously ... by battle experienced troops', they caused casualties among those troops who tried to 'close with the bayonet' as they had been exhorted to do.[20] In February 1917, the British army still advocated the importance of 'offensive spirit' to the training of all platoons: 'All ranks must be taught that their aim and object is to come to close quarters with the enemy as quickly as possible so as to be able to use the bayonet. This must become second nature.'[21] In response to the perceived over-use of grenades, the British army formally reaffirmed the priority of the rifle and bayonet—and therefore morale and aggression throughout the war: 'The bullet and the bayonet belong to the same parent, the rifle, which is still the deciding factor on the battlefield. One must work with the other. It is the spirit of the bayonet that captures the position, and the bullet that holds it.'[22] Not only did General Kiggell favour the bayonet charge per se but, as far as he was concerned, it was impractical to expect a citizen army to be capable of more complex tactical manoeuvres.[23]

Although British doctrine may appear reactionary here, the prioritization of the bayonet over the grenade in the assault was not absurd from a formation-level perspective. By going to ground and seeking to bomb their way forward, British troops did slow the momentum of an assault. Consequently, although it was individually rational to seek cover in grenade distance of the German trenches, it actually exposed the division to greater risk. It stalled the assault formation in the most dangerous area of the battlefield, the killing zone on which machine-guns, artillery, and mortars were all zeroed. Thus, the injunction that troops must dash through the killing zone in an act of collective élan to reach the parapet where they had the advantage of looking down upon their enemies to drive home the assault with bayonet and, perhaps, bullet at that point was not without reason. Yet, troops did not, of course, typically choose to go to ground in no man's land. They were forced to take cover in front of enemy trenches if they could not attain fire superiority, which typically occurred if the artillery bombardment had failed. At this point, the long distance recourse to grenade throwing was actually an attempt by assault troops to generate their own micro-artillery bombardment which would suppress the enemy (in the light of the failure of the fire plan), allowing them to assault. Once they had bombed their way forward, the charge might then occur. In reiterating the priority of the bayonet (whatever the circumstances), the training manuals failed to understand the tactical problems which induced the dependence on the grenade in the first place.

However, their reaffirmation of the bayonet—and the charge—as the key weapon is important. It demonstrates categorically that even towards the end of the war, high morale—evidenced above all in a moment of collective aggression—was critical on the battlefield, fire superiority notwithstanding. Even as late as 1918, the importance of the bayonet was emphasized in the British army: 'Every Platoon Commander must realize that the repeated act of putting his platoon over the bayonet-fighting course, if carried out with energy, determination and individual effort, has the direct effect of raising the fighting spirit of his command. In the Infantry the culminating point of all training is the assault.'[24]

The Canadian army units which were deployed to the Western Front always fought under British command, even when a full Canadian corps of four divisions was eventually formed in 1917. Consequently, especially given the heavy imperial influence of Britain on Canada and its army, which, as already noted, was drawn primarily from the anglophone community, these divisions adopted British military doctrine, including the proclivity for bayonet charges. Consequently, although the Canadian army was at the forefront of tactical innovations in the trenches, specializing in trench raiding which was first used by the Princess Patricia Canadian Light Infantry in February 1915,[25] it also employed the bayonet charge regularly. Although the assault on Vimy Ridge by the Canadian army employed a huge artillery bombardment and mines, both Brown and Red Lines were eventually secured by a bayonet charge.[26] Indeed, reports record that Red Line was secured by 'a mad charge' down the eastern slope of Vimy Ridge by 4th Division.[27] Inspired by a communal spirit of mateship, the Australians showed particular dash with their charges which were widely admired by commanders at the time as evidence of true offensive spirit.

The French and German armies before the First World War comprised large, mainly conscript forces for national territorial defence and they adopted notably similar infantry doctrine to the British. In particular, for both armies, the Boer War and the Russo-Japanese War in Manchuria were critical points of reference in developing infantry tactics at the turn of the twentieth century. For the French, military interpretation of these recent wars was inflected with a visceral sense of shame at the humiliation of Sedan where the French army had suffered an almost total collapse of command and morale.[28] The pre-war French army used these three conflicts as reference points for the development of a new military doctrine. Specifically and in direct opposition to the timidity and low morale demonstrated at Sedan, the French army developed a doctrine of *'l'attaque à outrance'*; all-out attack.[29] For the French army, a war could be won only through the assault, and the concept of *l'attaque à outrance* was intended to infuse every French soldier with a spirit of indomitable aggression. The concept of offence imposed upon commanders at all levels the responsibility to attack. In the absence of orders, commanders should always seek to attack and were required to do so; indeed, 'One defended only on receipt of an order not to attack or where attack was impossible.'[30] The concept of *offense à outrance* was plainly a response to the defeat at Sedan where the local aggression of tactical forces might have overcome the woeful incompetence of General MacMahon.

However, the concept of the offensive spirit was not only a tactic. It was also a manifestation of French nationalist self-conceptions in the early twentieth century. There were numerous publications preceding the First World War which promoted a distinctively French way of war in which the élan of the bayonet charge was primary. Indeed, the infantry tactics of the French army self-consciously utilized and indeed relied upon these patriotic sentiments. Specifically, the bayonet charge was seen as the appropriate tactical expression of a nation. 'Since Conscript armies were believed to represent their society and the spirit of the nation, it was necessary to develop a tactic which fitted the national temperament.'[31] Accordingly, military commanders became particularly interested in the human sciences at this point in order to refine their understanding of crowd dynamics and culture so that they would be better able to create the conditions of collective effervescence on which the bayonet charge depended.[32] The fact that French troops entered the Great War in nineteenth-century uniforms consisting of royal blue coats and red trousers emphasized the attachment to the nobility of offence; it was only in 1915 that the more tactical horizon blue uniform was adopted. These uniforms were deliberately designed to instil fear into the enemy confronted with an advancing mass, while actively encouraging cohesion among the infantry themselves who could all see each other very clearly and could be heartened by the stirring sight of the mass of bright colours around them. The blue and red uniform referenced the national flag and connected France's soldiers with the Napoleonic era.

Michael Goya[33] discusses the febrile work of Lieutenant Laure, *The French Offensive*, who, claiming that 'the vibrations of the offensive will have attained a paroxysm of intensity', appealed to his fellow officers:[34]

My comrades all of you, you who feel in your soldier's breast the heartbeat of all Frenchmen, I entreat you, let out with me this cri de guerre which inspires in you the instinct of your temperament: 'We want to conquer' and its corollary: ATTACK! Long live the manly, energetic, decisive and obstinate commander in his offensive spirit and whose magnificent propaganda seeks to 'nationalize such a doctrine'. Long live the break through! Long live the FRENCH OFFENSIVE![35]

The themes of masculinity and nationalism are to the forefront here, usefully indicating the close connection between the bayonet charge as a tactic and identity of the citizen army.

Laure's book, which accorded with a much wider flow of opinion, was certainly rhetorical but, in fact, French pre-war military doctrine accorded closely with its fundamental elements. Lieutenant Colonel de Grandmaison played a decisive role here, representing a whole cohort of officers before the First World War.[36] He was the chief of the army's Operations Branch before the war and organized two conferences at the Centre des Hautes Études Militaires in 1911 in which tactics were discussed. His book *Le Dressage de l'infanterie* (1909) synthesized his experiences as commander of the 1st Battalion, 30th Infantry Regiment, but also crucially drew on his interpretation of the Manchurian War.[37] There he emphasized the importance of aggressive spirit and fire in contemporary combat. Like many of his British contemporaries, Grandmaison interpreted the Japanese performance during that conflict as a vindication of high morale and close order mass assault. One of Grandmaison's axioms for young soldiers illuminated the relative priority of morale over fire: 'To conquer is to attack, to plant a bayonet in the heart of the enemy. The victory will go to those who accept death more easily.'[38] He concluded: 'The elegance of FRENCH MANOEUVRE is simple, direct and strong ATTACK.'[39]

In fact, French commanders were always well aware of the importance of fire superiority. *Reglement sur la conduite des grande unités* (1913) and *Reglement sur le service des armées en campagne* (1913) affirmed the requirement of fire superiority for an assault. However, the infantry assault was understood as a process by which the attainment of fire superiority facilitated the decisive bayonet charge; 'The artillery does not prepare attacks; it supports them.'[40] To that end, in the infantry regulations of April 1914, the bayonet was still described as the supreme weapon.[41] Intriguingly, Goya suggests that the social and institutional divide in the French army between aristocratic combat arm officers who graduated from Saint-Cyr and the technicians who trained at the Polytechnic for commissions in the engineers or artillery influenced the development of French infantry doctrine. The more influential Saint-Cyrians regarded the charge with its reliance on spirit and breeding as the appropriate expression of their ethos, the dependence on firepower as demeaning.[42] In the French army, the spirit of the bayonet was very prominent and explained many of the disastrous assaults in which it was involved especially in the early years of the war; 'many casualties suffered in the attacks from 14 to 23 August [1914] were unnecessary and came from foolish bayonet charges against entrenched enemy.'[43] Unfortunately, many of the attacks during the Nivelle Offensive on the Chemin des Dames in April 1917 assumed a similar form.

The German army of the First World War has been widely regarded as an innovative institution but its pre-war doctrine was startlingly similar to British and French thinking. For instance, the 1906 *Drill Regulations for Infantry* highlighted the importance of morale to victory. Specifically, victory could be won only through 'offensive spirit',[44] and the Regulations commanded: 'Forward against the enemy, whatever the cost!'[45] Naturally, the German army was well aware of the potency of fire on the modern battlefield, and fire superiority was seen as vital to the success of an assault. However, as with the British and French armies, firepower was only a means to an end. It produced a situation in which the assault, as the ultimate goal of the attack, could be accomplished; 'The attack consists of taking fire forward to the enemy, in extreme cases to the closest range. In the assault the defeat of the enemy will be achieved with edged [*blanken*] weapons.'[46] For Raths, this sentence from 1906 doctrine fused 'old and new images of war; the modern fire fight in open order on the one hand, the assault in thick closed lines on the other'.[47] The Regulations laid out a three-phase assault which paralleled French and British techniques. The units approached the enemy in formations as dense as possible to maximize control. An English summary of the 1909 Regulations noted that 'In no army in the world are smart, steady drill and rapidly manoeuvring in close order held in higher esteem or more assiduously practiced, than in Germany'; 'in action the Germans keep their men in close order as long as possible, preferring the loss of men to letting their commands get out of hand'.[48] As they came under enemy fire German troops dispersed into open or extended order, using the terrain to cover their advance, and moving forward in 80-metre dashes at most, thereby engaging in a fire-fight at a maximum of 800 to 1,200 metres from the enemy. Finally, when their fire was at its most intense, they mounted their assault.[49]

Echoing the sentiments of commanders like Grandmaison, German pre-war military writing was filled with eulogies to the bayonet to support this doctrine. Drawing on the Russo-Japanese War, Thilo von Trotta maintained the superiority of moral over material factors.[50] Colonel Freiherrn von Freytag-Lohringhoven recognized the power of new weaponry but even he saw the enduring relevance of moral factors.[51] In the correct circumstances, where fire superiority had been established, the bayonet assault could be successful especially since it was a simple manoeuvre. However, like the British and French, there are numerous occasions where the sanctification of the offence in general, and the bayonet charge in particular in German infantry doctrine and in pre-war military writing, precipitated poorly conceived and ultimately disastrous attacks. For instance, during the Battle of Tannenberg, the 41st Division 'advanced with the enthusiasm of inexperience, taking heavy casualties because of their reluctance to lie down between rushes'.[52] Similarly, while 43rd Hessian Brigade conducted a successful open order assault with light casualties on 9 September 1914 on Gerdauen, east Poland, the 44th Brigade mounted a close order assault relying on moral and offensive spirit. The latter attack failed with very heavy casualties.[53] Nevertheless, even after the First World War, the bayonet attack was seen as an important tactic. 'Good infantry must always look to the bayonet fight as the last resort, it is frequently indispensable to gain the decision. The final appeal to the bayonet, omitted in prior regulations, is found again in the new regulations.'[54]

The Italian army of the First World War replicated this preference for the bayonet charge. Most of the assaults which the Italians mounted especially on the Isonzo involved mass bayonet assaults, which were often unsuccessful and which were always very costly. Indeed, so futile were some of these uphill assaults against prepared Austrian positions that Austro-Hungarian defenders sometimes urged the Italians to retreat rather than to be massacred.[55] There is little doubt that General Cadorna favoured the bayonet charge because he was fully aware of the inadequacies of his army in terms of training and, in his view, discipline and morale. The bayonet charge was the easiest way of controlling his troops and of compelling them to assault. However, the belief in the bayonet charge was institutionalized into Italian military thinking. Italian military doctrine, *Frontal Attack and Tactical Training*, explicitly noted that 'stopping and lying down would be *a very serious mistake*' in the killing zone.[56] Accordingly, 'spirit and will' were crucial and Thompson notes the importance of the word 'slancio' (will) throughout Italian doctrine at this point.[57] Certainly, it was likely that many soldiers would be killed in such an assault but such a tactic minimized the time the formation as a whole spent in the killing zone and, at least, offered the possibility of success, even it was costly. Interestingly, Thompson explores the connections between this preference for the bayonet charge and deeper currents in Italian culture. The bayonet charge was not just one tactic among others but was 'the military expression of vitalist beliefs about nature and society'. The attack denied the dominance of technology over the human spirit and as such the bayonet assault was closely associated with the vitalist movement and with the futurism of Boccioni.[58] For Cadorna, the bayonet charge was not just a tactic, it was an existential expression of Italian nationalism.

Throughout the First World War, commanders often pushed the poorly trained infantry into mass bayonet assaults against prepared positions which had been inadequately disturbed or suppressed by artillery bombardments. Commanders did not trust their untrained citizen forces with any more complex tactical manoeuvres. In most cases, inspired by collective enthusiasm and the close proximity of their comrades, citizen soldiers were willing to charge forward against machine-guns and mortars with bayonets even at a terrible cost. In this way, the bayonet charge utilized the combination of a highly motivated but low-skilled infantry which often characterized the citizen army. With a poorly trained army, the mass assault of the infantry was easy to organize and, if the enemy had been suppressed or sufficiently demoralized, especially through the use of artillery, a bayonet charge could be effective. Indeed, there were occasions when it did work such as at Neuve Chapelle or the first day of the third Battle of Ypres after the blowing of the Messines Ridge or some of the assaults at Vimy Ridge. In Chapter 6, the innovation of infantry tactics away from the massed bayonet charge will be discussed, and it is imperative that the significance of these developments are recognized. Nevertheless, for all the innovations during the conflict, the mass assault of infantry driven home at the point of the bayonet was a regular practice right up to 1918, whether it was regarded as an ideal or not. Commanders self-consciously saw it as a means of overcoming battlefield inertia—the Marshall effect.

It might be assumed that the limitations of the bayonet and the mass assault demonstrated by catastrophic assaults by the French, the British, and the

Germans during the First World War would have led to its disfavour in the Second. Many troops recognized its shortcomings. The British and Canadians were deeply concerned about casualties throughout the Second World War, to the frequent frustration of the more aggressive Americans. In France, the First World War had cured the nation and its army of all desire for offensive action, and its doctrine of all-out offence (*attaque à outrance*) had dissolved into strategic and operational defensiveness.[59] In preparing for the Second, defensive fire took precedence over manoeuvre.[60] Yet, despite advances in mechanized warfare and in the application of firepower, the bayonet charge remained significant to military doctrine in the Second World War among western militaries. British publications in the 1930s eliminated naive appeals to the superiority of morale which were typical before the First World War. The British army retained the centrality of the offence but it too recognized the realities of the modern battlefield: 'Infantry cannot advance against even semi-organized resistance, unless that resistance is kept in subjection by fire, nor must it be launched against unbroken wire obstacles.'[61] The centrality of fire superiority was reiterated in all British publications and the concept of dense, close order assaults even once the enemy had been suppressed had disappeared entirely. Nevertheless, echoing First World War doctrine, *Infantry Training* still stressed the importance of the bayonet: 'The rifle and the bayonet are the principal weapons of the individual infantry solder. The first requirement of the infantry soldier is confidence in these weapons, based on his skill in their use.'[62] The manual continued: 'The bayonet is the weapon for hand-to-hand fighting, and its use, or the threat of it, finally drives the enemy from his position or causes him to surrender.'[63] Indeed, the relevance of the bayonet was recurrently affirmed during the Second World War. In order to disseminate lessons learnt from operations, the army published a monthly Army Training Memorandum. In August 1940, one of these Memoranda discussed the bayonet, confirming its ability to 'develop offensive spirit', especially since there was 'nothing peacetime or "pansy" about it'.[64] Moreover, it rejected the findings of a recent article which 'described the bayonet as obsolete' because few casualties had been inflicted in 1914 to 1918 by the weapon. The Memorandum noted that by contrast in Libya, the bayonet had been used effectively and it was the 'soldier's best method of offense and defence when he cannot shoot'. The Memorandum concluded: 'The bayonet often achieves success without inflicting casualties merely by the moral effect.'[65] British training literature repeatedly claimed that German troops generally fled when faced with the bayonet.[66] Advocating the potency of the bayonet charge, Brigadier Harding, commander of 138th Brigade in Italy, referred to numerous successful bayonet charges by his troops in an interview published by the War Office in June 1944.[67] In a work which explores the underdevelopment of British tactics, Timothy Harrison Place is clearly sceptical about the Brigadier's claims but in fact there are numerous cases of the successful employment of bayonet charges in the Second World War. At El Alamein, the British infantry advanced in dense bayonet waves in a manner very similar to the First World War, though their assault was successful. An officer from the Seaforth Highlanders recalled: 'a sight that will live for ever in our memories—line upon line of steel-helmeted figures with rifles at the high-port, bayonets catching in the moonlight, and over all the wailing of the pipes.'[68] Defenders were sufficiently distracted by the massive artillery bombardment and by the supporting tank assault to be overwhelmed. The British forces also employed the same tactic with success on numerous other documented occasions. On 19 August 1942, British and Canadian forces raided Dieppe. For the most part, the raid was a disaster, but No. 4 Commando achieved all its objectives, attacking and destroying an assault gun battery. The position was eventually taken by a bayonet charge.[69] Captain Pat Porteous, who led the charge and was awarded a Victoria Cross for it, recorded the action: 'It seemed a hell of a long way but I think it can't have been more than about eighty yards or a hundred yards.'[70]

The US Army certainly prioritized firepower in all their doctrine. However, bayonet training was also advocated especially at the platoon and company level in both the First and Second World Wars. It featured in the 1943 Field Manual *Bayonet*:

> The will to meet and destroy the enemy in hand to hand combat is the spirit of the Bayonet.... The full development of his physical prowess and complete confidence in his weapon culminates in the final expression of the spirit of the bayonet—fierce and relentless destruction of the enemy. For the enemy, demoralizing fear of the bayonet is added to the destructive power of every bomb, shell, bullet and grenade which supports and precedes the bayonet attack.[71]

General George Patton, who developed quite sophisticated fire and manoeuvre tactics, advocated the bayonet; 'Few men are killed by the bayonet; many are scared by it.'[72]

These statements might appear as mere rhetoric, intended to inspire an offensive spirit. Yet, there are many apparently unlikely cases of successful bayonet attacks by US troops in the European and Pacific theatres. The predicament of 3rd Battalion, 502nd Parachute Infantry Regiment, 101st Airborne, on the Carentan Causeway in Normandy on 11 June 1944 is instructive. Pinned down and exposed, individual soldiers refused to return fire despite the exhortations of their commander (see Chapter 3). The main enemy fire was coming from a farmhouse 'to the west of the road on ground that rose sharply from the marshes'.[73] At this point, their commanding officer, Lieutenant Colonel Robert Cole, ordered them to 'Fix bayonets' and to 'Charge'. The action was recorded prosaically in the US official army history:

In the morning of 11 June after attempts to knock this [the farmhouse] out with artillery had failed, Colonel Cole, battalion commander, ordered a charge which he and his executive officer, Lt. Col. John P. Stopka, led. Followed at first by only a quarter of their 250 men, Cole and Stopka ran through enemy fire. The charge gathered momentum as more men got up and ran forward. The farmhouse was not occupied but the Germans had rifle pits and machine gun emplacements in hedgerows to the west. These were overrun and the Germans killed with grenade and bayonet.[74]

Despite the fact that most failed to participate, Cole's advocacy of this communal action had a galvanizing effect on men who only moments earlier had failed to show individual initiative: 'The men began to roar as they charged, their own version of the Rebel Yell. The Germans fired and cut down some, but not enough... Those Germans who dodged the bayonets ran out the back way and fled to the rear.'[75] Cole's order illustrated the tactical benefits of the bayonet charge. There are many other examples. On 3 January 1945, near Mande-Saint-Étienne 10 kilometres north of Bastogne, a platoon 3rd Battalion 513 Parachute Infantry Regiment, 17th Airborne Division, were caught by nebelwerfer and pinned down. The platoon captain ordered 'Fix bayonets': 'All fourteen men jumped out of the ditch, formed into a line of skirmishers, rifles at high port and moved toward the German position. They began shouting "Geronimo!". Gabel screamed with the others. They got into the German lines.'[76]

The First Marine Division in the Pacific disparaged the suicidal frontal assaults of the Japanese.[77] At the Battle of Tenaru, no more than a few Japanese soldiers from 1,200 survived a banzai attack, killing only 44 of their US opponents. On Guadalcanal, the banzai charge became a 'symbol of Japanese stupidity' and was seen as evidence of their racial inferiority.[78] Yet, throughout the Pacific campaign the Marines demonstrated a not completely dissimilar affinity for the bold but bloody frontal assault. In his widely read account of the Pacific War, Eugene Sledge provides significant support that, for the mass infantry, the bayonet assault was a favoured and indeed potentially effective form of tactics. Instructively, his description of 1 Marine Division's assault on Peleliu Airfield by an entire Marine regiment on 16 September 1944 records one of the most famous bayonet charges in the Second World War.

'Let's go', shouted an officer who waved toward the airfield. We moved at a walk, then a trot, in widely dispersed waves. Four infantry battalions—from left to right 2/1, 1/5, 2/5 and 3/5 (this put us on the edge of the airfield)—moved across the open, fire-swept airfield... For me the attack resembled World War I movies I had seen of suicidal Allied infantry attacks through shell fire on the Western Front.[79]

Nevertheless, at an admittedly very heavy cost, the attack succeeded and usefully Sledge records the reason for its success. He noted that as he himself charged 'my only concern then was my duty and survival'. Trained and indoctrinated as a Marine, he was committed to the assault. Moreover, he noted that this motivation was evident among the troops much more widely:

'Because of the superb discipline and excellent spirit of the Marines, it had never occurred to us that the attack might fail.'[80] The Marines on Peleliu demonstrated little tactical refinement. However, their severe training had at least welded them into a force with sufficiently high morale that it was able to conduct simple, aggressive tactical manoeuvres like a bayonet charge.

General George Patton understood and indeed had observed the Marshall effect at work among his units. He noted the tendency of soldiers to 'hit the dirt' in the face of enemy fire. While individually rational, he noted that paradoxically the tendency to take cover 'has done much to increase our casualties'.[81] Observing this tendency among US troops, the Germans 'wait until we have arrived at a predetermined spot on which they have ranged rockets, mortars or artillery'. Then, 'the soldier, obsessed with the idea of hitting the dirt, lies down and waits supinely for the arrival of the shells from the mortars, rockets, etc. He usually does not have long to wait.'[82] Patton insisted that 'the only time it is proper for a soldier to drop is when he is caught at short range—under three hundred yards—by concentrated small-arms fire'.[83] By taking cover, soldiers increased the amount of time they were exposed to enemy fire in the killing zone and consequently, despite the apparent security afforded by the ground, they increased overall casualty rates as well as their own chances of being wounded or killed; 'the time you are exposed [to enemy fire] is reduced by the rapidity of your advance'.[84] Accordingly, while he recognized the moral potency of the bayonet, in the context of industrialized warfare, he advocated a tactical technique which represented a fusion of mass bayonet and fire tactics: 'marching fire'.

The proper way to advance, particularly for troops armed with that magnificent weapon, the M-1 rifle, is to utilize marching fire and keep moving. This fire can be delivered from the shoulder, but it is just as effective if delivered with the butt of the rifle halfway between the belt and the armpit. One round should be fired every two or three paces. The whistle of the bullets, the scream of the ricochet, and the dust, twigs, and branches which are knocked from the ground and the trees have such an effect on the enemy that his small-arms fire becomes negligible.[85]

Marching fire did not only have a moral effect on the enemy but on the advancing US troops; 'The fact that you are shooting adds to your self-confidence, because you feel that you are doing something, and are not sitting like a duck in a bathtub being shot at.'[86] Patton's instructions are written in the second person, to an individual soldier. However, his concept of marching fire could be effective only if it were performed collectively. A lone soldier attempting to march and fire would be quickly eliminated by enemy fire. But if whole platoons and companies assaulted in this way, assaults could be effective. Certainly, as with the bayonet charge, casualties were likely to be incurred especially in the initial phase, but as the attacking force advanced and began to attain fire superiority, casualties would be minimized. Patton did not see the need for complex fire and movement tactics in which sub-units coordinated their actions. Rather, each individual advanced together in a line firing at will to generate a genuinely collective effect. Marching fire may have been an unsophisticated tactical concept, and yet, in the light of a poorly trained infantry, it may have represented a realistic solution to the battlefield problem of inertia. In particular, it added fire to the moral and tactical advantages of the bayonet charge but, to all intents and purposes, it was a form of mass bayonet assault.

The German armies of the First and, especially, Second World War are often held out as the avatar of military performance in the twentieth century. As discussed, the German commanders up to the First World War shared the faith of their French and British rivals in the primacy of morale over matériel and, therefore, the importance of the bayonet charge on the battlefield. Indeed, one of the most successful and innovative infantry commanders of the war, Erwin Rommel, frequently employed the bayonet assault to great effect.[87] Following their successes with stormtrooper tactics, the Germans continued to make tactical innovations in the Second World War, exploiting the potential of their rapid firing *Maschinengewehr* 34 and 42.[88] There was decidedly less emphasis on the bayonet in Wehrmacht doctrine. Indeed, Allied armies noted that the Germans were not adept at bayonet fighting and did not like it: 'the German hates the bayonet and is inferior to our men with it.'[89] However, American military observers of exercises in the 1930s noted that once German infanteers reached assault distance 'they fixed bayonets and made the charge with a yell'.[90] The bayonet assault may have become a subordinate tactic for the German infantry, which preferred to exploit the high rates of fire from their machine-guns, but it was an unignorable element of their assault techniques.

Interestingly, the bayonet remained important even after the Second World War. In Korea, the French battalion deployed under the US 2nd Division consistently used bayonet tactics. The French battalion consisted of volunteer professionals commanded by their (pseudonymous) Lieutenant Colonel Monclar (in fact the much decorated and seventeen times wounded veteran Lieutenant General Magrin-Vernerey of the First World War) and was widely regarded as the finest infantry troops which the UN had at its disposal.[91] One reporter recorded that 'By any standards by which a soldier can be judged they are among the finest fighting men this war has seen. They truly go "to war as to a wedding". They keep their weapons clean. They endure cold, hardship, and the strain of ceaseless combat with an exuberance nobody else can.'[92] Perhaps as a result of his experiences in the First World War, Monclar believed in the power of the bayonet in both the assault and perhaps more surprisingly the defence, where they developed an unusual tactic. On Hill 543, the battalion dug two parallel trenches, occupying the second and leaving the first to the Chinese attackers. On finding an unoccupied trench, Chinese attackers would naturally seek cover in the first trench. At this point, the French battalion would mount a counter-attack driving the Chinese out with a bayonet charge. For this action and those at Chipyong-ni and Hongchon, the battalion received three Presidential Unit Citations which were awarded personally by General MacArthur.[93] The battalion's most famous accomplishment was, however, the seizure of Hill 851, the last pinnacle on Heartbreak Ridge, by bayonet charge on 13 October 1951.[94] Primarily because of these aggressive tactics, the French losses were proportionately the highest of all the UN forces serving in the Korean War, except for the American and Korean forces.[95] Yet, the French were also widely regarded as the most successful and competent military force there.

The Vietnam War offered less opportunity for bayonet charges with its dense jungles, inundated paddy fields, and often fleeting enemy. Even more than in Korea or the Second World War, the US Army and Marines relied on firepower and, above all, artillery and air support to mount attacks. Nevertheless, there were examples of bayonet charges in that conflict. Indeed, despite the training of 1st Division, there were limits even to 1/7's professionalism in the Battle of Ia Drang in November 1965. About two hours into the fight on the first afternoon, Captain Herren's Bravo Company aimed to assault the NVA forces in front of them to improve their position by clearing the creek-bed which ran along the eastern edge of the Landing Zone. They planned to use institutionalized fire and manoeuvre drills, but in reality the platoons were unable to coordinate their actions. 'The "fire and manoeuvre" plan was forgotten. By necessity, Bravo Company got on line and attacked toward a sizeable enemy force in the brush ahead of them.'[96] One of the lieutenants recorded that: 'We stood up and started the assault... we had men dropping all over the place. Finally, the assault line which had started out erect went down to our knees. And then down to the low crawl.'[97] The assault eventually became an outright bayonet charge. On Operational Eagle Claw on 15 February 1966, 2nd Battalion, 7th Cavalry assaulted an enemy position, initially using fire and movement but concluding the assault about 40 metres from the objective, 'the men clambered to their feet and charged forward at a dead run, yelling as loudly as they could'.[98] In July 1967, C Company 1st Battalion, 35th US Infantry, part of the 173rd Airborne Brigade, made a successful charge against enemy bunkers on a hill in the jungle to the southwest of Duc Pho. Exposed under enemy fire, the company commander decided that his only option was to mount a frontal assault with two of his platoons; 'Rising as one, shouting and screaming at the tops of their lungs, the men charged forward.'[99] Their commander noted afterwards: 'I still prefer to use our basic concept of finding and fixing the enemy.'[100] It was noticeable that it was often elite troops with the highest morale who resorted to the bayonet charge.

Despite the criticisms of analysts like Liddell Hart and its apparent obsolescence with the appearance of the machine-gun, artillery, and mortar, the successful use of the bayonet charge in the Second World War provides at least some support to the advocacy of the weapon by Du Picq and by military doctrine in the first half of the twentieth century. The bayonet charge overcame the problem of dispersal on the battlefield. It unified attacking forces into a common form of action in which each soldier provided mutual support and exercised moral pressure over others. Conditions clearly had to be favourable for such an attack to succeed. Typically in the siege conditions of the First World War, this was not the case. Assaulting against prepared defensive positions which protected the enemy infantry and slowed the assault with barbed wire, the bayonet charge was practically suicidal. In the sometimes more fluid conditions of the Second World War, however, when defences were often more improvised than on the Western Front and when enemy forces might be less prepared, isolated, or suffering under heavy artillery or air bombardment, an aggressive charge could frequently be successful. Indeed, given the relatively low training levels of the troops, the charge could be the most effective form of attack. As the Carentan Causeway demonstrated, soldiers who were incapable of attaining firing superiority—despite their supposedly elite training—might be willing to charge together at an enemy. It is easy to disparage the mass action of the bayonet charge. It was extremely risky but it represented a concrete means of overcoming the problem of dispersion presented by the modern battlefield. It was a proven, if unsophisticated and now derided, means of overcoming the Marshall effect.

The bayonet charge was a military tactic which was developed primarily in response to the need to cross the fire-swept killing zone quickly by means of a technique which a mass, often inadequately trained infantry could actually perform. It was, then, very substantially an understandable and coherent tactical response to a military problem: the Marshall effect. However, as military doctrine up to the 1940s but especially before the First World War demonstrates, there were evident cultural and political factors which recommended the bayonet charge to the army as an ideal. The bayonet charge was not simply an objective solution to a tactical conundrum. It was also explicitly seen as a form of assault which was culturally appropriate for a citizen army tasked with defending its nation. The bayonet charged embodied the concepts of nationalism and physically united the citizen soldiers into a mass body of equals all prepared to sacrifice themselves for the collective good of their nation. The bayonet charge was not then simply a tactical solution, it was a manifestation of national will and commitment. In the charge, the citizen soldiers physically realized the egalitarian community which the modern nation-state took as its ideal. All were united as equals in the defence of national honour.

Clearly, the bayonet is still issued to professional soldiers and, since the 1970s, it has been recurrently used by the infantry and is particularly favoured by the British army. During the Falklands War, bayonets were fixed for major assaults and a small number of Argentine soldiers were killed by them.[101] Before their assault on Mount Longdon on the night of 11/12 June 1982, B Company 3rd Battalion the Parachute Regiment was ordered to fix bayonets by their company sergeant major on their line of departure at the base of the mountain; despite their efforts, they made considerable noise which was heard by Argentine defenders, though they failed to react to it.[102] During the Battle of Goose Green on the morning of 29 May 1982, A Company 2nd Battalion the Parachute Regiment found its progress stopped in a gorse gully below Darwin Hill, a ridge line which ran across the narrow isthmus just over a kilometre north of the Goose Green settlement itself. When their commanding officer, Lieutenant Colonel Jones, came forward to ascertain what was happening and to re-inject momentum into the attack, the company mounted an assault as close to a bayonet charge as perhaps any professional soldiers have ever tried in the last fifty years, despite offers of assistance from D Company to outflank the position to the right.[103] The result was disastrous. Almost immediately on cresting the hill, the summit of which Argentines occupied by a series of bunkers, in an assault line with no fire support, Captain Dent, the second-in-command of the company, Captain Wood, the adjutant, and Corporal Hardman were killed.[104] Lieutenant Colonel Jones subsequently mounted his now famous Victoria Cross-winning lone assault up a gully to the right only to be overtaken by the same fate, shot from an unseen bunker to his right and rear. The scathing professional assessment of the entire action by Sergeant Blackburn, Colonel Jones's radio operator, was illuminating: 'It was a death before dishonor effort; but it wouldn't have passed junior Brecon.'[105] Illustrating the importance of firepower over morale and mass, the position was eventually unlocked when Corporal Abols destroyed a key bunker with a rocket.[106] More recently British forces have used bayonets in both Iraq and Afghanistan. On 14 May 2004, a company from the Argyll and Sutherland Highlanders (now 5 SCOTS) assaulted and cleared an insurgent mortar position on Danny Boy Hill in Al Amara bayonets fixed, killing approximately thirty enemy fighters in the process. Professional soldiers have fixed bayonets and individuals have used them. However, these cases cannot be said to amount to mass bayonet charges. With the partial and highly irregular exception of the assault on Darwin Hill by 2 PARA, wave tactics involving large units advancing *en masse* have not been evident in any of the other cases. Companies and platoons have used dispersed fire and movement tactics and, notwithstanding its moral effect, the bayonet was used by individual soldiers in the close fight. This was particularly noticeable on Mount Longdon where the complex geography of the position reduced the battle to a series of small actions often executed by small teams or pairs.

INDIVIDUAL ACTION

The bayonet charged solved the problem of dispersal and the inclination to individual passivity. However, the bayonet charge, although relatively straightforward, still presented coordination difficulties. Once the charge had started, it was generally more difficult for an individual to fall out of the line and shirk than to continue. However, soldiers had to be got to their feet and, if the assault was unplanned as it was on the Carentan Causeway, individuals might plausibly point to local circumstances as an excuse for non-compliance; they might shirk by claiming that they had not heard the order or could not respond through injury or their personal tactical circumstances. In classical reproduction of the collective action problem, individuals might simply wait to see what everyone else did before committing themselves to the charge. Indeed, as discussed in Chapter 3, this is precisely what happened to Lieutenant Colonel Cole on 6 June 1944 when his soldiers claimed to have failed to hear his order to charge. As each individual awaited the initiative of others, no collective action would take place.

Accordingly, the mass armies of the early twentieth century developed, sometimes quite accidentally, a second technique for encouraging activity. Instead of addressing the question of collective action, armies sought to circumvent the problem altogether. Individual initiative was adopted as a solution to the problem of inertia. In many cases, the individual action was entirely spontaneous and unexpected. As an attack stalled, one individual continued the assault alone, firing his weapon and throwing grenades as his comrades remained passive. However, there was an expectation that section and platoon commanders should lead from the front and, if an attack were failing, it was their personal responsibility to seize the initiative. The British army consistently recommended individual initiative. As Place records,[107] throughout the first part of the twentieth century, substantial elements within the British army were concerned that the institutionalization of tactical drills would rob soldiers and particularly officers of the possibility of individual initiative. Since combat was unpredictable and no two tactical situations were ever exactly the same, individuality was regarded as absolutely fundamental to successful military action. Indeed, the War Office was explicitly opposed to the attempts of Major Lionel Wigram, chief instructor at the Chelwood Gate Battle School, to introduce standardized tactics into the army; it saw 'tactical rules' as a menace to initiative.[108] The advocacy of individuality also accorded with and was probably a manifestation of the aristocratic culture of the British officer corps at the time in which every commanding officer acted as a country gentlemen with sole authority for his estate.[109] Although there was much discussion of fire and movement and tactical drills, individual initiative was recurrently stressed in British doctrine. For instance, the 1942 Instructors

Handbook introduced a number of techniques to overcome German tactical superiority at the platoon level and the publication suggested that inherent national characteristics like individual initiative might be utilized in combat: 'He [the British soldier] is equipped with a natural cunning which the individual German lacks.'[110] In line with this, the document stressed the importance of high morale which could be sustained and augmented by each individual: 'To generate high morale each man must be aware of his own genuine skill and power as a fighting unit; and feel his own importance to his comrades and his nation.'[111] In particular, individual rushes—personal charges—were advocated as a means of overcoming the final bound before the enemy position: 'If the hostile fire becomes so severe that sections are no longer possible, the advance must be continued by individual rushes, two or three or even single men from the advancing section moving forward at a time.'[112] When the majority of the attacking force was pinned down, a single individual might leap forward and initiate the collapse of a position. The British may have been unusual in their advocacy of individualism in warfighting but, on the battlefield, individual acts of initiative were a significant feature of the citizen army. The case of Lieutenant Masters at Hardt has already been noted. While the members of his platoon failed to take collective action, he jumped to his feet and charged the enemy. This individualism was replicated many times in the citizen wars of the twentieth century. Indeed, there are some celebrated examples of it. Alvin York and Audie Murphy represent two of the most renowned American examples in the First and Second World Wars and their actions provide useful individual examples of a wider social phenomenon.

Alvin York came from a poor farming family in Kentucky where he spent much of his youth hunting and shooting:[113] it was here that he developed his extraordinary individual marksmanship, shooting the heads off turkeys with a pistol at range.[114] York was eventually drafted in 1917 and was assigned to 328th Battalion 82nd Division and saw his first action in the Argonne Forest. On 8 October 1918, York's battalion were ordered to take Hill 223. At 6.10 a.m., the battalion advanced up a valley in waves with bayonets fixed but, without artillery support, they were immediately subjected to intense enemy machine-gun fire. York was positioned on the extreme left-hand side of the advance, as the assault faltered. In his autobiographical account, which illustrates his limited exposure to formal education, York and his platoon identified that the 'worstest machine-gun fire was coming from a ridge over our left front' and accordingly sought to outflank the position.[115] York's sergeant ordered the platoon to 'crawl back a little and try and work out a way down around on the left and then push on through the heavy underbrush and try and jump the machine guns from the rear'.[116] The platoon 'got around on the left and in single file advanced forward through the brush' towards the machine-guns.[117] Eventually, the platoon infiltrated the German rear, surprising medical and headquarters elements whom they tried to encourage to surrender. At this point, the platoon came under fire from the machine-guns on the hill above them (which had switched their fire rearwards); most of the platoon was killed or wounded at this point as 'thousands of bullets kicked up the dust all around us',[118] leaving only eight men alive and York in charge. Despite his exposed position, York began to engage the machine-gunners;

> I had no time nohow to do nothing but watch them there German machine gunners and give them the best I had. Every time I seed a German I jes teched him off [I just shot him]. At first I was shooting from a prone position; that is lying down; jes like we often shoot at the targets in the shooting matches in the mountains of Tennessee; and it was jes about the same distance. But the targets here were bigger. I jes couldn't miss a German's head or body at that distance.[119]

York was then charged by an officer and five men with fixed bayonets, all of whom he killed, shooting them in reverse order as he had learnt to shoot turkeys.[120] At this point, a German major surrendered the position to York, so that in the end York, having killed 25 enemy soldiers, was personally able to capture 132 Germans and approximately thirty-five machine-guns.[121]

Substantially as a result of his actions, the 328th Battalion was able to take Hill 223, described in the official report as a success 'of outstanding proportions', and York was himself awarded the Congressional Medal of Honor. York's endeavours remain a striking example of individual virtuosity. Yet, this virtuosity itself highlights the collective passivity not only of his own fellow US soldiers but also the German machine-gunners whom he single-handedly overwhelmed. Despite their great advantage in numbers and firepower, they collapsed in the face of the determined action of a single firer in a manner which Marshall would have recognized. The Germans themselves seemed to demonstrate the problems which citizen armies had in generating collective combat performance among their troops. Precisely because these were mass forces, they tended to encourage passivity and non-participation among large numbers of soldiers who explicitly or implicitly would leave it to others to take responsibility in combat.

Audie Murphy's military record in the Second World War bears some notable similarities to those of Alvin York both in terms of his combat performance and his biography. Like York, Murphy was raised in a poor, farming family in Texas during the Depression. He became an experienced hunter, claiming in later life that if he failed to shoot accurately, he 'didn't eat'.[122] Underweight, he was rejected by the Marines and the paratroopers but was eventually accepted by the US infantry and was assigned to the 3rd Infantry Division. He had been deployed to North Africa in 1943 and had fought in the Sicily and Anzio campaigns, distinguishing himself in each operation. He had then landed with the 3rd Division in southern France in 1944 as part of Operation Dragoon, fighting his way up into Germany to the Vosges mountains, during the course of which he was promoted to lieutenant.

On each occasion, Murphy demonstrated individual virtuosity in personally engaging with and killing the enemy while his colleagues remained passive. In the Vosges Mountains, on 25 January near Holtzwihr, Audie Murphy was involved in an action for which he received the Congressional Medal of Honor. Murphy's 30th Infantry Regiment was ordered to take the town of Holtzwihr. The regiment was positioned on a road at the top of a U-shaped valley leading to Holtzwihr, on either side of which woods extended in a long horseshoe down higher ridges to the village.[123] On 25 January, the 30th Regiment was attacked by a force of six tanks attempting 'an encircling movement using the fingers of trees for cover',[124] supported by a large number of German infantry. Almost immediately, one tank destroyer, accompanying Murphy's regiment, slid into a ditch, rendering itself useless, and soon after the attached second tank destroyer that had positioned itself on the road was knocked out by artillery fire and began to burn. As German tanks began to bypass Murphy's position aiming for a burning but not totally destroyed US tank destroyer, he recognized the threat posed by advancing companies of German infantry. He used the field telephone to order in artillery fire,[125] and then ordered his platoon to withdraw as a result of enemy artillery fire and machine-gun fire from the tanks. Alone he continued to call in fire while sniping with this carbine.[126] Finally, out of ammunition, Murphy withdrew to the tank destroyer, manning its 0.5 calibre machine-gun. The German infantry, assuming the tank destroyer was out of action, 'could not comprehend where this new hail of fire was coming from' and 'milled around, taking casualties all the time'.[127] At the same time, Murphy called in artillery strikes on the German tanks and infantry which were firing at him. Eventually, the German infantry mounted a series of assaults against Murphy's position on the tank destroyer which he repulsed with fire at ranges of as close as 10 yards;[128] Murphy recorded killing a number of Germans with his fire.[129] Murphy then called in an artillery barrage almost on his own position. Unable to dislodge Murphy, some German infantry bypassed him, eventually penetrating to rear areas where they were engaged by the battalion headquarters. The action was eventually halted when US air support arrived, forcing the Germans to withdraw, allowing Murphy to finally leave his position on the tank destroyer. Murphy's actions at Holtzwihr were clearly remarkable and have been the subject of much interest and admiration since that time. Yet, they display very similar features to York's some two decades before. In the face of intense enemy fire, Murphy's battalion began to fragment as soldiers within it were either killed, wounded, or morally incapacitated. At this point of crisis, Murphy acted autonomously to repulse the enemy. As in the Argonne, Murphy's German opponents demonstrated a similar incompetence to their US counterparts, unable to coordinate a collective response to this individual action.

These are extremely famous examples whose general validity might be questioned on the basis that they represent extraordinary—even unique—cases rather than a pattern, especially since the US Army and government deliberately sought to use York and Murphy for political ends, generating popular support for and demonstrating (especially in the case of York) the USA's contribution to the war. The representation of the actions of York and Murphy in the media deliberately emphasized the heroic dimension, informed by American ideals of rugged individualism. However, there are numerous other less well-known actions which display the same pattern. Thus, in *Night Drop*, Marshall describes how on D-Day Staff Sergeant Harrison Summers of 1st Battalion, 502nd Parachute Infantry Regiment single-handedly cleared a German barracks on the Reuville road, personally killing numerous enemy soldiers, with the minimal help of two other soldiers;[130] 'Summers rushed the buildings one by one, kicked in the doors, and sprayed the interior with his Tommy gun.'[131] By contrast, Marshall records how the other paratroopers 'responded somewhat sullenly to the order' (to assault),[132] and 'where they made any contribution, they stuck in the roadside ditch and provided some covering fire'.[133] On 7 June, Lieutenant Waverly Wray of the 505th Parachute Infantry Regiment, 82nd Airborne was fighting near Sainte-Mère-Église. Wray independently conducted his own reconnaissance patrol (colleagues describe him as 'moving like the deer-stalker he was') against the enemy, eventually finding eight officers at a command post. He tried to make them surrender and shot one of the officers who had drawn his pistol. Two Germans 100 metres away in a slit trench then fired at him, providing an opportunity for all the officers to scatter. Wray killed the retreating officers and the Germans in the slit trench single-handedly.[134] On his return to his lines, he then led the assault against the Germans and broke a German counter-attack which allowed Sainte-Mère-Église to be taken. He died on 19 September during the Nijmegen operation demonstrating similarly individualistic initiative: 'The last I saw of him, he was headed for the Germans with a grenade in one hand and a tommy gun in the other.' He was killed by a sniper, shot as he raised his head over the railway embankment.[135]

A similar pattern of individualism was evident among British forces. In his analysis of the Normandy campaign, Carlo D'Este noted the general lack of aggression and initiative displayed by British troops. He emphasizes that they typically fought well, especially in defence, but there was a notable tendency among soldiers in most regiments to want to follow. Consequently, British troops were very dependent upon their platoon and company commanders to lead assaults. The result was that casualty rates among platoon and company commanders in Normandy were extremely high. Mass passivity and individual action by designated leaders was almost a structural feature of the British army in Normandy. Drawing on D'Este's work, Max Hastings has also been critical of the performance of Allied armies and especially their infantries in the Battle of Normandy in 1944, noting how frequently acts of individual soldiers were critical to their tactical redemption. He cites the example of Sergeant-Major Stan Hollis, of 6th Green Howards, who having landed on Juno Beach on D-Day single-handedly eliminated two German pillboxes: 'Without hesitating, Hollis sprang to his feet and ran 30 yards to the German position, spraying sten-gun fire as he went, until he reached the weapon-slit, where he thrust in the barrel and hosed the interior with fire.'[136] He immediately advanced on the second bunker which he destroyed, returning with twenty-five prisoners for which he was awarded the Victoria Cross, the only one awarded on D-Day.[137] Over the next few weeks, he continued to perform similar acts of heroism.[138] His commanding officer emphasized his distinctiveness in comparison with typical citizen soldiers: 'He was absolutely personally dedicated to winning the war—one of the few men I ever met who felt like that.'[139]

The Canadian and Australian armies demonstrate a compatible reliance on individual heroism. At Vimy Ridge, as the 16th Battalion approached the Black Line, a German machine-gun firing from the left inflicted many casualties on No. 4 Company: 'fan-shaped around the position lay dead 16th men.'[140] At this point, the official report states that, just as a serious delay seemed inevitable, 'a series of bomb explosions were heard in the direction of the enemy and a 16th man, Private William Milne, sprang up from the shell hole close to it, signalling to his comrades to advance. He had crawled around on his hands and knees to within bombing distance of the enemy machine gun crew and with hand grenades had put every one of them out of action.'[141] He personally cleared the position and then called his company forward. He repeated the feat with another gun between the Black and Red line before he was himself killed; he was awarded a posthumous Victoria Cross. The recourse to individual heroism in place of platoon tactics was equally evident in the Canadian army in the Second World War.[142] During the Dieppe Raid in 1942, Lieutenant Colonel Charles Merritt led his battalion, which was taking heavy casualties, off Green Beach by personal example, walking casually across a bridge under German fire. Once across, he called back to his men, 'Come on over, there's nothing to it.'[143] Indeed, in the Second World War, there was some concern in the Australian army that battalion commanders were becoming too involved in the fighting itself at excessive risk to themselves. A number of commanding officers physically led assaults rather than simply commanding them. During the Parit Sulang retreat Lieutenant Colonel Charles Anderson led by personal example throughout the period 20 to 22 January including a bayonet charge.[144] Lieutenant Colonel Charles Assheton, similarly leading by example, was killed directing his machine-guns; 'Confronted with confusing and increasingly desperate tactical situations, and cut off from higher headquarters, COs resorted to instinct and fell back on the basic foundation of Australian command: lead from the front.'[145]

The German army of the Second World War has long been held out as among the most effective military forces that have ever existed. Its professionalism and determination have repeatedly impressed historians. Nevertheless, despite the potential distinctiveness of the German infantry in the Second World War, there remain numerous examples of almost entirely individual acts of bravery. German officers were expected to lead their units professionally and the Knight's Cross was often awarded to those commanders who instilled calm and determination in their troops, fulfilling their duties with skill. Yet, commanders were also often expected to lead from the front, as were Allied tactical commanders. Thus, Oberleutnant Martin Steglich, commanding I Kompanie, Grenadier Regiment 89, attached to 123rd Infanterie Division, was awarded a Knight's Cross for his actions in a defence against a major Russian attack near Zemena on 23 December 1942. His citation records that when a neighbouring battalion commander was wounded, 'Steglich immediately assumed responsibility for the sector' and that he 'was indefatigable in his efforts to bring order to choas' while also quelling 'defeatist rumours'.[146] Steglich showed exceptional command qualities but he also personally led the counter-attack: 'With loud cheers, Oberleutnant Steglich formed forward at the head of his men and closed the gap between the two units. His heroic bravery and acceptance of responsibility, witnessed by his decisive actions in knocking out the enemy tanks and succeeding in the counter-attack, secured the front line in his own hands.'[147] In leading the counter-attack personally, he effectively ensured that the counter-attack was not impeded by the natural reluctance of his troops.

Steglich showed leadership and individual bravery. However, some non-commissioned officers were awarded the Knight's Cross for acts of purely individual valour. For instance, SS-Unterscharführer Emil Dürr, fighting with 12th SS Panzer Division in Normandy, just south of Caen, earned a posthumous Knight's Cross when he personally destroyed a tank with a satchel charge (by which he was simultaneously mortally wounded).[148] Although the division had been fighting in Normandy since 6 June and was widely held to be the most fanatical German formation in the battle, Dürr 'became the first junior NCO of the division to receive the award' when it was eventually approved on 23 August.[149] On the night of 13–14 September 1942, Lieutenant Hans Sturm, of 473 Infanterie Regiment, 253rd Infanterie Division, personally stopped a Russian attack and the encirclement of his formation on the banks of the Volga at Rschew. Sturm's regiment was stationed at an old brickworks on the northern edge of the town when they were subjected to an intense barrage, followed by an attack on 3 Platoon. As company runner, Sturm was sent to ascertain the situation with 3 Platoon. Finding them besieged on his arrival, Sturm began to fire an MG 34, whose operators had been killed, at the attacking Russians. When the weapon ran out of ammunition, he threw grenades, allowing him time to reload the weapon which he then began to fire again, moving from position to position. He continued to fire even after he was wounded until a relief company arrived to repulse the attack. His regimental commander was in no doubt about the importance of Sturm's actions: 'This Sturm! He has single-handedly prevented the encirclement of the Regiment and halted the Russian breakthrough.'[150] While Sturm's actions may have been extraordinary, the passivity of 3 Platoon also needs to be recognized. Although they suffered heavily in the attack, there is no evidence that they were annihilated. Yet, Sturm acted entirely alone. The examples of Steglich and Sturm can be multiplied. Although the German infantry was typically able to rely on its crew-served machine-gun, individual virtuosity and concomitant passivity on the part of the mass was a common occurrence in the Wehrmacht and SS in the Second World War, as it was among their enemies.

A similar phenomenon was periodically observable in the Korean War, especially during the early phases. During the catastrophic early weeks in June and July 1950, when Task Force Smith and the entire US 24th Division and its Republic of Korea allies were driven back to the Pusan perimeter in the south, senior officers were forced not merely to command the fighting but engage in it themselves in a desperate attempt to get their troops to fight; 'It was not because colonels and generals had lost their minds that so many of them began to stand with bazooka teams or direct rifles.'[151] Without setting an individual example, senior officers had good reason to believe that US troops would scatter in the face of North Korean assaults. The French use of the bayonet charge on Heartbreak Ridge has already been discussed but that battle, like many others in Korea, also demonstrates the importance of individual action. One of the most obvious examples here is the performance of Herbert Pililaau for which he was awarded a posthumous medal of honour. On 17 September 1951, 23rd Infantry Regiment was ordered to take Hill 931, one of two prominent features on Heartbreak Ridge. C Company was tasked with the assault but stalled below the summit where Pililaau's platoon formed a defensive position. They were subjected to intense attack by Chinese forces which they fought off until Pililaau was left to cover the retreat of the main body. He fought off assaults with his BAR and grenades, until, finally running out of ammunition, he was overwhelmed and killed.[152]

Interestingly, although the technology of the US forces in Vietnam represented a major break with the Second World War, in some cases looking forward to the Revolution in Military Affairs in the 1990s, and, although US troops sometimes demonstrated high levels of cohesion there,[153] the individual–mass dynamic is also evident in that war. In the battle for Hue, 1/5 Marines had great difficulty crossing the road 'Phase Line Green' (Mai Thuc Loan) in order to assault the Citadel. After numerous attempts, one of the platoons from Alpha Company eventually secured some houses on the south side of the street on 14 February 1968. There they came under fire from two NVA machine-guns which wounded several Marines.[154] The riflemen in the platoon did nothing, stunned into inaction. However, one of the company's mortarmen, hearing the calls for help, dashed across Phase Line Green 'through a hail of fire from the tower before anyone could stop him. He made it to the death trap that was the corner house.'[155] 'The mortarman immediately took control of the situation, yelling at one of the most coherent Marines to tell him where the enemy positions were, and in what strength.'[156]

This lance corporal mortarman who had no business being there in the first place, calmly gathered up as many M-26 grenades from the dead and wounded Marines cringing in the corner, as he could carry and, without looking back or saying anything, charged out the side door and assaulted the main machine gun nest. The mortarman ran right up to within a few feet of it, tossing two M-26 grenades into the enemy position as he dived forward and hit the dirt. Both grenades hit their marks, and the first enemy machine gun nest was wiped out and silent for the first time in nearly a half hour. The mortarman didn't even stop to catch his breath. Scrambling back into the corner house, he grabbed some more grenades and charged out the back door into the teeth of the second machine gun position's fire. He took the second machine gun position out in the same way as the first.[157]

Warr emphasized the heroism of the mortarman's actions: 'I am certain that the Alpha Company mortarman's heroics were the turning point for the battle of Hue's Citadel fortress because it gave 1/5 its first "beachhead" across phase line green.'[158] Warr was asked to write up the Marine for a Congressional Medal of Honour. In the event, he was only awarded a Silver Star partly because the citation 'has not been well written'. 'It is to my everlasting shame that I accept full responsibility for this travesty. I am even more ashamed, however, that today I don't remember this young man's name.'[159]

It is very difficult to corroborate this incident definitively. According to the accessible records, no Marine from A Company was awarded a Silver Star for an action on 14 February. However, for actions on 13 February, Corporal Walter W. Rosolie, a squad leader with Company A, not a mortarman, was awarded this medal. Given that a number of other Marines whom Warr discusses and who were decorated feature on this list it seems probable that Rosolie was the Marine about whom Warr writes. The official citation differs somewhat from Warr's account but also seems to bear sufficient similarities to be taken as the same incident, although some doubt remains.

On the morning of 13 February 1968, Corporal Rosolie's unit was maneuvering along the Hue City wall when the Marines suddenly came under intense small arms, automatic weapons and rocket fire. During the initial moments of the fire fight, the platoon sustained numerous casualties. Realizing the seriousness of the situation, Corporal Rosolie fearlessly led an assault upon a hostile position located in a heavily fortified tower. Disregarding the enemy rounds impacting around him, he climbed the wall, threw eight hand grenades into the tower and fired a light antitank assault weapon through the doorway. Realizing his unit was dangerously low of ammunition, he withdrew to a covered position. Quickly resupplying his fire team with ammunition, he began maneuvering his men toward the enemy emplacement and, as the Marines approached the tower, he alertly observed an enemy explosive device. With complete disregard for his own safety, he skillfully disarmed it and directed its removal. Arriving at the tower a second time, he again threw hand grenades into the hostile position until forced to withdraw due to the intense enemy fire.[160]

Yet, even if Warr's mortarman was someone else, Rosolie's citation confirms the importance of individual heroism to the battlefield performance of the platoon in Vietnam.

Certainly, similar acts of individual initiative were evident in Vietnam.[161] In his very successful recent novel, based on his own experiences of Vietnam as a US Marine officer, Karl Marlantes affirms the point. His narrative reaches its satirical denouement with a Marine infantry assault on a hill called Matterhorn, which the Marines have previously fortified and then evacuated. Their NVA opponents subsequently use the Marines' own bunkers from which to kill them as they mount their attack. The climax of the attack comes when the central character of the novel, Lieutenant Mellas, is pinned down with his platoon in a killing zone on the slopes of Matterhorn. At that moment, Mellas has an epiphany. He sees clearly the tactical crisis which confronts him and the solution to it: 'He studied the bunkers. He saw the interlocking fire as if in a drawing. He saw the machine gun Jermain had attacked—and he knew. He floated back to a tactics class at the Basic School where a redheaded major said that junior officers were mostly redundant because the corporals and sergeants could take care of just about anything. But there would come a time when the junior officers would earn every penny of their pay, and they would know when that time came... He saw what would open the door through the interlocking fire, and it was right in from of him, shooting at him.'[162] Mellas organizes his machine-gunner to fire at the bunker in front of him and his grenade launcher to fire at a second bunker.

Then he stood up and ran. He ran as he'd never run before, with neither hope nor despair. He ran because the world was divided into opposites and his side had already been chosen for him, his only choice being whether or not to play his part with heart and courage... He ran because his self-respect required it. He ran because he loved his friends and this was the only thing he could do to end the madness that was killing and maiming them.[163]

As he ran, Mellas noticed to his surprise that his corporal and three new Marines had followed him and were running behind him.[164] Eventually Mellas reached the bunker in front of him and eliminated its occupants with grenades and rifle fire. As a result of his action, the position was unlocked. Clearly, some care needs to be taken with Marlantes's account. It is avowedly a piece of fiction. Yet, his narrative is explicitly taken from Marlantes's own experience, as Executive Officer of 3rd Battalion Fourth Marines, and specifically from a major operation that took places near the demilitarized zone between 1 and 6 March 1969. Indeed, as the author has himself confirmed,[165] the attack is taken directly from an action in which Marlantes participated and for which he was personally awarded a Navy Cross. The official citation demonstrates the close connection between his fictional but in fact autobiographical description and the actual event:

During the period 1 to 6 March 1969, Company C was engaged in a combat operation north of the Rockpile and sustained numerous casualties from North Vietnamese Army mortars, rocket-propelled grenades, small arms, and automatic weapons fire. While continuing to function effectively in his primary billet, First Lieutenant Marlantes skillfully combined and reorganized the remaining members of two platoons, and on 6 March initiated an aggressive assault up a hill, the top of which was controlled by a hostile unit occupying well-fortified bunkers. Under First Lieutenant Marlantes' dynamic leadership, the attack gained momentum which carried it up the slope and through several enemy emplacements before the surprised North Vietnamese force was able to muster determined resistance. Delivering a heavy volume of fire, the enemy temporarily pinned down the friendly unit. First Lieutenant Marlantes, completely disregarding his own safety, charged across the fire-swept terrain to storm four bunkers in succession, completely destroying them. While thus engaged, he was seriously wounded, but steadfastly refusing medical attention, continued to lead his men until the objective was secured, a perimeter defense established, and all other casualties medically evacuated. Then, aware that all experienced officers and non-commissioned officers had become casualties, he resolutely refused medical evacuation for himself.[166]

It would be an error to conflate the Mellas character with the author throughout the novel but there is clear evidence of a fusion of author and protagonist here.[167] The Mellas/Marlantes case represents another example of the way in which citizen armies used individual heroism to galvanize activity on the battlefield. Mellas/Marlantes devised a plan and then ran forward alone, assuming the responsibility he knowingly accepted as a junior commander. He ran forward alone like York, Wray, or Masters before him. Marlantes usefully illustrates the mass–individual dynamic of the conscript army even in the apparently highly technological context of Vietnam. Indeed, for him, it is so important that it constitutes the climax of his entire novel. The act of individual bravery represents the ultimate achievement for Mellas; a moment of existential vindication. According to Marlantes, the kind of bravery demonstrated by himself (and represented by Mellas in his novel) was not accidental. It was institutionally expected that platoon commanders led from the front and that their troops do not simply follow their orders but follow them physically. Indeed, he observed that in the Second World War, US platoon commanders had rank bars painted on the back of their helmets so that their subordinates could identify them more easily as they followed them.

It is important to be careful with these examples of individual activity. The differences between the battle conditions in each of these wars (and indeed in different campaigns in these wars) have to be recognized. Nevertheless, despite dramatic differences in the geography, climate, and enemy, a recurrent pattern seems to be identifiable. Specifically, combat seems to have been typically characterized by a dynamic dialectic between individual and mass. In this way, these examples seem to vindicate Marshall's point that while the majority of riflemen merely occupied the battlefield, a small number of virtuosos dominated the action. These active individuals carried the fight forward, overcoming the collective action problem which was typical on a modern battlefield. Indeed, it is worth noting the situational context in which these individual actions took place. They display an unignorable and recurring pattern. These individual soldiers took initiative when their platoon came under heavy fire and forward momentum was lost; they had succumbed to the Marshall effect. Soldiers were intent on taking cover and preserving themselves. At this point, an individual jumped up and assaulted the enemy position alone. In most cases, this soldier held a position of rank or responsibility: he was an officer, NCO, or specialist weapon operator. In other words, he had been recognized by the organization as bearing special responsibility for the platoon, and both he himself and his fellow soldiers were fully aware of this role. There was, then, an organizational expectation that this individual should act in extremis which was reinforced in the field itself by the eyes of his fearful subordinates actively looking to him as they lay under enemy fire. Karl Marlantes has noted that as a platoon commander he was explicitly expected to lead—physically—in these circumstances of crisis. The passive mass deferred to their leader and, with this weight of responsibility upon him, he was impelled to act. He picked up his weapon and charged forward firing and grenading the enemy positions.

Not only did individual action become an effective means of overcoming the collective action problem in a mass army but the way in which military units were organized, placing responsibility on a minority of designated individuals, actually encouraged individualism as a possibility. Citizen soldiers expected their NCOs and officers to take charge and exerted moral pressure upon them to do so, while this expectation itself actively encouraged and justified their passivity. Marshall highlighted the curious contradiction between the virtuosic action of the few and the passivity of the many but he failed to recognize that this mass–individual balance was a dialectic. The passivity of the mass necessitated the action of the individual, while the expectation that the individual would act itself encouraged massive passivity, especially since the troops were generally insufficiently trained to overcome the collective action problem through prepared coordinated action. Indeed, the existence of primary groups, united around common concepts of masculinity and ethnicity which imposed mutual obligations of reputation upon their members, seems to have been well adapted to encouraging this individual action from their leaders. The leaders felt obliged to lead from the front if they were to hold the respect of their subordinates, demand their loyalty and obedience, and sustain the cohesion of the group. The social constitution of the mass army in which tactical skill was generally poor and extraneous social factors played an important role in engendering cohesion promoted individualism as an organizational response to the problems posed by the modern battlefield.

6

Modern Tactics

Despite his detractors and his own unfortunate exaggerations, Marshall was broadly correct. The citizen infantry of the twentieth century faced a predicament on the battlefield which for the most part its soldiers were unable to resolve. They struggled to overcome the problem of inertia generated by firepower. In the light of the tactical under-preparedness of the platoon, armies typically encouraged collective activity through appeals to comradeship, patriotism which encouraged mass action (the bayonet charge), and individual example. This is a broad summary of the way that the mass armies of the twentieth century solved the problem of collective action posed by the mechanization of warfare. However, it would be utterly misleading to ignore the vital tactical developments which did take place during the early part of the twentieth century at the platoon level. Indeed, by the middle of the First World War, the outlines of every form of tactical procedure later refined and sophisticated by today's professional army were substantially in place, at least in selected units within the combatant's armies. In order to provide an accurate picture of the state of infantry tactics in the first half of the twentieth century, to defend against any charge of simplification, and to provide the context for understanding more recent tactical innovations among all-volunteer professional forces, a detailed understanding of the evolution of modern infantry tactics from the First World War through to Vietnam is necessary.

FIRE AND MANOEUVRE

During the Franco-Prussian War and American Civil War, western infantry still adopted close order assault tactics which were nearer to the methods employed by infantries in the Napoleonic Wars than those which would become necessary during the First World War. On 3 July 1863, the Confederate general George Pickett ordered a mass assault of nine brigades, some 12,000 men, against Union lines on Cemetery Ridge. With little artillery support, the Confederate infantry marched *en masse* up the hill towards their enemy in an assault reminiscent of the advance of the Imperial Guard at Waterloo on the evening of 18 June 1815. Although the charge has now become legendary especially in Confederate memory, it was a disaster; Union artillery and rifles inflicted up to 50 per cent casualties and, although some Confederate troops reached Union positions, they were driven back. The Battle of Gettysburg and ultimately the Confederate cause was lost. Seven years later, many of the assaults by French and Prussian troops assumed a similar character, with troops advancing in dense formations with little supporting fire, and were subject to very high casualties. Even during the Boer War at the turn of the century, the British army persisted with dense assault formations against a concealed and often entrenched enemy resulting in similar disasters at Colenso, Spion Kop, and the Modder River.

These wars proved instructive. The need to support infantry assaults with fire began to be recognized by the late nineteenth century as the effects of the repeating rifle, machine-gun, and artillery became clear. Specifically, the concept of fire and movement was incorporated as an essential feature of infantry assaults. Nevertheless, late nineteenth-century innovations were a long way from anything which might be recognized as modern tactics. Close order attack was still affirmed by German theorists Honig and Meckel for control purposes, and Prussian *Drill Regulations of 1888* advocated close order.[1] Accordingly, in the battalion assault, companies would be formed up into two or three dense skirmishing lines each led by their company commanders, with the commanding officer leading the attack.[2] The 1906 *Drill Regulations* continued to recognize the importance of fire superiority, which was to be achieved by companies advancing in lines by means of mutual fire and movement until they reached the firing line. They would then build up a rate of fire, replacing casualties in the line, with supporting waves until the enemy had been beaten down. At this point, the bayonet charge would be released.

The British army 1909 *Field Service Regulations* and *Infantry Training* (1914) described a similar method of assault. These manuals depicted infantry companies approaching the enemy positions in dense lines with the assistance of artillery, generating a firing line from which the assault would be made: 'the object of fire in the attack, whether of artillery, machine guns or infantry, is to bring such a superiority of fire to bear on the enemy as to make the advance to close quarters possible.'[3] *The Field Service Regulations* stated that the opportunity for assault 'will be known by the weakening of the enemy's fire and perhaps by the movements of individuals or groups of men from the enemy's position towards the rear. The impulse for the assault must therefore often come from the firing line.'[4] Once fire superiority was attained, as noted in the previous chapter, a bayonet charge would be mounted from the firing line, as specified in Infantry Training (1914). Clearly, the importance of fire superiority was recognized. Yet, these close order tactics—even with the full recognition of the importance of fire superiority—and especially the formation of company or even battalion strength firing lines began to look unfeasible after the Boer and Russo-Japanese Wars. Early experiences of the First World War demonstrated their inappropriateness to the Germany army. Indeed, the so-called *Kindermord* at Langemarck in September 1914 when German recruits were massacred by Allied fire proved their obsolescence.[5]

Consequently, as early as 1914, some units began to innovate with small-group fire and movement tactics. Instead of advancing in dense lines, troops began to disperse into smaller, irregular formations which exploited the terrain and advanced in rapid bounds. By 1917, the basic tactical unit became the platoon, typically consisting of three or four sub-units, the section, which supported each other's advances by fire. The discovery of platoon fire and movement tactics represented an important, perhaps seminal, historical moment in infantry techniques. In his study of classical warfare, Hans Delbrück describes what he regarded as a critical moment in military development, when the Greek phalanx was replaced by the Roman legion.[6] The Greek phalanx was a potent force on the battlefield as the victories at Marathon and Issus had demonstrated but, as a tactical unit, the phalanx had major shortcomings; it was highly immobile and inflexible. Moreover, in combat, the hoplites in a phalanx tended to squeeze together, with the potentially disastrous consequences of opening gaps at other points: 'Both actions, the squeezing and pulling apart of the phalanx, occurred in natural alternation. If the soldiers pressed together at one spot, a breach would probably develop at another place. Therefore a cure had to be created for both evils simultaneously.'[7] To address these weaknesses, the Romans developed a more manoeuvrable formation, consisting of 'maniples', each of approximately 120 soldiers;[8] 'the ten maniples of the hastati, each twenty men wide and six men deep in the normal formation were placed side by side with small intervals.'[9] 'Behind them the class of principles was formed up a second echelon', covering the gaps between the hastati maniples, and 'behind them were the maniples of triarii'.[10] The deployment of more numerous smaller units onto the battlefield gave Roman commanders the opportunity to manoeuvre and to outflank opposing forces and to avoid unexpected openings in combat; maniples were trained to close up gaps in combat.[11] Of course, both the possibility of manoeuvre and cooperation between maniples required much higher levels of training than were typical in the Greek phalanx.[12] The move from company lines to the platoon might be seen as the twentieth-century equivalent of that classical innovation. The creation of the platoon allowed the infantry to exploit the potential of the terrain and fire much more effectively than the large battalion phalanx. The platoon might be seen, in effect, as the modern maniple.

Platoon tactics were evident in a number of infantry units, often before its ratification in doctrine. Perhaps, most famously, Erwin Rommel records the development of platoon tactics in his admittedly elite 124th Infantry, 6th Wüttemberger Division. Although company skirmish lines feature in his narrative especially in the early battles, Rommel emphasized from the outset of his memoirs the importance of platoon fire and movement tactics. Thus, his first assault as a platoon commander on the village of Bleid is instructive: 'My attack plan was to open fire on the enemy on the ground floor and garret of the building with the 2nd Section and go around the building to the right with the 1st Section and take it by assault.'[13] The 1st Section formed itself into an 'Assault Detachment' which 'picked up a few timbers' in order to batter down the doors and straw 'to smoke out any concealed men':[14] 'On signal, the 2nd Section opened fire. I dashed forward to the right with the 1st Section.'[15] Significantly, Rommel stresses that this fire and movement tactic had been practised by the unit before the war. Describing a subsequent assault on Hill 325, he notes: 'We rushed forward by groups, each being mutually supported by the others, a manoeuvre we had practiced frequently during peacetime.'[16] The Assault Detachment became a regular feature of Rommel's tactics. Later, as the war developed and Rommel was promoted, he emphasizes the integration of light and, particularly, heavy machine-guns in the assault, providing the crucial fire support for the assaulting troops who relied primarily on speed and fieldcraft. Thus, in the assault of Mount Cosna in August 1917, he describes that, 'Under the quickly organized fire support of a heavy machine-gun platoon, it was possible to regain the last line of the combat outposts without suffering much in the way of casualties. The fire and movement of the assault squads were in complete unison here.'[17] In addition, Rommel records the introduction of specialist weapons; grenades, machine pistols, machine-guns, mortars, and assault guns begin to become more important than rifles and bayonets.[18]

Rommel was perhaps atypical as an officer in an unusually advanced formation. The German armies of the First World War did not commonly use the tactics which Rommel described in 1914 until much later in the conflict. Captain Rohr with Captain Reddeman under the supervision of Colonel Bauer played a decisive role here. In the summer of 1915, they began to develop a new special assault battalion.[19] This unit was based on an experimental pioneer battalion but, with the impetus of Laffargue's captured document, it was turned into an elite infantry organization. The unit specialized in the application of new weapons: grenade, assault guns, machine-guns, trench-mortars, flamethrowers, and artillery. Long skirmish lines were replaced with squad-sized stormtroops. These units were small autonomous tactical entities, with their own organic firepower. The NCOs who commanded them were empowered to take responsibility.[20] Battle missions rather than specific orders and tasks were given to these junior commanders.

They were oriented to the tactical objective but were allowed to innovate and adapt in order to attain it. In particular, the commander identified a 'Schwerpunkt', a centre of gravity or main effort, on which all efforts should be focused.[21] Stormtroops sought to use their new firepower to infiltrate enemy positions, using ground and weapons to neutralize strong points and identify weak points. They then aimed to roll up trenches through the use of grenades and flamethrowers.

Stormtroops were first used in early October 1915 by 2 Pioneer Company in the Vosges Mountain at Schratzmannle. However, their first major success came in 1917 where they led a major counter-attack against the British to close the Battle of Cambrai. As Rommel's memoirs show, the stormtroopers were not quite the unique development which they have sometimes been considered to be. Indeed, other units demonstrate the same acumen. Nevertheless, the stormtroopers institutionalized three critical innovations; they broke themselves down into small mutually supporting squads of between four to twelve men which were capable of independent action, they prioritized the use of fire and movement (and fieldcraft) to infiltrate defences, and they exploited new specialist weapons displacing the exclusive reliance on the rifle and bayonet.

This is particularly apparent from the tactical doctrine published in January 1917, *Ausbildungsvorschrift für die Fusstruppen im Kriege*. This document noted that new methods were required to fight in modern wars, stressing the importance of firepower and dispersal and therefore greater initiative on the part of junior commanders and soldiers, who had to be better trained. In particular, the platoon and section were identified as critical to combat performance. *Ausbildungsvorschrift* describes the method of assault in detail. One of the first requirements in trench warfare is adequate preparation: 'The preparation for such attacks must be very accurately carried out; every commander and soldier must know precisely the position, at which he leaves the trench, the route he should take, which leads to the correct position in the enemy trench.'[22] Once prepared, the basic form of attack remains the skirmish line and the wave,[23] but the line moves forward and is organized in different ways from those specified in German doctrine before the war. *Ausbildungsvorschrift* specifies that sections are to work in cooperation with their neighbours to support each other through fire and movement.[24] Thus, rather than forming long company lines, sections should form skirmish lines together following their commander: 'On the command "Attack!", the section commanders jump up. They form the skeleton of a skirmish line as quickly as possible.'[25] The sections and platoon seek to bound forward in this line. Nevertheless, while a skirmish line consisting of sections is described, *Ausbildungsvorschrift* also stresses: 'The formation which a platoon will adopt to work up to an enemy cannot be pre-determined. Everything depends on the terrain and the effectiveness of enemy fire. Where bounds of the whole platoon become impossible, sections or even individuals will work their way up. The form, means and speed of the movement depends on the conditions.'[26] *Ausbildungsvorschrift* stresses that 'all bunching under enemy fire must be avoided'.[27] The light machine-gun was critical to fire support during the assault phase: 'The suppression of the enemy and the support of the execution of the assault is the main role of the machine gun.'[28] Once the enemy trench has been reached by fire and movement, *Ausbildungsvorschrift* recommended the use of grenades to roll up the trench and provided detailed instructions of how different kinds of traverses were best neutralized by bombing; 'the section commander accompanies the bomber, gives directions and range and protects the section from enemy attack from the front with rifle or pistol.'[29]

Ausbildungsvorschrift für die Fusstruppen interestingly includes a comprehensive section on the assault battalion (or stormtroops) primarily created to instruct new assault methods.[30] Indeed, *Ausbildungsvorschrift* recommends that stormtroops should not be used as complete platoons or companies in order to preserve their strength and avoid attrition.[31] Rather it was better to exploit the specialist training of single assault squads.[32] Assault teams should provide intelligence by scouting before an assault and engage in common training with the infantry on replica areas to heighten the chances of success and raise the confidence of the troops:[33] 'In the attack the storm troops of the assault battle should lead the infantry against the most difficult positions, open up the break-in positions, roll up enemy trenches, take enemy block-houses and machine-guns and support the infantry to secure the position.'[34] Crucially, the stormtroops must be carefully trained and prepared; 'In the squads, conduct in trench warfare is practised to perfection.' The doctrine stresses throughout that a successful assault requires detailed intelligence: 'a requirement is precise knowledge of our own and the most careful reconnaissance of enemy positions with the help of maps and aerial photographs.'[35]

Despite the tactical successes of stormtroops in the First World War, German military commanders remained concerned about the performance of their troops in general and, in the 1920 and 1930s, they sought to improve the infantry's ability to engage in the platoon tactics which had become so essential on a modern battlefield. General Von Seeckt, the head of the Truppenamt (the General Staff had been dissolved by the Treaty of Versailles), saw that the key to future victory was mobility.[36] He was no great supporter of conscription since he maintained that mass conscription generated 'nothing more than poorly trained recruits'.[37] For him the army of the future was small, elite, and mobile and he initiated a comprehensive programme of research to collect and analyse the experiences of the First World War to create a new body of doctrine for such an army.[38] Von Seeckt issued the Reichswehr's new field service regulations, *Führung und Gefecht der verbundenen Waffen* (FUG) (*Combined Arms Command and Combat*), in 1921 and, over the next two years, *Ausbildungsvorschrift für die Infantrie* (AVI) (Training Regulations for the Infantry), *Ausbildung der Schutzengruppe* (A.d.S. Training Regulation for the Rifle Squad), and *Einzelausbildung am Leicht Maschinengewehr* (ALMG, Individual Training with the Light Machine-Gun).[39] These documents represented both a distillation of the First World War lessons, including stormtroop tactics, and a complete break with pre-war doctrine. The close order ranks and the concept of the firing line were quite absent from this generation of doctrine. *Führung und Gefecht*, for instance, affirmed the importance of manoeuvrability and all arms cooperation. Drawing explicitly on stormtroop tactics, FUG understood large-scale infantry assaults as a series of individual assaults by squads and platoons.[40]

Ausbildungsvorschrift für die Infantrie, the most important publication in terms of platoon tactics, was published in 1922 and then revised in 1925 and throughout the 1930s up to 1942; indeed, A.d.S and ALMG were effectively extracts from *Ausbildungsvorschrift*. *Ausbildungsvorschrift*, in particular, represented the most comprehensive account of platoon tactics and, while there were minor changes in weaponry (the MG 08/15 [Machinen-Gewehr, machine-gun] was replaced by the MG 34 in 1938) and the platoon was reorganized in 1938, there is great continuity from the original *Ausbildungsvorschrift* in 1922 and the final edition in 1942. For *Ausbildungsvorschrift*, in all its iterations, the dispersed conditions which pertained on the modern battlefield were seen to represent a significant challenge which only extensive training could overcome: 'The multiple effects of today's weapons limit the use of close order formations and the immediate influence of the commander in combat than previously. This demands the wider dispersal of combat groups and places increased demands upon the training of individual soldiers.'[41] In this environment, *Ausbildungsvorschrift* consistently identified firepower as essential to success: 'From the beginning on it is necessary to alert every soldier's understanding that the combination of all infantry weapons on a common target will bring success.'[42] Indeed, 'the superior readiness of the German soldier to fire must come to its full fruition in the assault'.[43] From 1922 to 1942, *Ausbildungsvorschrift* consequently stressed individual weapon handling, with which the manual always began, but it also described at length the importance of camouflage, concealment, observation, and movement to allow the platoon to bring its weapons to bear with maximum effect.

Since firepower was decisive, according to *Ausbildungsvorschrift*, the light machine-gun, either the MG 08/15 or, from 1938, the MG 34 or, later, the MG 42, was the linchpin of the platoon around which all the riflemen operate.[44] Far more text was dedicated to the maintenance, firing, and use of this weapon in combat than any of the personal weapons. In 1922, the platoon consisted of one or two machine-gun sections of four men each and at least two rifle sections of nine men, a platoon commander, and various platoon elements. By 1936, the German platoon consisted of a platoon commander, platoon troops (three runners, one for each section, and a bugler), and three thirteen-man sections, each with a section commander, a second-in-command, seven riflemen, a machine-gunner, and three assistants.[45] The 1941 *Ausbildungsvorschrift* specified that the platoon consisted of a platoon commander and headquarters (1–3), four sections of ten soldiers, and a light mortar group of a further two soldiers. Each of the four sections was organized around the MG 34 which was operated by a three-man team with the gunner (No. 1) being trained as a specialist, the No. 2 his assistant, and the No. 3 acting as an ammunitions carrier;[46] the six other riflemen and section commander supported the machine-gun team. The Wehrmacht platoon, as it entered the Second World War, had fewer soldiers but one more machine-gun than its Reichswehr predecessor, therefore. Moreover, although the gun team was highly trained, the whole platoon was dedicated to the support of its machine-guns, which were seen as critical to tactical success since that weapon generated by far the highest rate of fire.

In *Ausbildungsvorschrift*, the section (and by implication the platoon) attack was conceived as consisting of four basic stages; the approach, the fire-fight, working up, and the break-in (or assault).[47] The approach, typically using the cover of terrain and artillery support, often involved close order file formation. In section file, for instance, the MG 34 followed the section commander. However, as the platoon neared the enemy, normally at a range of 400–800 metres, the platoon would typically deploy into a line and the riflemen would often also engage the enemy; 'The line [*Schützenkette*] is used above all for firefights involving the whole section.'[48] There were a number of established platoon manoeuvres for the deployment from file to line, the soldiers going left, right, or splitting into two with the MG 34 in the centre of the platoon.[49] At this point, the fire-fight was initiated, 'the aim of which was the suppression and elimination of the enemy'.[50] Typically, the machine-gun engaged the enemy as ordered by the section commander while the riflemen stayed in cover; riflemen fired at this early point only if there was a possibility of good fire effect or insufficient cover for them.[51] The fire-fight continued until the purpose for which it was initiated has been achieved; 'for instance, the attack of a neighbour has been helped.'[52] However, if the section had been identified as the assault force, it prepared for the 'working-up' phase, and *Ausbildungsvorschrift* described the action of the section in the working-up and assault phase in detail. The section (and platoon) employed fire and movement to close the distance to the enemy, moving from cover to cover while the light machine-gun (the MG 34) provided support, all on the section commander's orders. During this phase, *Ausbildungsvorschrift* explicitly warned: 'Riflemen help each other to move forwards. Fire and movement must always be coordinated. While one part moves, the other fires.'[53] Dashes could be made individually or in smaller or deeper formations and the machine-gun might also need to be unloaded and brought forward; riflemen could make small improvised entrenchments if necessary.[54] *Ausbildungsvorschrift* illustrated the best way of crawling with a weapon, how to place an entrenching tool to hand, when prone, and specified that as a soldier got up, his rifle should be in his left hand, that he should draw his knee to his chest, pushing forward, to reduce the target he presents; he should not simply stand up.[55] Once the platoon has reached assault distance, orders are given and the platoon prepares for the break-in.[56] As the section approaches the enemy, 'the section commander must exploit every opportunity for a break-in even without specific orders'.[57] 'In the assault the section commander drives the whole section forward through personal example. He leads from the front. He overcomes armed enemy with shots from his machine-pistol or lays him low with bayonet or spade. Shortly before and during the assault, the enemy is engaged with every weapon with the highest increase in fire. The LMG assaults too, firing on the move in the middle of the section. With hand-grenades, machine-guns, rifles, pistols and spades and with a constant "hurra", the resistance of the last enemy is broken.'[58]

By the Second World War, the Wehrmacht had refined and developed an elaborated and detailed set of procedures which its infantry was trained to follow in combat. At every point, the responsibility of the individual soldier was emphasized not only to take action himself but above all to coordinate with others in the prosecution of small-group tactics which were regarded as fundamental to battlefield success. As German doctrine noted, the infantry were not merely the most numerous element of the army, they constituted 'its moral backbone'.[59] Battles could only be won by close-quarter combat, and the discipline and skills which platoons generated were essential to the cohesion of the Wehrmacht as an army.

The French army was one of the most poorly prepared forces as it entered the First World War. Its doctrine was closely compatible with the German and British armies but it benefited neither from the intense training regime of the German army nor from the professional status of the British army. The battles of 1914 and 1915 in which the French suffered appalling casualties in numerous poorly executed attacks demonstrated its weakness. However, Goya has usefully noted that these early years of the First World War might be seen as the equivalent of the economic cycles described by Schumpeter or Kondratieff; a long period of stasis was suddenly punctuated by a period of rapid innovation facilitated by tactical and technological adaptation.[60] Crucially, the French dispensed (in theory at least) with the idea of the mass bayonet attack and 1915 marked the end of the paradigm of *'offensive á l'outrance'*.[61] Instead of relying on morale, the French sought a scientific approach to battle in which machines and firepower replaced men.[62] The 9 May 1915 offensive in Artois was taken as the model of this new 'methodical' battle. In order to pierce the enemy front, it was necessary to mount successive slow attacks, each one of which was executed with the massive artillery superiority.[63] The dependence on artillery necessitated an increasing centralization of command over the fireplan but, by contrast, the methodical battle simultaneously demanded a new flexibility from the infantry: 'the machine gunner became the nucleus of less visible small offensive groups on the ground.'[64] The emergence of small-group fire tactics, in which the machine-gun provided covering fire for the infantry, reduced casualties significantly: 'In the seventy-seven days of engagement, each infantry company from the 13th Infantry Division lost on average one man per day in combat, as opposed to seven in 1914 to five in 1915.'[65]

Captain André Laffargue's famous *The Attack in Trench Warfare* written in 1915 (and translated into English in 1917), based on his experiences during the Artois offensive, was a seminal work of military innovation which represented the transformation of the French army in 1915. The question which exercised Laffargue was how to reduce casualties in a period of mechanized siege warfare, in which 'infantry units disappear in the furnace of fire like handfuls of straw'.[66] In order to avoid casualties, the indispensability of attaining fire superiority was emphasized by Laffargue and he outlined several methods by which this could be attained. It was the duty of the soldiers in the skirmishing line to suppress the enemy with fire; Laffargue recalls how at Neuville on 9 May 1915 his company overcame the enemy wire entanglements by a mutual process of fire and movement.[67] Laffargue had some important instructions about combat firing: 'In close combat, men fight much more by shooting at point blank and very often from the hip than with the bayonet. The man should therefore be trained to use his rifle in close fighting. First teach him to watch that part of the parapet and the loopholes on which he marches in order to forestall the shots of the enemy; then to aim rapidly, throwing the piece to the shoulder to get the first shot at the enemy who is aiming at him.'[68] Alternatively, the best shots could fire from concealed positions.[69] Eventually, around 200 metres from the enemy, sections can alternately move forward in rushes covered by the fire of the other section.[70]

The French army instituted many of the procedures which Laffargue described. In 1916, they published *Instruction sur le combat offensif des petits unités* and *Manuel du chef de section d'infanterie* which represented an important transformation in French infantry doctrine. These publications described a new platoon structure. The platoon, consisting of forty soldiers commanded by a platoon commander, was divided into two half-platoons, themselves further subdivided into two sections or squads each, though both manuals emphasized the need for tactical flexibility and the preparedness of commanders to form temporary formations, 'groupements momentanés', when the situation demanded it.[71] The four sections of the new platoon were organized as specialist sub-units and an assault platoon would typically consisted of two bombing (grenade) sections, one rifle-grenadier section, and a rifle section.[72] In the assault, the platoon would form itself into two waves, a rifle section and bombing section followed by another bombing section and the rifle-grenadier section, all soldiers approximately 4 to 5 metres apart.[73] While, the term 'wave' was still used, the *Manuel* emphasized that these waves, in fact, constituted a 'juxtaposition' of coordinated sub-units, capable of movement and direction, not simply dense lines which surged forward together, with following lines merely pushing against the first wave.[74] The rest of the company echeloned behind the assault waves in files. On leaving its line of departure, the assault platoon follows the creeping barrage until it meets enemy resistance, at which point it uses coordinated fire and movement tactics to allow its sections to outflank and destroy strong points.[75] Long or short tactical bounds or, if the terrain were suitable, infiltration methods would be used in this assault.[76] When the section reached assault distance, they would hurl a volley of grenades and charge forward with bayonets.[77] If the platoon were pinned down, it would continue to fix the enemy with fire from the front, while a neighbouring platoon conducted the assault.[78] As a result of this doctrine, the French considered themselves significantly more adept than the British army. The French disparaged the British attacks on the first day of the Somme which they regarded negatively their artillery preparation was seen as particularly poor.[79] Indeed, the British themselves recognized that they had fallen behind the French army in 'attack technique, formation and the co-ordination of rifle, grenade and automatic fire'.[80] After the war, the French army continued to emphasize dispersed, platoon tactics. Thus, Lieutenant Colonel Culmann's *Cours de tactique générale d'après l'expérience de la Grande Guerre* stressed that infantry must not manoeuvre in dense formations.[81]

The German stormtroopers and the Wehrmacht platoon are often held out as the acme of progressive tactics in contrast to the plodding British infantry and their generals. Indeed, Paddy Griffith notes that this myth has been partly created by those involved themselves; thus Basil Liddell Hart's interest in disseminating the idea of British tactical incompetence 'lay in propagating the idea that his own re-writing of the infantry manual in 1920 filled a void that had been left entirely unfilled throughout the war'.[82] Griffith lays a similar charge against General Ivor Maxse. Griffith has sought to overcome this stereotype of British tactical obtuseness and its corollary, 'the idea that all German soldiers were almost always incomparable tacticians'.[83] In fact, the British were tactically innovative. As in Germany, the importance of fire and movement was explicitly recognized in pre-war British doctrine, as already noted, although it was practised in company lines. General Maxse, who eventually took over command of the training of the BEF, and publications such as *Instructions for the Training of Infantry Platoons for Offensive Action*, advocated platoon fire and movement. Accordingly, the British army began to introduce new weapons, techniques, and tactical formations—above all the platoon—onto the battlefield. Although Griffith emphasizes the autonomy of British innovation and there is little doubt that local techniques could have developed quite independently of each other, there was much cross-pollination from Allied and indeed enemy forces. In particular, many of the tactical innovations were learnt from large-scale Canadian raids in 1915.[84] As a result, lineal tactics were replaced by a more flexible, manipular system which, Griffith suggested, was not very distant from the much vaunted stormtrooper model. To prove his point, Griffith illustrates a dramatic change between the organization of British assault formations of the 27th and 26th Brigades in 1915 at Loos and those at Longueval the following year.[85] At Loos, both brigades deployed their two battalions in six double company lines, three in the first wave followed by three more lines in the second. A year later, 27 Brigade added depth to their assault, organizing the first wave as four double company lines but dropping the second wave much deeper. 26 Brigade demonstrated a quite different concept. They organized their assault into two 'columns' consisting of eight platoon lines organized into four separate waves. By deepening the attack and, in 26 Brigade's case, thinning the number of men in each assault line, the British consciously sought to introduce greater tactical flexibility into the attack so that smaller groups, above all the platoon, could use the terrain, covering each other's movement with fire; 'An important improvement was that it was now quite a widespread practice to divide the attacking waves into several different functions, as between "fighting platoons", "mopping up platoons", "support platoons" and "carrying platoons".'[86] In order to coordinate these small-group tactics, British doctrine and training gave 'constant encouragement for junior officers and senior NCOs to think for themselves'.[87]

Doctrinal publications during the First World War clearly show the development of infantry tactics. Specifically, the British, like the Germans, moved away from mass lineal assaults to manipular platoon and section tactics. The British army produced a number of infantry training publications during the First World War, in order to disseminate tactical innovations, two of which are particularly useful in illustrating the development of military thinking, *Instructions for the Training of Platoons for Offensive Action*, 1917, and *Assault Training*, 1917. The *Instructions* emphasized the importance of progressive training; 'true soldierly spirit must be built up in Sections and Platoons'. This spirit 'is obtained by encouraging Section leaders to take a pride in their sections and in their work'.[88] The *Instructions* divided the platoon into four specialist sections of nine men: the bombing section, the rifle bombing section, the Lewis gun section, and the rifle section. In order to generate tactical cohesion, the importance of live training was stressed: 'No form of instruction with arms can be complete until it has been carried out with live ammunition under conditions as nearly as possible approaching those which would pertain on the battlefield.'[89] In stark contrast to the pre-war emphasis on the ability of indomitable morale to overcome firepower, the *Instructions* wisely advocated the centrality of fieldcraft to infantry tactics.[90] The appendix of the Instructions provided detailed diagrams of platoon formations and manoeuvres for a 'trench to trench' attack and an 'attack in open warfare' (see Figures 6.1a and 6.1b). In both cases the assault was mounted by two platoons of forty-five

Modern Tactics 141

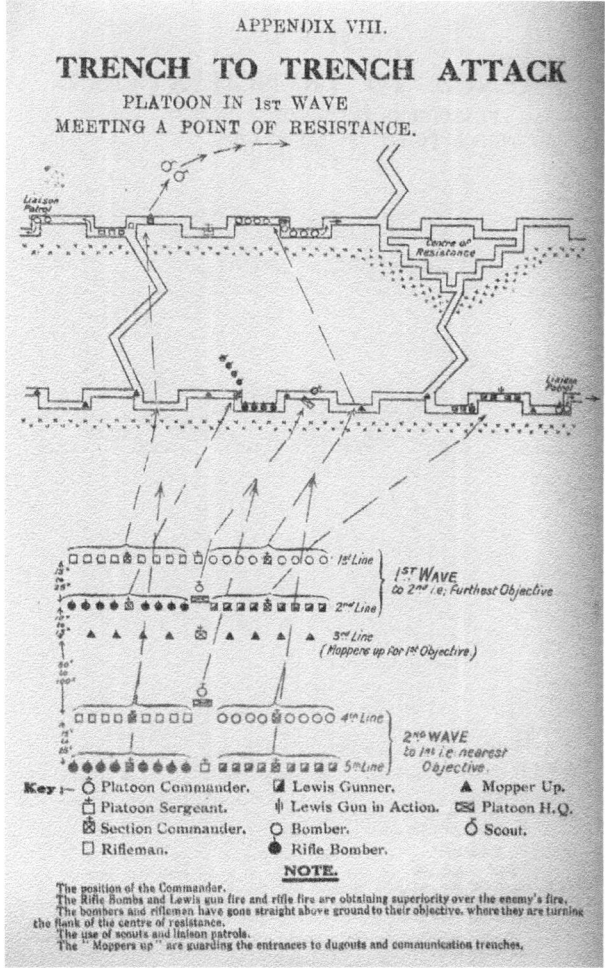

Fig. 6.1a. Trench to trench attack. Platoon in the first wave. The General Staff, *Instructions for the Training of Platoons for Offensive Action*, 1917 Appendix: VIII.

soldiers, a platoon commander and sergeant each. The trench assault is based on the example of assaulting a point of resistance, located in the second line of enemy trenches. The platoons organize themselves into two waves, the first consisting of three lines; a line of riflemen and bombers, followed by a mixed line of Lewis gunners and rifle-grenadiers, followed finally by 'moppers up'. The diagram depicts the riflemen and the bombers of the first line entering the first trench line and then infiltrating into the second, taking up a position to the strong point's rear flank. At the same time, as the first platoon is infiltrating, the second platoon assaults, sending its bombers, supported by its

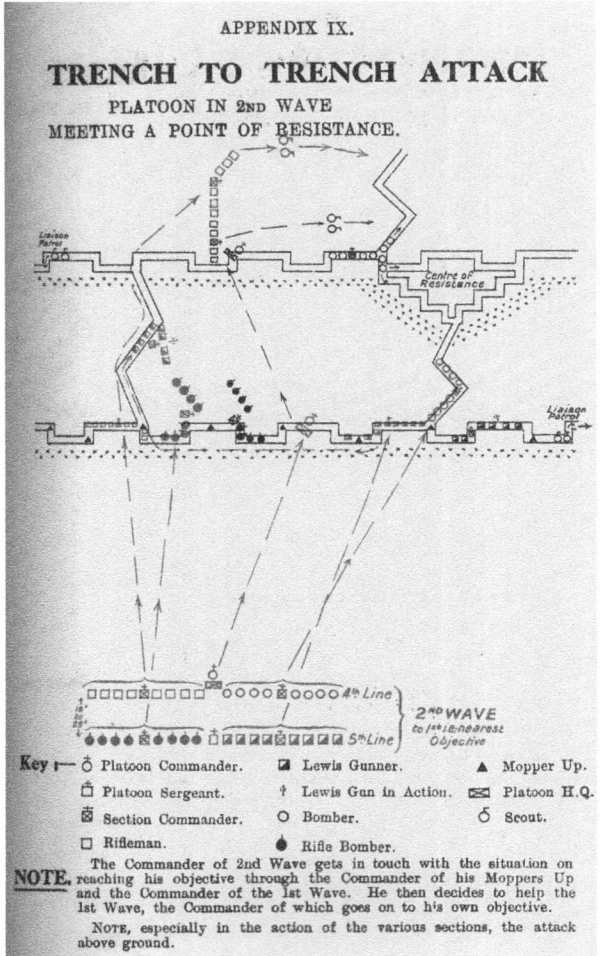

Fig. 6.1b. Trench to trench attack. Platoon in the second wave. The General Staff, *Instructions for the Training of Platoons for Offensive Action*, 1917 Appendix: IX.

Lewis gunners, riflemen, and rifle-grenadiers, to a forward left and right flank. The bombers from both platoons converge along adjoining trenches onto the strong point, taking it simultaneously, under cover from Lewis guns, rifles, and rifle-grenades.[91]

The diagram of the attack in open warfare demonstrates similar fire and manoeuvre tactics, emphasizing the importance of using depth and flanking by alternative illustrations of correct and incorrect assault methods (see Figure 6.2). In the incorrect assault, the platoon forms a simple firing line of two ranks which marches directly up to the enemy position.[92] The diagram

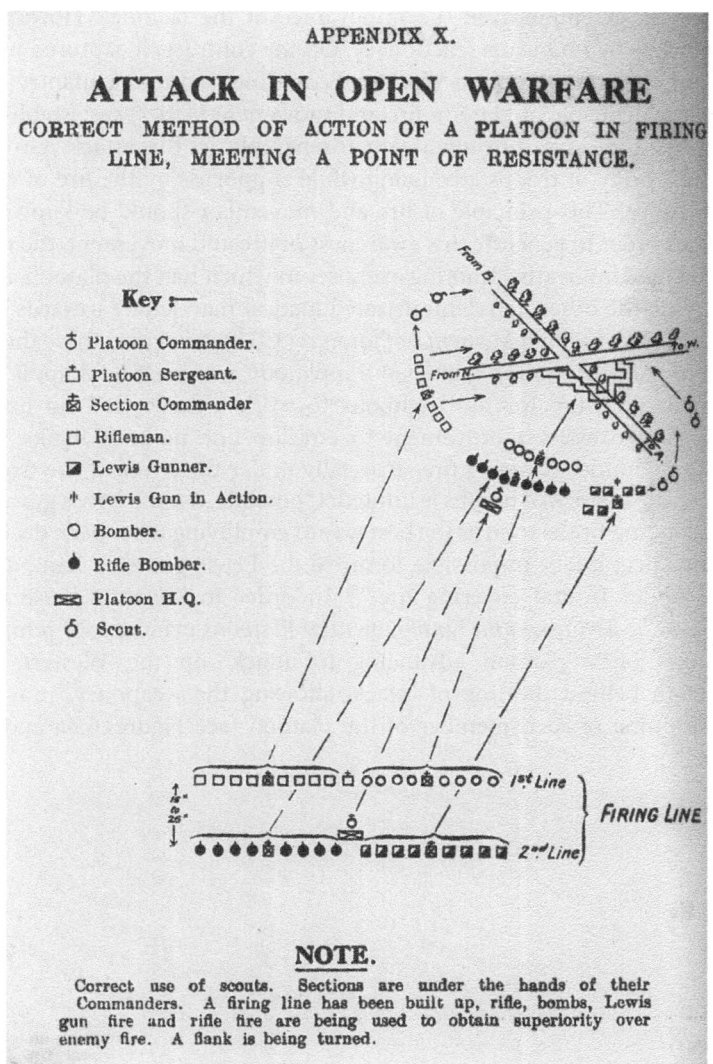

Fig. 6.2. Attack in open warfare. The General Staff, *Instructions for the Training of Platoons for Offensive Action*, 1917 Appendix: X.

'correct method of action of a platoon in firing line, meeting a point of resistance' depicts the use of a flanking force of riflemen on the left flank, Lewis gunners on the right, with rifle bombers firing over the heads of bombers in the centre.[93]

Assault Training prioritized offensive spirit and asserted that the 'chief duty and thought of all should be to kill as many of the enemy as possible'.[94]

Consequently, it emphasized the importance of the bayonet. However, the document was by no means reactionary. On the contrary, it captured many of the central changes which the Western Front imposed upon infantry tactics. Signally, the principle of platoon fire and movement tactics was established as paramount. 'Fire and movement are inseparable in the attack. Ground is gained by a body of troops advancing while supported by the fire of another body of troops. This principle of fire and movement should be known to all ranks.'[95] In order to generate this awareness of fire and movement, the manual provided some innovative training exercises in which half the platoon acted as enemy while the other half demonstrated tactical manoeuvre towards them.

The Training and Employment of Platoons (1918) developed these themes of fire and movement, noting that even if a platoon was itself held up, it should provide covering fire for its neighbour to work a flank: 'If held up, open covering fire to assist a movement to envelop one or both flanks; always move rapidly under covering fire, especially under the cover of fire from rifle bombers, whose stock of bombs is limited. If possible, use the Lewis gun section to open flanking fire as soon as the best way of employing it has been decided by reconnaissance; if it is impossible to move the Lewis gun to a flank, try and advance under frontal covering fire.'[96] In order to illustrate these tactical principles, *The Training and Employment of Platoons* printed four remarkable illustrations of 'a platoon advancing to attack' on the Western Front, drawn from behind the line of attack, showing the weaponry, movement, and positioning of each member of the platoon (see Figures 6.3a and 6.3b).

Fig. 6.3a. Plate I (a platoon advancing to attack) from *Instructions for the Training of Platoons for Offensive Action.*

Fig. 6.3b. Plate II (a platoon attacking a strong point) from *Instructions for the Training of Platoons for Offensive Action*.

The pictures were a major improvement on the more abstract diagram from the appendix of *Instructions for the Training of Platoons for Offensive Action*.

The plates clearly display a new form of dispersed tactics. The first plate shows a platoon advancing with a skirmish line of one section thrown out in front of three other sections all advancing in separate files or the sections advancing in file behind their own scouts. The third plate shows a platoon attacking a strong point. The rifle bombers and Lewis gunners lie in cover to a right flank while two sections of riflemen and bombers assault the trench. In the fourth print, the cooperation of platoons in depicted: 'The platoon on the right has been checked and its Lewis Gun put out of action.'[97] The Lewis gun section of the reinforcing platoon has taken up a position to the left flank while its riflemen and bombers advance to support the attack of the stalled platoon. As these publications accentuated, by the end of the First World War, British army tactics prioritized platoon fire and movement tactics. Platoons, divided into specialist sections, moved forward under mutual support and through the use of the terrain together to infiltrate German positions which would be mopped up (rolled up) by succeeding waves of platoons. The British had converged on a pattern of attack which accorded closely with the German concept. Indeed, since both were influenced by their common experience of the front and by Laffargue's pamphlet, such a convergence was understandable. The Canadian Expeditionary Force, always under British command and having developed many of the innovations through their raids, adopted British doctrine more or less wholesale. After major setbacks in 1915 and 1916, 1917 seems to have been a critical year for the Canadian infantry. In particular, as they prepared for their assault on Vimy Ridge in early April, they began to apply the *Instructions for the Training of Platoons for Offensive Action*. The Canadian infantry successfully incorporated these techniques into its assault on Vimy Ridge on 9 April 1917.

After the First World War, the importance of fire and movement at platoon level was emphasized very strongly in the British army. Basil Liddell Hart played an important role in writing infantry doctrine immediately after that conflict and was the author of important sections in *Infantry Training* (1921) including 'Use of the Ground', 'Use of Smoke', 'Fire and Movement', and 'Fire and Formations in Battle'.[98] In his memoirs (as discussed in the previous chapter), he disparaged the British affection for the bayonet, emphasizing the generation of firepower over brutal mass energy and morale. Trying to eliminate the old lineal concept of the advance, he replaced traditional terms like the 'firing line'—and especially 'the dreadful phrase "building up the firing line"'—and 'supports' with 'forward body' and 'manoeuvre body' 'in order to instil the idea of penetrating manoeuvre instead of lineal action'.[99] Liddell Hart stressed the autonomy of a platoon and the functional specialization of its sections. He envisaged a division of labour between the forward and main sections. *Infantry Training* was published in 1921 and a new version in 1922. Whatever the failings of practice during the First World War, the British had by its end—and in line with Griffith's claim—formally developed a refined concept of infantry tactics. In contrast to typical British practice in the First World War, the publication stressed the independence of the infantry.

> The bedrock of infantry tactics is the principle of fire and movement. Success, whether in attack or defence, largely depends on the skilful combination of these two elements... In attack, the task of the infantry is to close with the enemy's infantry and destroy it. To effect this, the infantry must use movement to get to close quarters, advancing from cover to cover, avoiding fire by quickness of movement and finally, by movement, endeavouring to work round the enemy's flank. With fire the infantry covers its own movement, by beating down the enemy's fire and forcing him to take cover. The main task of the other arms is to effect the result, but infantry must always be prepared to rely on its own fire to help it forward.[100]

Infantry Training maintained that 'the platoon is the smallest unit which can be divided into independent bodies each capable of fire and movement', and the document records the way in which the platoon worked its way forward and round to the flank, by the coordinated movement of the respective sections; 'the length of rushes must depend upon the ground, the enemy's fire and the physical condition of the troops.'[101]

Infantry Training underwent slow evolution in the next two decades but the principles which it promulgated were affirmed. Immediately before the Second World War, doctrine publications continued to demonstrate the point. *Infantry Tactics* and *Infantry Section Leading*, published respectively in 1937 and 1938, and *Infantry Training* of 1943 and 1944, all prioritized the use of fire and movement by the independent and functionally differentiated platoon. According to these publications, the platoon should organize one or more of its sections to neutralize the enemy with fire while the other advanced. Typically, as many diagrams depicted in *Infantry Training* (1944), the fire section—consisting of light and heavy machine-guns and mortars—was placed off to one flank while the assault sections, themselves given different specialist tasks, mounted the assault under its covering fire 'firing from the hip as they go in'.[102] British doctrine recognized the quasi-independence of the section itself which could mount its own small assault. In this case, the Bren gun (light machine-gun) would be placed off to one flank while the riflemen and grenadiers dashed forward for the assault; 'Every section is designed to provide its own covering fire within itself. It can, if necessary, rely on itself to get forward. This provision of covering fire is the primary task of the Bren gun in the attack.'[103]

Unsurprisingly, the evolution of American infantry tactics from the First World War accords closely with French, German, and British doctrine. Indeed, in many cases it simply adopted and reprinted British and French manuals. Thus, like European doctrine at the time, the 1914 Field Service Regulations advocated lineal assault and the concept of the firing line.[104] However, US Army doctrine of the First World War fully embraced the move to platoon tactics once the American Expedition Force (AEF) was deployed. Here fire was paramount but the concept of the firing line less applicable because troops moved forward in small section or squad groups. They formed bases of fire rather than firing lines. The May 1918 *Instructions for the Offensive Combat of Small Units* is an American army translation of the French army's *Instruction sur le combat offensif des petits unités*, though it includes the four prints of a modern platoon assault from the British *The Training and Employment of Platoons* published in February 1918. While the 1914 *US Army Field Service Regulations* (War Department, 1914) prioritized the rifle, the *Instructions* listed the rifle and bayonet, hand-grenades, rifle-grenades, automatic rifle, and machine-guns as vital to the generation of tactical firepower. The automatic rifle (the US equivalent of the Lewis gun) was given special emphasis. Although not as powerful as the heavy machine-gun, its mobility ensured that it was 'the accompanying weapon of the infantry'.[105] Specifically, in support of an attack, it provided 'sweeping fire' and 'marching fire', obliging 'the adversary to remain underground during the last rushes of the assault'.[106] The firing line, too, was fundamentally revised. The assault now consisted not of company lines but of platoons and half-platoons, the latter normally fighting on a front of about 100 to 125 metres.[107] The *Instructions* organized the platoon into small waves of differentially armed soldiers, with most of the bombers and riflemen and half of the automatic riflemen in the first wave, rifle-grenadiers and remaining riflemen and bombers in the second wave.[108] These waves consisted of half-platoons, led by the assault half-platoon; 'When a half-platoon encounters hostile resistance (machine gun, hand bombers, etc.), it endeavours to overcome it by combining movement and fire. The automatic rifles and the rifles take care of whatever rises above the ground, and thus oblige the defenders to conceal themselves, while the hand and rifle grenades take care of whatever is hiding below the ground.'[109] If a half-platoon were held up, it aimed to provide covering fire for a neighbouring unit to mount a flanking attack. There were examples when the US infantry made good use of these tactics. On 30 July 1918, 1st Battalion 47th Infantry was pinned down with little fire support. D Company pushed two automatic rifles forward, firing from the left flank, while the 3rd platoon rushed forward to overcome resistance.[110]

There is a clear line of continuity in US infantry doctrine from the 1918 Instructions into the Second World War. With the massive expansion of the US Army, the infantry platoon was reorganized on the German model. The Browning Automatic Rifle (BAR) squad was dissolved and the BAR was inserted into each rifle squad. The idea was to build fire and movement around that weapon with the squads divided into BAR and rifle teams which mutually supported one another.[111] One of the clearest statements of US infantry tactics was the 1944 Field Manual 7-10, *Rifle Company, Infantry Regiment*, which laid out all the fundamentals of small-group infantry tactics. The publication explained the now standard account of fire and manoeuvre; 'At the first firing position each attacking platoon seeks to gain fire superiority over the enemy to its front ... Further advances are made by successive rushes, or movements of individuals or small groups of the leading squads and platoons.'[112]

US commanders regularly emphasized the importance of fire and movement tactics. David Kenyon Webster, of the 101st Airborne Division, approved of the last speech of his regimental colonel, Robert Sink, before emplaning for D-Day. 'He didn't discuss the Allies or the Four Freedoms, Nazi tyranny or a Crusade in Europe. He talked of war as it is to soldiers: Kill or be killed. He talked of fire and movement.'[113] Balkosi's work on 29 Division demonstrates that the principle of fire and manoeuvre was fundamental to their tactics. The Divisional Commander, Major General Gerhardt, was unhappy with training in Britain before D-Day and insisted that a more open way of using weapons was adopted to encourage fire and manoeuvre: 'when the men were out on their own on the lonely moors, Gerhardt encouraged them to practice free shooting with pistols and carbines' and indeed demonstrated the importance of realistic live firing from his jeep, 'dropping hand-grenades out of the window onto the bogs, and howling with delight'.[114] His Deputy, Brigadier Cota, also used to drive around the training area on Dartmoor, 'giving instruction in tactics to units as small as twelve-man squads'.[115] There is clear evidence that US forces sometimes implemented fire and movement tactics very effectively during the Second World War.

DRILLS

Although the Greeks and especially the Romans perfected infantry drills for their era of warfare, the origin of recognizable modern infantry drills can be traced back to the late sixteenth and especially early seventeenth century, as the musket began to replace the pike as the dominant infantry weapon. At this point, in order to maximize the potential of firepower and, to a lesser extent, prevent fratricide, commanders began to institute set procedures in preparing and firing weapons. Gustavus Adolphus and Maurice of Nessau were important figures here, where each produced manuals which, in a strange anticipation of Taylorist principles, divided infantry practices into their smallest and simplest steps. The process of loading, aiming, and firing a musket was then instituted *en masse* through rigorous training. Gustavus Adolphus also introduced the concept of the counter-march, so that recurrent lines of infanteers would retreat to the back of the formation on firing their weapon, with the aim of maintaining a continuous fire. The better trained each individual musketeer and the more coordinated the lines of firers, the more likely it was to be successful in battle. Wellington's infantry perfected this technique of fire and, demonstrating high levels of discipline, were able to thin their formations out to two or three long ranks to deliver devastating fire at close range. The defeat of the Imperial Guards at Waterloo—advancing in their favoured column formation to maximize the effect of shock—was only the most famous example of this use of fire organized in lines. The formal military parade has its origins in these practical battlefield drills and, although Clausewitz disparaged mere parade ground training, the ability of a battalion to perform its drill, including manoeuvre evolutions, quickly and efficiently, thereby maximizing the fire it could generate, was fundamental to its survival.

The problem of the modern battlefield, as Marshall recognized, was that the lineal drills of the age of the musket became unworkable. Consequently, it might be claimed that infantry drill itself became irrelevant. Certainly, the fireswept battlefield of the First World War demonstrated the inadequacy of lineal infantry formation in contrast with fieldcraft and tactical skill; the use of terrain, camouflage, crawling, and individual initiative all became vital to tactical survival and success. As we have seen, one of the most important solutions which armies developed often quite spontaneously to overcome the Marshall effect was to rely upon, encourage, and reward purely individual action. With the introduction of the rapid-firing rifle and the machine-gun, the military automaton of the age of the musket seemed to have been replaced by the individual hunter, operating with no fixed method or means, cunningly improvising to the situation in hand. In fact, the infantry drill remained just as important in the age of the machine-gun as it was in Wellington's army. Periodically, armies recognized that allied with proper fire and movement doctrine and training, infantry drills became the central and most effective means of enhancing performance. Crucially infantry drills allowed soldiers to react individually and collectively to enemy fire, without having to go through the complex process of developing novel practices and coordinating their execution on the spot.

In the First World War, it is often presumed that the German army, and especially Captain Rohr's stormtroopers, pioneered tactical innovation. In fact, the French army recognized the importance of battle drills very early on. The *Manuel* of 1916 not only laid out a new tactical doctrine for the infantry platoon but also provided guidance on how troops were to be instructed. Accordingly, it recommended the construction of dummy trench systems in exercise areas behind the front on which various types of assault could be practised.[116] One of the central purposes of these repeated training evolutions was to instil in the troops 'l'automatisme au combat'; automatic reactions to combat—or battle drills, in short: 'the company must possess in advance a series of instinctive actions which can be executed [*jouer*] in the absence of any commander and without reflection; this essential equipment [*bagage*] can be acquired only through frequent repetition.'[117] The 1918 *Instruction* stated that successful attacks required speed, one of the central methods of attaining which was by 'the instantaneity, the automatism in the conception and execution of local manoeuvres'.[118]

Nevertheless, despite French developments, it is certainly true that Rohr's stormtroopers played a key role in demonstrating the potential values of new techniques including battle drills. They exploited the terrain, specialist weapons, and tactical acumen (crawling, camouflage, or speed) to infiltrate enemy positions. They adhered to the principles of fire and movement throughout, developing basic drills for individual specialist weapon use. For instance, Ernst Jünger records an argument he had with his subaltern during the famous counter-attack at Cambrai, when stormtroops (of which Jünger was one) were first used in significant numbers. 'He wanted to be first, and insisted that I supply him with bombs, rather than throw them myself.'[119] Jünger, assuming the prerogative of command, was grenading British positions himself but his subordinate cited established drills to correct him: '*One man* to throw! And after all I was the instructor at storm troop training!'[120]

By the Second World War, battle drills (*Kampfweise*) had become a central part of German infantry doctrine. Dr Reibert's *Deinstunterricht* which draws heavily on *Ausbildungsvorschrift* demonstrates the institutionalization of drills in the Wehrmacht in order to conduct fire and movement. Dr Reibert highlights the importance of weapons training, especially on the MG 34, which receives extensive treatment.[121] Citing *Ausbildungsvorschrift* he notes: 'Combat training and training in field service are the most important branches of training. The goal is skilled conduct on the battlefield as well as the purposeful and safe use of weapons in combat. Good weapons and shooting-technique training are a requirement.'[122] Specifically, this weapons expertise was not primarily developed through parade drill and formal firing ranges but by realistic training exercises which inculcated combat drills: 'Exercise in combat drills is the thing which indelibly impresses on the soldier the right way of moving on the battlefield and the independent, purposeful use of his weapons in combat.'[123] In this way, the German army sought to inculcate set drills and, indeed, Captain Dawson, of No. 4 Commando, noticed that reaction times of German troops demonstrated that they must have been using pre-arranged drills in combat.[124]

The institutionalization of drills is observable among British forces. Fire and movement conducted by platoons was established as a fundamental infantry principle early on in the First World War. General Maxse provided detailed instruction, which he called 'methods', about the particular means by which his troops were to assault the enemy.[125] He believed that if both officers and troops had a clear common idea of what to do in particular circumstances, they would be able to react quicker and more coherently than if procedure had to be improvised under fire. Thus before the assaults on Thiepval and the Schwaben Redoubt in September 1916, he issued specific instruction to his division:

As regards attacks formations ... the doctrine in the 18th Division is as follows:—(a) teach, drill and practise a definite form of attack so that every man shall know it thoroughly. On this basis of theory and knowledge common to all, any brigade, battalion or company commander varies his attack formation to suit any condition which may be peculiar to his front or objective.[126]

As many of the major battles demonstrated, the British army in the First World War did not institutionalize these drills particularly effectively.

In the Second World War, there was a similar attempt to institute battle drills in the British army but there was also substantial resistance to the concept of the drill. The War Office in particularly regarded drills with suspicion since they putatively robbed soldiers of the initiative which was essential on the modern battlefield.[127] However, the *Instructor's Handbook on Fieldcraft and Battle Drill* (1942) and *Infantry Training* (1944) both stressed the importance of established battle drills. *The Instructor's Handbook* was particularly interesting on the theme of drills. It claimed that mechanized warfare had dispersed forces and, therefore, made the role of small-group tactics ever more important to the overall outcome: 'in this war of dispersal, it is not an exaggeration to say that an army is as good as its section and platoon commanders.'[128] In order to maximize the effectiveness of its sections and platoon, the *Handbook* recommended the institutionalization of 'battle drill'. These would act as a 'firm basis on which to develop his individual initiative much in the same way that the young cricketer is taught the basic principles of stroke play on which he develops his own style'.[129] Throughout the publication, it was stressed that battle drills were not dogmatic but were to unify practice, ease coordination, and accelerate the tempo of tactical action on the field of battle itself. Interestingly, the *Handbook* saw the section battle drill as the 'modern equivalent of close order drill which 150 years ago was the way the soldier fought'.[130] However, drill served an important unifying purpose: 'You are ensuring that every subaltern, serjeant, corporal and private soldier has a clear idea of an ideal plan photographed on his mind. He will know what is being aimed at, what battle is all about and what everyone is trying to do—things he seldom knows now.'[131]

The *Handbook* proposed that this drill was first practised on the parade square with a flag representing the enemy position. The section then mounted a series of mock assaults, pincer, left and right flanking, with the Bren and rifle group mutually covering each other's movements.[132] This drill then led on to the platoon drill. *Infantry Training* (1944) took up the theme of battle drills, breaking down the assault into a sequence of simple actions and specific tasks for the Bren and rifle groups. Indeed, the pamphlet includes a spreadsheet for infantry training drills in the assault, in columns, named Drill No., Word of Command given by, Word of Command, Action taken and by whom.[133]

Drill No.	Word of command given by	Word of command	Actions taken and by whom
5	Sec. comd.	Rifle group—stop. We will kill all enemy in that post. Right (or left) flanking. Rifle group follow me. Halt. Enemy left (or right). 200 covering fire	Group moves round rear of Bren group to bound. If more than one bound has been ordered by instructor this will be short of 3 o'clock to post, as 3 o'clock is assault position in right-flanking. Halt. Left (or right) turn and shout:—'Down crawl observe sights fire'.[134]

Place has provided the best account of the institutionalization of battle drills and he demonstrates that much of this doctrine was actually implemented. He points to the important foundations laid by General Maxse at the end of the First World War where 'codified tactical drills' were established.[135] Building on Maxse's work, Place identifies key figures in the officer corps who provided impetus for the battle drill programme, among the most important of whom were Lieutenant General Harold Alexander (who had commanded 1 Corps after Dunkirk), Major General Utterson-Kelso of 47th Division, and Lieutenant General Paget.[136] Lieutenant General Alexander's '1st Corps Tactical Notes' and Utterson-Kelso's own mimeographed book, *Battle Drill*, articulated their thoughts on the importance of drills in print.[137] Alexander and Utterson-Kelso expanded on the concept of the corps and divisional battle schools which already existed to establish new schools, specifically to develop and disseminate battle drills: the central battle school at Barnard Castle and 47th Division's school at Chelwood Gate. The chief instructor there was Major Lionel Wigram, who became acquainted with battle drills through his brother-in-law, Lieutenant John Jokelson, who was an instructor at 1 Corps's school in Lincoln.[138] The course at Chelwood Gate, lasting two weeks, proved very successful and involved practical work, lectures, and tactical exercises without troops. Battle drills were sometimes deliberately unrealistic but were intended to demonstrate the fundamentals of platoon tactics. Thus, part of the course involved battle drills on the parade ground. Sections and platoons would run through a mock assault, putting the Bren gun to the side to cover riflemen's assault and standing to attention to indicate that they were 'firing'.[139] The parade drills were easy to parody but they were also practised in more realistic conditions, once they had been mastered. Indeed, at Barnard Castle, there were three deaths during live firing exercises.[140] Students expected to return to battalions and pass on what they had learned: 'the result is an unbelievable increase of standard of training of the "man in the section".'[141] In autumn 1941, the 2nd Canadian Infantry Division relieved the British 55th Division defending the Sussex coast and, as a result of its association with 47th Division commanded by Major General Utterson-Kelso, began to adopt battle drill.[142] Subjected to tedious and apparently irrelevant training from 1940, the Canadians embraced battle drill with enthusiasm; indeed, the Calgary Highlanders became 'fanatical disciples' of it.[143] Canadian platoons and companies learnt a series of automatic responses to contact with the enemy, eventually establishing their own battle school after the closure of Chelwood Gate in 1942.

Before 1942, the British army had organized assaults by means of formal orders which could be cumbersome and might be impossible in the instance of an unexpected action. Precisely because the battle schools sought to develop standard tactical practices which were recognized by all, it was possible to retain tactical coordination without recourse to a long orders process. At battle school, the rapid call of 'right flanking', 'pincers', or 'stay and cover' sufficed.[144] As Major Hickson recorded:

> True it is a flexible drill which can be altered to suit circumstances but the orders required are stereotyped, and such that not only NCOs but private soldiers can understand at once, so that he knows from the start what he is doing, which he rarely did before. Verbal orders are reduced to a minimum; within five minutes of sighting the enemy, a platoon is off like a pack of long dogs.[145]

Significantly, progressive officers who saw the values of established and codified drills were supported by experienced soldiers in the field. For instance,

Lieutenant Colonel Victor Turner, 2 Rifle Brigade, who won a Victoria Cross at the Battle of El Alamein in North Africa on 23 October 1942, emphasized the importance of drills: 'It's therefore extremely important that in peace time you should get the drill, as it were, of conducting a battle, or exercises so that when it actually comes to the fighting [and] you are enmeshed in the fog of war, your actions are so automatic that you can adapt yourself to the unexpected and carry on without losing your head.'[146]

The battle schools did not train all infantrymen. Selected individuals, typically officers or NCOs, attended the course and then passed on their instruction to their units, on their return. This seems to have improved tactical skills, but the institution of battle drills was particularly marked among those specialist soldiers who underwent collective battle training together; specifically, in Britain, the commandos. During the Second World War, British forces began to raise commando units in 1940 following Churchill's call for the creation of an elite fighting force capable of taking the fight to the Nazis on continental Europe. Although some of the initial training and operations of the commandos were haphazard and amateurish (as Evelyn Waugh recorded), by 1941 a Commando Training Base had been established at Lochailort and subsequently moved to the famous 'Castle Commando' at Achnacarry.[147] Under Lieutenant Colonel Charles Vaughan, this location became an important milieu of military innovation for the Allies. Not only did the British commandos train here, from which emerged Britain's marine and airborne elites as well as their Special Forces,[148] but American Rangers also went through the stern training and selection process which Vaughan created.

Achnacarry bore some similarities with the regular battle schools such as Barnard Castle and Chelwood Gate. It prioritized battle drills, but the training at Achnacarry was far more intense and realistic than anything which took place elsewhere. In addition to an arduous physical regime, including a demanding assault course (normally completed with simulation explosions and overhead fire), candidates at the school underwent extensive live firing training. There was no range target shooting. All shooting took place in realistic exercises against human-form targets, typically under simulated fire conditions. The intensity of the training was important but so was the fact that troops conducted their training together. Typically, commandos would be trained and selected in batches which would then be kept together as fighting units; whole Royal Marines battalions were converted into commando units in this way (with significant wastage of men). Consequently, the members of commando units were unified around common drills which they had trained together to perform, increasing their cohesiveness under fire. Central to the syllabus at Achnacarry was the development of battle drills, from getting off the landing beach as quickly as possible to assaulting enemy positions with covering fire. Major Henry Hall was one of the founders of Special Training Centre Lochailort and noted that as a result of the training 'Whatever you did, you did it automatically, subconsciously. The answer to whatever attack you were up against would be an instinctive reaction.'[149]

Moreover, although the discipline was strict at Achnacarry, there was an active attempt to embrace tactical innovation among the commandos. Thus, following their training, commando units sought to learn from experience and develop their tactics: 'After practising their skills, the commandos started to notice that their responses in attack and defence situations were starting to consolidate into set moves.'[150] This was particularly apparent in No. 4 Commando immediately before the ill-fated Dieppe raid, which started to develop a set of tactical drills:

Discussions within No. 4 Cdo led to American football where set drills are used during games. The first battle drill revolved around the smallest unit—the sub-section. Within each sub-section, the Bren Unit would be the fire unit and the Tommy gunners would be the movement group. A typical drill was the response to enemy fire. Once the section came under fire immediate cover was sought and the source of enemy fire was located. Having done this, the Bren gun would engage the enemy position and the assault group would find the best route to the enemy position and attack once within range with grenade and guns and bayonets.[151]

The battle drills of No. 4 Commando played a crucial role in their assault on the Dieppe Battery at Varengeville in August 1942. In a well-known report, Martin Lindsay of No. 4 Commando underscored the importance of their training in producing these effective reactions. 'Waterloo may or may not have been won on the playing fields of Eton; it is certainly much truer to say that Operation "Cauldron" was won on the training fields of England.'[152] The officer recorded the 'remarkable features of the training programme' as the 'accuracy with which the nature of the various actions was foreseen' and the 'soundness of the training programme, which resulted in the soldier's meeting of the sudden events of the day with the confidence of a highly trained athlete hearing the expected starting pistol for his race'.[153]

It is a model of 'fire and movement' tactics. Frontal fire pinned the enemy to the ground while the assault troops moved round their flank to the forming up position, the assault itself being preceded by a final crescendo of fire. The principle of this attack and that of the battle drill taught at the School of Infantry are the same.[154]

Interestingly, as part of the collective training programme, the notes specified: 'fire and movement on the range; battle drill with live ammunition, bayonet training and unarmed combat ... Practise withdrawal, first as a drill, then with smoke, then with smoke, opposition and casualties.'[155] It was noticeable that, on the operation itself, the withdrawal was done in small groups leap-frogging back exactly as they had practised.[156] The author of the notes on the Varengeville raid emphasized the importance of training and the development of automatically followed drills. Although Canada's 1 Division were certainly unfortunate during the Dieppe Raid, many of their boats being struck before they had landed, the success of No. 4 Commando was not a matter of luck. They avoided the disasters that befell the Canadians substantially as a result of their training and battle drills. They were not the only force to put their successes down to their drills. Colonel Bill Darby, commander of the 1 Ranger Battalion, which stormed the Point Du Hoc near Omaha Beach on D-Day, also emphasized the importance of severe and realistic collective training which induced appropriate drills: 'The achievements [on D-Day] were due entirely to our training at Achnacarry.'[157]

Battle drills were fully institutionalized into US doctrine after the Second World War and featured prominently in manuals like the 1959 version of FM 7–10, *Rifle Company, Infantry and Airborne Division Battle Groups*.[158] In Vietnam, US infantry, especially in the better-trained divisions, regularly used these drills in practice. During the Battle of Dau Tieng, Marshall describes how US riflemen 'flopped down evenly like a well-drilled fire team and as methodically as if they had been doing it in a test exercise at Fort Benning, blazed away'.[159] In his memoirs of his combat experiences, Marlantes has also emphasized the importance of battle drills to the training and performance of Marines in Vietnam. The Australian Task Force in Vietnam was one of the most effective battalions to fight in the war and its successes were substantially due to the inculcation of battle drills during intense training. Their most famous battle took place at Long Tan in 1966 and will be discussed at greater length in the following chapter. In that fight, D Company 6 Royal Australian Regiment (RAR) was most heavily involved, and relied on drills to prevent itself being annihilated. Before deployment, the company was subjected to a very stern training regime by Major Harry Smith, the company commander. The hard and realistic training which the Company underwent before deployment seems to have been critical to their performance at Long Tan. Specifically, 6 RAR's artillery battery commander stressed the importance of training and drills in the field: 'battle drills are an important part of our training, the whole aim being to learn to act and react speedily whilst under pressure, in such a way as to reduce mistakes and errors to a minimum . . . The battle drills are practised and practised until they become second nature—a reflex.'[160] 6 RAR in general and D Company in particular seem to have developed great proficiency in their battle drills so that at Long Tan even the national servicemen were able to support each other and respond collectively to the very serious threat which confronted them. For instance, at a late stage in the battle, Lieutenant Sabben, commander of 12 Platoon, described how one of his machine-gunners was 'hit in the chest and rolled away from the gun. His gun number two took over the gun and the next digger along took over the belt-feeding duties. It all looked so like a rehearsed manoeuvre that I just stared for a moment, not comprehending that Kev [the machine-gunner] had indeed been wounded.'[161] The importance of the battle drill was widely recognized and periodically practised in the mass army.

BATTLE PREPARATION

Although drills and training were critical to cohesion, during the First World War it was realized that combat performance could be significantly improved by mission-specific rehearsal. Here the troops prepared in detail for the particular tasks which they expected, preparing themselves for the distinctive conditions which they expected to meet. The modern battlefield posed serious coordination problems for assaulting troops. Where infantry had previously been able to simply advance together behind their leader against normally very obvious objectives, with flags and banners often helping to maintain cohesion, troops were now dispersed and hidden from each other, attacking camouflaged positions which it was very difficult to discern. In order to sustain any level of coordination in this context, it became imperative that all soldiers recognized the terrain and had a clear understanding of their missions so that they could continue to conduct the assault even in the light of almost inevitably unexpected difficulties. Although it was certainly not universal practice and, indeed, it was typically the preserve of specialist forces on prepared operations, armies began to condition their assault troops.

In France, battle preparation was identified early on as critical to success. Despite his importance, many of Laffargue's recommendations were not, in fact, original, but much of his novelty lay in his emphasis on intelligence and rehearsal: 'During the days which precede the attack, a minute study of the hostile trenches should be made.'[162] In particular, the position of machine-gun emplacements had to be identified and specific plans for their elimination developed. Captain Laffargue emphasized the importance of adequate battle preparation. Assault troops needed not only to understand their objectives clearly and collectively. Specifically, the assaulting infantry had to have a clear understanding of what they were trying to achieve and how they were going to conduct the assault: 'Let us prepare our business down to the slightest detail in order to conquer and live.'[163] Laffargue continued:

> Before the attack, the physiognomy of the terrain of the enemy's defences should be well impressed on the memory. The position should be known not only from the front, but in profile. This study is of the greatest importance, particularly *for the troops of the second line*, because the greatest cause of stoppage in an offensive against a fortified position is the incomplete knowledge of the position.[164]

It was insufficient that this knowledge was limited to company and platoon commanders who were to be issued with 'very detailed maps'. 'Before the attack on 9th May, I had recopied for each non-commissioned officer the part of my map concerning the zone of attack of the company, entering on it all the known information.'[165] According to both the *Manuel* and the *Instruction*, preparation was of prime importance and the success of an assault depended upon it. Both documents stressed that attacking formations had to know the exact lay-out of enemy positions derived from aerial photographs, observation, reconnaissance, and the interrogation of prisoners.[166]

After the First World War, the importance of battle preparation continued to be stressed in training manuals and tracts on infantry tactics in France. For instance, Lieutenant Colonel Culmann's *Cours de tactique générale d'après l'expérience de la Grande Guerre* discusses the preparation of the infantry at length. He recommended that assaulting battalions should be taken out of the line a week before the planned assault, only the pioneers remaining to continue their reconnaissance and obstacle-clearing work.[167] The battalion should be allowed a significant period of rest and reflection so that at the time of the assault, the soldiers 'have nothing to regret'. Culmann emphasized the importance of establishing close fraternal bonds of comradeship so that the regiment was like a 'family'.[168] This would enable the caring officer to 'harvest the fruits of his labour' on the battlefield because his soldiers would be willing to risk themselves for him. In addition to this moral preparation, the battalion must be submitted to intense military instruction: 'Each and every soldier is instructed in minute detail what he has to do on the day of the attack, him and his neighbours.'[169] Significantly, models were to be a major part of this process of instruction and Culmann recommended that models should consist of a relief map of the sector at a scale of approximately 1:5000: 'On a selected spot, in the centre of the camp, all the defensive positions, friendly and hostile, in the attack sector are represented, faithfully reproduced. The trenches and saps are staked out, highlighted [*décapé*] and marked off with white tape. Writing indicates the names of all bunkers, shelters, machine-gun and mortar emplacements etc which are to be taken.'[170] Culmann continued: 'On this model, which depicts the situation [*parle aux yeux*], the officers show their cadres and men the manoeuvres they will make, the defiles to avoid, the probable pockets of resistance which will have to be reduced, the objectives to take, the terrain to occupy.'[171] These models should depict the French and German positions and troops in different colours and should show company, battalion, and regimental movements 'until all know their role'.[172]

Perhaps unsurprisingly, among the first troops to use rehearsals were Captain Rohr's Assault Detachment. His troops used detailed maps at a scale of 1:5000, recording dug-outs and gun emplacements in detail, in planning operations. By referring to these maps, the troops were familiarized with the ground over which they would assault. On the basis of these maps, full-scale models of enemy positions would be built which would be used for real-time live rehearsals.[173] This technique eventually became established as standard practice and stormtroopers trained on a dummy trench system with rehearsals up to divisional level, as a matter of course. Jünger described the use of such dummy trenches both at company and divisional level: 'Several times I had the company practise attacks on complicated trench networks, using live hand grenades, to turn to account the lessons of Cambrai. Here too there were casualties.'[174] 'After several battalion and regimental drills, we twice rehearsed an entire divisional breakthrough, on a large site marked with white ribbons.'[175] Rommel's 124th Infantry Regiment also used to employ rehearsals extensively, which Rommel recorded in his memoirs: 'The attack moved through smoke and noise of battlefield with the same precision shown in rehearsals of the past few days.'[176] These practices persisted in the Reichswehr after the war. Indeed, during the 1920s, the US Military Attaché in Berlin repeatedly noted the quality of the models, build on 'sand tables', which German soldiers constructed during exercises which he observed; 'the [relief] map I was shown on one [sand-table] was the most carefully made and realistic war-game map I have ever seen. The scale was approximately 20 inches to the mile. Wood, streams, roads, bridges, houses, individual trees were all shown in most painstaking detail. Plowed fields, standing crops and stubbled fields were represented, as different coloured earths.'[177]

Rehearsals were conducted by British forces during the First World War before major assaults. Summarizing the performance of his division at the Somme, General Maxse concluded: 18th Division's performance on the Somme demonstrated that

Training is everything. If British men know what is wanted of them, they can accomplish marvels; the difficulty lies in getting them to understand, each unit its particular job. This was done very thoroughly in the 18th Division on model trenches which we dug at Cavillon representing (from aeroplane photos) the exact Bosche trenches each company and platoon had to take.[178]

As assaults in 1917 and even 1918 demonstrated, General Maxse's injunction about the importance of detailed training and rehearsal was not always followed. Poorly prepared attacks did take place. However, as Inspector General of Training, he formally instituted a new regime of tactical preparation and practice which was closely compatible with and, as Paddy Griffith would argue, the equal to German stormtrooper techniques. Liddell Hart's 1921 *Infantry Training* similarly emphasized the importance of preparation. Crucially, to conduct successfully coordinated small-group tactics, it was essential that there was a collective understanding of the mission: 'To ensure success in an attack, every platoon, every section, and every man must know what its or his objective is.'[179] For a raid in which surprise and therefore a higher degree of coordination was necessary, the manual specified that preparation had to be elaborate: 'Raids must be planned in great detail and when out of the line the raiding parties should, if possible, rehearse the operation on a facsimile of that portion of the enemy's trenches which it is intended to raid.'[180] British doctrine in the Second World War continued to emphasize the importance of collective preparation. *The Instructor's Handbook*, for instance, which was prepared by Wigram, recommended that special preparation was advisable when attacking pillboxes and fortifications. 'A drill and very careful rehearsals (if possible using a model) will be necessary.'[181]

Learning from their allies, the US Army recognized the importance of battle preparation quickly. The 1918 *Instructions* provided precise details about how to prosecute an assault in the new conditions which existed on the modern battlefield. However, in order to be able to coordinate these complex tactical manoeuvres, the *Instructions* recognized that troops needed to have a better collective understanding of what they were trying to achieve. Accordingly, the *Instructions* also included some important observations about the importance of preparation and rehearsal.

Attacking troops must be 'put in condition'. When circumstances allow, rehearsals of the attack in all kinds of weather should take place in rear on terrain where there can be reproduced the hostile position to be captured—each unit receiving a mission which is identical in every respect with that which will be assigned to it in the combat. It is well to show the men large scale relief maps (in cement, for example) constructed in the open.[182]

Intelligence of precise enemy locations was regarded as critical and consequently an assault would be preceded by a period of covert observation.

Knowledge of the enemy's situation is an element of supreme importance for the commander in arriving at his decision. Commanders of small units must thoroughly study the documents of every sort which are transmitted to them by their commander. These consist of vertical or oblique photographs, battle maps, relief maps, sketches and intelligence bulletins. But the preparation of these documents necessitates the collaboration of the troops and must be constantly verified by them.[183]

The Canadian Expeditionary Force began the war badly but sought to learn from their experiences and, in particular, began to develop new tactics as a result of raids in 1915 and 1916, which assaults became a crucible for adaptation. It is possible that the first major raids were carried out by 1st and 2nd Battalions of the Gerwhal Rifles on the night of 9–10 November 1914, and drawing on this experience and probably on their instruction from 2nd Royal Welsh, Princess Patricia's Canadian Light Infantry mounted the first major Canadian raid on 28 February 1915.[184] As a result of its raids, the Canadian army, like the other combatants, developed a series of practices to improve performance. Reproductions of enemy trenches were dug in fields to the rear on which the raiders practised bombing attacks; ladders were tested to cross ditches and soldiers were organized into specialist groups, assigned specific functions. Before the assault, scouts escorted the raiders over the ground.[185]

According to Rawling, the Canadians developed four lessons from raiding: the need for careful training and rehearsal, artillery in cutting off the Germans, first-class scouting, and correct selection of soldiers for the specialist roles.[186] A large raid was a very different operation from a major attack but, on the basis of their raiding experiences, standards in the Canadian infantry improved throughout the war and, indeed, when Canadian divisions were committed to the Battle of the Somme in 1916, platoon tactics had substantially evolved. Before Vimy Ridge in April 1917, the Canadian troops underwent a rigorous tactical training programme which culminated in 'tape courses'. These tape courses, erected in fields behind the lines, represented the exact lay-out of German trenches and pillboxes on Vimy Ridge and units would advance over them, behind an artillery simulated by umpires waving flags, so that they became completely familiarized with their objectives.

Rehearsals were common in the Second World War especially before major or specialist operations and, as in the Great War, involved the drawing of detailed maps or the construction of models to demonstrate the operation to troops followed by rehearsals on terrain which replicated reality. The raid on Dieppe was a case in point. No. 4 Commando developed their plan for the raid from detailed aerial and maritime photographs, illustrating the beach and the objective. On the basis of this data, the Commando developed a plan on a large-scale map, marking all the features which they would encounter. This plan was then rehearsed extensively. No. 4 Commando chose Lulworth Cove on the Dorset coast as the coastline most like the one they would encounter near Dieppe. They then submitted themselves to fitness training to get off the beach and up the cliffs at Lulworth.[187] Martin Lindsay, in his *Notes from the Theatre of War No. 4 Destruction of a German Battery by No. 4 Commando*, emphasized the significance of these rehearsals on topography similar to those on which the actual operation would take place to the cohesive performance of the unit.

D-Day was perhaps the most rehearsed operation in the course of the war primarily because of its complexity, the huge risks involved, and the absolute requirement for success. As Lieutenant General Truscott recorded, to aid the process of planning and preparation, relief models of 'the entire invasion coast' were constructed with great difficulty since the project required 'several hundred model-makers for a number of months' but 'there were only a handful of specialists in England, and apparently no more in the United States'.[188] With the help of these devices, all the assault formations underwent extensive preparation which was particularly pronounced among elite units. The British army's D Company Oxford and Buckinghamshire Light Infantry were selected for a glider assault on the important bridges across the Orne River and canal just south of Sword Beach, on which No. 4 Commando would land a few hours later on D-Day. They would be the very first Allied troops in action and, like No. 4 Commando, they used detailed maps and photographs to develop their plan.[189] In order to rehearse the mission, they identified two equivalent bridges over the Exe River and canal just south of Exeter in Devon to practise their mission. A month before the operation, the troops practised the mission for six days at that location, preparing for untoward eventualities such as the failure of some of the gliders to arrive or an incorrect landing sequence or the death of officers on the ground.[190] Similarly, as part of their rehearsals, American troops were explained their D-Day missions by reference to models of their landing zones. Balkosi records the preparations of the 29th Division: 'Officers showed their men rubber models of Omaha Beach that simulated the coastal terrain with remarkable accuracy.' The 29ers were told the exact time and place their outfits would land and what they had to do to get off the beach: 'They studied aerial photos and 1:7,000 maps of their sectors, which indicated enemy positions as small as individual machine gun pits.'[191] Indeed, the models were transported with American forces across the Channel, although on Utah and Omaha beaches, some American forces landed so far from their designated destination that they were off their own models.[192] Nevertheless, the use of these models seems to have oriented troops on D-Day even though the landings were disorganized and came under very heavy fire. Even under the conditions which confronted them on Omaha, the troops were able to identify the draws up which they would leave the beach even though the formations which advanced up them were formed ad hoc on the beach.[193] Although the Omaha assault was chaotic on most sectors, the fact that the divisions involved had been engaged in extensive rehearsals on similar beaches in Devon seems to have helped the troops to adapt.

The rehearsal was an important development for the infantry platoon. Orienting troops to common objectives, with which all were intimately familiar, and ensuring that everyone knew both their own and their comrades' role, significantly increased levels of cohesion in combat. Since troops were prepared for the terrain over which they would fight and the likely enemy strengths against them, and they had practised their manoeuvres, they were able to perform better on the real operation. Preparation served a further purpose: it articulated the general principles of fire and movement tactics and the generic drills which platoons had learned to perform to the specific circumstances of the operation. Through battle preparation, drills were refined and adapted to the mission to maximize the chance of success. Through battle preparation, the operation itself might become a large-scale drill, with sections and platoons requiring minimal coordination or guidance as they simply executed the manoeuvres they had practised. Many of the most successful platoon actions conducted by citizen soldiers from the Canadian raids or Operation Michael in the First World War to actions in Vietnam were the result of ingrained drills specifically tailored and practised for the operation in hand.

During the First and Second World Wars, the infantry did not always perform according to doctrine. However, by the middle of the First World War, the centrality of fire and movement tactics, prosecuted not by infantry companies and battalions operating in lines, supervised not by officers, but by independent, mutually supporting platoons and sections commanded by junior officers and NCOs, was accepted by all the major combatants, as this survey of western tactics demonstrates. The First World War was a crucible of change here. Despite the casualty rates of that war, often poor strategic and operational command (especially on the French and British part), and inadequate execution of tactics by divisions on all side, the fundamental principles of modern infantry tactics came into being. Armies began to try and develop procedures for overcoming the Marshall effect not through the use of bayonet or individual virtuosity—or by appeals to extraneous social criteria—but by platoon tactics. The common method which all forces converged upon was to turn the infantry phalanx of the nineteenth century into a platoon maniple. The most advanced doctrine before the First World War envisaged long lines of infantry advancing to the enemy to gain fire superiority until the moment of the charge. By 1917, the infantry sought to reduce the assault line. Small numbers of troops in platoons and sections surged forward, using the terrain and mutually covering fire, to infiltrate enemy positions. This remained the ideal throughout the Second World War (and beyond). At the same time, drills and rehearsals were developed to ensure that infantry soldiers would be able to execute these tactics in combat. Doctrine can be misleading since it is not always implemented. However, the evidence shows that in many cases, the infantry did employ precisely the kind of fire and movement tactics prosecuted at the platoon level which formal doctrine described. The institution of platoon fire and movement tactics did in some cases overcome the predicament which the mass army faced on the twentieth-century battlefield.

7

The Persistence of Mass

The citizen armies of the twentieth century developed many of the techniques which are still used today and which are indeed central to professional forces. Moreover, while there were clearly pockets of tactical excellence among the infantry during both world wars, Korea, and Vietnam, it is important to recognize that the infantry was still a mass formation. Accordingly, while certain units, especially elite ones, carefully selected, trained, and rehearsed, could solve the Marshall problem through genuine tactical cohesiveness for the most part, the infantry continued to rely on mass and morale in the twentieth century and generally performed poorly. It was capable only of the simplest collective tactical actions; the bayonet charge typified this approach and mass exhortations about masculine or ethnic pride featured heavily in preparing troops for these kinds of action. The job of commanders was to inspire the enthusiasm of their troops or to force it upon them by fear of shame and punishment. Alternatively, the individual firer was relied upon to provide the impetus which was lacking among fearful and poorly trained troops. For the most part, the citizen infantry was poor and did not follow the preparations outlined by elite forces nor were they able to engage in complex fire and movement drills once in combat. They were badly trained mass formations which displayed fragile cohesiveness and low levels of participation and performance on the battlefield from the First World War to Vietnam.

Clearly, to question the performance of the citizen soldier and to argue that they generally performed poorly is highly contentious. National and personal pride especially in the Great Generation which fought the Second World War naturally bristles at the suggestion that citizens who were willing to fight and, in many cases, die for their country were less than competent soldiers. Such an interpretation seems to defile their memory and undermine the nobility of their sacrifice. Indeed, even when national pride is not a factor, the combat performance of the citizen soldier is presumed often on the basis of a cursory reference to the historical record. Perhaps because the four conflicts discussed here and especially the world wars were such major events in which millions died, the effectiveness of the armed forces at every level—including the infantry platoon—is presumed. It is certainly true that the modern armies which fought these wars generated a huge amount of firepower and killed and maimed many hundreds of thousands of soldiers and civilians in the process. Their organizational potency is not in doubt. However, the specific question posed here addresses not the general ability of these armies to generate massive combat power but the effectiveness of their smallest unit—the infantry platoon—to fight and, specifically, to attack. There is no contradiction in suggesting that the industrial army was capable of prodigious destructive power and at the same time claiming that its platoons were relatively weak. Indeed, it has been widely acknowledged that the most potent offensive element of a modern army has never been its infantry but its artillery. From 1914 on, most casualties have been caused by the guns of the artillery.

Nevertheless, even with these qualifications, in order to sustain the claim that the citizen infantry typically demonstrated low levels of cohesion and performed poorly against this natural resistance, it is necessary to examine the historical record in greater depth. Clearly, the case has already begun to be made with the discussions of Marshall, comradeship, and mass tactics, but it is necessary to establish the case more systematically. It needs to be shown that even those armies presumed to be successful also displayed deficiencies at the level of the infantry platoon. If it can be shown by other scholars that some citizen armies were effective, then the general claim advanced here about the mass army in the twentieth century is severely damaged if not disproven outright. It must be shown, therefore, that low levels of performance were endemic to citizen infantries generally, and were due not to putative weaknesses in national character or indigenous military culture or specific commanders but to structural factors which were organic to the mass army itself. It is necessary to prove, in short, that Marshall was right not just about US riflemen in the Second World War but citizen riflemen in general. Clearly, in order to do this, it is not only necessary to demolish cherished beliefs about the Great Generation of the Allies but also to question the concomitant myth that they fought against a particularly effective force in the form of the Wehrmacht and SS, which was fundamentally different from them. It is, in short, necessary to dissect the performance of all six armies which are part of this study over the period 1914 to 1973 (the end of the Vietnam War). Although the distinctiveness of the Wehrmacht will eventually be questioned, the examination will discuss its performance later as the hardest case in terms of mass twentieth-century armies. It is easier to begin with less problematic and less difficult examples in order to confirm the general applicability of Marshall's observation before extending it to all mass armies.

The French army claimed that it had discovered the secret to the attack in trench warfare after 9 May 1915. Nevertheless, although it had doctrinally established plausible procedures for the difficult conditions of the modern battlefield and, as Montdidier showed, it was sometimes capable of implementing them, the French infantry engaged in many poor attacks.

On 27 September 1915, for instance, the French 254th Brigade deployed a deep column of assault, with waves 300 metres apart. The 19th Battalion formed one dense skirmish line; 600 bayonets assaulted a front of 800 metres. 'Enthusiasm ran high for this was the third day of the great French offensive... an offensive that would bring Germany to its knees': 'So at the appointed hour this brigade of 6000 high-hearted and determined men stood up and at the word of command fixed their bayonets... and marched forward in quick time and in step to an entrenched enemy with machine guns.'[1] When the artillery lifted, however, the assault waves were subjected to the fire of the enemy's artillery, machine guns, and rifles; 'the leading wave went down... the others were blown apart.'[2] The 1917 mutinies were the direct result of the badly executed assaults by General Nivelle on 17 April on the Chemin des Dames which once again exposed the French infantry to unsuppressed enemy fire.

French infantry tactics showed little improvement in the Second World War. The French humiliation in 1940 has been widely taken as evidence of the inappropriateness of the concept of the methodical battle. In fact, as the Allies showed later in the war, the concept of a methodical battle in which firepower predominated was not irrelevant even in the era of armoured warfare. At the operational level, the central error was the plan for the bulk of the French army to advance into Belgium on the pre-war presumption that Belgium would fight, leaving France's centre exposed to a totally unexpected thrust through the Ardennes.[3] That mistake was so serious that the French were never subsequently in a position to apply the principles of methodical battle because they had allowed the Germans freedom of manoeuvre.

Various scholars have suggested that the French collapse in 1940 was due to their adherence to outmoded First World War doctrine. Their dependence on defence and the methodical battle do seem to have infected commanders and troops alike. However, probably more important than this doctrinal underinvestment or the poor handling of the battles of 1940 was the extremely poor infantry training evident throughout the inter-war period. The period of conscription was reduced by 1927–8 laws so that soldiers served for eighteen months; they were consequently weak in the fundamental skills of soldiering. National service was subsequently extended in the late 1930s but, by then, it was too late to affect the performance of the French army. Moreover, as Doughty records, there was systematic disorganization in the system of conscript training so that training units were constituted solely for the purpose of reserve duty. The soldiers in them never eventually fought together.[4] They were sent on to different formations where further problems were encountered. Kiesling notes that the programme of biannual incorporation in which new conscripts were inserted into existing formations ensured that levels of soldiering never improved. More experienced soldiers had to repeat basic training recurrently for the benefit of the new conscripts.[5] Recognizing the lack of instruction which many of their soldiers had received, the Second Army on the Meuse attempted to improve training standards by setting up training schools. However, the requirement to keep four divisions in the line meant that it was almost impossible to put the divisions through a training programme.[6] As a result, two regiments were rotated through training at a time, which effected some improvement, but overwhelmingly the French infantry were extremely poorly prepared for combat.

This poverty was demonstrated by 55th Division at Sedan on 6 May 1940. Lieutenant Drapier of 9th Company 3/147th Infantry noted that most of his soldiers' weapons were dirty and that they were unfamiliar with their equipment.[7] This was a widespread phenomenon; 18 per cent of riflemen in the average regiment had never fired a rifle.[8] Drapier highlighted the incorporation of new, poorly trained company teams into the existing formation as central to the collapse of his unit. 331st Infantry Regiment has subsequently been held out as one of the prime examples of French military failure; the soldiers panicked and fled.[9] Indeed, Doughty describes their performance as 'cowardly'.[10] General Lafontaine, commander of 55th Division, explained the poor performance of his formation; 'Without combat experience, they were surprised by the violence of fire and by the use of new combat procedures.'[11] French infantry had been woefully trained.

It is easy to dismiss the French capitulation at Sedan as an aberration but it may be more useful to see it as an extreme example of potential frailties of all mass conscript armies. However developed infantry doctrine was in theory, the requirement for pure mass to combat the huge armies of enemy states placed a severe limitation on how proficient the infantry could become. Ultimately, it was impossible for the mass infantry to be adequately trained and under-performance was all but inevitable for most combat formations, as Marshall suggested. The French catastrophe at Sedan usefully stands as an example of how poorly an infantry could perform in this era; it is not an exception but an extreme case of a trend.

Griffith has stressed the potential of the British infantry. Certainly, they were often robust and could perform well. However, both in the First and Second World Wars, the infantry often demonstrated low levels of cohesion in combat. Indeed, the infantry, as many commentators noted, depended primarily on artillery. In the First World War, British troops conducted some successful attacks through platoon fire and movement, but often companies and battalions simply advanced under the protection of the creeping barrage. Their success lay almost exclusively in staying as close to the creeping barrage as possible. German observers, Jünger and Balck among them, noted (often with envy) the superiority in matériel which allowed the British to advance in this way. Novels and memoirs of the First World War identify, if accidentally, how this dependence upon the artillery developed. British infantry training throughout the war was very poor. Even a committed soldier in an elite regiment like John Dunn noted the poverty of training; still in May 1917, he describes that training was 'most elementary' due to a lack of specialist officers.[12] Indeed, even when he describes a training programme quite neutrally, its content is instructive. For 14–16 December 1917, the elite Welch Fusiliers organized their training into four elements. On Friday morning, the companies conducted close order drill, extended order drill, bayonet training, and saluting drill, each lasting forty-five minutes. On Saturday, the programme was much the same, although extended order drill was replaced by practising artillery formations (i.e. dispersing).[13] It is true that later, under a new commanding officer, better training was introduced, but the standard of training among the British infantry was generally poor.

Indeed, even Griffith notes that British infantry could be very badly organized. He cites the example of 1/5th Warwickshires who assaulted Ovillers 'shoulder to shoulder' on 16 July 1918 and were simply lucky to advance against a position where no enemy machine-gun was firing.[14] Furthermore, despite Maxse's advocacy of small-group tactics, Lieutenant Colonel Frank Maxwell, who won a Victoria Cross in the Boer War, commanding two battalions, had to lead the assault line against the Trones Wood on 14 July 1916:

I formed a line with fragments of the Northamptons and two companies of my own... After infinite difficulty, I got it shaped in the right direction and then began the advance, very, very slowly. Men nearly all much shaken by the clamour and din of shell-fire, and nervy and jumpy about advancing in such a tangle of debris and trenches etc... I immediately found that without my being there the whole thing would collapse in a few minutes. Sounds vain, perhaps, but there is nothing of vanity about it really. So off I went with the line, leading it, pulling it on, keeping its direction, keeping it from its hopeless (and humanly natural) desire to get into a single file behind me, instead of a long line on either side. So I made them advance with fixed bayonet, and ordered them, by way of encouraging themselves, to fire ahead of them into the tangle all the way.[15]

The use of walking fire to suppress the enemy and instil confidence was a perceptive move by Maxwell, but the manner of the attack, in which he, as an acting regimental commander, personally led the assault, does not quite match Griffith's own claims.

Other evidence also suggests that Griffith's argument about innovation is to be treated with care. John Dunn, the Royal Welch Fusiliers' medical officer, frequently records old-fashioned extended line tactics. On 22 April 1917, Dunn observed the 'swathes' of Royal Fusilier and Middlesex dead, 'lying where the machine-guns swept the brow of the rise—all shot in the head or chest'.[16] James Edmonds, the official army historian, drew similar conclusions. In his 1916 retrospect, he noted the 'perfunctory battle training of the troops was based upon tactical principles sound for the most part, but lacking in some essential details and in proper anticipation of the difficulties with which the infantry would have to contend'.[17] He continued:

The new British infantry, unlike that of the old Regular Army, had not been taught to combine fire and movement to the best advantage; it had begun to rely too little on the rifle and too much upon the bomb, it was not well practised in the use of ground; and whilst inclined to be unduly sensitive about its own flanks, did not sufficiently appreciate the need of helping adjacent formations.[18]

The battalions had insufficient knowledge of tactics and, consequently, 'an admirable anxiety to do the right thing too often founded in ignorance of what was the right thing to do'.[19] 'The best formation, as was proved later in the War, consisted of small groups each trained to use ground and covering fire to the best advantage, and to work on its own initiative whilst affording support and assistance to other groups.'[20] In fact, in the 1917 volume of the official history, Edmonds more or less repeats the point; a lack of training and experience resulting in costly lineal tactics was still evident.[21] He cited the report of Major General Pereira, commander of 2nd Division: 'Our troops are excellently trained to advance under a barrage and in this respect very much superior to the Germans; they suffer in comparison when called upon to carry out a manoeuvre in semi-open warfare.'[22] Military historians have concurred with this judgement. In assessing the military effectiveness of the British army on the Western Front, Paul Kennedy observed that 'the British were not particularly effective at the "sharp end" of battle fighting' and that the British tactical record on the Western Front was 'not a good one'.[23] Overwhelmingly, as many commentators have stressed, the British—and French—infantry relied on artillery support even in the final phases of the assault. Where the Germans were constrained by matériel and possessed a forward-thinking officer corps, the British (and French) looked to technical and quantitative solutions; they demanded more and more artillery and (eventually) tanks. In this way, fundamental tactical innovations were not required.

Especially as a result of the work of Maxse and Liddell Hart on infantry doctrine after the First World War, it might be assumed that in the Second World War, British infantry would perform well. Yet, despite the efforts of those like Alexander or Wigram, the evidence points to a repetition of the First World War. The British infantry remained dependent on artillery. As Place demonstrates,[24] the poverty of the British infantry could not be put down to a complete neglect of tactics as some critics have claimed. British drills were developed and the battle schools trained some officers who returned to units. There was a problem with training and exercises in particular which did not support the battle drills and schools nor disseminate and inculcate their procedures. Ultimately, Timothy Place puts down the poor performance of the British infantry to a lack of appropriate close-quarters battle training and an overdependence on artillery due to a false doctrinal dichotomy between minor tactics and organized fire support.[25] For Place, the British wrongly thought of artillery and infantry as alternatives, not as complementary. Yet, however good the infantry–artillery coordination, at a certain point in the battle, the infantry would need to operate independently to take enemy positions through their own organic fire.[26] Unfortunately, lacking infantry skills, artillery was 'the standard recourse of infantry commanders whatever the scale of enemy resistance... Emphasis on the need for infantry to keep close to the falling shells was a constant refrain.'[27] David French has emphasized the point. Throughout his work on Churchill's army, he reaffirms the poverty of British infantry training and the subsequent dependence of the infantry on artillery support. There was a woeful lack of attention to the last 300 yards of the assault in which the infantry necessarily has to operate independently. There were examples of excellence in the British army in the Second World War, the commandos at the forefront, but the general image of the British infantry, stolid in defence, determined and robust, but lacking flair and initiative, was often accurate. British infantry remained overwhelmingly dependent on artillery, as Marshall predicted in a conscript army.

In his assessment of the performance of British infantry in Normandy, Carlo D'Este is condemnatory.[28] Inadequately trained for the bocage and platoon tactics, in general, they were sluggish and reluctant, relying on their officers to lead them and on their artillery to defeat the Germans. The British army reports on operations in Normandy similarly record the poor quality of training the troops had received and expressed concerns about the uniformity of drill.[29] In addition to the problems with the 51st Highland Division and especially the infantry brigade of 7th Armoured Division, 6 Duke of Wellington Regiment had to be removed from the line after a new Commanding Officer appointed on 26 June wrote a long report condemning the battalion. Montgomery was incensed, sacked the Commanding Officer, and disbanded the regiment.[30]

Other Allied forces demonstrated a similar pattern in both conflicts. The Canadian Expeditionary force was first deployed on the Western Front in 1915 but, as Bill Rawling noted, 'it was hardly a well-prepared formation', having received only a few weeks' training consisting primarily of bayonet fighting and mass assaults.[31] The infantry was not assisted by the poor performance of the Ross Rifle with which they were armed, but their inadequate preparation became obvious in their first major engagement at the 2nd Battle of Ypres in April 1915, where they suffered 6,000 casualties; the Canadians were 'not yet ready for the kind of war which had been thrust upon them'.[32] In particular, their limited tactical training and simple linear tactics were inappropriate against an enemy armed with rifles and machine-guns.[33] As a result of their experiences of raiding and their adoption of the new British infantry doctrine, the Canadian Expeditionary Force became a more competent force. Yet, it is important not to overstate the extent of these changes.

Vimy Ridge, regarded as one of the most successful assaults of the war, is a good example of reality of combat performance even in the later part of the war after the apparent institution of platoon tactics. The four Canadian divisions which assaulted Vimy Ridge on 9 April 1917 utilized platoon tactics and they had been trained to a much higher level than their predecessors in 1915 and 1916, especially in the possibility of fire and movement. These tactics were used against machine-gun nests on Vimy Ridge so that when 14th Battalion, advancing as part of the 3rd Brigade, came under fire from four machine-guns located in the Eisener Krenz Weg trench, Lewis gunners advanced, firing from the hip to suppress the enemy, while two of the positions were destroyed by hand-grenades.[34] Nevertheless, the assault was, according to Humphries, predominantly lineal; it was a mass advance with troops organized into platoon and section lines. Andrew Godefroy's analysis of the disastrous assault by 4th Division on Hill 145 and the Pimple across admittedly very difficult ground provides evidence of this. Within minutes of its assault, two of its brigades had disintegrated. Particularly notable here was the poor performance of 87th Battalion which was stopped by as few as two machine-guns from 5th Company, 261st Infantry Regiment, which annihilated the battalion's entire right flank.[35] The 4th Division's war diary recorded: 'The two lines of 87th were stopped dead, losing 60 per cent of their strength from machine-gun fire.'[36] The ease with which these casualties were inflicted by German machine-gunners seems to prove that, for all its advances, the Canadian army of the First World War was still a mass force; under-trained and inexperienced, it necessarily resorted to simple lineal tactics which made it potentially very vulnerable to fire from close range.

Shane Schreiber has argued that the Canadian Expeditionary Force became the 'shock army of the British Empire' used during the Hundred Days in 1918 to crack vital points of the German defence[37] and, by the end of the war, the Canadian army seems to have been a very effective force. However, in the light of Humphries and Godefroy's analysis, some caution is required. At the company and platoon level, the Canadian divisions were as good as any other Allied infantry and probably better than most. However, the use of the Canadian Corps against strong points may not have been evidence of obvious tactical superiority but of their unity and cohesion at the divisional level. While British divisions rotated battalions in and out of divisional battle orders, General Currie insisted that the Canada Corps fought as an entity and its divisions remained intact.[38] This stability is likely to have given the Canadian divisions and the corps itself an advantage in terms of the efficiency of its fire support, logistics, and command and control. It seems probable that it was these higher-level organizational advantages which recommended it to its British commanders during the Hundred Days. It was a ready-formed and available formation of a size capable of executing large-scale assaults. The formational advantages of the Canadian Corps in no way deny the improvements to tactics and training evident in the Canadian army from 1915 to 1918, and some historians attribute their success to their being 'unhindered by such prejudices' and tradition in contrast to the British army,[39] but it should warn against presuming that the mass citizen army which Canada fielded in that conflict constituted a professional force or could imitate the performance expected of a professional army.

The First World War and Vimy Ridge in particular has become an important founding myth for the Canadian nation and, while collective national memories are typically at least partially invented, there is some evidence that the pride which Canadians have attached to their army's performance in the Great War is not entirely misplaced. Even the most avid Canadian nationalists might be dismayed by the performance of the Canadian army in the Second World War. Indeed, the official historian of the Canadian army, C. P. Stacey, questioned its performance in Normandy, endorsing the views of Major General Charles Foulkes, General Officer Commanding 2 Division:

> When we went into battle at Falaise and Caen we found that when we bumped into battle-experienced German troops we were no match for them. We would not have been successful had it not been for our air and artillery support. We had had four years of real hard going and it took about two months to get that Division so shaken down that we were really a machine that could fight.[40]

In his assessment of the reasons for their disappointing performance and following Foulkes, Stacey emphasized the lack of battle experience as a major issue: 'The lack of battle experience undoubtedly had its due effect within the Canadian formations. They did well, but they would certainly have done better had they not been learning the business as they fought.'[41] He noted: 'it is true that all had undergone exceptionally long and careful training; but no training is entirely a substitute for experience of battle.'[42] Yet, in highlighting the inexperience of the Canadian divisions, Stacey himself had to admit that many of the German formations which they faced and which caused them such problems had not been in battle either. The 12th SS Panzer Division, drafted from the Hitler Youth, was formed only in 1943 and had never fought a battle before 7 June 1944. Yet, it was a formidable force in Normandy. Similarly, other German divisions had not fought before.[43] Some of the experienced British divisions, as already noted, fought woefully in Normandy. In the end, it was the Canadians' lack of appropriate training rather than experience which seems to have been fundamental: 'We had probably not got as much out of our long training as we might have.'[44]

Terry Copp has sought to rehabilitate the Canadian army, arguing that the German forces enjoyed the benefit of defending prepared ground which the Canadians had to take and, less plausibly, that Allied numerical supremacy especially in the air 'meant little at the tactical level' where the 'enemy had to be drawn into combat, not simply struck by high explosives'.[45] Copp suggests:

'Perhaps it is time to recognize the extraordinary achievements that marked the progress of the Canadians across Normandy's fields of fire.'[46] It is certainly true that the Canadian divisions, on the eastern flank of the bridgehead along with the British, confronted the highest concentration of German Panzer forces around Caen, and it is also now widely accepted that a number of Canadian commanders at brigade, divisional, and corps level were inadequate. Much of the sluggishness of Canada's operations around Caen, including the poorly handled Operations Spring, Totalise, and Tractable, can be attributed to them; neither Major General Foulkes nor Keller, respective commanders of the 2nd and 3rd Divisions, were considered competent generals by their superiors.[47] Copp himself is critical of brigade, divisional, and corps commanders' handling of certain parts of the campaign;[48] 'it does not seem possible to argue that any of the three Canadian divisional commanders passed the test of battle.'[49] It was notable that Lieutenant General Crerar, 1 Corps Commander, had no experience of major operations at all.[50]

However, even with these caveats the performance of the Canadian infantry in Normandy has not been regarded as impressive. Indeed, Copp is himself somewhat disingenuous about some of the evidence. Thus, in an attempt to refute the criticisms of Max Hastings and Carlo D'Este, Copp noted that, during Operation Epsom, when 6th Duke of Wellington Regiment of the 49th Division collapsed, 'the performance of 3rd Canadian Division on 8 July did not draw any criticism from Crocker or Dempsey'.[51] Perhaps not on that day. However, as Copp himself records, on 5 July 1944, Lieutenant General John Crocker, Commander 1 Corps, wrote a formal letter to Dempsey, his army commander, documenting his dissatisfaction with Major General Keller and the Canadians under him. This letter was forwarded to Montgomery who endorsed its assessment.[52] After the success of D-Day, 3rd Canadian Division had become 'jumpy and excitable'.[53] Crocker also claimed that the Canadians had failed to press their advance into Caen on 9 July.[54] Against Copp's claim, British commanders were critical of the Canadian performance. Copp attempts to refute Crocker's complaints in early July by arguing that the Canadians were disadvantaged by Crocker's 'penchant for isolated battalion or brigade attacks on strongly held enemy position'.[55] It is very difficult to know whether the Canadians would have performed better had Crocker used them differently. Yet, it would be difficult to dismiss British commanders' criticisms as self-serving. The British acknowledged the Canadians' efforts on D-Day very clearly and Montgomery was even more critical of some of his own divisions, whose commanders he removed immediately. Although it is understandable why Copp would want to salvage the reputation of the Canadian army in Normandy, the evidence suggests that there were significant shortfalls in its performance. Certainly, this is the position of most other military historians.

The first Canadian troops arrived in Britain in 1940, fortunately missing the Battle of France and the evacuation from Dunkirk. Apart from the disastrous raid at Dieppe which all but destroyed the 2nd Division so that it had to be completely rebuilt, the Canadian 2nd and 3rd Divisions spent four years training in England before D-Day, while the 1st Division fought in Africa and Italy. Nevertheless, despite their enthusiasm for battle drills, there were serious shortcomings in the training of the Canadian division. In February 1942, Montgomery remarked on the 'backwardness' of 3rd Division. In fact, Montgomery saw the principal weakness not at the level of the soldier or small unit but at the higher level because they were commanded by officers who did not understand the art of war and could not handle their formations, and he sacked a number of officers, including the corps commander, General McNaughton, in December 1943 for precisely this reason.[56] However, even at the small unit level, Montgomery and others seem to have had reservations about the quality of Canadian training. Montgomery noted that most Canadian commanders were unable to conduct proper troop training;[57] he regarded seven out of twenty-five commanding officers as unacceptable. Certainly, John English claims that the training schedules of Canadian units in the approach to D-Day 'put to rest the myth that Canadian soldiers overtrained for four years in Britain'.[58] Despite the enthusiasm for battle drill among some battalions, the Canadian army landing in Normandy seem to demonstrate the typical weaknesses of the mass army. It was inadequately trained to perform complex tactics at the platoon level. Lacking the battle experience which their predecessors had earned in the First World War, Canadian troops were hesitant due primarily to 'unimaginative training' and consequently had to learn 'in action what their training in Canada and Britain had not taught them'.[59]

Consequently, there were substantial problems with line infantry units. Indeed some engagements were actively disastrous, showing a complete failure of the basic infantry tactics. On 20 July as part of Operation Atlantic, the South Saskatchewan Regiment of General Keller's 3rd Division was ordered to gain the crest of the Verrières Ridge to the south-east of Caen. The Saskatchewan had to advance up an open slope with inadequate support and were subject to intense German fire from concealed positions on the ridge. In the face of an enemy counter-attack, they fled, streaming back towards the Essex Scottish Regiment lines, behind whom they were ordered to withdraw. At this point, two companies of the Essex Scottish Regiment began to withdraw, at least one of them out of control. There was, in effect, a panic.[60] Indeed, Copp himself describes the period as a 'debacle'.[61]

However, possibly the most notorious infantry action during the Normandy campaign was the attack of Montreal's Black Watch, 5th Brigade, 2nd Infantry Division, on Verrières Ridge east of May-sur-Orne on 25 July 1944. This assault, part of Operation Spring, demonstrated blatant problems at the level of command and control. The entire operation was over-ambitious and not well conceived. As a result neither the Black Watch's forming-up point at Saint-Martin-de-Fontenay (to the north) nor their start line at May-sur-Orne was secured by the Calgary Highlanders even though they claimed the area was clear; indeed, the acting commanding officer of that regiment was under the misapprehension that he was in May-sur-Orne when he was in fact in Saint-Martin.[62] Two regiments of the German 272nd Division were secreted in the environs of Saint-André and Saint-Martin, exploiting the cover of mine-workings or well-concealed positions to the south-eastern fringes of Saint-Martin.[63] Consequently, as soon as the Black Watch began their advance, at 3.30 a.m. on 19 July, they came under heavy fire while approaching Saint-Martin in which their commanding officer was mortally wounded and senior company commander injured.[64] Major Phillip Griffin took command and advanced to his start line at May-sur-Orne on a compass bearing. Griffin realized that the attack would have to be delayed and successfully re-scheduled the artillery barrage. He then consulted with this brigade commander, Brigadier Megill, outlining his attack plan. Essentially the attack would proceed as planned, with the Black Watch advancing up the slope of the Verrières Ridge across open wheat fields. Megill thought the plan a 'dicey proposition', because the enemy still held May-sur-Orne, the right flank of the start line, and could potentially enfilade the attack; but Griffin convinced him that Calgaries could infiltrate into May behind the Black Watch to neutralize the enemy there.[65] Megill eventually agreed and the attack went ahead at 9.10.[66]

Perhaps predictably, the 300 remaining men of the Black Watch, advancing in company lines, two up, one back, were subjected to intense enemy fire from the moment they approached their start line and throughout the advance. Only 60 reached the crest of the ridge, including Griffin, only to be ambushed by concealed German tanks and machine-guns. Only fifteen men returned, Griffin himself having been killed on the ridge with many others.[67] The Black Watch suffered 307 casualties in the assault, the worst single loss of any Canadian battalion in the Second World War except for Dieppe. Poor command especially at the divisional and brigade level has been identified as central to this catastrophe but the Black Watch's death march up the Verrières Ridge illustrates—in its extremity—features which are typical of the mass army. Even though they were aware of the inadequacy of their artillery support, the Black Watch attacked a concealed, entrenched enemy by means of lineal assault. The 300 men of the battalion organized themselves into long company lines and, making little attempt to use fire and movement or to infiltrate, simply advanced up the ridge into unsuppressed enemy fire as many of Kitchener's battalions had done on the Somme. It was a testament to their courage and morale that even sixty reached the crest, and provides some support to Montgomery's claim that Canadian soldiers were 'probably the best material in any armies of the Empire'.[68] However, the Black Watch's attack formation was perhaps surprising. Although he may have wanted to deflect attention from his own performance as II Corps Commander, Lieutenant General Guy Simons noted the tactical deficiencies of the Black Watch: 'The losses were unnecessarily heavy and the results achieved disappointing. Such heavy losses were not inherent in the plan nor in its intended execution. The action of the Black Watch was most gallant but was tactically unsound in its detailed execution.'[69] Nevertheless, Simonds seemed to be well aware of the limitations of the Canadian infantry. In his 1 July 1944 guidance, he stipulated that the infantry should lean on the barrage. Revealingly, he stipulated that the attacking troops should not stop or even open fire until the objective was reached because if troops adopted fire positions it was 'extremely difficult to get them on the move again'.[70] The assault on Verrières Ridge was a disaster, but the general level of tactical competence in Canadian infantry regiments on the admission of one of Canada's own generals was poor. Given their level of training, it could not be hoped that they would execute fire and movement tactics against an enemy as determined and well equipped as the Wehrmacht and SS divisions which confronted them.

The poor performance of the Italian army and its reliance on mass tactics in the First World War has already been discussed. It is important to recognize that the Italian army did innovate in the Great War, in a manner closely compatible with its allies and enemies. While it is true that General Cadorna continued to rely on wave attacks supported by massive artillery bombardments,[71] the need for specialist assault troops eventually became clear: 'The first step towards a new technique of assault came in 1917 with the creation of small units of picked men—the *Arditi*', although in fact their introduction did little to change the Italian reliance on clumsy assault tactics.[72] The Arditi consisted of small units of selected and highly trained soldiers who specialized in assault techniques in a manner closely consistent with German stormtroopers. They were armed with a variety of weapons including clubs and maces for close combat and relied on a formidable aggression and fearlessness.[73] During the Battle of Vittorio Veneto in 1918, the Italian counter-attack after Caporetto, the Arditi featured prominently, taking a critical bridge. The Arditi originally consisted of small units, then separate battalions, until they were finally formed into divisions.[74] The Arditi were capable of small-group tactics. The Alpini regiments, raised locally from mountain villages and fighting in the Dolomites, have also been identified as highly competent.

John Sadkovich has sought to demolish the myth of Italian military incompetence (and the genius of the Germans and especially Rommel), plausibly pointing to the excellent performance of the mechanized Ariete Division in 1941–3.[75] The airborne Folgore Division has also been highlighted by military historians for praise.[76] In the western desert, Italy's infantry divisions were always hugely disadvantaged by their lack of mobility. In this war of movement, footborne Italian infantry could fulfil only a static defensive role. As a result, they were at great risk of being cut off and encircled by the mechanized British forces; they were unable to keep up with withdrawing forces and were left behind in Tripolitania.[77] In this situation, separated from their own forces for fire support and supplies, resistance was more or less useless. The mass capitulation of Italian infantry, especially during Wavell's drive west in late 1940 and early 1941, might be seen not as a deficiency in the competence and morale of Italian infantry, still less a defect in national character (despite Erwin Rommel's recurrent accusations), but a reflection of the nation's poor industrial base, its inadequate military equipment, and poor command decisions which exposed infantry to the dangers of encirclement.[78] In a similar predicament, citizen soldiers of all other armies would also have surrendered, as they often did.

Nevertheless, despite these caveats, it would be peculiar to claim that the Italian infantry displayed high levels of cohesion in either the First or Second World War, excepting specialist troops like the Arditi, the Alpini, and the Folgore Division. They did not. Their cohesiveness in combat demonstrated the typical fragility which attended citizen armies throughout this period and was probably worse than most other combatants. Again, if the Italian infantry did perform as poorly or worse than their allies and enemies in both conflicts, this was not due to national character. Rather, as most military historians have emphasized, the poor performance of the Italian infantry was due almost entirely to a complete lack of training. The Italian armies of the First World War were possibly the worst trained and equipped of any except the Russians in that conflict.[79] Compounding the problem of a lack of training, officers were drawn from only a small pool of educated upper-class Italians, which comprised some 10 to 15 per cent of the population. Yet, only the dullest of this group entered the army.[80] There was subsequently a scarcity of officers at the front. At the same time, the Italian army lacked a solid body of non-commissioned officers.[81] It seems probable that it was because of this lack of training and an absence of adequate junior commanders that General Cadorna resorted to extreme disciplinary measures.

The performance of the Italian infantry was equally problematic in the Second World War, Sadkovich's apposite remarks about German propaganda notwithstanding. In the inter-war period, troops received 'very poor' training.[82] Italian troops still predominantly practised close order drill and had few areas in which to train. As a result there was little live firing training or instruction in modern tactics.[83] Tactical exercises consisted of artificially choreographed scripts, and the non-commissioned officers remained poorly trained and insufficient in number. In 1940, 93 per cent of army officers were reservists or recently promoted non-commissioned officers.[84] The invasion of Greece and the assault on France in 1940 were disastrous, and even a sympathetic analyst such as Macgregor Knox describes this first campaign in Africa as a 'debacle';[85] General Graziani's attack on Egypt had failed entirely and he had lost Cyrenaica in two months, with the British capturing 115,000 prisoners of war, 1,290 guns, and 140 tanks.[86] As Knox has noted, formations like the Ariete mechanized division became highly proficient after 1940, but infantry tactics remained inadequate. The result was a number of notorious collapses even after 1940, including the capitulation of the 'Wolves of Tuscany' and the Bari divisions.[87] Moreover, even after the Germans reinforced North Africa, the Italian infantry often compared unfavourably with the Afrika Korps, notwithstanding the near racism of German commanders towards the Italians. Throughout El Alamein, Italian infantry performed better when stiffened with German infantry,[88] yet a differential remained. When the Australians and Highlanders crossed Boxes J and L, they encountered a mixed force of Italian 62nd Infantry from the Trento Division which had abandoned its unconsolidated positions during the preliminary bombardment; the German 382nd Grenadier Regiment stayed to fight.[89] While the Allies overran all but one company of the Italian regiments, the 1st Battalion of the German regiment held out until dawn. On 25 October, the Highlanders advanced from Box L on 382 Grenadier Regiment's exhausted 2nd Battalion which was almost wiped out; the 3rd Battalion from the Italian 61st Regiment surrendered.[90] In the First and Second World Wars, the Italian army seemed to represent the problems which characterized a mass citizen army at its most extreme. Inadequately trained (and, especially in the Second World War, lacking a sense of national mission), the Italian infantry was able to generate only very low levels of cohesion in combat and was generally capable of only the simplest mass tactics; when mass tactics, enforced by the harshest sanctions, were not available as they were not in the Second World War, units typically performed weakly and broke quickly. In this way, the Italian infantry was not the aberration which near racist stereotypes promulgated by their German allies and British enemies suggest. They demonstrated the typical characteristics of a conscript army operating in more difficult economic, political, and operational circumstances than others.

Despite pockets of excellence, when American troops demonstrated high levels of cohesion, the US infantry was similarly compromised. The United States entered the First World War very late and, therefore, had to learn lessons from their allies. Moreover, as a result of Woodrow Wilson's determination to maximize his influence over any post-war settlement as well as the ambitions of the appointed commander, General Pershing, the United States military rushed precipitately into the creation of an American Expeditionary Force as a completely independent army, when French and British allies recommended that US divisions be incorporated into existing Allied corps.[91] The USA lacked the command experience to coordinate large-scale independent manoeuvres with adequate artillery and logistical support and, therefore, necessarily inflicted higher casualty rates on their troops. This unpreparedness was compounded by General Pershing's stubborn concept of how the Americans should fight and his unfounded belief in their natural superiority.[92] His training methods differed markedly from his allies, prioritizing aimed rifle fire and stressing open warfare; 'It was my opinion that victory could not be won by the costly process of attrition, but it must be won by driving the enemy out into the open and engaging him in a war of movement.'[93] On Pershing's concept of war, forms of infantry tactics which had been proven to be inadequate on the Western Front were regarded as optimal: 'My view was that the rifle and bayonet still remained the essential weapons of the infantry.'[94] Acquiescing to Pershing's demands for independence, the AEF was assigned quiet sectors and initially they played a negligible role in the campaign. However, from May 1918, they were drawn into some significant battles including an assault on Cartigny and the containment of a German advance at Chemin des Dames.[95] But as they become involved in larger battles, the tactical shortcomings of the US soldiers and their inadequate training became ever more evident. After the 4th Marine Division's assault on Belleau Wood, their commander, Major-General Lejeune, observed that 'the reckless courage of the foot soldier with his rifle and bayonet could not overcome machine-guns well protected in rocky nests'.[96] Their German opponents made similar assessments of the AEF. German officers noted that, although 'the American soldier is courageous, strong and clever', 'the manner in which large units attack is not up-to-date and leadership is poor' and that 'they were unskilfully led, attacked in dense masses, and failed'.[97]

Indeed, Paul Braim has emphasized the AEF infantry's lack of training and competence. One infantryman reported that he had never seen a rifle or pistol till they arrived at his division.[98] Another commented: 'Our training did nothing to equip us to take care of ourselves in combat. Training in the US was unrealistic. Training in France was non-existent.'[99] Braim records the fact that infantry soldiers were incapable of performing basic functions like loading their weapons, never mind complex tactical manoeuvres.[100] The result was that 'the "green" troops moved fast until they hit the first effective enemy fire; then, according to friendly and enemy observers, they tended to "mill about", remaining in the killing zone, taking no action to silence the fire'.[101] According to David Trask, the AEF improved considerably during the course of 1918. As the transposition of British doctrine (cited above) in US army manuals demonstrated, the First Army began to take 'advantage of tactical innovations that had served the British and French armies well during the battles of July–October'.[102] Despite Pershing's advocacy of the mass bayonet charge, they introduced 'special training for infantry "assault teams" to prepare them to engage machine-gun nests successfully while other infantry elements bypassed them'.[103] However, David Trask concludes that despite these developments, the AEF's performance in the First World War was characterized by the mass expenditure of unskilled conscript forces in badly organized and executed attacks, for which the infantry had not been sufficiently trained.

Despite obvious exceptions in the Second World War, these patterns recurred. Poorly trained US conscripts performed crude mass tactics. Indeed, there is very significant evidence which supports Marshall's assessment of lack of training and subsequently poor performance. The debacle of the Kasserine Pass was an early example but the problem of infantry performance persisted throughout the war. In his work on 29 Division, *Beyond a Beachhead*, Joseph Balkosi aims to refute Max Hastings's 'astonishing' claim that 'Few American infantry units arrived in Normandy with a grasp of basic tactics'.[104] However, the actual work shows the difficulties which the 29th Division experienced. Despite the fact that the division was formed from existing National Guard units and subsequently trained for two years for D-Day, their performance in Normandy was mediocre. On D-Day, they landed incorrectly on Dog Red sector, Omaha Beach, where they were subjected to intense German fire. However, despite these problems, it would be difficult to praise their performance. On Dog Red, there was chaos and inactivity. When Brigadier General Cota, deputy commander of the division, came ashore in a supporting wave, he found his division cowering under the sea-wall, making no attempt to return fire or to extract themselves from the situation through the assaults they had planned and practised in Britain. He personally commanded an assault of about company strength with fire and manoeuvre up the bluff behind Hamel-au-Prêtre just to the west of Le Moulins draw.[105] He then led six men down to take the Vierville draw to the west.[106]

The shock of Omaha, for troops who had never experienced combat, could explain their inactivity on D-Day. Yet, even on the following day, when the situation had stabilized, the infantry continued to perform poorly, demonstrating low levels of cohesion in combat. Near Saint-Laurent, Brigadier General Cota found a 'group of infantrymen' led by a captain (probably of platoon strength) which was being held up by fire from a house. They failed to react. Brigadier General Cota had to demonstrate fire and manoeuvre tactics to the group. Cota conducted the assault himself, grenading the house, instructing the young captain: 'You've seen how to take a house. Do you understand? Do you know how to do it now?'[107] This unpreparedness was repeated. The 2nd Battalion 115th Infantry failed to dig in on the evening of 9 June outside Saint-Lo at Le Carrefour. They were ambushed and destroyed; 50 killed, 100 wounded or captured, the rest scattered.[108] The 29th Division was not exceptional. The 90th Infantry Division's conduct in the Normandy campaign was the source of consternation to General Omar Bradley, who sacked two successive commanders who did not seem capable of stabilizing the formation.[109] Up until the end of the campaign, the division consistently failed to achieve its missions, and its very high casualty rates suggested a low level of tactical competence especially since it was not exposed to the intense or prolonged fighting of some divisions.

Although the US Army eventually prevailed in Korea with the assistance of its allies in preserving South Korean sovereignty, there was considerable concern about the performance of its troops in the 8th Army throughout the campaign.[110] Specifically, from the rout of Task Force Smith in June 1950 to the end of the war, there was recurrent concern about the tendency of American troops to 'bug out' (i.e to run away in the face of enemy fire or assault). The 24th Infantry Division, which was stationed in Korea at the start of the war, was the subject of intense criticism. A sergeant in the 34th Regiment, whose performance was so weak that it had to be withdrawn from the line completely after a disaster on 14 July,[111] was 'disgusted that so many of his men hadn't fired', in many cases because their rifles had jammed due to poor handling or maintenance including incorrect assembly; due to poor training and preparation the 24th Division as a whole performed badly.[112] The all-black 24th Infantry Regiment of the 25th Division was particularly prone to 'bugging out' and, while Ridgeway undoubtedly opposed segregation for political reasons, the desegregation of black and white divisions in Korea was also a product of operational necessity. General Kean broke up the 24th Infantry Regiment, reassigning personnel in it to ten other units because 'he could spare neither the time nor the effort to bring marginal units up to the rather low standard of the US Army performance at the time'.[113] Due to poor leadership and lack of motivation (itself substantially due to civilian and military discrimination), the 24th Regiment might have been an extreme case but many white units also recorded a 'dismaying number of "bug outs"'.[114] The infantry troops of the 45th Division had a tendency to go to ground in the attack rather than 'press on to Chinese positions which would provide cover'.[115]

The 45th Division was seen to improve during the Korean War but it 'never reached the levels of performance desired by General Styron' and its performance was 'unfulfilling and confusing compared to its triumphant World War II service'.[116] The inadequate performance of the US Army in Korea was a result of the fact that its citizen soldiers, most of whom had been conscripted into service, were inadequately trained for the situation which confronted them. Indeed, some commentators have argued that only the British and Turkish conscripts were fully trained for this war.[117] The lack of training was compounded by a system of individual replacement which further undermined cohesion; without a sense of mutual obligation to each other and a fear of dishonour by their peers, it was easier for US Army troops to run away. US soldiers in Korea were often heavily dependent on artillery, and when artillery support failed, collapse was very common. The Chinese communist forces were highly critical of US infantry: 'Their infantry men are weak', 'they must have proper terrain and good weather', specializing only in 'day fighting'.

The Chinese further observed that they were completely lost without the use of their mortars and became very afraid when they were cut off from the rear.[118] Of great concern to the US Army was the apparent susceptibility of US soldiers to brainwashing in North Korean and Chinese prisoner-of-war camps. The low morale of US Army prisoners of war was seen as a continuation of the general poverty of their combat performance where a lack of collective competence and determination was often very evident.[119]

Despite the declaration of General William Westmoreland, Commander Military Advisory Command Vietnam (MACV), in a speech to Congress on 28 April 1967 that 'Our soldiers, sailors, airmen, marines and coast guardsmen in Vietnam are the finest ever fielded by our nation',[120] and the evident professionalism of certain formations, such as the 1st Cavalry Division in the first year of the war, Vietnam is rightly remembered for the poor performance and eventual collapse of the US Army. Notwithstanding Marshall's assertions and Glenn's more evidenced appeals to high levels of performance, there is substantial evidence that cohesion among the American infantry in Vietnam was as fragile as it had been in the Second World War at the best of times. Of course, by 1970, cohesion had begun to collapse. Yet, long before 1970, low levels of cohesion were often detectable among the infantry, evidenced by its dependence on supporting fire.

In the early years of Vietnam, infantry soldiers from the 1st Cavalry, especially, were highly trained and performed well, but overwhelmingly the performance of troops in combat demonstrated the same weaknesses which attended troops in the First and Second World Wars. Like their First and Second World War predecessors, the conscript forces in Vietnam demonstrated a dependence upon artillery and air support. The strategy of 'search and destroy' was to use the infantry to bring the enemy forces to battle in order to destroy them by indirect fire from the air and from fire support bases; 'the infantry finds the enemy, the air and artillery kill them', as one general noted. While the attempt to maximize the destructive potential of supporting fire, especially given the US competitive advantage over their opponents in this area, might have been a valid military decision. In the light of the inadequate training and patchy performance of the US infantry, it led to a dependence on this fire and an active fear of moving away from the immediate environs of a fire-support base.[121] At the battle of Ap Bau Bang II, Colonel Alexander Haig, commander of the 1st Battalion 76th Infantry (and future Secretary of State under Ronald Reagan), explicitly identified cluster bomb units (CBU), dropped from the air, as the decisive element in his victory; 'with the arrival of the air, tactical air and especially the ordnance, the CBU ordnance was the main factor.'[122]

The dependence on supporting fire indicates potential shortcomings in infantry training and these gaps were manifest early in the campaign. Lieutenant Colonel Harold Moore also provides some interesting evidence here.

1/7 Cavalry was one of the most highly trained units ever to be deployed into Vietnam but the whole of the 1st Cavalry Division, as an elite unit, underwent some of the best preparation of any formation deployed. Nevertheless, Moore contrasts the level of training 1/7 received with their sister battalion, 2/7; 'The 2nd Battalion, 7th Cavalry was the same mix of draftees, good NCOs, green lieutenants and good company commanders as 1st Battalion, 7th Cavalry. But it did not have the same intense airmobile training that we had gotten in the 11th Air Assault Test.'[123] Moreover, in order to make up the numbers, it was 'filled with people from the 101st, the 82nd Airborne some from Fort Lewis'; 'they were just a bunch of strangers to each other'.[124] In their baptism of fire at LZ [Landing Zone] Albany, only a few days after the battle of LZ X-Ray, the lack of training became evident. The battle at LZ Albany shared many of the same features as Moore's own fight; 2/7 were inserted onto a landing zone in close proximity to NVA forces in order to bring them to battle and to try and destroy them. However, surprised by the intensity of the fighting, 'some of the soldiers in a state of shock, were running and firing at anything... Throughout the action, groups of cavalrymen rushed toward the nearby patches of open grassland, mistaking them for the landing zone and thereby actually running deeper into the ambush. Their terrified troops were either frozen to the ground or crawled on their elbows, pistols and rifles propped up in wildly shaking hands.'[125] Perhaps, significantly, the troops received very little briefing or time for battle preparation before the operation. The commander of the battalion, Colonel McDade, noted ruefully: 'It was right back to 1950 or 1944 Europe. All we got were verbal orders: Go here. Finger on map. And we just marched off like we were in Korea.'[126]

In fact, the training of 2/7 Cavalry at LZ Albany was more typical of the level of performance of US troops in Vietnam than that of 1/7 a few days earlier. Shelby Stanton, in direct contrast to Westmoreland, claimed that the 1st Division in Corps Tactical Zone III to the west of Saigon had experienced 'difficulties' which were 'reminiscent of its first experiences in North Africa in World War II', including the Kasserine Pass in 1943.[127] In 1966, 88 per cent of all fights were initiated by the Vietcong (VC) or North Vietnamese Army (NVA) and the 'green' American soldiers were faring poorly as a result.[128] 1st Infantry Division was 'painfully assimilating jungle warfare experience'.[129] Indeed, when Major General William DePuy assumed command in March 1966, he was forced to relieve so many subordinates that the army Chief of Staff expressed concern. Despite a temporary loss of morale, 1st Division eventually mastered the tactical problems of fighting in Vietnam and, according to Stanton, the division began to live up to the reputation which it had earned in the Second World War.[130] Nevertheless, while 1st Division's performance may have improved, by the end of 1966, there were major concerns about the competence of the US Army. On 16 May 1966, at the Battle of Vinh Thanh, 2nd Battalion 8th Cavalry inserted Company B into LZ Hereford to assault a mountain overlooking their base which was known to be an enemy position. A squad of soldiers was overrun and the company engulfed by enemy forces.[131] On the same operation, Marshall records the decimation of Charley Company which was scattered by enemy fire when caught in an ambush.[132] Marshall's description of the assault of LZ Bird suggests significant non-participation by a number of its occupying troops: 'The American Army ended 1966 campaign on a sombre note and similar incidents continued to plague its performance for the duration of the Vietnam War.'[133]

Westmoreland himself admitted that disciplinary problems began as US troops started to withdraw in 1969,[134] and Moskos also dates the fundamental collapse of the US Army to the final phase of the war in 1969. For Westmoreland, such disciplinary problems were the inevitable result of the US drawdown; nobody wanted to be the last soldier to die on a mission which Washington had now signalled was no longer of critical strategic importance. Westmoreland's excuse is perhaps generous, given the nature and extent of the disciplinary problems to which he referred: mutiny, fragging, drug use, and atrocity. Yet, in fact, while extensive drug use became endemic only later in the campaign, there is clear evidence of the collapse of military discipline and of course tactical competence before 1969. For instance, in his account of the First Wolfhounds, Alfred Bradford records that after Tet in the summer of 1968, the unit collapsed under an inept commander.[135] Following a series of defeats, including the incompetent air insertion of a company on top of an occupied bunker complex, 'platoons of forty were down to ten, sometimes led by PFCs [Private First Class], and the survivors were drinking beer, smoking dope and going crazy'.[136] Soldiers inflicted wounds on themselves, 'refused to obey orders', and in one incident a private tried to stab his company commander. Officers had to 'face down a mutiny every time they went out'.[137] Bradford's unit recovered under the command of Colonel Reece.

Other units and formations did not. The most notorious here is the Americal Division, stationed in Tactical Corps Zone I in the north. The Americal Division had originally been raised during the Second World War on New Caledonia to support the Marine offensive on Guadalcanal, its name denoting it was an American army division raised on that island. The division was reconstituted for Vietnam, with two brigades, the 198th and the 11 Infantry Brigades, which respectively performed police duties in the Dominican Republic and acted as Pacific reserve in Hawaii. They were not fully trained or equipped before deploying to Vietnam. Indeed, Sam Stanton is condemnatory about the division: 'The Americal Division suffered from grave command and control problems, stemming from poor training and a lack of leadership, from the division down to the platoon.'[138] Indeed, 'some elements of its 11th Brigade (Light) were little better than organised bands of thugs'; 'dereliction of duty, ignored regulations and hoodlum activity were more common place than the army had ever imagined'.[139] Even Westmoreland admitted that the division and 11th Infantry Brigade in particular were poor: 'at President Johnson's direction the 11th Infantry Brigade had deployed to Vietnam before completing its training... Although I committed the brigade in a quiet sector so training might continue, just over a month later the troops were caught up in the enemy's Tet offensive.'[140] Not only was 11th Brigade's combat performance poor but one of its sub-units perpetrated the My Lai massacre in March 1968, when 500 civilians were murdered. Illustrating the poverty of the division, the platoon commander at the centre of the atrocity and eventually convicted for it, Lieutenant Calley, was incompetent and unfit for command. Indeed, 'had it not been for educational draft deferments, which prevented the Army from drawing upon the intellectual segment of society for its junior officers, Calley probably never would have been an officer'.[141] The Americal Division with its notorious 11th Infantry Brigade may have been extreme but the problem of leadership was widespread, as Westmoreland suggested. Stanton notes not only that 'by 1969 US soldiers in Vietnam usually represented the poorer and less educated segments of American society' but also that the 'drastic reduction' of graduate officers from the university ROTC reduced the quality of officers across the army, while 'the non-commissioned officer corps had suffered an alarming decline in quality'.[142]

Instructively, in 1965, 1st Cavalry Division were among the best-trained formations in Vietnam but their level of performance declined substantially. During their notorious assault on Hamburger Hill in A Shau Valley in 1969, which was a poorly executed frontal assault, a captured North Vietnamese after battle report stated that American troops in this formation had poor fire discipline, were easily frightened at night, tended to bunch up when brought under fire, and could be best attacked when dismounting from their helicopters.[143] In his history of the division, Stanton noted that the soldiers began to inscribe their helmets with slogans and short calendars, indicating their DEROS, and reflecting this decline of morale, the division suffered a number of near mutinies. In 1970, there were 5 fragging incidents and 72 shootings in six months in the division.

Although they operated in Tactical Corps Zone I, often independently of the US Army, the US Marines demonstrated a similar trajectory. With some exceptions, they performed competently in the early part of the war and indeed introduced a number of significant innovations like the Combined Action Program, County Fair, Golden Fleece, and Stingray operations. During the Battle of Hue, Marine infantry employed fire and movement to secure the city. Yet the same problems of recruit quality and morale eventually undermined the force: 'In a Corps infected with sociopathic recruits and led by too many cynical, demoralised NCOs and apathetic officers, the signs of social disintegration grew.'[144] Millett concluded that the Marines' performance was both 'very good and very bad'. At their best, they demonstrated professional standards of tactical competence and military discipline, but towards the end of the war, fraggings and drug use were common. There were forty-seven fraggings in 1 Marine Division in 1970.[145]

The examples discussed up to this point represent perhaps easy and uncontentious cases where few would dispute that an inadequately prepared citizen infantry struggled to perform in combat. However, in order to sustain the claim about the citizen infantry's performance, it is necessary to examine genuinely hard cases and to examine the combat performance of citizen armies which are widely presumed to be very good. There are two forces which stand out in the twentieth century in this context; the German army and, especially, the Wehrmacht of the Second World War is the obvious example here. Less well known is the performance of the small Australian Task Force in Vietnam, which despite the context of the war and the collapse of the US Army around them demonstrated high levels of cohesion throughout the campaign; but it also constitutes a useful counter-example by which the Marshall thesis can be assessed. As Paddy Griffith has emphasized, the German army of the First World War has been consistently held out as the paragon of tactical acumen. It is certainly true that the stormtroops demonstrated new skills. However, the stormtroops represented an infinitesimally small part of Germany's armies. By February 1917, fifteen assault battalions and two independent assault companies had been formed.[146] One battalion (600 troopers) was attached to each German army (consisting of 250,000 men);[147] 2nd Assault Battalion was assigned to the Third Army, 3rd Jäger Assault Battalion to the Second Army.[148] In the assault, these storm battalions were separated into small storm squads of approximately twelve men which led the attack. In his description of the battalion attack, Wilhem Balck depicts four of these squads (approximately 50–60 men in total) going forward independently ahead of the rest of the troops organized in four waves each of between twenty and twenty-four men.[149] Very small numbers of stormtroops infiltrated to be followed by denser skirmish lines. In his work on stormtroop tactics, Gudmundsson claims that German doctrine misrepresents itself. German doctrine at the end of the First World War described the advance of the line infantry in 'skirmish lines'. Gudmundsson insists that such skirmish lines no longer existed; he presumed that German doctrine really meant 'lines of squads', and he apostrophizes the term 'skirmish lines'.[150] He maintains that the manipular tactics of the stormtroop had become universal. His enthusiasm for German initiative outruns the evidence. Wilhelm Balck himself affirms that skirmish lines existed[151]—his diagram of a battalion assault is explicit. Indeed, both Jünger and Rommel discuss the presence of skirmish lines. Jünger as a company commander (and stormtrooper) ordered their use on a number of occasions even late on in the war after the putative stormtrooper revolution. Indeed, drums and bugles were still used to coordinate their movement.[152] Mass lineal tactics was a recurrent feature of

German infantry operations throughout the war, despite the development of stormtroopers.

Moreover, not only were stormtroopers with their advanced tactics a small part of the German army but the German infantry right to the end of the war could perform very badly. Paddy Griffith, keen to eliminate the stormtrooper myth and to promote British innovation, isolates several examples of poor German tactics. During the final offensive of the war, the Germans 'habitually' displayed a total neglect of tactics which resulted in very severe casualties.[153] In March 1918, during the supposedly revolutionary Operation Michael offensive, British soldiers were surprised to see mass German infantry shambling around 'in cricket match crowds';[154] they presented lucrative targets to machine-gunners and the artillery. This was not an isolated incident. On 15 July 1918, north-east of Fossoy, 'the German infantry and machine-gunners came on at a slow walk and as steadily as though on parade. An officer walked at their head swinging a walking stick.'[155] Indeed, the great assaults in which Ernst Jünger was involved in March 1918 may have been led by stormtroops but Jünger records the density of the formations behind those leading troops: 'No man's land was packed tight with attackers, advancing singly, in little groups or great masses towards the curtain of fire. They didn't run or even take cover if the vast plume of an explosion rose between them. Ponderous, but unstoppable, they advanced on the enemy lines.'[156] After recovering from a wound received in this assault, Jünger was reassigned to training of shock troops:

I had come to understand in the course of the last few engagements that there was an increasing rearrangement of fighting strength in progress. To make an actual breach or advance, there was now only a very limited number of men on whom one might rely, who had developed into a particularly resilient body of fighters, whereas the bulk of the men were at best fit to lend support.[157]

It was perhaps no surprise that German infantry attacked *en masse*. Yet, the inconsistent performance of the German infantry was not limited to the First World War. Terry Copp, for instance, has questioned the concept of German military superiority 'as an article of faith among NATO armies' propounded by Basil Liddell Hart, Charles Stacey, Trevor Dupuy, and Martin Van Creveld.[158] Despite evidence that the German army generally outfought the Allies in the Second World War, German troops often failed to demonstrate the professionalism for which they were reputed. For instance, on 16 December 1944 during the Ardennes offensive, a platoon of eighteen US infantrymen from the Intelligence and Reconnaissance Platoon of 394th Infantry Regiment, 99th Infantry Division were ordered to occupy forested high ground slightly north-west of Lanzerath, overlooking a road, in order to plug a gap between 394th Regiment's lines and the forward location of the 14th Cavalry Group.[159] This position placed the Intelligence and Reconnaissance Platoon on the main effort of the 1 SS Panzer Corps and, indeed, the entire Sixth Panzer Army.[160] Armed only with their own weapons, they were able to repulse three major assaults, inflicting approximately 300–400 casualties and destroying 1 Battalion Fallschirmjäger Regiment 9, because the German paratroopers, from an elite formation, repeatedly advanced *en masse* across snow-covered fields.[161] One of the US soldiers involved in the action commented: 'Whoever's ordering that attack must be frantic. Nobody in their right mind would send troops into something like this without fire support.'[162] Stephen Ambrose records compatible defensive incompetence by two German infantry companies near Zetten in Holland in 5 October 1944 which were routed by a charge by another platoon of paratroopers from E Company 506th Parachute Infantry Regiment, led by Captain Richard Winters, which was executed only because the Americans found themselves in an untenable tactical position.[163]

Such performances should not be surprising. They were a typical feature of the German army, as a citizen military, in both wars. Indeed, as these conflicts proceeded, David Glantz plausibly claims that the problem of poor training and inexperience worsened. As they prepared for Stalingrad, German divisions often received replacement soldiers with as little as two months' training.[164] Adelbert Holl, who fought as an infantryman at Stalingrad, highlighted the problem in his unit II Battalion, 276 Infantry Regiment which was engaged in the struggles around the mouth of the Tsavitsa in the north of the city centre and later used in the assaults on the Barrikady Factory to the north of the city. Not only were the 'new replacements mainly men from other regions of our Fatherland', principally the Sudetenland, and, therefore, Holl implies, not so committed, but they were not prepared for combat having only received eight weeks' training.[165] The result was that they panicked in combat and fled.[166] This subsequently caused the death of one of the young soldiers who was abandoned in his foxhole during a night attack and was killed by a Russian grenade.[167]

Major General Hans Kissel, who was decorated in both world wars and commanded a Wehrmacht division on the Eastern Front in the Second World War, fully recognized the shortcomings of the German army. He accepted, following Marshall, that 'there were also numerous "nonshooters" in the German infantry', although he believed that the problem was less severe in the Wehrmacht because of higher levels of training based 'on team like collaboration' in the form of 'fire combat in the framework of the field squad' or, in defence, 'squad nests' (usually around the machine-gun) instead of individual foxholes.[168] Indeed, while Kissel's article on panic was intended as a general theoretical treatise on the social dynamics of combat, he drew on a number of examples which he had experienced on the Eastern Front: 'In the Russian Campaign of 1941–5 innumerable panics occurred on both sides.'[169] Specifically, Kissel identifies a panic at Moscow where German troops abandoned their position in the light of the threat from the direction of Terayevo.[170] Kissel records the rout of Hungarian battalions under his command in autumn 1944 but he also noted that at 'Gawaiten-Gumbinen, the picture of the battle with the invincible Soviet defender was so extremely different from the peacetime exercises that the troops were not sufficiently able to withstand the effects of their adversary's fire'.[171] Kissel's picture of the Wehrmacht depicts a force which was not fundamentally different from the conscript armies it faced; it too was susceptible to poor battlefield performances in which non-firing and panic were endemic. Indeed, even when they were not guilty of gross tactical incompetence, the German army was never as professional as some of the panegyrics would imply. In his investigation of the 'Blitzkrieg' legend, Karl-Heinz Frieser usefully observes that only sixteen German divisions in 1940 were well equipped and motorized; the Battle of France was ultimately won with ten armoured and six mechanized divisions.[172] The remaining 127 divisions, by far the largest part of the Wehrmacht, were of rapidly declining quality, relying on horses for transportation and consisting of often quite poorly trained troops. Indeed, Frieser notes that during the campaign when the Wehrmacht was at the height of its powers in the west, 45 per cent of its soldiers were over 40 years old while 50 per cent had had only a few weeks' training.[173]

An article published in *Die Wehrmacht* in 1940 and translated by US Military Intelligence records a successful battalion attack in Belgium in detail for which the commanding officer, Major Albrecht Lanz, was decorated. The attack took place in the vicinity of Theilt during the crossing of the Lys River. On 24 May 1940, the battalion attacked the village of Gothem but were beaten back by enemy fire. Two days later, on 26 May, the battalion attacked again across high and very dense grain fields. Once again, the attack became broken and confused, with the platoons disintegrating under enemy fire in the thick vegetation. At this point, Major Lanz went forward, organizing the remnants of his companies into a stormtroop of two and a half platoons. The assault group advanced to the edge of Denterghem and attacked a farmhouse in which the enemy defences were concentrated, forcing their surrender.[174] The article was intended to communicate the inevitable confusion of battle and the importance of leadership, which was evidently demonstrated by Major Lanz. Yet, other inferences about the quality of the Wehrmacht might have been drawn from it. Indeed, the Military Attaché who filed the report drew alternative conclusions: 'an experienced German infantry unit during an initial assault occasionally became confused, lost or even disintegrated, in spite of the thoroughness of its organisation and training.'[175] It might be possible to go further. Not only did this battalion become disorganized in the grain fields (on an initial and subsequent attack) but it seems difficult not to draw the conclusion that the Marshall effect overcame many individuals and sub-units as it sought to attack. Disoriented by heavy fire, it seems that many soldiers understandably took the opportunity of availing themselves of the considerable cover which the tall grain afforded. The result was that in an apparently fully trained battalion barely a sixth, some eighty soldiers, took part in an attack. It seems plausible to claim that this level of confusion and subsequent participation was probably fairly typical of the Wehrmacht. Indeed, this low level of participation was only to be expected because as a mass conscript force, the Wehrmacht too was often incapable of training its mass soldiery to a sufficient level so that they could execute modern infantry tactics.

Similar patterns were evident during the Battle of Normandy, even though the German high command recognized that an Allied invasion of northern France was inevitable from 1943 and they, therefore, theoretically had months to prepare their divisions for it.[176] Moreover, it was obvious to all in the German forces that the success or failure of this invasion would be utterly decisive in terms of the outcome of the war. There was, therefore, every motivation for soldiers and their commanders to oppose the landings vigorously. Yet, while there were a number of German divisions which performed well in Normandy including Sepp Dittrich's 12th SS Jugend Division, German officers were candid in their assessment of some of the infantry. Thus, Oberstleutnant E Maurer, a staff officer in charge of training in the 253 Infantry Division, noted some shortcomings in the three infantry divisions (709, 253, and 91 Luftlande Division) deployed on the Cotentin Peninsula north of Carentan before D-Day. Maurer praised the 709 Infantry Division under General von Schlieben: 'The fighting troops and their commanders demonstrated excellent morale, were disciplined and combat-experienced soldiers, whose construction of positions and combat posts was laudable.'[177] He was condemnatory about 91 Luftlande Division commanded by Lieutenant General Wihelm Falley, 'which was not particularly battle-worthy, despite the fact that they had good personnel at their disposal and were equipped with excellent modern weapons'. For Maurer, the problem was that the majority of the soldiers were drafted from the air force and 'had experienced only an accelerated infantry training and lacked essential combat experience'.[178] Maurer's concerns about the Luftlande Division might be dismissed as Wehrmacht bias but his assessment of his own division, 253, was pointedly realistic. Thirty per cent of his division were from *Volksliste* III and many were not German but Polish and spoke little German. In Maurer's view, their performance was, consequently, 'good as long as their commanders and NCOs fulfilled their duty immaculately'.[179] Accordingly, he summarized the combat power of 253 as adequate to good in places.[180] Lieutenant General Kurt Badinski, commander of the 276 Infantry Division, recorded similar problems. While many men in his division were experienced and combat-proven soldiers, new draftees of 17 to 22 were ordered to Normandy by Hitler 'in the middle of their training', where they spent most of their time constructing the Atlantic

Wall.[181] He noted that 'rockets [*Ofenrohre*] and Panzfausts were not used to their full defensive effect, because the crews could not be adequately trained as a result of a shortage of training munitions'.[182] Oberst Friedrich Freiherr von der Heydte, commander of Fallschirmjägerregiment 6, concurred. While he (perhaps self-interestedly) emphasized the hard four-month training period during which his own regiment had been forged after its creation in January 1944, he recorded the low assessment of the troops in Normandy by the commander of another regiment; 'The troops, which were available at the beginning of the Invasion to defend against the Allied Landings, could not be compared to soldiers deployed to the Eastern Front. Combat morale was meagre, the mass of men and NCOs had no combat experience; among the officers' corps one had the impression that the majority of officers consisted of those who were no longer needed in the East because of inadequate aptitude, wounds or illness.'[183] Von der Heydte accepted that the pessimistic view of this officer reflected his own experiences of exercises in May, and might have been entirely personal. Yet, von der Heydte went on to note the very poor equipment which was issued to some of the troops, including weapons which were decades old.[184] Notably, he concluded his description of preparations for the invasion with a comment which Lieutenant General Erich Marcks, commander LXXXIV Corps, had made to the regiment after an exercise: 'Bunkers without guns, ammunition dumps without ammunition, minefields without mines and a uniformed mass, with scarcely a soldier among them.'[185] Indeed, the leitmotif of von der Heydte's report is a furious criticism of the 17th SS Panzer-Grenadier Division 'Götz von Berlichingen' with whom Fallschirmjägerregiment 6 had to fight for most of the campaign. He noted that while the SS Division was excellently equipped, 'the majority of the leadership had neither training nor combat experience' and 'regimental, battalion and subunit commanders generally lacked elementary tactical knowledge'.[186] Von der Heydte recorded the failure of the division in the counter-attack on Carentan on 13 June 1944. In the face of heavy losses, the morale of the SS troops sank and increasing numbers of individual troops and even whole sections retreated. A panic was only prevented when the adjutant of Fallschirmjägerregiment 6 ordered the troops to stop, threatening them with his pistol.[187] Later that day, he made the assessment that 'even a single infantry attack by the Americans' would be enough to induce uncontrolled panic in large parts of the SS regiment.[188] Clearly, von der Heydte's testimony is potentially biased. For instance, by pointing up the weaknesses of the SS formation, he might highlight the quality of his own regiment and, indeed, distract attention from the fact his paratroopers also failed to retake Carentan and might not have always performed outstandingly at other moments in the campaign. Yet, von Heydte's assessment is corroborated by others. Antony Beevor, for instance, described the 'Götz von Berlichingen' Division as 'the weakest and worst trained of all the Waffen-SS formations in Normandy'.[189]

These shortcomings were often recognized by Allied troops. For instance, the British platoon commander Sidney Jary observed in Normandy that the Germans were not always as professional on the battlefield as their post-war reputation would suggest. Jary noted that it was in the interests of German veterans to exaggerate their performance: 'It is, of course, very much in their own interest to encourage the theory and myth that, although superior, as fighting men, they were beaten only by numerically superior forces and firepower.'[190] Jary's experiences of the German army did not confirm its reputation: 'In my experience this was not so. In many attacks the prisoners we took outnumbered our attacking force and German units who would continue to resist at close quarters were few indeed. Unlike us, they rarely fought at night, when they were excessively nervous and unsure of themselves. Where we patrolled extensively they avoided it. I can remember only one successful German patrol and not one successful night action. If our positions had been reversed, I doubt if they would have performed better than we did.'[191] Jary appositely identified the basis of their combat superiority. 'The RAF undoubtedly dominated the skies above us, but it was my experience that German infantry and armour had far superior firepower to ours. Their guns and mortars could also produce a devastating display.'[192] The 'spandau' (the MG 42) featured regularly in Jary's account and seemed to be the weapon which threatened his platoon most frequently. Jary implied that apart from their MG 42 (and their mortars), German infantry were by no means special. Despite their own difficulties in Normandy, US soldiers observed a similar phenomenon themselves. While the Germans feared Allied artillery, they were disconcerted even more by aggressive infantry. A successful infantry officer observed: 'We have learned to keep moving forward. If there is anything the Germans hate it is close fighting.'[193] A fellow officer concurred: 'Move forward aggressively. The German is a poor marksman under the best conditions. In the face of heavy fire and an aggressive enemy his fire becomes highly ineffective.'[194] The reasons for this ineffectiveness in marksmanship and in close fighting were not only a product of the lack of training which many German officers recorded but can be traced back to the doctrine and training which the Reichswehr and Wehrmacht developed in the inter-war period.

As the Germans themselves discovered in Stalingrad, they were not always proficient, at the platoon or section tactics which they themselves had pioneered in the First World War. The problem here seems to have been the emphasis which German doctrine placed upon the MG 34 and 42. While these weapons were fearsome on the battlefield and were intended to generate fire superiority at a range under which riflemen could assault the enemy, in reality, German riflemen often became dependent upon their machine-guns. In defence, riflemen supported the weapon delivering ammunition for the gun and with fire and, in assault, there was a tendency for the riflemen to infiltrate the position with support of the machine-gun rather than employ section fire and movement tactics. Indeed, American military attachés observing German exercises up until the declaration of war in December 1941 noted the infrequency with which riflemen actually used the weapons even in the assault very clearly. At a demonstration at the infantry school of a battalion attack in 1939, the assistant Military Attaché, Major Black, recorded that: 'Under cover of this fire each platoon moved by groups (sections), each group covered by the fire of its light machine gun, by bounds of approximately 25 yards. Riflemen took no part in the firefight except to support the advance of the light machine gun squad.'[195] Observing a demonstration at the Infantry School in January 1940, the Military Attaché once again emphasized the role of the light machine-guns providing covering fire under which sections advanced: 'The main concern of the German Army infantry in this demonstration was to get the light machine guns forward, to get the firepower advanced. In other words, the advance was executed by small teams, a group of riflemen built around a machine gun.'[196] As late as April 1940, the Military Attaché, Colonel Peyton, and his assistant, Major Hohenstal, were allowed to observe the final pre-deployment exercise of the 382rd Infantry Regiment of the 164th Infantry Division which involved a series of assaults on positions in the environs of Königsbrück in Saxony. The description of the exercise is particularly useful because it involved a fully trained formation about to deploy and, therefore, probably represented German infantry tactics at their most refined. Initially, it appears as if this division did indeed engage in advanced fire and movement platoon tactics. Peyton, for instance, noted: 'German soldiers in attack advance in pairs and their coordination and cooperation are excellent, and one might say almost automatic. They are thoroughly trained in taking advantage of terrain and advancing each other by fire covering'.[197] In fact, Peyton is describing only the actions of the specialist reconnaissance troops as they initiated contact with the enemy before the main assault took place. Earlier in his report, Peyton explicitly observed that these reconnaissance teams 'worked in pairs'.[198] By contrast, the main body of infantry attacked in waves supported by light and heavy machine-guns, the volume of fire from which Peyton found 'impressive'.[199] The final assault on the machine-gun nests was described as a 'rush' 'preceded by hand-grenade attacks' with bayonets fixed.[200] Although, in other reports, riflemen were observed rushing forward in pairs,[201] there is no specific mention of section fire and movement tactics; the machine-guns provided the fire support and the infantry dashed forward. In the ideal training situation, German riflemen advanced by bounds under the cover of their machine-guns and, having thrown grenades, rushed forward in the final assault once machine-guns had suppressed the position. Clearly, with weapons as good as the MG 34 and MG 42 and operators as well trained as they were, this was a very effective form of tactics—especially for a citizen infantry in which training was limited—but even among the best-prepared German forces at the start of the Second World War, presumptions about the professionalism of the rifle sections and platoons need to be avoided. On the contrary, it was not surprising that US and British infantrymen in Normandy found the German platoon very susceptible if it was possible to obviate the fire superiority of the MG 42.

In the ruins of Stalingrad, this dependence on the machine-gun exposed the weakness of the Wehrmacht, illustrating that it was not so very different from the citizen armies which it fought. There, as the machine-gun's range became less relevant, the German infantry were thrown back onto their small-group tactics. It is particularly noticeable that in Stalingrad the Germans favoured Russian weapons over their own. In particular, they found the Russian submachine-gun far more reliable than their own machine-pistol and often armed themselves with it. In the dust and filth of Stalingrad, the Russian submachine-gun was less prone to jam.[202] Indeed, Adelbert Holl's 7 Company, II Battalion 276 Regiment had its own supply of 'special weapons': captured Russian anti-tank rifles, sub-machines which were more effective than the traditional armoury of the Wehrmacht infantry platoon and company in these bewildering urban surroundings.[203]

Yet, of course, it was more than just their weaponry which troubled the German troops and commanders in Stalingrad. As David Glantz has noted, in October 6th Army were able to use mechanized manoeuvre to achieve their objectives in the city; thus between 14 and 22 October, Heim's 14 Panzer Division and Oppenlander's 305 Infantry division demolished the Russian Guards Regiment in Spartanovka,[204] and the assault of Seydlitz's forces against the Tractor Factory was successful.[205] However, 'it was also the last occasion when Paulus's Sixth Army was able to exploit surprise and employ manoeuvre successfully'.[206] Thereafter the war became a brutal attritional struggle in a complex and nightmarish urban environment. This was a war for which the Wehrmacht were not prepared: 'Stalin's decision to defend the city deprived the Germans of their traditional advantages of mobility, manoeuvre and precise, overwhelming, and deadly artillery fire and air support.'[207] At the level of the platoon, the intense street fight was disorienting in the extreme. It was described as 'an insatiable abyss of chaos'; 'it was a nightmare world for the grenadiers; was that movement a piece of sheet-iron swaying in the wind or a stealthy Russian just waiting for his opportunity to cut some throats?'[208] The soldiers freely confessed 'this was another kind of war, which we had to wage here'.[209] Captain Uli Weingärtner of the 94th Infantry Division recorded; 'The hardest fighting was around the Grain Elevator. Other brick and concrete buildings were defended tooth and nail by "Ivan". It is a completely new way of fighting for us. You've got to expect a burst of fire out of every hole or gap in a wall.'[210] Not only did the Russians typically fight to the death, fortifying every structure, but they used the sewer system to infiltrate the German lines and attack them from behind. In this chaotic environment, German riflemen were reliant upon their own individual reactions and the rapid synchronization of fire and movement between and within squads, using grenades and personal weapons. They were not nearly so proficient at this close-quarters battle technique. In the face of this environment, the 6th Army found that it had to adopt new tactics, especially as they sought to clear the heavily fortified and intensely defended Barrikady Factory district. Every assault on every building had to be meticulously planned and supported by a huge weight of indirect fire from artillery and mortars and direct fire from tanks and assault canons.[211]

However, so difficult was the assault of the Barrikady Gun Factory, and the subsequence clearance to the Volga, that the Wehrmacht returned to First World War methods. Five experienced pioneer battalions were assigned to the 6th Army to be trained as specialist storm battalions; three were assigned to 305 Infantry Division and two to the 380 Infantry Division. A further unit, Sturmkompanie 44, had been formed earlier. These units were given some specialist weapons such as satchel charges, assault canons, and flamethrowers and practised assault troop tactics in the ruins of nearby villages; a Pioneerschule had been established near Kalach for this purpose.[212] On 11 November, the day of the major assault beyond Barrikady Factory towards the Volga, the troops from the pioneer battalions crawled ahead to take the Apotheke with 'well practised movements' under the support of artillery, machine-guns, and assault canons.[213] Perhaps the most successful assault came when pioneer assault troops took the Kommisarhaus on 13 November. This building lay just east of the factories and represented a formidable obstacle because of its position and its architecture: it was built in the style of a mock castle with thick high walls and its defenders had further improved its defences. The Kommisarhaus was built on a U-shaped design, whose courtyard faced the German positions across the street. Instead of opting for the obvious assault direction, attacking the external wings of the building, the Sturmkompanie charged the main door in the middle of the central building. They correctly surmised that not only would the flanks of the building be subjected to enfilading fire from other Russian positions but the occupants had blocked the entrances to the ground floor of the wings, presuming that this was where any assault would come from: 'the pioneers and grenadiers of Sturmkompanie 44 dashed past the entrances of the southern keep and worked their way along the inner wall of the practically enclosed forecourt, hurling grenades through windows as they went. This daring move caught Klyukin (the Russian officer commanding the defence of the Kommissarhaus) and his men off guard, for there were now precious few defenders positioned at the windows overlooking the courtyard. The pioneers swiftly formed up outside the portico of the central entrance. Others, kneeling down, covered the main group by aiming their weapons at the many leering windows.'[214] The Sturmkompanie broke in and, once again surprising the Russians, made their way immediately to the first floor, seeking to clear the building from top to bottom, outflanking the defenders from above.[215] The fighting was fierce but, eventually, with significant casualties, the Kommisarhaus was seized. Most of the fighting in November around the Barrikady assumed this form with a series of deliberate attacks by specialist troops, though by 25 November the German forces were exhausted. So heavy were the casualties in the pioneer battalions that they had to be amalgamated, which Marks plausibly claims signified the culminating point of the German offensive in Stalingrad.[216]

The attacks in November in the Barrikady district demonstrate that the Wehrmacht did adapt to the horrific environment which confronted them; the assault on the Kommisarhaus was an almost ideal, though costly, operation. Yet, the very fact that they had to create specialist assault troops to break into and clear each building seems to provide evidence that the normal German platoon was not always so resolute and skilled as is often asserted. In particular, it does seem to suggest that away from the key platoon weapon, the MG 34 and 42, the German infantry became ordinary citizen soldiers just like their enemies. It has been widely claimed that Allied infantry were dependent upon their artillery in both the First and Second World Wars. It might be argued that the Wehrmacht's infantry, although generally better trained especially in the early stages of the war, displayed a similar dependence on crew-served fire support weapons, as Marshall suggested: its machine-guns. Interestingly, the British army's *Instructor's Handbook* criticized the German soldier as 'stereotyped'; 'he is cunning but has a predisposition to do the same thing in the same way.'[217] In a document aiming at generating cohesion and morale among British troops, there seems to have been propaganda value to this denigration but the reasons for the criticism were instructive. The British noticed that the Wehrmacht placed 'great emphasis on the firefight' whereby 'fire superiority on a narrow front was chosen as a critical objective'.[218] Specifically, the German platoon recurrently relied on their ability to beat down, pin down, or blind the enemy with light machine-guns, which were much more numerous than in the Allied infantry companies.[219] The German army doctrine of 1933, *Truppenführung*, is explicit on the centrality of the machine-gun to infantry tactics.

> When advancing, the squads deploy at irregular distances and intervals with sufficient depth, based on the terrain conditions and enemy reaction. They make maximum use of dead space and cover from enemy fire. In the face of strong enemy fire, the advance continues by bounds and rushes, in either large or small groups. During lulls in the fighting, riflemen seek cover to limit the effect of enemy fire as much as possible. Light machine guns open fire when in effective range, and riflemen continue advancing under this covering fire. When closing with the enemy positions, the infantrymen also engage with fire . . . The advance against the enemy is executed through the careful synchronization of fire and movement. Exposed advancing units must not lack fire support. While they work forward, adjacent elements suppress the enemy, especially with light machine guns in conjunction with heavy weapons.[220]

Truppenführung has significantly little to say about platoon and section tactics themselves, heavily implying that these were local matters determined primarily by the ground. Immediate tactical success for small infantry groups depended upon the ability of machine-guns to suppress the enemy as they moved forward.

At its most effective, it might be argued that the Wehrmacht infantry had developed a doctrine which was compatible with the small-group fire and movement tactics which are evident in the professional force of the twenty-first century. However, in practice, it prosecuted platoon tactics by generating firepower from small numbers of specialist troops, the machine-gunners, on whom the less well-trained riflemen were very dependent. In this way, the Wehrmacht did not really represent a professional force, as Von Seeckt conceived of it. It was still characterized by a mass of conscripts energized by small numbers of experts operating a very effective weapon. The machine-gun suppressed while the infantry charged forward in files or lines. The fact that the machine-gun was organic to the infantry, rather than part of a supporting combined arms organization as the artillery was for the Allies, has perhaps deceived commentators into presuming that therefore the Wehrmacht infantry as a whole was hugely more impressive than the Allies. This may be an illusion. Indeed, in 1940, the US Military Attaché observed the advantages afforded the German army by prioritizing the machine-gun and positioning it as far forward as possible: 'One of the most difficult problems to solve in the attack is to provide the very close coordination required between the infantry–artillery teams in the smaller units. It was observed, that to a large extent, the necessity for this coordination in the German Army has been eliminated. The artillery was concentrated on the final objective and the fire of the infantry supporting weapons provided the close support to assist the infantry in its advance.'[221] Certainly, American military attachés throughout the inter-war period were respectfully critical of the German army which they did not think was any better than the US Army and, on some points, inferior. In 1925, for instance, the US Military Attaché acknowledged that the Reichswehr was, due to its equipment and training (and above all its machine-guns and gunners), 'the best one hundred thousand soldiers on the continent of Europe', but, at the same time, he pointedly noted: 'The legend of superior German efficiency is not concurred in. There is no reason why the American army should not be superior.'[222] In 1940, Colonel Peyton also recognized shortcomings in the Wehrmacht. On the basis of his observation of the 382rd Infantry Regiment training and exercising in April 1940, he thought their bayonet training was 'very mediocre' and 'distinctly inferior to our own bayonet training' while their hand-to-hand fighting was 'distinctly inferior to what can be expected from American soldiers'.[223] By organizing the infantry around its best infantry weapon, the Wehrmacht may have maximized the performance of its troops. Despite the fact that many of them were conscripted, they became a formidable opponent in close combat on open terrain especially during the day. However, the skilful use of its machine-guns did not fundamentally transform the organizational dynamics within the force. It was still characterized by a poorly trained mass, mitigated by small areas of excellence; above all, its machine-gun teams. In this way, the Wehrmacht, despite its reputation, may not have been fundamentally different from the armies which it faced.

The Vietnam War is correctly remembered for the defeat and collapse of the US Army but while European forces—even Britain refused to send troops—did not participate in the conflict, other Allied forces participated, most notably the Australians who deployed a battalion-sized task force, given responsibility for Phuoc Tuy Province to the south-east of Saigon from 1965, eventually withdrawing in 1972, when the USA themselves left the country. The performance of this task force represents a very useful case study for examining cohesion among the citizen army. The Australian infantry in the First and Second World Wars emphasized the bonds of masculinity, of mateship, to encourage participation and although, especially in the First World War, mateship often produced a notably offensive spirit among Australian troops, their performances accorded with those expected of a citizen army at this point. Tactical performances were generally characterized by a reliance on commanders to lead men forward. In Vietnam, the Australian Task Force often displayed a higher level of performance. Although the Australian Task Force was in the country for some seven years, the most dramatic and important event of the campaign occurred on 18 August 1966 when 6th Royal Australian Regiment fought the Battle of Long Tan against the NVA D445 Regiment, which battle had attracted the most commentary and analytical attention. In fact, this battle usefully illustrates the nature of the Australian army at that time as a partly conscripted force, especially since it displayed a surprisingly high level of competence and cohesion in that fight.

The Battle of Long Tan was initiated on the night of 16–17 August when the Australian base at Nat Dui was mortared by the NVA. Consequently, on 18 August, Delta Company were ordered to patrol out to the east of the base to locate the mortar firing points. As they advanced through a rubber plantation some 3 kilometres east of the base, members of 11 platoon D Company surprised a patrol of six NVA soldiers (whom they mistook for Viet Cong). They fired at and wounded at least one of these soldiers whom the company, led by 11 Platoon, pursued. 11 Platoon rapidly ran into a major enemy force and were enfiladed from the left flank with heavy fire. 4 Section was annihilated almost immediately, every member killed or wounded, and the platoon was subjected to a company size assault, in which their commander, Lieutenant Sharp, was killed; fifteen were killed or seriously wounded in the contact. 10 Platoon moved forward to assist 11 Platoon and they too quickly encountered heavy enemy forces but were able to break up a major assault on

11 Platoon by surprising the enemy from the rear; but they were then subjected to an assault. 12 Platoon held in reserve were moved forward. Desperate fighting followed but eventually, the company commander was able to concentrate the three platoons, including the remnants of 11 Platoon of this company, into a perimeter as the NVA Regiment began to encircle it. D Company held out with small arms fire and artillery support until 1 Armoured Personnel Carrier Squadron from Alpha Company eventually broke through to D Company and repulsed the attacking force.[224] The battle lasted for some three hours and was the most intense contact in which the Australian Task Force had been or would be involved: 14 Australians were killed and 245 Vietnamese dead were found on the battlefield the next day, although it is likely that 6 RAR inflicted higher casualties than this.

The performance of 6 RAR and especially D Company has often been held up as an evidence of the skill and professionalism of the Australian conscript army. Indeed, in his work on the battle, Alex McAulay claimed that Long Tan demonstrated 'the legend of ANZAC upheld'.[225] Ultimately a single Australian company of some 100 soldiers held off and, with the assistance of artillery and an armoured personnel squadron, heavily defeated an NVA regiment of perhaps many times their own number. D Company demonstrated impressive tactical expertise and determination during the fight. Sergeant Bob Buick, who took over command of 11 Platoon after Lieutenant Sharp was killed, recorded that 'I saw no one falter, each supporting his mate and defending his patch of dirt to the death'.[226] In fact, when D Company revisited the battlefield the next day, they found their comrades lying dead, killed in their firing positions, still over their weapons. D Company do seem to have demonstrated a very high level of performance especially for a (partly) conscripted force.

There seem to be a number of explanations for their high level of performance. Although there were a large number of national servicemen in the company and the battalion, the majority of soldiers, all but one of the officers, and all the senior NCOs were regular professionals; D company was one of only two of the battalion's companies designated to accept national servicemen.[227] Throughout 1965 and early 1966, as they prepared for their tour, D Company became more professional with the arrival of more regular NCOs. This pattern continued throughout the war. In 1971, Delta Company 4 RAR consisted of 55 per cent regulars and 45 per cent national servicemen.[228] Moreover, there was a very high level of personnel stability in the battalions with national servicemen remaining for their entire two-year service within the same unit. Instructed and commanded by professionals who made up the majority of the company, it does not seem implausible to suggest that these citizen soldiers were likely to have been all but equal to their professional peers. The high ratio of regulars to national servicemen and the stability of their service seem to be important factors in explaining the performance of 6 RAR at Long Tan and that of the Australian Task Force in Phuoc Toc in Vietnam. In October to November 1971, the commander of D Company, 4 RAR believed that the standard of tactical performance dropped when the conscripts who had been with the company for two full years returned home to be replaced by newcomers, who were not 'battle hardened' and did not have the 'knowledge and experience' of those they replaced.[229]

6 RAR and the Australian Task Force in general seemed to be advantaged over their American allies (and over most citizen armies in the twentieth century) by the high proportion of regulars and by the low personnel turnover in the battalions. There was, therefore, time to foster a sense of unity and to instil the correct individual and collective skills to prevent Long Tan from turning into the massacre which was often the outcome when the NVA ambushed conscript American forces. The Battle of Long Tan and the performance of the Australian Task Forces seem to prove the general claim about the fragility of citizen armies rather than to disprove it.

Although an advocate of mechanized warfare, Basil Liddell Hart recognized the potential of well-trained infantry and had a clear vision of how they could operate as early as 1926. The infantry 'must become light infantry of the Peninsula pattern, agile groups of skirmishers who will exploit to the full the tactics of infiltration and manoeuvre'.[230] Liddell Hart was explicit about how to achieve this: 'this high standard will become practicable, particularly if we save more time for tactical training by bringing our drill up to date—basing it on modern tactical movement instead of on those of Waterloo, as is the present drill.'[231] He dismissed the idea that infantry should be just 'cannon-fodder' but rather viewed them in a sociological light, recognizing its collective potency: 'No military spectacle appeals to me as much as the sight of a light infantry or a rifle regiment at drill or on the march, because it seems to represent a collective intelligence rather than a collection of automata... the infantry soldier needs to revive the tradition of the Peninsular skirmisher but carry it to a higher pitch.'[232] Individual tactical skill would be fused into cohesive collective action in this new infantry through established practices and drills to produce a powerful and flexible military body.

The mass citizen armies from 1914 adapted, in theory, at least to the new conditions of the modern battlefield. In some cases from the Great War to Vietnam, there were significant examples of precisely the 'collective intelligence' which so inspired Liddell Hart among the West's citizen armies. The publication of new doctrine and the generation of new kinds of forces demonstrated that the old lineal tactics of the late nineteenth and early twentieth centuries were recognized as obsolete. Infantries, at least in theory, broke up the company firing lines into small, more flexible platoons which mutually supported each other in an assault maximizing use of the cover provided by the terrain. The platoon began to be recognized as the prime unit of infantry combat. As commanders recognized and feared, flexible small-group tactics demanded much higher degrees of coordination and training and to that end, drills and advanced battle preparations were developed. By 1915, some troops on the Western Front were beginning to demonstrate new techniques and developing new doctrines which continued to be elaborated in that conflict, the inter-war period, and in the Second World War. The principles of platoon tactics were recognized and they were sporadically practised by the most highly trained infantries. From the German stormtroopers of the First World War to the commandos of the Second and on to the most highly trained US troops in Vietnam, it is possible to trace the germination of infantry tactics which overcame the Marshall effect. It is, in effect, possible to see the development and evolution of military processes which would be at the very heart of the professional all-volunteer armies of the twenty-first century.

Although the germs of effective infantry tactics found their origins in the mass conscript armies of the twentieth century, the vast majority of infantry was poorly trained. Platoon cohesion was, consequently, weak. Typically, in the absence of adequate training for individual platoons, armies relied on high morale to inspire and encourage their troops; they appealed to their sense of masculine honour or to their patriotic duty to attempt to obligate soldiers to each other in combat. Despite his methods, Marshall's instincts about the infantry were broadly correct. Required to perform complex fire and movement tactics in the face of mechanized warfare, platoons were overwhelmed by the task and simply failed to perform from the Western Front to Vietnam. The average platoon under fire was not able to solve the Marshall effect. Overwhelmingly, the inadequately trained platoon in the major conflicts of the twentieth century relied not on small-group tactics which were too difficult to execute but on supporting fires and, above all, artillery. Indeed, it might be plausibly argued that in Korea and Vietnam, the same dependence on supporting fires was even more evident. When they were left to independent action in close quarters with the enemy, the infantry typically fell back on mass tactics or individual virtuosity. It is important to recognize that the combat performance of the citizen platoon is not only recurrent over the period 1914 to 1973 (when western citizen armies were engaged in major wars) but that this pattern of under-performance was broadly similar across all armies at this period. Even the much admired Wehrmacht, though advantaged by radical political indoctrination and an excellent machine-gun, displayed similar social dynamics under fire as the derided Italians. Crucially, it is important to recognize that there is no hidden source of evidence to support common presumptions about how the citizen army must have performed. All the major western combatants in four major wars have been assessed. Excepting the case of highly trained specialist troops, in the citizen armies of the twentieth century, the phenomenon of the Marshall effect was a widespread and recurrent—almost universal—sociological phenomenon.

It is important to recognize why this effect was a more or less structural and, therefore, unavoidable feature of the mass army. In the era of total war, it was essential that states generated mass armies if they were to prevail primarily because their competitors were doing so. No matter how skilled an army was in the industrial era, if it was too small it would be simply overwhelmed by its enemies. Napoleon was a brilliant commander but he fully recognized the importance of sheer numbers. Military commanders in the twentieth century also fully recognized the indispensability of mass. In the 1930s, Charles de Gaulle, supported by Paul Reynaud, published a number of pieces which argued for the creation of a professionalized and mechanized French army. De Gaulle presciently observed that the reductions in the duration of conscript service, excessive concentration on frontier defences, and expenditure cuts had diminished the French army's combat readiness.[233] While General Maurice Gamelin concurred with some of de Gaulle's criticisms he dismissed the consideration of a professional army as inappropriate after 1935: 'Perhaps it might have been viable had it been created in the early 1920s as a means to preserve the Versailles system (when it would have faced a Reichswehr of only 100,000 men without armour or air support).' But, Gamelin contended with good reason, 'a small professional army could not be trusted to guarantee French security in the radically altered and more dangerous circumstances of the mid-1930s'.[234] Since Germany had reinstated conscription in 1935, France was similarly obliged to maintain a mass army. Indeed, Erwin Rommel, a putative genius of manoeuvre warfare, became fully aware of the requirement for mass in Normandy in 1944. There he noted that in order to unlock the Allied bridgehead, with its huge firepower and vast artillery batteries, a genuinely massive counter-force was required:

At one time they [German senior commanders] looked on mobile warfare as something to keep clear of at all costs, but now that our freedom of manoeuvre in the West is gone, they're crazy after it. Whereas, in fact, it's obvious that if the enemy once gets his foot in, he'll put every anti-tank gun and tank he can into the bridgehead and let us beat our heads against it, as he did at Medenine. To break through such a front you have to attack slowly and methodically, under cover of massed artillery, but we, of course, thanks to the Allied air forces, will have nothing there in time. The day of the dashing cut-and-thrust tank attack of the early war years is past and gone.[235]

The German 7th Army with some Panzer divisions from the 15th Army was simply too small to oppose the invasion. Mass could be defeated only by mass. As the breakout in Normandy loomed, Rommel correctly concluded: 'The fighting has shown that with this use of material by the enemy, even the bravest army will be smashed piece by piece, losing men, arms and territory in the process.'[236]

In Europe, Britain's convenient geographic isolation from the Continent and its subsequent global and imperial orientation allowed it to maintain a small army throughout the eighteenth and nineteenth centuries, when mass armies became the norm elsewhere. Yet, the British way in warfare did not in any way question the need for mass. The Royal Navy was truly massive throughout this era, as the Naval Defence Act of 1889 established at the turn of the twentieth century, when it specified that the Royal Navy had to be bigger than both of its two immediate rivals put together. Even the army recognized the requirement. Thus, while a diminutive army was adequate for most of the small colonial wars of the Victorian era, a dramatic expansion and the calling up of volunteers from the reserves was required for the Boer War of 1899–1902 when the British army faced a large and well-equipped opponent.[237] Moreover, once Britain became involved in the continental struggles of 1914 to 1918 and 1939 to 1945, it too discovered that it required a truly mass army. In neither war was it able to choose to remain small and professional. In the twentieth century, mass citizen armies were requisite, precisely because every other state raised them. Mass had become a historical necessity. Yet, once a mass citizen army had become a requirement, it became all but impossible to train this force, and especially the infantry with the most demanding and dangerous role, effectively. It was simply too big.

Of course, this training problem was compounded for the infantry by the new circumstances of war. Napoleon's *Grande Armée* comprised a predominantly unskilled and untrained infantry, but column tactics reduced the demands on the troops while producing an effective assault method against all but the most disciplined opposition. Mechanized warfare of the twentieth century, by contrast, demanded initiative, independence, and complex fire and movement tactics at the platoon level. The lethality of the killing zone had multiplied dramatically. Platoons in the mass armies of the twentieth century required sophisticated and coordinated tactics but, precisely because armies were necessarily so huge, there was very little possibility of infantry troops receiving adequate levels of training to execute such practices in combat. The problem of training was compounded not only by the size of recruit cadres but also, once a war had started, by the pressures of time to regenerate and expand forces and by the need to replace casualties. Even the best-trained citizen armies, like the Wehrmacht in 1940, began to lower training standards as casualties began to mount, notwithstanding the endurance of primary groups. Here again the fact of total warfare set a deeply problematic but unavoidable dynamic in process. As a result of massive industrial firepower, high casualties were all but inevitable so the problem of replacements and training them became unavoidably acute. Of course, once casualties started to be incurred and untrained replacements rushed to the front, a vicious cycle of ever lower training and tactical capability with concomitantly increasing casualties was initiated. In the face of these ineluctable problems, citizen armies typically prioritized their artilleries, while their infantries understandably—and indeed rationally—relied on appeals to masculinity and nationalism to generate morale and recourse to the mass tactics of the charge or to individual acts of heroism. The dynamics of infantry combat in the citizen army was not contingent, accidental, or avoidable, therefore. Their generally poor performance was not resolvable within the institutional context of mass armies and total industrial war. On the contrary, it was a necessary and organic part of the mass army. It would take the professionalization of the army to implement small-group tactics fully and to generate the training systems which these specialist forms of collective practice demanded. In short, it would take a massive reduction in the size of the infantry and the introduction of professionalism before the infantry platoon could generate genuinely high levels of cohesion.

Training is manifestly central to the execution of complex platoon tactics and, accordingly, this account up to this point has emphasized the inadequacy of the training which the citizen soldier received to explain the pattern of poor combat performance. The fact that western powers had developed refined infantry doctrine which their platoons were, however, normally unable to execute in combat, despite expectations that they should do so, indicates that a shortfall in training provides a central explanatory role. Nevertheless, it is important to recognize that while the lack of training was often decisive, it was not the only factor in generating the kinds of performance which were typical of the mass army in the twentieth century. Although Korea and Vietnam were important exceptions, mass armies (especially in Europe) were primarily developed to protect the territorial integrity of the state. They were designed to fight interstate wars, in extremis, for national survival itself: the First and Second World Wars were precisely the kind of existential struggle for which the mass citizen army was conceived. Accordingly, notwithstanding the difficulty of training a mass force, it seems plausible to suggest that especially in the first half of the twentieth century at least and possibly even in Korea and Vietnam, a 'structure of feeling' might have been in place which did not merely mean that mass armies fought as they did because they were inadequately prepared but because cultural and political presumptions suggested that they should fight in the way they did.

Edward Luttwak's concept of 'post-heroic' warfare may be relevant here.[238] In that article, Luttwak explores a potential contradiction between modern western concepts of warfare and the reality of post-Cold War conflicts in the 1990s. Clausewitzian warfare prioritized decisive victory by conventional military means; the central purpose of the armed forces was to defeat the armies of hostile states, thereby imposing a political settlement upon them. Luttwak is interested in overcoming the tension between a Clausewitzian predilection for potentially absolute war and the muddy and conditional realities of the new wars in which the West was increasingly involved in the 1990s. For Luttwak, precisely because these new wars offered no prospect of decisive Clausewitzian victories, western powers have not seen the sense in losing soldiers fighting them. They became casualty-averse. Luttwak's strategic point is not relevant here but what he implies is that in the Clausewitzian era, and above all in the First and Second World Wars, the idea of sacrifice was central to the prosecution of war. Political and military leaders and populations themselves were willing to make massive sacrifices precisely because they saw war in existential terms. Precisely because the nation was in danger, the army's reliance on individual and mass heroism—and sacrifice—was regarded as appropriate, desirable, and even necessary.

In this context, the bayonet charge, a favoured tactic throughout the period and especially in the First and Second World Wars, was not simply a product of inadequacy. It was rather a positive existential choice. It was seen as the ideal expression of national will. Indeed, before the First World War, it is quite clear that the bayonet charge was not simply a tactic. It was seen in moral terms, expressing the communal will of the nation. In this way, it demonstrated the indomitable spirit of the nation to both enemies and citizens alike and communicated the willingness of a society to incur even very heavy casualties in the pursuit of its goals. It signified the absolute acceptance of sacrifice. By contrast, not only were platoon tactics difficult to perform but the careful use of fire, movement, and concealment to suppress and outflank a knowingly smaller opponent could not inspire the same fervour as a dynamic bayonet charge into the face of the enemy. Bayonet charges were thrilling in the way in which an orchestrated platoon assault often was not. Even in the Second World War, the bayonet was infused with intense moral significance. Armies themselves seemed to be aware of this. The award of medals involves complex institutional politics but it was noticeable that while Lieutenant Colonel Cole received a Medal of Honour for his bayonet charge at Carentan on 11 June 1944, Lieutenant Winters, from a sister regiment, the 506th, leading a company assault on an artillery position at Brecourt Manor on 6 June in which he skilfully utilized fire and movement tactics, was awarded the lesser Distinguished Service Cross, even though both were recommended by Colonel Sink, their commander in the 101st Airborne Division, for the Medal of Honour. Although professional, the French battalion in Korea favoured the bayonet. The use of sophisticated modern tactics—which would actually reduce casualties—seems to have been seen as less brave and virtuous in a mass army fighting a war of survival than the knowingly sacrificial act of a bayonet charge.

It would seem to be strange to claim that after the First World War, western powers—excluding perhaps Nazi Germany—were beguiled by the notion of sacrifice and were attracted by forms of attack which were likely to incur high costs. Certainly, after 1918, western powers (including to a lesser extent Nazi Germany) were more concerned about casualties, especially in comparison with authoritarian regimes like Russia and Japan. One of the main causes of dispute throughout the Normandy campaign was that Bernard Montgomery could not accept the same level of casualties for the British and Canadian armies as Omar Bradley. The experience of the First World War and the losses which had been built up since 1939 severely hampered him. Even then, in the Second World War, the Americans sought to replace the soldier with firepower primarily as a way of reducing casualties. This approach was institutionalized in Korea and Vietnam. Nevertheless, political and military leaders for most of the twentieth century were still willing to accept casualty levels which would be untenable today. Even in Vietnam, fighting a war of choice, very high levels of casualties were accepted by the US government; 58,000 US soldiers were killed. When, during the Korean War, the disastrous Battle of the Imjin in April 1951, in which 622 British soldiers were killed, wounded, or captured due to command incompetence, was reported to Parliament by the Minister of War, the members of the House cheered. Even though Korea represented a war which had only an indirect relevance to the defence of Britain, casualties signified military prowess and inspired sentiments of national pride. Of course, while these were very large figures in absolute terms, relative to the size of the armies at the time, they represented sustainable casualty rates. In this context, western societies seem to have made a virtue out of a necessity. The bayonet charge was a tactic which their under-trained armies could not only perform but which was actively seen as a noble expression of national will.

A similar claim might be made about the importance of individual action on the battlefield. Although individual heroism normally indicated not only a failure of training—platoons could not coordinate their actions—but also active passivity if not outright cowardice on the part of the majority of the troops, there are indications that it too might have been valued in the citizen army. It seems that rather than dwelling on the negative side of individual heroism, which was necessitated because of mass passivity, heroism was held up as an ideal not only of national character but also of masculinity. Combat leaders were supposed not merely to command their troops, coordinating their movements in combat, but to act as moral exemplars for them. In this context, it was not simply enough to command an assault. The moral authority of leadership depended on the willingness of a commander to lead it himself. The cohesion of his unit relied on his embodiment of masculine norms which could be demonstrated only by his performance of manly acts of heroism on the battlefield. In order to sustain the unity of a platoon, it was necessary for its leader to be heroic. Individual heroism was actively encouraged and, indeed, required. It seems possible that in the First and Second World Wars, especially in the British, French, and German cases, this investment in the individual actions of the leaders was a reflection of a more deferential and hierarchical culture. None of this suggests that the prime cause of low levels of participation in combat was not the lack of training. However, it seems plausible to suggest that the social dynamics displayed by the citizen platoon in combat was not only the result of a necessary lack of training but also possibly the product of an often unacknowledged and active preference for individual heroism and mass bayonet charges in the armed forces and wider societies of the time. The culture of western societies especially before the 1960s may have actively and surprisingly encouraged a preference for mass assault and individual heroism over modern tactics, even though the latter was fully recognized and advocated by the army.

8

Battle Drills

THE PROFESSIONAL ARMY

The western armies which fought the major wars of the twentieth century, from the First World War to Vietnam, included professional soldiers especially in senior enlisted and officer ranks. However, these armies were citizen armies, consisting for the most part of civilian volunteers or conscripts. In the case of the First and Second World Wars, citizens were plausibly enlisted on the basis of national survival and, in many cases, recognized it as their duty to fight. In the case of Korea and Vietnam and a series of other minor wars in which Britain and France, in particular, were involved, the concept of national mission was less convincing but nevertheless western states still enjoyed the authority and legitimacy to demand that citizens perform military duties, even if there was increasing reluctance and outright opposition to forced military service by the time of Vietnam.[1] Although many citizens volunteered for military service especially in the First and Second World Wars, it would be wrong to confuse the decision to enlist with that of the professional soldier. Few citizens understood service as a vocation. Rather volunteering was motivated by a sense of national duty and was often a method of exercising choice over the services, arm, and comrades with which a recruit would serve, knowing that subsequent drafting was either very likely or certain. Consequently, the basis of recruitment and the moral commitment of the citizen soldier to the armed forces was quite different from that of the professional soldier. From the 1960s, the armed forces began to undergo a profound historical transformation; they started to professionalize. Most social scientists have understandably focused on the implications of the decline of conscription to civil–military relations and the changing contract between the citizen and state. The abolition of conscription clearly has very significant implications for the political settlement. However, professionalization also has deep but often overlooked implications for the armed forces as an organization, fundamentally altering the relations between military personnel, the basis of institutional solidarity, capabilities, and operational performance. Professionalization does not simply involve a change of employment contract between the soldier and the armed forces. It represents a profound transformation of the associative patterns within the armed forces and the solidarities displayed within military units.

Canada was the first western force to professionalize. Immediately after the Second World War, as conscripts were demobilized, the Canadian Armed Forces reverted to an all-volunteer professional basis. The rapid abolition of conscription was substantially a product of Canada's favourable geo-strategic position. It faced no realistic territorial threat and could effectively free-ride on the United States' strategic defence. Consequently, Canada had no requirement for a large military force, especially since conscription was a contentious political issue between English- and French-speaking communities. Accordingly, in the post-war decades, partly to demonstrate its contribution to NATO and the international community, the Canadian armed forces claimed to specialize in peace-keeping; this was one of the few missions which Canada's tiny but deployable military could conduct. It was not until their involvement in Afghanistan, especially since 2006, that the Canadian forces were once again confronted with the serious prospect of combat. Some commentators have observed that from 1945 until 2000, Canada was effectively a professional force in name only. While its soldiers were full-time career service personnel, the level of competence was low. Granatstein and English have both argued that the abuses perpetrated by Canadian airborne soldiers in Somalia in 1993 were the product not so much of hypermasculinity but of inadequate professionalism.[2] They reflected the poor standard of training, discipline, and morale in the Canadian army, which was, they claim, typical throughout the post-war period. In effect, although the first 'professional' army in its terms of service, it might be possible to argue that the Canadian military professionalized along with most other western nations after the end of the Cold War in the 1990s and the 2000s.

Britain is unusual in Europe in that, following the Sandys Defence Review of 1957, which represented a major restructuring of defence, national service (conscription) was abolished in 1960 and the last cohort of conscripts left the armed forces in 1963, nearly forty years before most other European countries. The United States fought and lost Vietnam with a conscripted force. That campaign demonstrated many weaknesses in American political and military command but an ignorable difficulty was that posed by conscription itself. The conscripts who had fought in Vietnam were, as discussed, insufficiently trained or motivated to participate in such a campaign. Moreover, the presence of large numbers of conscripts, especially since they were disproportionately drawn from lower socio-economic groups, politicized the campaign, poisoning civil–military relations especially from 1968 until the American withdrawal in 1973. Accordingly, identifying conscription as a major problem, the draft was abolished in 1973 as an all-volunteer professional military was established. Although the Australian Task Force performed relatively well in Vietnam, there was substantial controversy about the retention of national service and the deployment of conscripts to this campaign. Accordingly, following the end of the Vietnam War, Australia, imitating its nearest allies the USA, Britain, and Canada, abolished national service in 1973.

Canada (in a strange way), the UK, USA, and Australia were at the forefront of military professionalism but they have not been completely exceptional. On the contrary, the pattern of professionalization was widespread in all western nations. From the 1970s, sociologists began to record the decline of the mass citizen army and emphasized the process of fairly dramatic reduction which has occurred since the end of the Cold War.[3] Tracing the steady decline in the numbers of conscripts and the length of their service, Michel Martin accurately predicted in 1977 the eventual appearance of professionalism in France, though the process took longer than he perhaps anticipated. It was not until the poor performance of French forces in the Gulf War of 1991 that political and military leaders began to consider the possibility of an all-volunteer force seriously. During that conflict, France could deploy only 14,000 soldiers to form the Daguet Division, even though their army at a strength of 280,000 was more than double the size of Britain's army, which deployed 40,000 troops. Moreover, these troops were of little operational value since they were improperly trained and equipped to integrate with the United States forces.[4] The performance of French troops in the Gulf led to the publication of the 1994 *Livre blanc* which outlined military reform.[5] In 1996, Jacques Chirac announced that the French military would be converted to an all-volunteer force with the abolition of military service by 2002.[6] The French military has now been fully professional for over a decade. The Italian army went through a similar process of reform; having provided a large conscript force for the defence of NATO's southern flank, it eventually abolished conscription in 2002. Italy abolished conscription in 2000.

The gradual decline of conscription was notable in Germany from the late 1960s, and professionalization of the Bundeswehr accelerated from the mid-1990s. One of the most radical phases of professionalization following Peter Struck's Defence Policy Guidelines in 2003, the Bundeswehr was reduced yet further to 252,000 by 2010, the majority of which was professional. Nevertheless, although an advocate of professionalization, Peter Struck affirmed the principle of conscription which he and the majority of political and military leaders regarded as essential to the maintenance of civilian control over the military. His successor Franz-Josef Jung confirmed its status. The consensus was that conscription ensured a close connection between civil society and the armed forces; conscription maintained political oversight and sustained moderation by ensuring that the Bundeswehr is populated by individuals from the entire spectrum of German society. The fear was that without this civil–military connection, extremists would infiltrate the Bundeswehr. The fear of a radicalized Bundeswehr is understandable given the history of Germany and there seemed to be no prospect of establishing an all-volunteer Bundeswehr. Nevertheless, following the credit crunch of 2008 and a severe contraction of the public budget, as well as increasing pressure from allies, Angela Merkel's (now disgraced) Defence Minister, Karl-Theodore zu Guttenberg, announced in the summer of 2010 that conscription would be abolished in 2011. Despite all the debate about conscription over the preceding decade and the regular assertions of its indispensability to both the Bundeswehr and the Bundesrepublik by ministers, politicians, and generals, the announcement was surprisingly uncontroversial. It was neither preceded nor followed by the intense political or public debate which had previously attended the issue. Professionalization was accepted without question, dissent, or much discussion and, in July 2011, the Bundeswehr became an all-volunteer force.

THE CONCEPT OF PROFESSIONALISM

The professionalization of western armed forces represents a historical change which reverses a 200-year trend, when states generally preferred mass citizen armies. It represents a profound transformation for the armed forces, not only at the level of civil–military relations but at the most intimate and brutal level of the combat soldier. In order to appreciate the nature of this transformation and to begin to understand the impact of professionalization on the cohesiveness of the infantry platoon, it is imperative in the first instance that a general coherent concept of professionalism is established. Military professionalization refers, in the first instance, merely to a contractual change in the employment status of soldiers and their individual relationships to the armed forces. Professionals volunteer for military service and are remunerated financially for the work they perform during their career, the length of which is not tied to any specific national mission such as a war but to contracts which they have signed for set periods ultimately with their respective defence ministries. In the 1970s, partly as a response to the professionalization of the US Army, Charles Moskos sought to identify the main differences between a conscript and an all-volunteer force, developing the well-known 'institutional-occupational' categories to define the respective military types. For Charles Moskos, the defining feature which distinguished the institutional or citizen army from the occupational or professional military was primarily economic. While the institutional (citizen) military is legitimized by normative values of national defence and civic responsibility, 'transcending individual self interest', the occupational force is organized around a 'marketplace economy'.[7] In the occupational military, there is 'no difference between cost-effectiveness analysis of civilian enterprise and military services', 'military compensation should as much as possible be in cash, rather than in kind or deferred' and 'military compensation should be linked directly to skill differences of individual service members'.[8] In the occupational military, which Moskos regards as retrograde, service personnel are oriented to money and are rewarded and promoted on the basis of their specialist individual skills. Somewhat counterintuitively the professionalized, occupational military becomes increasingly civilianized according to Moskos, even though there are no longer any citizen soldiers in it. Specifically, Moskos's account heavily implies that the emergence of an occupational military as a result of professionalization has increased individualism and undermined corporate commitment, which has been typically essential to military performance. Soldiers simply work for the military, feel less commitment to their regiment or division (and more commitment to their employment specialism like civilian workers), and, therefore, Moskos implies, are less willing to risk themselves on operations.[9]

Moskos admits that the divide between institutional and occupational militaries is not always clear: 'The institutional versus occupational thesis seeks to identify an overarching trend while still recognising that military systems are differentially shaped, depending upon a country's civil–military history, military traditions, and geopolitical position. Moreover, I/O modalities interface in different ways even with the same national military system. Differences exist between military services and between branches within these services, I/O modalities may also vary along internal distinctions, such as those between officers, non-commissioned officers, and lower ranks; between career and single-term military members.'[10] It seems highly likely that, as Moskos implies, it is not possible to sustain a monolithic division between institutional and occupational categories; soldiers in any army are likely to define themselves and be motivated in various ways. Nevertheless, Moskos maintains the fundamental veracity of his 'institutional/occupational' distinction. For him, professionalization refers to the influx of the market forces into the military with a fundamental change in relation between service personnel and the armed forces (turning them into no more than employees and employers) and between service personnel themselves.

Moskos's concept of an occupational military usefully highlights the organizational principle which respectively underpins the citizen and professional army, and he is surely correct to emphasize the profound implications of professionalization. Yet, the suggestion of the organizational superiority of the institutional (citizen) model seems to be highly dubious. Like Moskos, other scholars have also observed that professionalization almost necessarily involves very significant transformations in the structure of the armed forces. Professionalization involves the end of the mass army; it is above all a process of 'downsizing'.[11] Thus, today's professional armed forces are significantly smaller than their forebears in the Cold War; indeed, in most European countries, they are the smallest they have been for over a century, and even longer, in the case of Britain. However, the armed forces should not be understood merely as in numeric decline. Against Moskos, professionalization does not seem to represent a weakening of the armed forces. All volunteer forces are qualitatively different from the mass armies of the twentieth century. Yet, as they have become professionalized, they have specialized. Indeed, as they have moved to a professional model, resources have focused on particular forces within the military; investment has been directed to selected forces which have actually increased in size and capability. Professional forces are not simply smaller (and better-paid) versions of conscript armies. The concept of downsizing is not wrong—forces are shrinking—but it may not capture the full dynamic of professionalization. The professionalized army, while significantly smaller than the citizen army, is, in relative terms, substantially more powerful. Professionalization involves not so much a reduction of capability but its concentration: a condensation of military power rather than its mere diminution. In contrast to Moskos's implication, that the occupational army is weaker than the institutional force, the evidence suggests the opposite; that the professional force increases the commitment and performance levels of the armed forces. Certainly, a belief in the greater effectiveness of the all-volunteer force is the critical reason why political leaders and military commanders have favoured it over the last few decades. Indeed, in his analysis of the rise of Special Operations Forces (SOF), which (as we shall see) are at the heart of the process of professionalization, Rune Henriksen has usefully elucidated the logic of concentration: 'With limited budget, political will and manpower, Western militaries are being forced to make sure every individual who is willing to fight is not just an able individual, but a very capable one. This is in part the reason for the relative doctrinal surge of SOF in postmodern militaries.'[12] Professional forces have been increasingly favoured by western polities because they offer greater military capability for relatively less money.

The increase in military competence which is typically involved with professionalization can be perhaps best understood and explored if the process is situated in wider studies of professionalism. The emergence of professional status groups has been of long-standing interest to sociologists and it was central to the work of both Max Weber and Émile Durkheim from the end of the nineteenth century. In his famous definition of status groups Weber claims that status groups always aimed to monopolize certain 'ideal and material goods or opportunities'.[13] Indeed, Weber employed an economic example to illustrate the point: the status group typically emerges 'when the number of competitors increases in relation to the profit span' and 'the participants become interested in curbing competition'.[14] In order to do this, any would-be status group has to exclude non-members.[15] The status group has to form itself into a unified entity closed to outsiders. The members of the group have to recognize the special relationship which binds them to each other to the exclusion of others; they have to recognize their *collective* interests.

As Weber stresses, in order for a group to emerge, its members have to recognize that they share something in common but what they share is not predetermined. On the contrary, the group has to select certain criteria of group membership which all consciously recognize. Perhaps surprisingly, Weber argues that membership criteria are ultimately arbitrary: 'Usually one group of competitors takes some externally identifiable characteristic of another group of (actual or potential) competitors—race, language, religion, local or social origin, descent, residence, etc—as a pretext for attempting their exclusion. It does not matter which characteristic is chosen in the individual case: whatever suggests itself most easily is seized upon.'[16] As a result of these identifications, the members of the status group develop a special relationship with each other, on the basis of which they are prepared to assist each other in the generation of collective goods while explicitly ensuring that non-group members (who lack the requisite identification) are excluded. Indeed, it is a duty of status group membership to ensure that outsiders are excluded. On the basis of arbitrary social criteria, status groups monopolize opportunities of every kind. As Reinhard Bendix has perceptively observed: 'Weber's approach conceived of society as an arena of competing status groups, each with its own economic interest and orientation toward the world and man... This emphasis upon the struggle among different social groups was at the core of Max Weber's personal and intellectual outlook on life.'[17] Weber further noted that historically the most successful status groups have sought to protect their monopolies by legal sanction: 'if the monopolistic interests persist, the time comes when the competitors, or another group whom they can influence (for example, a political community) establish a legal order that limits competition through formal monopolies.'[18] It becomes a 'legally privileged group'.

Weber gives numerous examples of 'cooperative [i.e. monopolizing] organization' such as fishermen 'taking their name from a certain fishing area, associations of engineering graduates monopolising positions, villagers protecting their commons, associations of shop clerks, knights, university graduates, craftsmen and ex-soldiers'.[19] Weber does not specifically mention the professional military as a status group but since he defined a standing army as enjoying 'the monopoly of legitimate violence' and he recognized that both knights and ex-soldiers constituted status groups, it is plausible to suggest that the military and especially the officer corps could be seen as a professional status group in his eyes. Indeed, in his writing on German capitalism, he noted the huge cultural and political influence which the Prussian Junkers, as a military and landowning elite, exerted over German society. Weber certainly regarded status groups as irrationally manipulative; they actively sought to undermine free competition by the erection of arbitrary and self-interested cultural barriers. Yet, status groups were not defined merely by the privileges which they enjoyed and protected, though they were not insignificant. Significantly, Weber was well aware of the role of training and education in the development of successful status groups: 'This monopolistic tendency takes on specific forms when groups are formed by persons with shared qualities *acquired* through upbringing, apprenticeship and training...if in such a case an association results from social action, it tends toward the *guild*.'[20] The maintenance of a guild requires the irrational reference to arbitrary criteria so that outsiders are excluded, but Weber implies that guilds can sustain themselves only if they possess some genuine and socially useful expertise. Thus, Weber emphasizes the importance of training and qualifications to would-be members: 'Only those are admitted to the unrestricted practice of the vocation who (1) have completed a novitiate in order to acquire the proper training, (2) have proven their qualification, and (3) sometimes have passed through further waiting periods and met additional requirements.'[21] Typically, these qualifications become the decisive criteria of membership; they must be earned in order to join the group.

Weber also highlighted the role of what he called 'status honour' in the maintenance of these groups. Members of status groups were expected to comport themselves in distinctive ways, sanctioned by the group. They were to adopt the status lifestyle of the group. Weber was well aware of the purely exclusive function which these lifestyles served; they distinguished the group from outsiders. However, status honour referred not merely to appropriate comportment. It denoted the individual member's obligation to defend and contribute to the group's collective interests: to cooperate willingly with fellow group members and to defend the group's monopoly from outsiders.

Weber's theory of the status group is, despite its deceptive and often overlooked simplicity, profound. However, with the exception of his work on bureaucracy and his lectures about the vocations of science and politics, Weber rarely applied this theory of status groups to an empirical exposition of a particular profession and, in fact, his work on bureaucracy ironically lacked the sociological orientation which the theory of status groups suggested. Weber saw bureaucracy as a form of rationalization in which individual clerks followed general rules, rather than as a status group of administrators (a *noblesse de robe*) with its own status honour, lifestyle, and monopolizations. Fortunately, Randall Collins, especially in his work on credentialism, has explicitly sought to demonstrate how the theory of the status group is relevant to understanding professionalism in modern western society. For Collins, modern credentials such as school diplomas and university degrees are not only designed to inculcate skills so that appropriately trained and capable individuals are able to fulfil socially necessary functions. Collins shows that credentials are substantially about limiting access to particular economic opportunities. Professional lawyers, doctors, and academics all create credential systems not because these necessarily ensure that the most capable candidates are accepted and promoted or that the social function of the profession is fulfilled, but to sustain the group's monopoly as a whole:

'People are actively concerned with the process of gaining and controlling occupational power and income, not merely (or even primarily) with using skills to maximize production.'[22] Credentials limit access to these opportunities and ensure that only those who are members of the professional status group (the 'qualified') can exploit these opportunities. Collins notes cases where individuals have masqueraded as doctors or lawyers without qualifications, indicating that credentials are not strictly required to perform a role. In other cases, it is far from clear that the objectively best individual candidates for certain kinds of profession are selected. By questioning the link between credentials, skills, opportunity, and remuneration, Collins usefully places professionalism within its proper competitive context. Status groups have to maintain themselves and their exclusive monopolies constantly; credentials are a central element in this process. Paralleling Collins's work, Andrew Abbott has shown how professions operate not alone but in a system, with a multiplicity of status groups interacting and competing with each other.[23] The standing and role of each status group is a product of these interactional dynamics between the professions, each seeking to defend its own jurisdictions while promoting itself over others.

However, although Collins's original writing sometimes suggests that the skill of professional status groups is a fiction, designed merely to justify monopolies, he does not dismiss the importance of expertise in sustaining them. As he notes: 'this is not to say that no one is involved in productive work.'[24] Andrew Abbott similarly demonstrates that while political machinations are critical, the specialist expertise of each status group is critical to its success in its struggle with others.[25] In order for a status group to monopolize an opportunity over any length of time, it must be able to perform tangible services and functions which others cannot match. While it is true that rare individuals are able to fake their way into a status group periodically without the credentials, this seems to be primarily because the status groups' expertise already exists so that it can be learnt on the job by enterprising individuals. In his more recent work on the sociology of philosophy, effectively one form of arcane professional expertise, Collins fully recognizes the importance of identifiable forms of expertise.[26] In this work, Collins explores the way in which philosophical movements emerge to dominate intellectual debate. Collins consistently demonstrates the contingent and political factors which underpin the success of these movements; the genius of major intellectual figures lies not simply in their own brilliance but crucially in fertile network connections and alliances which germinate and disseminate their work. However, the great philosophers are able to exploit the networks in which they are advantageously positioned because their work synthesizes a variety of strands of thought and represents an answer to major contemporary issues in ways which resolve critical problems or at least suggest fruitful lines of research. Geniuses and their followers are not illusionists; they develop forms of knowledge, skills, and expertise which are real. It is precisely the coherence of their thought and the utility it has for others that assure them their privileged positions.

Professionalism has typically arisen within status groups as they have monopolized certain opportunities in industrial society. The naked and self-interested politics of this process is evident. However, monopolizing particular activities and dedicating themselves to them, professional groups have developed skills which have concrete and productive effects. Crucially, other groups find these skills useful and, indeed, the more dependent other groups, the state, or the population as a whole become on the expertise of a professional group, the more powerful they become. Expertise is, then, almost always crucial to the maintenance and consolidation of any professional status group; professional groups generate and protect unique forms of knowledge and skill. Accordingly, professionalism might then be defined as specialist expertise which is generated and sustained by a unified group. Indeed, it might be possible to generalize and to suggest that professionalism may be most easily defined as systematic and collective attention to detail in a specific domain of activity by this group. Professional status groups maintain their monopolies by refining, expanding, and unifying their expertise, not least because political debates with competitors can be most successfully won by pointing to concrete and non-replicable skills.

While Charles Moskos was sceptical about the professionalization of the military, which he saw as potentially undermining its moral strength, other prominent social scientists, equating professionalism with precisely the specialist expertise which is found in the more general literature from Weber onwards, have stressed its importance to the performance and reputation of the armed forces from the mid-twentieth century. Thus, in the opening paragraph of *The Soldier and the State*, one of the most important works on the military profession and civil–military relations in the twentieth century, Samuel Huntington suggests that 'a profession is a peculiar type of functional group with highly specialized characteristics' and he maintains that, defined in this way, 'professionalism' is 'characteristic of the modern officer in the same sense in which it is characteristic of the physician or the lawyer'.[27] For Huntington, in contrast to Moskos, professionalism consists of expertise ('specialized knowledge and skill in a significant field of human endeavour'),[28] responsibility (the service is 'essential to the functioning of society'), and corporateness ('the members of a profession share a sense of organic unity and consciousness of themselves as a group apart from laymen').[29] The question of monopolization is less evident in Huntington's definition of professionalism than in the work of Weber and Collins but his emphasis on skills and the provision of essential services to others is consistent with wider sociological accounts of professionalism.

Huntington then applies these three definitions to the military profession. The question of corporateness (the third characteristic) will be discussed in

Chapter 10 but the first two characteristics—expertise and responsibility—are of immediate relevance here. Huntington notes the wide diversity of roles and skills in the officer corps which consists of 'engineers, doctors, pilots, ordnance experts, personnel experts, intelligence experts, communications experts'.[30] 'Yet, a distinct sphere of military competence does exist which is common to all, or almost all, officers and which distinguishes them from all, or almost all, civilians. This central skill is perhaps best summed up in Harold Lasswell's phrase "the management of violence".'[31] In order to manage violence successfully, officers have three fundamental duties: '(1) the organization, equipping, and training of this force; (2) the planning of its activities; and (3) the direction of its operation in and out of combat.'[32] Huntington contrasts modern military professionalism with historical forms of military command: 'Before the management of violence became the extremely complex task it is in modern civilization, it was possible for someone without specialized training to practice officership.'[33] By contrast with the rise of industrial warfare, 'no individual, whatever his inherent intellectual ability of character and leadership, could perform these functions efficiently without considerable training and experience'.[34] Huntington describes three central capabilities which a successful officer requires. First, officers need to understand warfare in general and, intriguingly, he draws a parallel between this 'universal' knowledge and other professions: 'Just as the qualifications of a good surgeon are the same in Zurich as they are in New York, the same standards of professional military competence apply in Russia as in America and in the nineteenth century as in the twentieth.'[35] The professional officer understands the utility of military violence, the dynamics of war, and the frictions of battle just as a doctor comprehends the workings of the human body, its pathologies, and the likely effects of different kinds of treatments. Secondly, the military officer requires an understanding of the development of war and a proven knowledge of his own specific area of expertise: 'A military specialist is an officer who is peculiarly expert at directing the application of violence under certain prescribed conditions.'[36] Thirdly, echoing Andrew Abbott's concept of the system of professions, the officer requires some understanding of cognate forms of expertise: 'To understand his trade properly, the officer must have some idea of its relation to these other fields and... The fact that, like the lawyer and the physician, he is continuously dealing with human beings requires him to have the deeper understandings of human attitudes, motivation, and behaviour which a liberal education stimulates.'[37] In terms of responsibility, the military professional applies this skill in 'the management of violence'[38] to ensure 'the military security of his client, society'.[39]

The result is a distinctive pattern of motivation which Huntington regards as constitutive of professionalism. Against Moskos's concept of the occupational type, the professional officer is not a mercenary driven only by 'economic incentives'. On the contrary, 'the motivations of the officer are a technical love for his craft and the sense of social obligation to utilize this craft for the benefit of society. The combination of these drives constitutes professional motivation.'[40] For Huntington, military professionalism consists of a diverse body of expertise, including detailed and highly technical knowledge in numerous domains, united by its focus on the management of violence and underpinned by an individual and collective ethos not of financial remuneration but of vocational excellence. Huntington's point here is interesting, for while the formal definition of professionalism is purely economic—individuals are paid for the service they provide—professionalism, in fact, necessarily involves an existential commitment, even a passion, which quite transcends financial motivations. Indeed, in his work on command, Bryan McCoy explicitly describes professionalism as a passion.[41] True professionalism is a vocation. There are evident sociological reasons why professional groups are oriented in the first instance to expertise, rather than money, especially at the collective level. Their existence as a status group and all the social and economic benefits which derive from it are dependent not in the first instance on the money they make—their wealth—but on the skills which they have. Financial remuneration is doubtless critical to most status groups, but to pursue only money is for a profession to deny its status as such. One of the most damaging criticisms for a profession is that it is simply using its privileges to earn money. Accordingly, it might be possible to suggest somewhat paradoxically that true professionals need to be amateurs (in the proper sense of the word) in their orientation to their employment: they have to love their work precisely because it is the skills which they bring to it which sustain their status. Huntington does not explicitly define military professionalism as the dedicated attention to detail, but his account of this pursuit of excellence implies precisely this orientation. The professional officer corps refines every aspect of its specialist knowledge while uniting it in a coherent understanding of the management of violence in general and sufficiently trained to employ that knowledge.

In his work *The Professional Soldier*, published only three years after *The Soldier and the State*, and addressing very similar themes, Morris Janowitz described military professionalism in terms closely compatible with Huntington: 'The officer corps can also be analysed as a professional group by means of sociological concepts. Law and medicine have been identified as the most ancient professions. The professional, as a result of prolonged training, acquires a skill which enables him to render specialized service.'[42] In line with his argument for the civilianization of the military and, therefore, a version of subjective civilian control of the armed forces, against Huntington's ideal of objective military control, Janowitz lays out five hypotheses which qualify the character of the military profession. First, 'authoritarian domination' has been replaced by 'greater reliance on manipulation, persuasion and group consensus' due primarily to the 'technical character of modern warfare' which requires complex military teams, each member of whom 'must make a technical contribution'.[43] Secondly, 'the new tasks of the military require that the professional officer develop more and more of the skills and orientations common to civilian administrators and civilian leaders' and be more politically oriented.[44] Thirdly, reflecting its civilianization, officer recruitment has become less elitist and 'more representative of the population as a whole',[45] with a concomitant '"democratization" of outlook and behaviour'. Fourthly, a concept of a military career has appeared. Finally, professional officers have become more politically attuned to the political ends for which they fight and less nationalistic and ethnocentric in their outlook.[46] It is not relevant to determine here whether Janowitz was correct about the civilianization of the professional soldier either in the 1950s when he was writing or in the present. However, there are evident parallels with Huntington's work which are extremely useful in establishing a definition of professionalism. Like Huntington, Janowitz regards a set of specialist skills which are developed to provide a service for society as essential. Indeed, in his concluding passage on professionalism, Janowitz begins to describe a professional ethos which echoes Huntington's account very closely: 'the military professional is unique because he is an expert in war-making and in the organized use of violence.'[47] He adds: 'The style of life of the military community and a sense of military honour serve to perpetuate professional distinctiveness.'[48] The military profession is characterized by a distinctive form of expertise but the genesis, maintenance, and refinement of this expertise relies not on technical knowledge but on a collective ethos: a vocational commitment to excellence which will ensure the reputation of the armed forces in civilian society. Like Huntington, Janowitz's professional soldiers display an amateur enthusiasm for their career. This passionate pursuit of improvement is facilitated by the professional soldier's full-time employment by the state, but the ethos quite transcends its economic underpinnings. Indeed, it would be quite possible to think of historic and indeed contemporary examples of 'professional' standing armies which did not display a professional ethos at all. Here officers were or are uninterested in military science or in the efficient management of the armed forces; troops under their command would be badly trained and disordered. Accordingly, following the sociological literature on professions and Huntington's and Janowitz's work on the officer, professionalism is not used to refer simply to soldiers who volunteer to be employed on a full-time basis by the state. Professionalism refers to the highly distinctive ethos of a status group, dedicated to the maintenance and improvement of its expertise, in order to sustain its privileges and monopolies.

Janowitz and Huntington, of course, discuss only officers. They rightly recognize that in understanding the structure and workings of the armed forces as a whole and, in particular, in elucidating the relationship between the military and civil society, the officer corps and, above all, the commanders of the forces play a decisive role. In one sentence in his chapter defining professionalism, Huntingdon mentions the firing of a rifle and the command of a rifle company as an example of the management of violence. The end of his book concludes with a notorious passage which equates the US Army's West Point, training future platoon commanders, with Sparta and American civil society with Babylon,[49] but he is overwhelmingly concerned with higher command, as is Janowitz. Not only is this officerly focus justified in terms of their interests but, in the 1950s, when they were conducting their research, all western armed forces still retained conscription and, thus, professionalism was limited to a small cadre of career officers and non-commissioned officers. However, there is no reason to think that Janowitz and Huntington believed that their concept of professionalism applied only to the officer corps. As his famous 1948 article demonstrated, Janowitz was deeply interested by combat soldiers and had analysed an army which was widely regarded as the most professional of all the conscript forces, namely the Wehrmacht. Combat soldiers, in their sections and platoons, are engaged in perhaps less intellectually complex problems than those which exercised generals, but it is evident that their work implies that professionalism can be found among the infantry. The individual and collective skills of the infantry soldier are more practical than those of the general. Yet, the tactical decisions of section or squad and platoon commanders correspond with those of generals at the operational and strategic levels. Moreover, the successful execution of platoon tactics requires a high level of individual and collective skill.

A successful infantry platoon demonstrates the central features of professionalism recognized by Janowitz and Huntington. The platoon is involved in the management of violence, to which end it develops a range of specialist skills which improve its performance in combat. The genuinely professional infantry soldier is not just a paid functionary but an individual committed to soldiering as a vocation, ideally fascinated by every dimension of tactics. Like the generals who ultimately command them, infantry soldiers are ideally attuned to every detail of their performance, which they attempt to improve. Attention to detail, training, and preparation are, consequently, central characteristics of the professional infantry soldier as they are of the officer corps. It is certainly true that these skills may be more practical than the intellectual activities more often associated with professionalism. Yet, the fact that those skills are closer to those of a professional mountaineer than of a lawyer should not detract from the fact that the infantry possess specialist knowledge and skills. In Chapter 6, it was noted that the citizen infantry aspired to and could sometimes attain professional standards if training cycles were adequate and personnel remained stable. These conditions were rarely met especially in wartime due to the structural constraints generated by total industrial war and the requirement for mass armies. The smaller all-volunteer force, by contrast, fundamentally alters the associative dynamics in the infantry platoon.

Precisely because the force is so much smaller and, as a professional force, it is expected to be more proficient, the levels of individual and collective skill are able to be increased through intensive training. Moreover, in contrast to the citizen army, the fact that everyone in a platoon is a full-time professional with a substantially higher level of individual experience and training improves the collective performance. In effect, the fact that everyone is a professional alters the nature of solidarity in the platoon, changing the possibility and dynamics of combat performance.

PROFESSIONAL DRILLS

The professional expertise of the infantry soldier is diverse. Contemporary US Army doctrine defines individual infantry skills as consisting of five basic capabilities: shoot, move, communicate, survive, and sustain.[50] This list appears simple. Yet, each of these capabilities requires numerous individual and collective skills which differ according to the environment in which the soldier is operating. Although the principles remain the same, patrolling and, indeed, just surviving in the Arctic requires a completely different set of skills from operating in the jungle. The professional soldier learns and develops all these skills, and any one of these capabilities might be used to explore the question of professionalism in the military. However, in the current context, the enquiry has focused on the platoon attack as the simplest and most brutal collective activity in which combat soldiers are employed. This study is interested in the execution of battlefield tactics. Clearly, in different environments and against different enemies, tactics differ, even at the platoon level. Accordingly, the term tactics typically presumes an idea of analysis and judgement in the application of military violence. However, although tactics therefore, necessarily, exceeds mere drill since it always includes judgement, at platoon level there is very considerable overlap between battle drill and battle tactics. The platoon or section attack, using fire and movement, are plainly a drill, and yet they are also a tactic. Moreover, successful battlefield tactics almost always requires a platoon to be able to perform a basic repertoire of battle drills effectively. Platoon commanders are only able to make a judgement about tactics insofar as their platoons can execute them. The battle drill is, then, central to the platoon attack and to all platoon tactics in fact, as Chapter 6 has demonstrated, and, consequently, in order to explore the distinctive character of professionalism within the infantry, the following examination focuses on the drill.

In doing so, the analysis follows much existing scholarship. In his intervention into the cohesion debate, Hew Strachan has considered the classic explanations of cohesion, the primary group, political ideology, and punishment, but has emphasized the importance of the battle drill which can be inculcated only through training: 'Training creates the psychological capacity to elongate the peak phase [of combat effectiveness] and to surmount the low points of the later phases. It does so in part through the inculcation of battle drills, a set of procedures, so that when exhaustion makes rational thought impossible, or when fear has taken over, individuals react without thinking.'[51] Strachan illustrates the development of training and battle drills through the history of stormtroopers in the German army and General Maxse's innovations in the British army in the First World War and the subsequent practices of the British and German armies before and during the Second World War. He explores the ways in which training had a psychological effect on the individual, preparing him for the experience of combat and noting that good training reduced the number of psychological casualties.[52]

However, while the psychological and individual impacts are significant, the importance of training for Strachan lies in its collective effects. Crucially, hard training generated a high level of unity and cohesion: 'Men were taught automatically to adopt certain drills which were designed to create self-confidence and give them a plan of action to be carried out when in battle.'[53] Through the institution of battle drills, soldiers collectively committed themselves to an automatic repertoire of activities, which everyone knew (and knew that everyone else knew) and which they trusted their comrades to perform. For Strachan, the effective performance of infantry from the First World War to the Falklands was ultimately the product not of primary group cohesion (Moskos's quasi-mystical bond), political ideology, or punishment but these battle drills. Strachan does not ultimately dismiss the concept of the primary group. The tactical cohesion of small units of soldiers was essential for effective military performance. However, cohesion was founded on the immediately relevant performance of military practices—tactical drills—not on friendship or generic social solidarity, which brings Strachan to a profound conclusion about the nature of professionalism itself: 'A professional soldier may pass his whole career without ever confronting it [combat]. But that does not relieve him of the need to prepare for it, nor erode the centrality of training for all he does.'[54] Here Strachan makes an important point which connects closely with Huntington and Janowitz's definition of professionalism as a complex of skills. For Strachan, especially at the level of the infantry platoon, the decisive skill seems to be the battle drill whose fundamental purpose is to ensure appropriate and replicable individual and collective responses in the face of enemy fire. The battle drill seems to have become the core of professionalism at the level of the infantry platoon, embodying the central individual, communal, intellectual, and physical skills which are requisite for success at this level of combat.

Of course, the battle drill has a much longer history than the First World War. Battle drills have been central to the training and performance of armies since at least the late sixteenth and early seventeenth centuries with the reforms of Maurice of Orange and Gustavus Adolphus. All nineteenth-century armies had battle drills based on lineal manoeuvre supported by fire, although by the American Civil War, their utility had become questionable. Chapter 6 described how modern battle drills were discovered and recurrently practised in the First and Second World Wars and beyond, and many historians, like Strachan, have emphasized their importance. There is a clear line of continuity in the development of battle drills since the First World War. Nevertheless, from the 1970s, as western forces professionalized, battle drills have become notably more refined. Contemporary infantry doctrine certainly bears a great similarity to early tactical manuals, therefore, but it is distinguished by the detail and comprehensiveness of its battle drills. In particular, battle drills have sought to standardize every aspect of infantry performance. Indeed, contemporary doctrine emphasizes this aim of standardization; it deliberately seeks to reduce infantry tactics to a set of predictable and regulated practices. 'Infantry doctrine expresses the concise expression of how Infantry forces fight. It is comprised of principles; tactics, techniques and procedures (TTP); and terms and symbols.'[55] Standardized procedure is explicitly recognized as being the best antidote to the disorientation of combat: 'One of the defining characteristics of war is chaos. TTP are the counter-weight to this chaos.'[56] Furthermore the development and application of detailed and prescriptive doctrine is seen as central to professionalism: 'Doctrine provides a common language that professionals use to communicate with one another. Terms with commonly understood definitions are a major component of the language. Symbols are its graphic representation. Establishing and using words and symbols of common military meaning enhances communications among military professionals in all environments, and makes a common understanding of doctrine possible.'[57] Specifically, through the creation of a common lexicon of actions and commands, contemporary doctrine aims to improve the cohesiveness of the armed forces. Every member of the force is oriented to the same suite of practices, which are cued by universally recognized symbols.

The battle drill has become one of the central practices of the professional infantry and it has been reduced to a set of simple common terms, represented by verbal symbols. Thus, in contemporary US infantry doctrine, all the elements of a platoon attack are prescribed: '2-1. General. A battle drill is a collective action executed by a platoon or smaller element without the application of a deliberate decision making process. The action is vital to success in combat or critical to preserving life.'[58] The publication describes 'Battle Drill 07-3-D3991—React to Contact' in detail:[59]

TASK STEPS AND PERFORMANCE MEASURES:
b. Immediate Assault
  (1) The platoon and the enemy simultaneously detect each other at close range.
  (2) All Soldiers who see the enemy engage and announce 'contact' with a clock direction and distance to enemy, example, 'Contact three o'clock, 100 meters.'
  (3) Squads in contact immediately assault the enemy using fire and movement.
  (4) The platoon destroys the enemy or forces them to withdraw.
  (5) The platoon leader reports the contact to higher headquarters.[60]

Twentieth-century doctrine described the platoon assault in terms of fire and movement. However, the prosecution of the final assault was described in general terms, leaving it to the platoon or section commander to improvise as they saw fit. Current battle drills break the process down into a series of standardized stages. For instance, in place of general comments about assaulting with all weapons firing and speed and aggression which featured in First and Second World War doctrine, current US infantry doctrine, Field Manual (FM) 3-21.8, *The Infantry Rifle Platoon and Squad*, provides clear and demonstrative directions about how to assault a trench.
  Entering the Trenchline
  7-228. To enter the enemy trench the platoon takes the following steps:

- The squad leader and the assault fire team move to the last covered and concealed position near the entry point.
- The squad leader confirms the entry point.
- The platoon leader or squad leader shifts the base of fire away from the entry point.
- The base of fire continues to suppress trench and adjacent enemy positions as required.
- Buddy team #1 [the US divides its fire-teams of four soldiers into two 'buddy' pairs] (team leader and automatic rifleman) remains in a position short of the trench to add suppressive fires for the initial entry.
- Buddy team #2 (grenadier and rifleman) and squad leader move to the entry point. They move by rushes or by crawling (squad leader positions himself where he can best control his teams).
- Buddy team #2 positions itself parallel to the edge of the trench. Team members get on their backs.
- On the squad leader command of COOK OFF GRENADES (2 second maximum), they shout, FRAG OUT, and throw the grenades into the trench.

- Upon detonation of both grenades, the Soldiers roll into the trench, landing on their feet and back-to-back. They engage all known, likely or suspected enemy positions.[61]

There is a clear parallel between this account and western infantry doctrine in the First and Second World Wars. Infantry doctrine at that point certainly implied this sequence of events and, in practice, experienced combat soldiers would probably have performed a very similar set of actions, but the specific details of individual and squad actions were often omitted from doctrine. Here, US doctrine provides not only a comprehensive account of the assault broken into discrete stages but it even specifies how troops should roll into the trench with their backs to each other so that they are mutually protecting each other. In addition, there is complete clarity throughout the doctrine as to the roles and functions of each member of the platoon. Such a level of detail is absent in previous doctrine. Indeed, the final section of the assault of western infantry doctrine was typically described in emotive rather than instructive terms emphasizing the morale of the troops rather than focusing on their performance. For instance, German infantry doctrine in the Second World War recommended that the final assault was initiated with a 'Hurra',[62] and Allied doctrine was similarly exhortatory rather than prescriptive. In the professional infantry, drills have been developed in response to an increased number of situations and these drills have been dissected and refined into the micro-actions of the assault.

Drills have been developed to improve individual performance, of course, aiming to inculcate the novice soldier with the knowledge of the experienced combat veteran. However, the refinement of drills has simultaneously involved their standardization so that all soldiers are trained to perform the same acts. The standardization of drills is important to improving the competence of the infantry since coordinated action is essential to combat performance. Standardization obviates the need for individual improvisation and, therefore, idiosyncrasy, deviancy, and malcoordination. All infantry soldiers are trained to the same highly detailed pattern. This standardization of process is perhaps most obviously demonstrated in infantry doctrine by the use of mnemonics, which are a prominent feature of contemporary doctrine. Mnemonics aim to assist soldiers and platoon commanders in remembering particular drills and they have proved effective in this way. However, of course, the mnemonics are effective only because drills have been reduced to a series of divisible procedures. Thus although the USA's Field Manual *The Infantry Rifle Platoon and Squad* is a discursive document, involving much descriptive text, its fundamentals could be reduced to a series of easily remembered mnemonics. Similarly, in Britain, contemporary infantry training demonstrates the importance of set drills, crystallized into simple mnemonics very clearly. In British doctrine, section squad attacks are reduced to the 'six section battle drill' PREWAR (preparation, reaction to effective enemy fire, enemy location, win the fire-fight, assault, reorganize). The preparation phase referring to the phase of a section attack is itself further broken down into a secondary mnemonic, PAWPERSO, which stands for protection, ammunition, weapons, personal camouflage, equipment, radios, special equipment, and orders. During battle preparation (P), the section commander gathers his section and puts out sentries to ensure security as they check they have the correct ammunition, weapons, and equipment. He then gives them their orders and checks that communications are working. Assuming that contact with the enemy is made, the section reacts to enemy fire, seeking to take cover and return fire (R). After this initial reaction, the section must locate the enemy (E), which is achieved by the observation of movement, smoke, or muzzle flashes and communicated by standardized means which involve the identification of axes that are obvious to the section. At this point, the section commander issues a fire control order (including the direction and weight of fire) in an attempt to win the fire-fight (W). The section commander then conducts an estimate to determine whether the section should mount a left, right, or frontal attack. He issues quick battle orders which will normally involve the establishment of a final position for one of the fire teams to create a base of fire, while the other team, normally led by the section commander, mounts the assault (A). Once the assault has been successfully executed, the section reorganizes itself beyond the objective (R). In training, the two mnemonics (PAWPERSO and PREWAR) are repeatedly emphasized, and indeed both demonstrations and instruction are organized around these mnemonics so that every action is identified and confirmed and all members of the section are explicitly united in what their reactions should be. British instructors stress the importance of these battle drills to coherent battlefield performance: 'When it goes bang, the world gets smaller. You forget you are part of a section. But if it goes off in the next 20 metres, you have put in the back of the lads' mind the six section battle drill.' As a result of repeated training, soldiers will know that 'if something goes bang, they take cover' and from that cover they begin to initiate the six-section battle drill. The PREWAR mnemonic allows the commander and his soldiers to coordinate their reaction in time and space with minimal requirement for discussion and deliberation. The actions of each individual are phased and ordered into a commonly understood sequence oriented to a clear objective. Following an instruction lesson on the section attack and the six-section battle drill at the Braunton Burrows training area in north Devon, a Royal Marine sergeant concluded: 'PAWPERSO and PREWAR; as long as you are doing that you can't go wrong. It is a guide. You can put in your own artistic licence but, for this course, this is how it will go.'[63]

Battle drills have become very important to the US Marines and they have similarly utilized the acronym as a means of remembering and unifying standard procedures across units. One of the most obvious and, indeed, important battle drills upon which they have relied in recent campaigns in

Iraq and Afghanistan is a procedure for assaulting urban areas, called the 'phase line battle drill'. An acronym is used as a mnemonic and stressed throughout training to ensure that the phase line battle drill has been internalized by corporals and platoon commanders so that procedures are uniform across the Corps. Accordingly, the phase battle drill involves four elements known as the RIGS process. RIGS stands for the four key elements of an urban assault: Reconnaissance, Isolation, Gain a foothold, Secure. In practice, the RIGS process involves a platoon commander examining the objective building immediately in front of his position which he has been assigned to assault. He reconnoitres the building (normally by simple observation) to determine the most likely entry point: R. The building is then isolated from its surrounding structures: I. The aim here is to ensure that the assaulting troops are not hit by fire from other buildings as they approach the target (typically across a street or an open area) when they are very vulnerable. The Marines identify three ways to isolate a building for the assault; by terrain (when the approach is obscured by the contours or by structures like walls or other buildings), by fire (supporting Marines suppress potential enemy by firing at the surrounding buildings and creating a tunnel through which the first buddy teams advance), or by obscuration (i.e. smoke). The assault itself is typically initiated with a decisive close-range use of a rocket or a burst of machine-gun fire into the building which is to be attacked to disorient the enemy and to signal to the platoon that the assault has started, intensifying and unifying their efforts on the objective. Once the building has been isolated, the first 'buddy' pair dash into it through a window or door. They secure a foothold in that room (G), supported by following teams, who all enter through the same entry point. Once they have a foothold in the building, the assault troops clear the structure from room to room (S), setting up the platoon for the next phase line battle drill and the next RIGS process.[64]

The phase line battle drill is a central part of the Marines' urban operations curriculum and features heavily in their training of platoon commanders (see Figure 8.1). On the final exercise at the Infantry Officer Course, the graduates of which will go on to command Marine infantry platoons, the students spend three days practising the phase line battle alone. On 4 June 2011, nearing the end of the final assault, one of the platoons had secured a building on the edge of a pre-designated phase line, called Phase Line Green; machine-gunners positioned themselves on the roof behind a balcony periodically firing their weapons at enemy forces. Significantly, Nicholas Warr's book of the same title (cited in Chapter 5) had been required reading on the course and was occasionally referenced by the instructors in their debriefings. There was a pause in this building but it was not merely a delay. The platoon was self-consciously preparing itself for a phase line battle drill. Indeed, on being questioned by adjacent troops about what was happening, the platoon commander shouted out that he was just about to initiate a phase line battle drill. In the event, the end of the exercise was called before the drill could be executed, but it was

Figure 8.1. Phase line battle drill. Range 220, US Marine student officers, on Phase Line Green, preparing for a phase line battle drill, 4 June 2011.

noticeable that even among this inexperienced group of officers the concept of the drill had been institutionalized. It had become an established and communal routine. There was some irony to the fact that on the Phase Line Green of this exercise, student officers were about to execute an established drill in which they sought to coordinate fire with manoeuvre to maximize the chances of success. In Warr's book, his platoon had simply walked out into the street, defined as Phase Line Green, to be enfiladed and destroyed by enemy fire.

Indeed, USMC instructors who had fought in Iraq and Afghanistan emphasized the importance of drills to the student officers. One captain who had been an enlisted Marine stressed how ingrained the drills had become so that they were executed eventually by instinct. As an officer, he found drills equally important, but they had changed from being primarily physical into mental drills.[65] Indeed another of the instructors used a sporting analogy to illustrate the instinctiveness of drills: 'It's like plays in [American] football, you have got to repeat them over and over until they are instinctive.'[66] Drills have become not only more refined and comprehensive in the professional western military but greater efforts have been taken to ensure they are executed in a standardized manner. The mnemonics which platoon and section commanders are forced to learn and re-learn are not theoretical formula but are supposed to become habitualized forms of collective practice, which trained troops will instinctively enact.

Chapter 5 explored how individual heroics, which were a recurrent if not institutionalized feature of the performance of the citizen army, were the product not primarily of personal characteristics but situational dynamics. The passivity of the unprepared mass citizenry soldiery required designated individuals to engage in autonomous actions. The citizen army formally recognized the importance of collective drills but was typically unable to execute these complex actions in combat; they could not overcome the collective action problem. Individual bravery has not disappeared from the twenty-first-century battlefield. There are numerous examples of individual heroism of a type which has clear parallels with that found in the citizen army, especially in rescuing the wounded. However, as a result of the inculcation of battle drills into everything which a soldier now does, the dynamics of combat seem to have changed quite significantly. Specifically, the mass–individual dialectic seems have been increasingly replaced by a group dynamic of collective action and teamwork. The drills—through acronyms and training—seem to have become so impressed onto the mental and bodily responses of professional soldiers that their instinctive and first response to fire is to conduct established battle drills with their colleagues. A collective approach seems to have become ironically natural for them.

There are some notable examples of the instinctive employment of battle drills in contemporary combat. In his widely read work *War*, Sebastian Junger records the tour of 2nd Platoon, Battle Company, 173rd Airborne Brigade in the Korengal Valley in Kunar Province eastern Afghanistan in 2007 to 2008. Junger's account is not always unproblematic since, as will be discussed in Chapter 10, he has a tendency to impose his own preconceptions about cohesion onto his material. However, his descriptions of the tactical actions executed by the 2nd Platoon are highly useful as they are some of the most recent, most direct, and best evidenced accounts of professional combat which are currently available. Even more helpfully, his narrative is supported by a documentary film, *Restrepo*, which provides a visual archive to support his text. One of the most important passages in that book describes the so-called 'Gatigal ambush' which occurred during Operation Rock Avalanche on 25 October 2007, when 2nd Platoon sought to penetrate further south into Korengal Valley to set up a new combat outpost. Junger records the importance of instinctive performance of battle drills as a result of training to this platoon's survival in this ambush. The Gatigal ambush is a very useful example of contemporary infantry tactics because there was no involvement of supporting fires such as artillery and Apache. It represented a nearly pure case of small infantry combat in which a group of some twenty insurgent fighters ambushed a similarly sized unit of American soldiers at close range. While the insurgents had the tactical advantage, both groups were equipped with similar weapons; automatic rifles, machine-guns, grenades, or rocket-propelled grenades. Because of the full moon, the typical US advantage of night-vision goggles was also negated. The fight was as close as it is probably possible to get in combat to an equal confrontation.

A soldier called Josh Brennan was on point, followed by a machine-gunner named Eckrode, Staff Sergeant Gallardo, and Specialist Sal Giunta. Junger described the terrain in detail: 'The soldiers walk single file along the crest of the spur spaced ten or fifteen yards apart. The terrain falls off steeply on both sides into holly forest and shale scree.'[67] As they descended from a crest, the platoon walked into an 'L shaped ambush' and were soon engulfed in fire from their front and side. These four soldiers were immediately isolated by the weight of fire from the rest of the platoon and could not be assisted by Apache as the enemy were too close; it was believed that the insurgents were attempting to isolate this fire team in order to take a prisoner. Brennan and Eckrode were badly wounded immediately but Gallardo instinctively recognized the danger and sought to link up with the two injured men. He was hit in the helmet but, recovering quickly, fought forward with Giunta and Private First Class Casey with hand-grenades, 'sprinting between the blasts': 'Even enemy who are not hit are so disoriented by the concussion that they have trouble functioning for a second or two.'[68] The team reached Eckrode and then Giunta went forward on his own to where Brennan should have been. Giunta saw two enemy fighters dragging Brennan down the hillside. He fired his weapon at them, killing one and forcing the other to flee, and then ran towards Brennan, whom he was able to drag back up the hill to his colleagues.

On this basis, Junger is able to make some apposite observations about the fight which are directly relevant to the question of cohesion.

> Stripped to its essence, combat is a series of quick decisions and rather precise actions carried out in concert with ten or twelve other men. In that sense it's much more like football than, say, like a gang fight. The unit that choreographs their actions best usually wins. They might take casualties, but they win. That choreography—*you lay down fire while I run forward, then I cover you while you move your team up*—is so powerful that it can overcome enormous tactical deficits. There is a choreography for storming Omaha Beach, for taking out a pillbox bunker, and for surviving an L-shaped ambush at night on the Gatigal. The choreography always requires that each man makes decisions based not on what's best for *him*, but on what's best for the group. If everyone does that, most of the group survives. If no one does, most of the group dies.[69]

Crucially on the Gatigal, the 2nd Platoon was sufficiently well drilled that Gallardo, Giunta, and Casey collectively reacted all but automatically. Without the need for consultation, they followed their anti-ambush training. Junger emphasizes the importance of the speed of their collective reaction:

> Giunta estimates that not more than ten or fifteen seconds elapsed between the initial attack and his own counterattack. An untrained civilian would have experienced those ten or fifteen seconds as a disorienting barrage of light and noise and probably have

spent most of it curled up on the ground. An entire platoon of men who react that way would undoubtedly die to the last man.[70]

Indeed, that is precisely the way in which citizen soldiers typically responded to ambushes from the First World War to Vietnam, going some way to explaining the very high casualty rates among the infantry in those conflicts. At the Gatigal, Giunta and his two colleagues instinctively followed their drills, aggressively fighting their way forward toward the enemy together (in contrast to the natural individual instinct to seek cover or to run away). Significantly, these three soldiers also primarily used grenades rather than fire—without any apparent consultation with each other; the decision seems to have been automatic. Yet, this was a quite complex tactical decision. There seem to have been two reasons for the selection of grenades over firing, both of which these soldiers learnt in training. At night, grenades, unlike shooting, do not give away the thrower's position and grenades have the ability to suppress enemy in a wider area. Having selected the appropriate weapon for the situation (within 15 seconds of being attacked), the three soldiers used grenades to implement fire and movement tactics, 'sprinting between the blasts' as the enemy were suppressed.

The Gatigal ambush has become a particularly celebrated infantry action in recent times. Yet, there are many other examples which affirm the centrality of the drill to tactical performance today. One of the most useful accounts is Colonel Bryan McCoy's deliberately didactic description of the ambush in which his unit 3rd Battalion, 4th Marine Regiment was involved in Al Kut Iraq on 3 April 2003 as the 1 Marine Division advanced on Baghdad. As the battalion drove down Highway 6, they were engaged from the right from as close as 30 metres by a large Iraqi force including two T-62 tanks and three armoured personnel carriers in prepared positions, between the road and the Tigris River. Colonel McCoy was himself in one of the leading HMMWVs (high-mobility, multipurpose, wheeled vehicle) which was caught in the killing zone and subjected to intense fire. In the opening exchanges, McCoy was forced to fire his weapon but he was also able to call up Kilo Company following the lead group, ordering it to 'attack the seam between the Tigris River and the palm grove to get behind the enemy in order to unhinge his right flank'.[71]

What happens next is pure violence, yet elegant in its harmony. Thirty-five US Marines of Kilo Company's 3rd Platoon rush out of the gloomy confines of their AAVs [Armoured Amphibious Vehicles] and into the teeth of the enemy fire. They know nothing of the enemy's strength or disposition. All they know is that this is a 'contact right' battle drill, and this is what we do in a 'contact right'. Private First Class Dusty Ladendorf, one of the platoon's riflemen, is less than a year out of high school. In an after-action review he makes this comment on the firefight: 'You come out of the back of the track and just do it like you were trained. Execute your battle drill, take cover and fire, cover your buddy's move, and move yourself when he covers you. Find the enemy, close in on him, and kill him.'[72]

For McCoy, his unit's ability to turn around a potentially lethal ambush (which was well conceived by his Iraqi opponents) into a resounding victory was evidence of the absolute centrality of battle drills to combat performance. Indeed, in his subsequent discussions of leadership, he proposed five habits which he regarded as essential to combat performance, the fourth of which was 'battle drills'.

Battle drills were essential not only to the counter-ambush at Al Kut but to other actions. For instance, on 21 March 2003, the unit's mortar platoon came under fire from a position 300 metres to its left flank. The platoon had established a drill of 'hip shoots' by which they returned fire rapidly. This drill involved one of the mortar men placing a stake in the ground some 50 metres from the mortar base plate in line with the target. The mortar crew would aim their shots from this stake, adjusting their range as they fired. On this occasion, the Marine tasked to place the stake found that the 50-metre mark was on an asphalt road into which he could not insert the stake. Instead, he lay prone on the road throughout the fire-fight holding up the stake with his hands. McCoy noted: 'a battle drill is a battle drill', and so instinctive had it become that this Marine did not question his role but merely improvised how best to achieve it in the circumstances. McCoy concluded: 'Without the "lighthouse" of battle drill, the platoon could not have performed as well as it did.'[73] Indeed, for McCoy the performance of Kilo Company and particularly Private Ladendorf was to be explained exclusively in terms of the drill. 'Your body is on automatic pilot. You yell "frag out" per the battle drill and throw from the prone at the bunker aperture. Your aim is true; the bunker is silenced. In the Al Kut palm grove, PFC Dusty Ladendorf did not suffer from dislocation of expectation that causes men to freeze; he knew what to expect and he knew how to react.'[74] The importance of battle drills to competent combat performance may ease the pressures of leadership but they do not eliminate the need for command. First, as McCoy himself demonstrated in Al Kut, commanders still have to decide which drill is required and, perhaps more significantly, in training they must prioritize which battle drills are to be the most important: 'As task force leaders, we deliberately set about forecasting what we believed would be the most crucial battle drills, then rehearsed those drills hundreds of times under increasingly challenging conditions.'[75] Commanders remain, therefore, important in cuing battle drills. However, by contrast, in the First and Second World Wars, individual junior commanders were recurrently central to maintaining the initiative of the attack by charging forward themselves. As the Gatigal and Al Kut ambushes show, professional soldiers now seek in the first instance to enjoin their colleagues to prosecute tactical manoeuvre together, and they are able to do this precisely because they know that their fellow soldiers will be oriented to the same set of battle drills. Battle drills have become ingrained into the very individual and collective comportment, expectations, and reactions of the professional soldier and the platoon on the battlefield. Strachan's observations seem to be entirely accurate. Indeed, many examples of this collectivization of action—and indeed of bravery itself—will be described in this and the following chapters.

However, in order to emphasize the profound social effects of the institution of battle drills on behaviour in combat, it might be useful to examine a case where the individual action of a leader of the type which was typical in the citizen army—and the concomitant passivity of his subordinates—seems to have been most apparent. Yet, significantly, even in actions of apparently total individualism, the reference to the collective seems to remain evident in the professional army. In extreme cases, there are examples of individuals acting as virtuosos in a way which seems to be entirely compatible with their twentieth-century peers, but even here the collectively institutionalized drill is present. One of the most striking and well-known examples of this highly individualistic and yet paradoxically collectivized action occurred in the Second Battle of Fallujah in November 2004. Having handed back control of the city to local security forces after their assault in April 2004, two Regimental Combat Teams from 1 Marine Division were ordered to retake the city with the attachment of two US Army units, 2nd Battalion, 7th Cavalry and 2nd Battalion, 2nd Infantry Regiment, 1st Infantry Division. The plan was for the assault force to clear the city from north to south with 2/2 Infantry assigned to left flank of the Marines' Regimental Combat Team 7 which was to clear the eastern side of the city, through Askari District, while Regimental Combat Team 1 cleared the Jolan District along the Euphrates River. On the third day of the operation on the night of 10 November 2004, A Company 2/2 Infantry were back-clearing the district and were ordered to clear a block of twelve buildings near Main Supply Route Michigan, a major highway which ran west to east through the town.[76] The first nine buildings were unoccupied but when Staff Sergeant David Bellavia, of the 2nd Battalion, 2nd Infantry, and his squad entered and cleared the tenth house, they came under intense fire. Bellavia had previously retreated from the house, but recognizing its importance to the safety of his squad and platoon, he decided to take the house alone, consciously seeing it as a challenge to his manhood. He was also incited by the news of the death of his command sergeant major, to whom he was intensely attached; the presence of a reporter, as an external observer, also does not seem to have been irrelevant to his decision.[77] Bellavia alone re-entered the house containing five insurgents, all of whom he personally killed, including one with a knife during a hand-to-hand struggle. Bellavia was awarded a Silver Star and his citation emphasizes the individualistic nature of his actions:

At this point, Sergeant Bellavia, armed with a M249 SAW gun, entered the room where the insurgents were located and sprayed the room with gunfire, forcing the Jihadists to take cover and allowing the squad to move out into the street. Jihadists on the roof began firing at the squad, forcing them to take cover in a nearby building. Sergeant

Bellavia then went back to the street and called in a Bradley Fighting Vehicle to shell the houses. After this was done, he decided to re-enter the building to determine whether the enemy fighters were still active. Seeing a Jihadist loading an RPG launcher, Sergeant Bellavia gunned him down. A second Jihadist began firing as the soldier ran toward the kitchen and Bellavia fired back, wounding him in the shoulder. A third Jihadist began yelling from the second floor. Sergeant Bellavia then entered the uncleared master bedroom and emptied gunfire into all the corners, at which point the wounded insurgent entered the room, yelling and firing his weapon. Sergeant Bellavia fired back, killing the man. Sergeant Bellavia then came under fire from the insurgent upstairs and the staff sergeant returned the fire, killing the man. At that point, a Jihadist hiding in a wardrobe in a bedroom jumped out, firing wildly around the room and knocking over the wardrobe. As the man leaped over the bed he tripped and Sergeant Bellavia shot him several times, wounding but not killing him. Another insurgent was yelling from upstairs, and the wounded Jihadist escaped the bedroom and ran upstairs. Sergeant Bellavia pursued, but slipped on the blood-soaked stairs. The wounded insurgent fired at him but missed. He followed the bloody tracks up the stairs to a room to the left. Hearing the wounded insurgent inside, he threw a fragmentary grenade into the room, sending the wounded Jihadist onto the roof. The insurgent fired his weapon in all directions until he ran out of ammunition. He then started back into the bedroom, which was rapidly filling with smoke. Hearing two other insurgents screaming from the third story of the building, Sergeant Bellavia put a choke hold on the wounded insurgent to keep him from giving away their position. The wounded Jihadist then bit Sergeant Bellavia on the arm and smacked him in the face with the butt of his AK-47. In the wild scuffle that followed, Sergeant Bellavia took out his knife and slit the Jihadist's throat. Two other insurgents who were trying to come to their comrade's rescue, fired at Bellavia, but he had slipped out of the room, which was now full of smoke and fire. Without warning, another insurgent dropped from the third story to the second-story roof. Sergeant Bellavia fired at him, hitting him in the back and the legs and causing him to fall off the roof, dead. At this point, five members of 3d Platoon entered the house and took control of the first floor. Before they would [sic] finish off the remaining Jihadists, however, they were ordered to move out of the area because close air support had been called in by a nearby unit.[78]

Bellavia's action is clear evidence of the persistence of actions defined as heroic by an individual identified as holding moral responsibility for his soldiers in the professional army. Yet, there are some notable differences from apparently similar cases of heroism in the citizen armies. In contrast to York, Murphy, or Marshall's Sergeant Summers, Bellavia personally insisted on assaulting the house alone, but his men were not passive. When Bellavia explained that they were going to take the house, his squad concurred immediately.[79] In order to effect an entry, 'the four men moved to the front gate and stacked along the nine-foot wall'.[80] Bellavia then ordered the men not to stop to give aid should one of them be hit, even if it was Bellavia himself: 'I don't care if I'm hit and screaming Jesus—leave me. Do not look down, do not look back. Continue to move forward and shoot. Kill the threat, or we will all go down.'[81] On this command, 'Ohle, Maxfield and Bellavia formed into a three-man wedge and charged through the garden up to the front door, with Lawson following close behind'.[82] Bellavia's squad then followed his instructions to provide cover and support from outside the building and followed him into the building as he fought his way from the bottom to top floors, although he cleared the rooms alone. Indeed, the last insurgent he killed had already been badly wounded by one of his own SAW gunners who continued to fire at the body after the insurgent had finally been shot by Bellavia on the roof and had toppled over onto the ground.[83] Moreover, while Bellavia's decision to enter the house was partly motivated by personal psychology, it was also a rational tactical decision. The house was small and he was concerned that the more soldiers who entered the property, the easier the target they presented to the insurgents. His platoon also lacked training in close-quarters battle techniques (see below) and, having failed to practise the complex clearing drills required in buildings, were more likely to obstruct and shoot each other than to prevail over the insurgents in the house. Finally, Bellavia had previously encountered 'Building-Contained Improvised Explosive Devices' in Fallujah, of a type which had been used to eliminate an entire Special Operations Forces team in Baghdad, and there was every chance that this house would be similarly booby-trapped. Bellavia's definition of the situation in Fallujah and the understandings and social dynamics which informed his actions seem to have been quite different from his forebears in the Second World War. Then, junior commanders had stormed forward alone out of necessity, driven by the mass–individual dialectic. In Fallujah, Bellavia seems, even in this moment of manifest individual bravery, to have been oriented by professional concepts, closely if indirectly supported by subordinates, all acting on his command. Paradoxically, Bellavia's house-clearing heroics might best be seen as evidence of collective expertise.

As the Gatigal and Al Kut ambushes and Bellavia's actions at Fallujah demonstrate, battle drills have become central to the conduct of professional soldiers on the battlefield even when they are apparently acting alone. They have enabled collective performance to replace the individual–mass dynamics of the citizen army. The successful institutionalization of the drill has fundamentally changed the nature of the professional force and its combat performance. The drill seems to have changed the nature of cohesion in the professional force. The refinement of the battle drill in doctrine and its internalization in training has facilitated a higher level of individual and collective performance from the infantry platoon. Platoons are now standardly able to execute set procedures in the face of a diversity of situations. They display precisely the knowledge and skill which Huntington and Janowitz identified as definitive of the military professional.

CLOSE-QUARTERS BATTLE

Battle drills have become central to the performance of the professional platoon and it is likely that they will only become more important in the immediate future. In the light of Iraq and Afghanistan, western armies have increasingly begun to emphasize complex urban operations as the most likely and the worst environment in which they will have to fight. Not only do western armies believe that these urban operations will involve house-to-house fighting but they will also feature irregular forces mingled with a civilian population. Although, in fact, this kind of fighting was always a subordinated feature of conventional wars in the twentieth century, operations since the end of the Cold War in the Balkans, Somalia, but above all in Iraq and Afghanistan have persuaded western militaries that their infantries have to be trained to conduct operations in this difficult and confusing environment. This new environment is demanding new skills and new battle drills from the infantry.

It is often assumed by commentators and military practitioners that the urban environment is the more likely scenario for future conflict because increasingly large numbers of people live in cities and consequently this is where political struggles will take place. In addition, western forces are likely to confront small numbers of irregular forces which will choose to fight in the city in order to negate western advantages of firepower, air superiority, and surveillance. These arguments for the new salience of urban combat are not implausible. However, the increasing likelihood of urban combat may itself be a product of western military professionalization and downsizing. In the twentieth century, western Europe was highly urbanized by global and especially by historic standards. However, with some notable exceptions including Stalingrad and Berlin, most of the ground fighting in the First World War, Spanish Civil War, and the Second World War took place in the countryside; this is also true of the Korean and Vietnam Wars. Urban populations during these wars were very large. Moreover, major and, especially, capital cities were politically and often economically critical. However, twentieth-century warfare developed as a campaign of fronts, which often encompassed cities, but fought primarily in the countryside simply because of the massive size of armies at that time. While these armies fought *for* cities, they were simply too big to fight *in* cities. They struggled for cities on huge fronts, spanning hundreds of miles, occupied by million-man armies which were perforce located mainly in the countryside. It is noticeable that even the major urban battles of these conflicts were typically part of a wider campaign most of which was fought outside the town; Stalingrad was only the most obvious example of this. Most of the Wehrmacht's Army Group South (B) and even a significant proportion of General von Paulus's 6th Army was not deployed into the relatively small city of Stalingrad but occupied the countryside around it (and was subsequently enveloped by gargantuan Soviet armies). Army Group B comprised some seventy-four divisions in total, of which only the twenty-four divisions of the 6th Army and 4th Panzer Armies were committed to the fight in Stalingrad itself. The rest were deployed on the Russian steppes. However, even in the case of the 6th Army, only five or six of its eleven divisions were involved in urban combat in September and October at any one time while the rest defended German lines to the north and west of the city. In the last ten years, armies have been and are likely to be drawn into cities not because cities themselves are more strategically important than they were. It is because armies are much smaller now. The military participation rate has declined hugely and the battle-space has become far less dense; combatants are able to manoeuvre on the new battlefield in a manner quite impossible in the twentieth century. As a result armies no longer just fight *for* cities but find that they are forced to fight in cities themselves. This has produced a potentially interesting, not to say ironic, dynamic for military transformation. The very reduction of western forces, primarily because of professionalization, has simultaneously altered the likely battlefield for those forces and compelled them to adopt new and more advanced combat techniques. By promoting the complex urban environment as a likely battlefield, professionalization seems to have demanded greater professionalism. The development of urban combat skills has involved the development, dissemination, and inculcation of a whole suite of specialist battle drills which are transforming the capabilities of the average infantry soldier.

The dissemination of close-quarters battle drills over the last ten years has a longer and very interesting history which goes back to the 1970s and the emergence of the Special Forces at that time. On 5 September 1972, eight members of the Palestinian terrorist organization Black September took twenty-one Israeli hostages at the Munich Olympics, shooting two in the initial break-in. The German authorities were totally unprepared for such an eventuality and, in a chaotic fire-fight at the airport as the terrorists tried to escape, all nine remaining hostages were killed. The Munich disaster was in fact part of a much wider growth of terrorism across Europe. The Provisional IRA in the United Kingdom, the Red Brigades in Italy, ETA in Spain, Action Directe in France, and the Red Army Faction in Germany all emerged at this time to present a significant challenge to governments and the security forces. In response to this new threat and galvanized by the tragedy at Munich, western governments began to develop specialist counter-terrorist units capable of conducting hostage rescue missions. In the United Kingdom and the USA, the SAS and Delta Force were at the forefront of these developments; in France, a number of Special Forces groups emerged but the Groupement de Sécurité et d'Intervention de la Gendarmerie Nationale took on the role of counter-terrorism and remained most advanced in terms of urban combat techniques, while Grenzschutzgruppe (GSG) 9 became Germany's foremost anti-terrorist unit. Close transnational connections developed between these groups so that they shared tactics and procedures, trained together, and in some cases conducted joint operations together; two members of the SAS were part of the GSG 9 team which resolved the Lufthansa Flight 181 hijack in Mogadishu in October 1977.[84] The central problem of hostage rescue was how to effect a rapid and unexpected entry into a structure (typically a building or a plane) to overwhelm the terrorists through the use of accurate but restrained force so hostage casualties were minimized. The Special Forces developed distinctive techniques to be able to conduct such assaults. They became adept at breaching structures and eliminating hostiles with accurate fire.

For approximately three decades after 1972, these close-quarters battle techniques remained the preserve of the Special Forces. However, as a result of operations in Afghanistan, and especially Iraq since 2003, the Special Forces monopoly began to be broken. A crucial event here was the 2004 US Marine assault on Fallujah. Following the murder of two US contractors in the city, political pressure was exerted on the US forces to suppress the insurgency in the city and to re-establish control of it. In the event, the US Marines adopted a conventional approach to this urban operation, which had some parallels with their recapturing of Hue in 1968 after the Tet offensive. Maximum firepower was brought to bear with the result that the retaking of Fallujah caused many civilian casualties and much collateral damage. The Marines employed most of their weaponry against the city and, following conventional infantry battle drills, squads of Marines entered buildings with traditional clearing practices. Grenades were thrown into open windows or doors, followed by the spraying of rooms with automatic gunfire. Numerous civilians were killed in this way and the Marines took unnecessary casualties throughout the operation. It was in the course of this fighting that David Bellavia conducted his house clearance.

Following these experiences, the US Marines and US Army made a substantial investment in the development of close-quarters battle techniques, developing new tactics and training methods. It is true that the Special Forces and elite forces such as the Rangers, the Airborne, and selected US Marine units have developed the greatest expertise in close-quarters battle. However, these techniques now feature as a standard part of infantry doctrine, are described at length in the US Army's *The Infantry Rifle Platoon and Squad*, and are an increasingly normal part of the repertoire of skills required by even regular line infantry. Since urban operations in the last decade have almost exclusively involved environments where non-combatants have been present, one of the main aims of close-quarters battle (CQB) tactics has been to replace the maximum use of firepower, regarded as appropriate in the past when facing a mass enemy force in an urban area, by rapid, accurate, and discriminate shooting. Indeed, even in the high-intensity context, the use of maximal firepower has often been ineffective. When David Bellavia cleared the insurgent-held house in Fallujah, he fired the entire magazine of a 5.56 SAW machine-gun into one of the rooms on entering the building: some 500 rounds. Despite going 'cyclic' with his weapon, he did not hit a single one of the insurgents who had barricaded themselves inside the house. There is some suggestion than even if confronted with a conventional enemy force (as the US Marines effectively were in Fallujah), precision clearance techniques might be favourable over traditional methods which are now often satirized as 'spraying and praying'.

A similar dissemination from Special Forces is evident in other western military. In Britain, as in the USA, elite forces, the Royal Marines and Parachute Regiment, have been particularly favoured in this process due primarily to their close association with the Special Forces, but the process is very widespread with new skills being adopted by all infantry. Despite their small size (some 7,000), the Royal Marines have been an important leader in the dissemination of close-quarters battle (CQB) skills in Britain, identifying the requirement to improve urban combat skills from 2007. In response to the concerns of their own Marines and substantial public outcry, the US Marines began to explore new, more discriminate techniques for urban combat and began to develop their own CQB techniques from the US Special Forces. The US Army was simultaneously developing its own techniques. At the same time, British forces were increasingly involved in non-permissive urban operations in Iraq and in Afghanistan (where a principal role was clearing compounds). In Britain, the Royal Marines, principally because of a close institutional link with the US Marine Corps, have been at the forefront of these developments: 'In 2006 and 2007 British Forces in Iraq and Afghanistan sustained the most intense period of close combat since the Falklands resulting in 134 UK fatalities with hundreds more badly injured—enemy losses were colossal. In the wake of this fighting, elements of the Royal Marines began to question the relevance of the urban combat (FIBUA, OBUA) training they had received. The tactics were perceived as being a relic of the cold war focussed on destroying a Soviet armoured enemy in a German village.'[85] The Royal Marines, like the US Army and Marines, began to look for possible ways to develop their shooting skills: 'In recognition of the lessons identified on operations and in response to Project Odysseus (a review of Army Operational Shooting Policy), instructors at CTCRM [Command Training Centre, Royal Marines, Lympstone] newly returned from theatre reviewed urban combat training and departed on a fact finding mission to NATO allies and Special Forces in order to create a best practice urban combat instructor course that would reinvigorate and modernize close combat.'[86] The Royal Marines were substantially advantaged in the adoption of CQB since one of their units, the Fleet Protection Group, tasked to protect the Royal Navy's nuclear bases in Scotland, had received considerable CQB training as part of this role. Their close links with the US Marines and independence were all important in allowing them to innovate with CQB.

The new close-quarters battle techniques can be observed in any of the major western forces, which are the focus of this study. The US Army run specialist CQB courses at various training establishments in the United States; the USMC have developed a Basic Urban Skills Training course and a more advanced CQB course. The Canadian army runs a Platoon Weapons

Level 4 course ('Gunfighter') at the Combat Training Centre Gagetown, New Brunswick. The British army currently runs an Afghan specific compound clearing course which includes elements of CQB, while the Royal Marines Platoon Weapons Specialist Level 2 course (for corporals) includes a four-week CQB package. The German, French, and Australian armies all run similar courses or include elements of CQB in their infantry training. Indeed, the French are investing heavily in this capability at their new Centre d'Entraînement aux Actions en Zone Urbaine (CENZUB) (see Chapter 9). Any of these courses might be used to illustrate the introduction of CQB battle drills and their significance for battlefield cohesion. However, as a result of location and access, the Royal Marines Platoon Weapons Course, run from the Commando Training Centre at Lympstone, constitutes the primary evidential base for the following discussion. This is not because it is more important or better than other courses being run in Britain or other western countries but because the evidence gathered there has been accumulated through longest and most in-depth exposure to the course. The course lasts four weeks and includes a two-week shooting package in which candidates are trained in individual marksmanship, followed by a tactical phase in which they learn to operate as an assault team within the urban environment. Most of this collective training takes place at a specially constructed compound at the Royal Marines' Commando Training Centre, Lympstone. This venue was excellent for research purposes since there was a gantry over the facility in which all the action could be viewed from above. One of the problems of researching urban combat is that it is simply very difficult to see anything because the action takes place in enclosed rooms which are dangerous to be in when anything other than blank rounds are used; even then, the experience is disconcerting and the possibilities for observation of teamwork are limited. There is another advantage to using the Royal Marines. Drawing on existing practices and, in particular, US doctrine, since 2009, the Royal Marines have been compiling their own close-quarters battle doctrine, *The Close Quarters Battle Instructor: course manual*, a long and detailed description of all the techniques which they have developed. Consequently, it has provided the best and most elaborated examples of the practices which are central to this study. It should be emphasized, however, that the Royal Marines course should be taken to stand for an important transnational process of change and, throughout the following text, parallels with the other western militaries will be highlighted. Especially given the close connection between these forces, it is not surprising that the CQB training and drills are internationally extremely similar; indeed, with very minor national differences they are all but identical. In addition, while undoubtedly advantaged because of their elite status, the techniques practised and taught by the Royal Marines are widely evident in the line infantry regiments in the UK. They are evidence of higher levels of professionalization across the armed forces not merely within small, selected elites.

In the 1970s, the Special Forces prioritized accurate shooting in their hostage rescues because they had to minimize civilian casualties. Accuracy remains central to CQB. Thus, the US Army and Marines note in their infantry doctrine that 'if there are known or suspected noncombatants within the building, the platoon may conduct precision room clearings'.[87] The Royal Marines similarly understand CQB substantially in terms of precision shooting over weight of fire.

> Put simply MUC [Modern Urban Combat, i.e. CQB] is a concept for operating in complex terrain in a controlled manner. It can be escalated or de-escalated to cover the full spectrum of conflict from COIN [counter-insurgency] to warfighting, and is designed to minimize friendly casualties—especially from fratricide—and inflict maximum losses on the enemy, while exercising judgement and reducing collateral damage.[88]

The requirement for accuracy—although an apparently relatively minor alteration—has involved a series of quite profound changes in terms of techniques, drills, and training, however. CQB, consequently, involves a series of distinctive individual drills, quite different from traditional platoon and section tactics developed primarily for rural warfare. From these new individual drills, critical collective drills have been developed, which will be discussed below. Traditionally, infantry soldiers learnt to shoot at static targets, which represented an enemy at a distance of 300 metres. Royal Marines noted that in the past, 'markmanship training lacked any shoots closer than 100 meters and was orientated around the tick box mentality of passing APWT (Advanced Platoon Weapons Training)'.[89] At this range, British soldiers were taught to hit any part of the main body of their opponent; the middle of the old Figure 11 target. Partly as a result of this approach to shooting, the British army has become deeply concerned with the accuracy of its combat shooting. In recent live combat shooting tests in daylight, only 5 to 20 per cent of rounds hit the target; at night the figure fell to 4 per cent. In their post-operational reports, British brigades in 2010 and 2011 both claimed that their shooting in Helmand had not been accurate enough.

The same preference for traditional range shooting was evident in the Canadian forces until 2005 when the new Platoon Weapons 4 ('Gunfighter') course was introduced. Before that time, Canadian soldiers were taught to shoot and tested to shoot on ranges of 100 to 400 metres on Platoon Weapons Levels 1 to 3. The tests on these ranges were stringent, culminating in the Platoon Weapons 3 test. On that test, soldiers had to fire at 400 metres in a timed shoot, then run forward to the 300 where they would have to fire again at targets which presented themselves only for a few seconds. The process was repeated at 200 and 100 metres and soldiers were expected to reach a high standard in terms of accuracy.[90] It would be wrong to think that traditional range shooting was easy or irrelevant; not unreasonably armies in the twentieth century estimated that they would most likely be fighting in open terrain (as they were for the most of the conflicts of that century) at the limit of their rifles' ranges. However, in the new urban environment, this type of range shooting is regarded as inadequate and unrealistic. As one Canadian CQB instructor noted: 'Now in Afghanistan, targets are much closer and you must be thinking 360 degrees. You must be looking all around. After shooting or when you are moving you must look around behind you.'[91] Consequently, soldiers have to be prepared to engage targets at extreme close range from 25 down to 5 metres and, in contrast to traditional shooting, CQB marksmanship focuses on small areas around the middle of the face (especially the eye area) and around the heart in the middle of the chest, since these are most likely to incapacitate opponents most effectively. Soldiers are also taught to shoot quickly in reaction to the appearance of unexpected targets from potentially any direction, from unusual firing positions, and, in particular, on the move. Accordingly, infantry soldiers in modern urban combat are taught to move differently and, in fact, apparently quite unnaturally—adopting quite strange body postures and movements—in order to maximize the chances of precision shooting in this demanding environment.

The Royal Marines' *The Close Quarters Battle Instructor: course manual* documents these new body positions and techniques which are required to ensure this accuracy. The document begins with a long and detailed section including diagrams and photographs of ideal firing positions for the individual in the urban environment. Correct hand movements in relation to the primary and secondary weapons (rifle and pistol) are described at length, so that reloading the weapon or clearing stoppages is executed in the most efficient way, which represents at least an improvement of the techniques which soldiers have been taught. In some cases, approved CQB weapon manipulation represents a form of remedial training where old habits have to be eliminated. The importance of correct weapon manipulation is particularly obvious in the case of the pistol as a secondary weapon. While soldiers are typically highly trained and extremely experienced in the use of a rifle, having been taught to use a rifle from initial training, they are typically unfamiliar with pistols, although they have often been issued one for a variety of duties. The correct pistol grip is, consequently, perhaps less obvious and natural than holding a rifle. According to the CQB manual and current practice, pistols are ideally held at arm's length (so that the shoulders brace the weapon), with the left hand clasped firmly around the right hand which holds the grip itself.

It is perhaps obvious that hand positions need to be correct in order to shoot accurately but in fact an ideal body position as a whole needs to be adopted for precision firing. To this end, western forces have adopted the concept of a standing combative or fighting position where the marksmen stand square on to the target, knees bent, slightly hunched over their weapons with elbows tucked into their sides, presenting the body armour forward. This body position both provides the most stable platform for shooting—it is self-consciously derived from the basic sporting position which tennis players, goalkeepers, or batsmen adopt as they are about to receive the ball—but it is self-consciously aggressive, forcefully projecting forward against threats while maximizing the protection of the body armour (see Figure 8.2a). It is also the position which a boxer adopts in a fight. When practising the combative position without weapons Marines will clench their fists like fighters. The US Marines assume the same combat body position, hunching forward, legs apart in a position which they describe as a strong sporting position. Echoing the UK forces, the US Marines also emphasize that unlike the traditional shooting stance of 'blading' (where the soldier stands sideways on to the target), the combat body position maximizes the protection afforded by body armour.[92] The same stance has become quite universal, however. The Canadian army also teaches this same body position with feet apart, knees bent, and shoulders hunched over the weapon. Not only is this position more stable and aggressive, but by hunching forward over the weapon, whether it is a rifle or pistol, as it is held tightly to the chest pointing downwards, it minimizes the chances of shooting colleagues accidentally. By hunching over the rifle, the barrel of the weapon remains pointing safely at the ground close to the soldier's feet even when turning to face new threats.[93]

In order to adopt and maintain good firing positions, the placement of the feet is surprisingly critical. After each burst of fire, Marines engage in a distinctive rocking motion, inclining their head and upper body to the left and right. Feet positioning is the basis of a stable platform for firing and correct placement can allow a soldier to fire accurately while in cover around a corner or through a door or on the move. One of the most interesting discussions of foot position relates to pivots, when the firer needs to turn through 90 or 180 degrees to face a new threat. The natural reaction to an unexpected threat is to twist quickly around to it moving head, body, and feet together. This instinctive twisting movement is suboptimal in a close-quarters environment; it does not provide a stable firing position and delays identifying whether there is a threat present or not. In close-quarters battle, 'the purpose of learning the turning and pivoting techniques are to enable the students to quickly bring the weapon to bear on a threat in a direction other than to the direct front in the most efficient and aggressive manner possible'.[94] Accordingly, the head is turned first in order to ascertain what is in that direction and therefore, if necessary, to 'acquire the threat'. In order to turn quickly into a new firing position, the manual identifies a sequence of movements. For a 90 degree pivot, the manual instructs: 'drive the eyes, head and shoulders toward the threat'. Then 'drive the torso and hips toward the threat by turning the foot in the direction of the threat [i.e. the foot nearest the threat] toward the threat by turning on the heel [i.e. lifting the toes and swivelling on the heel] and driving off the toes of the rear foot'.[95] The same movement is used by the

Canadian army with instructors explaining the rationale for leading pivots with the head in close-quarters battle: 'You turn with the head first because when you come around a corner, people could be very close to you. You need to identify and acquire them and work out whether they are a threat. You then turn towards them, spinning on your feet, and either relax, leaving your weapon down or you engage them.'[96] The 180 degree pivot is used most often on entering a room when 'button-hooking through a door'.[97] The soldier enters the room facing in one direction but needs to sweep the room, finishing in a position where he is pointing to the corner adjacent to the door through which he has just entered. The manual records a three-step movement in images, supported by text to describe the movement. The soldier puts one foot against the door frame and then steps into the room with the following foot. At this point, 'the head and shoulders are pushed forward with the rear foot

Figure 8.2a. Demonstrating the 90 degree pivot, Straight Point Range, Devon, 10 May 2011. The demonstrator begins in the alert combat position, hunched with weapon pointing downwards.

into the turn, providing solid shoulder support for the presentation of the weapon as with a 90 degree pivot'.[98] At this point the weapon is brought up to the 'Alert' and the 'rear foot will rise' and be brought forward so that the soldier can adopt the combative position.

The manual builds upon these basic foot positions to give instruction about how to shoot on the move. On entering a building, soldiers are instructed to move aggressively towards their opponent firing as they advance and potentially transitioning to a pistol if necessary. In order to reduce barrel movement, soldiers move slightly slower than normal walking pace, drop their centre of gravity by bending their legs more, and, in order to minimize bouncing, 'the foot should be rolled from heel to toe, with a smooth transition from toe to heel'.[99] This technique is sometimes called the 'combat glide' and is a highly distinctive body movement, the soldiers advancing steadily and smoothly with

Figure 8.2b. He pivots with the feet with weapon down.

Figure 8.2c. Assuming the combat position, he brings up his weapon to engage the target.

knees bent and shoulders hunched so that the upper body barely moves. The manoeuvre could be seen very clearly in both Canada and America. On the Canadian army's gunfighter programme, the same peculiar crouched padding, as soldiers advanced firing towards the target, was practised repeatedly (see Figure 8.3). It was noticeable that when experienced directing staff demonstrated the technique, their foot placement was very careful and methodical so that the rifle barrel hardly moved at all.

The emphasis on foot placement crucially sensitizes combatants to their tactical positioning. Close-quarters battle inside buildings and structures is understood to be predominantly about 'angles'. From a tactical perspective, buildings consist of a series of corners or doorways which are decisive in any fighting which takes place within them. Corners and doorways provide cover around which shots can be fired in relative safety but they also necessarily therefore obstruct vision. It is impossible to see around a corner or through an opening without exposing the eyes and head. One of the most dangerous moments in close-quarters battle involves looking around a corner or doorway in order to check for threats. Accordingly, success in the urban environment involves perfecting this art of looking around corners while maximizing the

Figure 8.3. Canadian soldiers practise the combat glide on the 'Gunfighter' course.

cover they provide. It is here that 'angles' become important. With care, it is possible to increase the angle of vision into a room or down an obscured corridor in a manner which puts an opponent at a permanent disadvantage. Foot placement becomes critical in exploiting these angles. By placing the right or left foot on an imaginary line extending back from the corner or the doorway, it is possible to maximize cover and the angle around which the soldier can see. Careful foot placement does not negate the problem entirely, however. At some point, soldiers will have to expose themselves in order to see around the corner at all. To this end, soldiers are taught to 'pop' corners and doorways. 'Popping' is an established concept in 'close-quarters battle' which involves a coordinated and definable set of actions known as 'close drill'. With their feet positioned on the imaginary line of cover, maximizing the angle of their vision, they simultaneously look around the corner and bring their weapon to bear. Consequently, hostiles within the room are not presented with a target—a head—but with the muzzle of a weapon, down which the combatant observes the space. 'Popping' is designed to ensure that combatants are covering themselves from threat and, if presented with a threat as they turn the corner, they are immediately in a position to fire.[100]

On the range in the initial phase of the CQB course, Royal Marines were taught a distinctive rocking movement which they were instructed to perform after every shoot; they would fire a sequence of shots at the target and then rock to the left and right with their weapon at the ready in the firing position.[101] This was known as 'shooting and scanning'. The purpose of this motion was not initially obvious until the course began to exercise in structures. The rocking movement prepared the candidates on the course to be able to 'pop' corners. Once positioned on a corner or at a doorway with their feet aligned, combatants rock left or right with their weapon aiming out in front of them; each rock brings their head and weapon out of cover, opening up the angle a small amount, allowing them to observe more and more of the room. Each rock is, in effect, a pop. As they begin to clear a room, combatants 'pop' the corner with a rock and then move their feet a small amount, rock again, repeating the same action until they have covered the entire room or corridor. The Canadian army uses the term 'popping' and they have the same concept of gradually opening up the angle of vision round a corner with a series of small rocks which they term 'pizza-slicing' in order to denote the arcs of fire which the soldier gradually opens up.[102] Popping is a difficult movement which students sometimes failed to perform properly. This was clearly demonstrated during the CQB course at Lympstone when students conducted assaults against live enemy, armed with simunition rounds,[103] and played by their fellow course students. On several occasions, students popped doors or corners with the distinctive rocking motion but, allowing their head to be exposed before their weapon was ready to fire, they were struck on the mask or helmet with a simunition round; the instructors, shouting from the gantry, declared them dead.[104]

Many of the drills which are taught as part of close-quarters battle procedures are individual. However, soldiers never operate alone (unless all their colleagues have been injured or killed). They operate as part of a platoon, squad, assault team, or, in the urban environment, a 'stack'. A stack refers to the formation which an assault team makes as it enters a building. Typically, the assault team will form a line outside a door or in a corridor before its assault; it will literally stack up behind the first soldier. The coordination of this stack is critical for effective performance in the urban environment and the ideal is for the stack to act as a team rather than four separate individuals. Crucially, therefore, the members of their stack remain in close physical proximity with each other so that each individual weapon supports and covers the others. In effect, the ideal is that the stack operates as a single weapon system. In the urban context, collective performance certainly depends upon individual skills but it is the coordination of the stack which is decisive.

Close-quarters battle represents a quite different tactical predicament from open warfare. In open warfare, the platoon is typically tasked with eliminating a small enemy position usually consisting of a bunker or a trench. There are typically only one or two major threats with which the platoon has to deal. The sections are all oriented to this single objective which they have to neutralize through fire and movement. Precisely because the terrain is normally open and the sections can move fairly freely across the ground maintaining their assault formations to find cover where they need it, the coordination of the platoon is relatively straightforward. Of course, as the experiences of the citizen army demonstrate, in the confusion of combat, even this relatively simple coordination becomes very difficult. Nevertheless, the fundamentals of a platoon attack in open warfare are simple. The platoon commander has the choice of mounting a frontal or a left or right flanking attack, typically using one section to support, one to assault, and one in reserve. In conventional combat, there is a single threat (a position) or a few closely contiguous ones (two or three positions) which are neutralized by the selection of one of three options by the platoon commander.

The urban environment poses a quite different challenge. Fundamentally, it represents a complex coordination problem of a quite different order from conventional tactics. On entering a building, an assault team is confronted with a multiplicity of threats which must be neutralized if the mission is to be successful and casualties avoided. There are numerous rooms which may contain enemies or booby-traps; the corridors and rooms lead off in multiple unseen directions all representing dangers. It is possible to be enfiladed from next-door buildings. Stairwells, cellars, attics, and furniture can all conceal dangers. Moreover, because the environment is complex and the threats numerous, they cannot be neutralized by any individual alone. The threats which are typically simultaneous have to be addressed by the assault team together. There is no one unified objective as there is in open warfare. An assault team is presented with a vexatious collective action problem. Accordingly, close-quarters battle requires an assault team to identify all the threats, to prioritize them in terms of immediate danger, and to assign team members to their neutralization. CQB demands therefore high levels of coordination within the teams so that all necessary actions are completed, often simultaneously. House clearances are, therefore, complex collective action problems which need also to be done at speed in a highly confusing and dangerous environment. The coordination problem is compounded by other factors, quite different from open warfare. Buildings necessarily channel assault teams down corridors or stairwells, separating them as sub-teams clear rooms individually. Unlike rural warfare, the sections and fire teams can quickly become separated and dispersed, unable to see or hear one another. It is precisely because of the complexity of the urban environment and above all the bewildering collective action problem which any building can pose an assault team that collective drills have become so central to CQB. The drills become a means of standardizing an assault team's responses to the problems it faces, thereby accelerating its reaction time and furnishing it with a wide repertoire of possible reactions to special problems. While complex, there are only so many configurations which a building can have. By establishing drills for a multiplicity of possible configurations which become automatic, the assault team is in a favourable position to improvise on the basis of known existing drills in the face of a new environment.

The difficulties of coordination and the concomitant importance of established drills in overcoming the collective action problem were graphically illustrated in the tactical phase of the CQB course at the Royal Marines Commando Training Centre in July 2011 when students were working in the initial training phase in the compound. In urban combat, it is inevitable that, at some point, the assault team will have to enter a building or a room. This is one of the most dangerous moments for the assault team because the entry point constrains physical movement, channelling bodies in particular ways, and is also typically defended by opponents who exploit the vulnerability which entry imposes on the attacker. The dangers of room entry are fully recognized by CQB specialists who call the area immediately inside a doorway the 'fatal funnel'. Not only do hostiles within the room know that their opponents must cross this area to enter the room (and can therefore aim their fire towards it pre-emptively) but typically the doorway or entry point casts a shaft of light into the room in which the soldier is silhouetted. Accordingly, the most basic and important collective drill is the entry method; the so-called 'five-step entry'. The five-step entry involves a number of interlocking elements which are intended to minimize the risks of entry and to resolve the problems of collective action. There are a number of variations of the basic assault manoeuvre, with team members assuming different positions in the room. However, each variation includes five basic requirements; 'a. Clear the doorway. b. Clear the immediate area. c. Clear your corner. d. Sweep your arc of fire. e. Establish a dominant position.'[105] Demonstrating the importance of teamwork over the individual, the success of the five-step entry perhaps surprisingly involves slowing down individual actions. In particular, assaulters have found that by slowing down the actions of the No. 1, they actually speed up the ingress of the whole team. As the Manual states: 'Speed is relative and the key is to be smooth and under control while operating within an enclosure.'[106] The point here is that maximum individual rapidity while apparently desirable in combat is actually counter-productive in the close-quarter environment since it threatens to fragment the team. If the No. 1 moves too fast, team-mates will be unable to follow with the result that the No. 1 will become separated from them; the ultimate danger in close-quarter battle is to become isolated in a room. Consequently, reduced individual speed can in fact accelerate the movement of the assault team as a group. This is what is meant by speed being relative and why the aphorism that 'Slow is smooth, smooth is fast' is so central to close-quarters battle.[107] In a well-executed five-step entry, the first member of the stack moves smoothly through the door to one corner, No. 2 'buttonhooks' round to the other, allowing his team-mates to follow in behind to stand along the wall. From this wall, the team forms a base line and advances forward to the 'dominant position' in the middle of the room.¹⁰⁸ The stack members are assigned to clear particular sections of the room in front of them (see Figures 8.4a and 8.4b).

Contemporary US doctrine on room clearance is very similar, stressing the importance of coordinated action to neutralize multiple simultaneous threats:

7-195. For this battle drill to be effectively employed, each member of the team must know his sector of fire and how his sector overlaps and links with the sectors of the other team members. No movement should mask the fire of any of the other team members.

7-196 On the signal, the team enters through the entry point (or breach). As the team members move to their points of domination, they engage all threats or hostile targets in sequence in their sector. The direction each man moves should not be preplanned unless the exact room layout is known. Each man should, however, go in a direction opposite the man in front of him. For example:

- #1 Man. The #1 man enters the room and eliminates any immediate threat. He can move left or right, moving along the path of least resistance to the point of domination—one of the two corners and continues down the room to gain depth.

Figure 8.4a. Stacking, Royal Marines CQB course. Royal Marines stack before entry into a cellar through which they will clear the basement. The No. 2 has 'flared' out and is about to throw a flash-crash distraction device. The assault commander stands in the foreground, counting down the assault on his figures from five; two seconds remain before the assault. A second assault team is ready to effect a second simultaneous entry onto the ground floor through the front door.

Figure 8.4b. Marines stack in the compound before breaching their way into a room.

- #2 Man. The #2 man enters simultaneously with the first and moves in the opposite direction, following the wall. The #2 man must clear the entry point, clear the immediate threat area, and move to his point of domination
- #3 Man. The #3 Man simply moves in the opposite direction to #2 man inside the room, moves at least 1 metre from the entry point, and takes a position that dominates his sector.
- #4 Man. The #4 Man moves in the opposite direction of #3 man, clears the doorway by at least 1 metre and moves to a position that dominates his sector.[109]

The French army also emphasizes the importance of drills or TTPs as its officers typically call them in urban operations and they have developed compatible entry drills to their NATO allies. The French platoon and section are structured somewhat differently from the US, Canadian, or British equivalent. The French platoon consists of one anti-tank section and three rifle sections. Each section consists of seven soldiers, two three-man fire teams (*trinomes*) and a section commander. The section commander acts as the controller and link-man between the two *trinomes*. The *trinomes* themselves consist of two rifleman and a team leader who directs their activities. In urban combat, the section advances as a group, one *trinome* leading, followed by the commander in front of the second *trinome*. On reaching an entry, the section stacks up on an adjacent wall and the first *trinome* commander sends in his two riflemen. They effect a standard urban entry, assaulting the room, they clear the corners first before establishing themselves in a position away from the 'fatal funnel'. French doctrine specifies that these soldiers should stack very closely to each other and enter the room together. The reason for this simultaneous entry with the No. 2 man up against the No. 1 is partly moral. The No. 1 feels physically supported by his team-mate. However, there is a practical purpose to this close proximity. The body armour and indeed body of the first soldier protects the second so that even if the No. 1 is shot, the No. 2 can eliminate enemy within the room. Crucially, by going in together, the dangerous process of room entry does not have to be effected twice. The No. 1 may have indeed become a casualty but at least the No. 2 is established in the room; the commander of the *trinome* follows his team in once the room has been neutralized. In training, French assault pairs enact a distinctive action as they enter a room. They stand very close together, then the No. 1 rocks back into the No. 2, coiling the pair for the assault. The No. 2 will shout 'Avant' or simply push the No. 1 forward and they spring into the room together. Once they are through the door the No. 2 will go the opposite way to the No. 1. Some instructing staff call this opening out from the very close stack into a dominant position, the 'papillon' (the butterfly), because the assaulting soldiers open out in alternate directions like a pair of wings. At CENZUB (see Chapter 9), the centre for urban operations training in France, the importance of these low-level drills at section and platoon levels is repeatedly practised and emphasized. Most of the training takes place with laser detector technology so that troops playing the opposing force can simulate killing and wounding the French troops. It is repeatedly emphasized that minimizing casualties requires excellent individual and team drills at the level of the *trinome*, section, and platoon.[110]

The five-step entry method is a simple collective drill which requires little coordination once it is learnt (see Figures 8.5a–d). The actions of each member of the stack are more or less standardized: No. 1 enters, clearing the blind corner, No. 2 follows to clear the open corner, Nos. 3 and 4 follow engaging threats directly ahead. All move quickly away from the fatal funnel whatever the room shape, until they have occupied the dominant position. On confronting a door, the assault team simply goes through this entry drill, automatically, until the team is stood in the dominant position in the room. Indeed, illustrating the point, the students on the Royal Marines CQB course learnt the entry method very quickly. After a couple of hours of practising in

Figure 8.5a. The five-step entry sequence. Nos. 1 and 2 enter and clear the corners.

Figure 8.5b. No. 3 enters and clears in depth.

Figure 8.5c. No. 4 enters and team establishes dominant position.

Figure 8.5d. Two-man dominant position.

teams, they were able to enter rooms quickly and effectively. In some cases, students made individual errors such as not clearing their corner or standing in the fatal funnel, for which they were criticized by the instructors, but overwhelmingly their entry drills became proficient very quickly. Consequently, when confronted with simple rooms which contained little furniture and were of a regular square or oblong shape with no hidden areas, they quickly demonstrated a facility with clearance drills. In these cases, the collective action problem was very simple. There were few threats with which to deal and the order of threats was relatively obvious. The five-step entry drill could be followed automatically. The team burst into the room, cleared the corners, and were then presented with an empty room everything in which they could see. Effectively, the standard entry drill alone was in and of itself sufficient to clear this room.[111]

However, there was one room in the compound which proved especially difficult (see Figure 8.6). This room demonstrated both the difficulty of coordination and the importance of drills as a means of overcoming the collective

Figure 8.6. A room clearance demonstration by directing staff. Note that the directing staff on the back wall are using the body armour to body armour drill to cover potential threats, behind the sofa and the hidden corners. This is one of the rooms which caused students difficulty. The directing staff had opened the door in the foreground and reorganized the closed wooden barricade beside the directing staff to the left, so that it had a door leading into a small oblong alcove. The doors to the alcove and to the adjacent room were, therefore, opposite each other, presenting equal and simultaneous threats.

action problem. Initially, the instructors had organized it as a simple square room containing a large sofa, chair, and television which the students found straightforward to clear. The standard five-step entry was totally adequate to the minimal threats which the room presented. However, later in the day, the directing staff reorganized the room so that there was an open doorway to an adjacent room on the right wall, opposite which on the left wall was another door into a small recess. The furniture also remained. Having entered the room unproblematically before the reconfiguration, the student teams now faltered badly. In particular, they were disoriented by the open doorway into the next room to their right which, in fact, represented the major threat, since an enemy was most likely to hide in that space and shoot from it. They could not work out how to cover this space and to clear the room which they were in. This was compounded by the fact that opposite the open doorway on the right was a closed door into a small alcove on the left, beyond which was a hidden corner which also had to be cleared. Teams were at a loss of how to assign the task of covering three threats—the open door on the right, the closed door on the left, and the corner beyond—while clearing the room before them. They visibly hesitated on entering the room and became uncertain about how to continue and were then incapable of coordinating their actions properly. They were unable to prioritize threats and to assign team members to those threats in order that they could clear the room simultaneously. Typically, the teams forgot to cover the open doorway (even though it was the greatest threat). Students advanced into the room, crossing the 'fatal funnel' created by the doorway on the right and exposing themselves to fire. In some cases, they turned their backs to the open door entirely as they dealt with the closed door to the alcove. The directing staff had to position the students physically in the room, pushing and pulling them by the body armour so that each individual was addressing the correct threat. Students gradually began to realize that the best way of resolving the dual problem of the open and closed door was for two of the team to advance and turn their backs to each other with their weapons ready, so that they were mutually protecting each other's backs while countering the main threats from left and right. Yet, it was the only obvious means of overcoming the collective action problem which this room presented them.[112]

Later that day, following from this experience, the students were taught the 'body armour to body armour' drill (see Figure 8.7).[113] This involved two assaulters moving forward together and simultaneously turning away from each other, their backs touching, each addressing a threat to their front and in that way protecting their team-mate's back. They were explicitly seeking to protect each other not only with their weapons but with their body armour. This simple but unnatural technique, especially for those trained in open warfare where troops would never stand together, solved the problem of how to confront two simultaneous and opposite threats without exposing

Figure 8.7. Body armour to body armour drill.

the back of any team member to a threat. The students practised the body armour to body armour drill and became proficient not only at the drill but also when to use it (although some individuals and teams still struggled to utilize this drill to solve more complex room clearance problems).[114]

There are established drills for room and building clearance which soldiers or Marines are taught to follow together and which can be adapted for a variety of contingencies. Consequently, the most practised close-quarters battle troops, and above all the Special Forces, have refined drills to such a point that in the face of almost any structure, they resort to a set drill which they have already conducted in training. Nevertheless, while drills have become institutionalized and while the Special Forces, in particular, seek to gather comprehensive data on the structures which they are to assault, they often lack details about the interior lay-out. Designs may be available for many buildings in western cities but, in Iraq and Afghanistan, the Special Forces have normally had little intelligence about the micro-geography of interiors.

This lack of knowledge about interiors is even more pronounced among the conventional infantry who almost never enjoy the level of intelligence supplied to the Special Forces. Moreover, even if accurate mapping were available of interiors, it is difficult to predict which rooms hostiles may choose to defend and, of course, as the Bellavia example shows, they often change the structure of the interior in order to improve their defensive positions.

In the light of these uncertainties, no matter how highly drilled an assault team has become it is regarded as essential that it should retain its flexibility. It should be able to adapt itself to the special requirements of the building and enemy contained within it. In the section attack drill, the options are more or less limited to a left or right flanking decision for the commander while the individual rifleman has to make only a few independent decisions, while he maintains his position in relation to his fire team. Such standard approaches are inadequate to the urban environment and to the coordination problems it poses. The complexity of human structures enforces tactical adaptability on the assaulting troops, forcing them into novel and unexpected configurations, while facing numerous threats. After clearing an initial room, the stack may have to split and clear rooms separately, re-forming later, or join with other stacks. At any point, there are likely to be unexpected recesses, cupboards, or siderooms to have to check. Teams will often have to work away from designated commanders. As a result, although the stack order is established at the entry point, it is very unlikely that the stack will retain this structure throughout the assault and it is vital that each individual is empowered to respond to a threat as they encounter it, if it is a threat to the team and if no one else can respond to it or has seen it. To that end, a concept of Initiative-Based Tactics (IBT) has been developed by western forces, pioneered by the USA. Initiative-Based Tactics refers to the requirement for adaptability and flexibility in the urban environment. Accordingly, each individual has to be prepared not only to conduct individual drills properly but to execute the drills associated with each element of the stack from No. 1 to the last man. Crucially, the members of the stack are trained to take their cue from the No. 1 while No. 2 (whoever it is) commands the stack at that point. They are taught to recognize that whatever the No. 1 does as he effects an entry into the room is correct and to follow the No. 2's instructions. The No. 1.'s actions must be followed, dictating what the following members of a stack will do. If the No. 1 goes right on entering a room, the No. 2 must go left.

Initiative-Based Tactics refers not only to flexibility but also to the interactional dynamics of the group. Every member of the team is taught to conduct himself in a way which maximizes the performance of the group as a whole. Specifically, Initiative-Based Tactics features three 'primary rules' which 'must be followed at all times':[115] soldiers must 'a. Cover all immediate danger areas', 'b. Eliminate all threats' (with precision marksmanship), and 'c. Protect your buddy': 'At all times firers should be looking to protect team members during the clearance of the structure, adopting a proactive approach throughout the clearance will aid enormously in the protection of the team.'[116] The US Marines have adopted a similar set of principles.[117] They have established standard procedures for clearing rooms of courtyards, with which all are familiar. Consequently on entering an area, the No. 1 will make a decision of which way to go and his team-mates will follow him. Initiative-Based Tactics is an institutional method to generate the highest level of collective action in assault teams by enabling individuals to take pre-emptive action without being commanded. Indeed, throughout the CQB course at the Commando Training Centre, the instructors would remind the students when they had overlooked a threat: 'IBT, IBT'. They were actively encouraging individuals to take pre-emptive action if they were in the best position to address a threat. If every individual was so motivated, the team as a whole had a much better chance of addressing the multiple threats it faced.

CQB requires a high level of individual skill both in marksmanship, tactics, and communication and great individual and collective flexibility. Unlike conventional infantry tactics where the members of a fire team maintain their positions relative to each other during an assault, Initiative-Based Tactics presumes almost total turbulence in roles and positions in the assault team; the No. 1 could end up as a No. 4 and vice versa. However, somewhat ironically, it also demands that all the members of the team are thoroughly united around a common set of procedures so that each individual can fit seamlessly into the proper role at each phase of a clearance operation, even though they could not anticipate before the assault that they would be in this role. The Royal Marines CQB handbook underscores the point: 'IBT (initiative based tactics) are enclosure-clearing tactics that are driven by the actions and initiative of the individual firers. Baseline knowledge of the techniques and fundamentals is common to all, but the application of these skills is directed by challenges of the particular scenario and the courses of action that the firers choose to follow.'[118] There are clear reasons why Initiative-Based Tactics requires even greater standardization of practice. While Initiative-Based Tactics liberates the individual to address an immediate threat, it could easily lead to chaos in an assault team. Individuals could invoke Initiative-Based Tactics to operate quite independently of each other with disastrous effects on the team. In this case, Initiative-Based Tactics would not solve the collective action problem posed by room clearances but compound the problem. Accordingly, for Initiative-Based Tactics to improve collective performance, individuals need to utilize it in line with team practices and expectations. Although they might individually take an unpremeditated action in a building, the way they neutralize a threat is in line with the team's methods, expectations, and its prioritization of threats. The only way for an individual to know when to employ Initiative-Based Tactics properly is by intimate understanding of the team's assault techniques and how it typically ranks and addresses threats. Only with this knowledge can they know that their individual action contributes to the collective performance of the team rather than undermining it.

There are many examples of this collectively oriented initiative being displayed in training. For instance, during the British army's Compound Clearance Course, which is run by the Operational Training and Advisory Group at the Stanford Training Area, which uses the two mock Afghan villages, trainees are forced to rely on Initiative-Based Tactics exclusively during one phase of the training where they practise clearances with no preparation or orders. They are simply organized into assault teams and told to attack a compound immediately which contains a number of enemy (other members of the course). Simunitions are used throughout to generate realism so that attackers and defenders seek to hit each other and to eliminate each other from the exercise. The assault teams choose a point and method of entry but once they are inside the structure they are forced to adapt by the unfamiliar building lay-out and by enemy action. In one compound, an assault team, consisting of a number of Royal Marine corporals who had done the CQB course at Lympstone, conducted a simultaneous assault of the upper and lower floors; a typical CQB tactic. As the upper team cleared the first storey, they entered a final room which led out onto an L-shaped balcony. One of the Royal Marines assumed a position in cover by the door leading out onto the balcony and called for support. A colleague stacked immediately behind him, at which point, the first Marine called, 'I'm going long; you go short.' They then moved decisively out on the balcony together in a combat glide, weapons at the ready, the No. 1 clearing round the corner to the end of the balcony, while the No. 2 'button-hooked' to the left and cleared the area immediately beside the door.[119] It was a very basic procedure consisting of only two soldiers and an easily visible if large space to clear. However, it demonstrated Initiative-Based Tactics at work where the members of the stack had adapted to the environment and conducted complementary clears simultaneously and without preparation. Indeed, in a highly drilled team, the instruction which the No. 1 gave might not have been necessary. His decision to go long might have automatically generated a corresponding decision to go short by his No. 2.

Individual competence is manifestly essential especially in the light of the high risk of fratricide and the complexity of the environment. In order for an assault team to execute Initiative-Based Tactics effectively, however, all the individuals have to have an intimate knowledge of the corpus of CQB tactics and each of its manoeuvres. The adoption of completely uniform body postures and movements by all members of the assault team seems to aid Initiative-Based Tactics. Since all the members of an assault team assume the same posture in the light of different circumstances, the individuals are able to read each other's postures to ascertain the threat without the need for verbal communication. Every member of the team responds in physically the same way to a danger and consequently the following members of the stack will automatically be able to respond to and to adopt an appropriate stance to this signalling. In such a highly trained unit where body postures are shared, it becomes irrelevant which soldier is the No. 1 and which the No. 4; they are effectively interchangeable. Citizen soldiers frequently fought in towns and cities from the First World War onwards and some became proficient at urban combat. However, contemporary CQB tactics are markedly different from the past. CQB tactics have been refined to a very high level, reduced to a series of detailed and elaborate drills, which have been described at length in doctrine. At the same time, professional western forces have attempted to inculcate these new methods through extensive training techniques. These changes represent a very significant rupture from the citizen armies of the twentieth century where mass was institutionally and historically more important—and indeed necessary—than precision.

Social scientists have defined professionalism in terms of the generation and monopolization of specific skills and expertise. The development of infantry tactics and particularly the dissemination of close-quarters battle seems to represent a significant improvement in the performance of the platoon. It seems to represent the refinement and standardization of skills, fundamentally changing the way teams of infantry soldiers operate in battle. It seems plausible to suggest that, on the basis of these new performance levels, the infantry has become highly proficient in combat in a way which typically eluded the citizen army. In particular, in the development of infantry tactics, a sometimes extraordinary attention to detail has become evident: feet, elbows, thumbs, and angles have all, perhaps surprisingly, become central elements of infantry tactics. As a result of this scrutiny of minutiae, the infantry has been substantially able to overcome the Marshall effect which bedevilled the citizen armies of the twentieth century.

Nevertheless, it is important to avoid a facile ameliorism here. The performance levels of the infantry platoon do seem to have been substantially augmented by the institutionalization of ever more precise, comprehensive, and standardized drills, now practised on an increasingly transnational basis. Yet, it would be historically inaccurate to claim that today's army as a whole is simply better than its citizen predecessor or that mass armies could or should have somehow performed like a professional force. As discussed at the end of the last chapter, in drawing conclusions about the citizen army, mass brings with it its own unavoidable deficiencies but also its own strengths. Clearly, it is impossible to compare the contemporary professional army, equipped at the formation level with quite different capabilities from its twentieth-century predecessor, but it does not seem obvious that today's small professional western militaries would be a match for a genuinely mass army in battle, even though in relative terms the professional force would be far more capable. It would seem likely that it would make a similar discovery to the one which Rommel made when fighting the Allies and above all in Normandy. Future conflicts are typically unpredictable but it is not at all obvious that the downsized western force would be the optimal force posture against an opponent like the Chinese who retain a genuinely mass army. However, of course, precisely because the professional army is so much smaller than its mass forebear, such an improvement in performance has both been entirely necessary to compensate for the loss of size and possible because of this diminution; armies have been capable of training their smaller infantries to a higher level. Thus, the introduction of CQB does seem to represent an increase in skill for the infantry; soldiers and Marines are now able to shoot more accurately and conduct urban assaults in new ways.

Nevertheless, while individual skills may have improved, especially in terms of the dissemination of CQB methods, it is difficult to ascertain whether it represents an optimal solution at the institutional level for current operations. CQB techniques were developed for the Special Forces tasked specifically with hostage rescue for which these methods are vital. The primary role of the regular infantry, like the Royal Marines, is battalion- and brigade-sized manoeuvre in all terrains, including urban. Even if on future operations the infantry is tasked to fight in towns and cities, it is not at all clear that the development of advanced CQB is necessarily the optimal organizational skill. Operating at the battalion and brigade level, established section and platoon techniques might be adequate, especially given the prodigious costs of developing and maintaining CQB skills. It might be argued that with a large infantry force, it is impossible to sustain CQB techniques at a sufficient level among enough of the troops to make their promulgation worthwhile. Perhaps a better strategy might have been to introduce tanks or heavy vehicles into infantry tactics, as the US Marine Corps does, while maintaining a basic standard of soldiering. Indeed, despite the claim that most of the population lives in cities, it is by no means certain that therefore future wars will necessarily take place there. The struggles over natural resources which many envisage as the most likely cause of future war could easily take place in very remote locations. The Canadian army is currently worried about a potential Russian attempt to seize land in the Arctic. In these environments, CQB is likely to be irrelevant beyond improving the marksmanship of individual soldiers. Nevertheless, despite these uncertainties, western armies have invested heavily in CQB. This does not seem to be a decision strictly based on rational calculation of optimal ends and the efficient use of organizational resources to maximize operational results. On the contrary, in the light of inevitable uncertainty about future conflict, such rational decision making would seem to be almost impossible.

At the beginning of this chapter, the work of Weber and Collins was employed to develop the concept of military professionalism laid out by Huntington and Janowitz. Huntington and Janowitz emphasized the importance of the special expertise which the officer corps display to justify their position and determine optimal relations with civil society. Weber and Collins certainly do not dismiss the importance of expertise and skill in the constitution of professionalism. However, they are rather more interested in the monopolizing strategies of professional status groups which seek to sustain their political and economic positions through the preservation of privileges. With this concept of credentialism, Collins represented the most extreme version of this argument. Credentials had almost nothing to do with skills acquisition but with exclusion and self-defence. A similar argument might be made about CQB. It might be seen as a new form of credentialism within the armed forces. In the light of the difficulties and expense of training soldiers to become proficient at CQB and the uncertainty about its future utility, it might be possible to claim that the acquisition of CQB is not primarily about improving operational effectiveness but more a matter of status assertion and monopolization. While an advance in capability seems evident with these heightened levels of individual and collective skill, a status demonstration may have encouraged armies to adopt CQB tactics. At one level, CQB tactics represent just another series of skills and capabilities to which the infantry can plausibly point to protect themselves from cuts. CQB becomes a way of maintaining their monopoly over ground combat. Yet, CQB is not just one capability among many others. Specifically, CQB is a tactic in which the Special Forces have specialized since the 1970s. It is therefore a combat method closely associated and indeed synonymous with troops who enjoy the highest status in the armed forces today. Close-quarters battle techniques have an allure for the public, politicians, and military commanders—and indeed for soldiers themselves. Whatever the actual military value of this expensive capability, the performance of these highly trained assault teams with their distinct movements and precise shooting has deep cultural resonances which are reflected in public imagery, films, and video games. While an operational requirement may have been identified for the dissemination of CQB methods to regular infantry, the high status of CQB and cultural allure does not just seem to be irrelevant to infantry regiments as they struggle to defend themselves from being cut in a highly competitive budgetary environment. CQB not only represents an extension of the infantry role for current and future operations but it allows the infantry to enjoy some of the prestige of the Special Forces by appropriating some of the latter's traditional skills. The professionalization of the infantry is not simply about objectively improved performance, although that is not irrelevant. It has also involved a status element, then. On the basis of these new skills, infantries are seeking to protect their resources and privileges. The infantry is acquiring new professional skills not just because these are most likely to be the best adapted to future conflict but because these skills are most likely to increase the status of the infantry.

9

Training

RECENT SCHOLARSHIP

The battle drill has been proved to be central to combat performance both in conventional open warfare and in the new context of urban combat. By uniting soldiers around a set of established drills, the cohesiveness of the platoon has been significantly enhanced. The battle drill has enabled soldiers to respond both automatically and collectively to the threats which confront them, thereby reducing the need for coordination and the chances of fragmentation. Yet, it is not enough merely to recognize relevant drills even though they might be extremely refined and theoretically effective. Modern platoon battle drills were fully recognized as early as 1916 but, crucially, the mass armies of the twentieth century were rarely able to train their troops sufficiently so that they were able to execute these drills in combat. The theory existed but soldiers typically reverted to appeals to morale and to mass tactics to overcome the Marshall effect. Citizen armies often relied upon and favoured motivational methods to encourage performance which were most closely aligned with the strategic and cultural requirements. Accordingly, it is here in the domain of training that a profound divide appears between the citizen and the professional army. In the all-volunteer armies of the early twenty-first century, this deficit between theory and practice has been substantially overcome so that the professional platoon is typically able to enact their drills in combat. Moreover, as the dissemination of close-quarters battle drills indicates, in many cases professional platoons are now capable of executing drills which are more complex than those recommended to their citizen predecessors. The ability of professional soldiers to perform these drills has little to do with their qualities as individuals; there seems to be little evidence to show, as individuals, they are any braver, more determined, or tougher than their forebears. Indeed, in many cases, the suffering and sacrifice of citizen soldiers in the conflicts of the twentieth century might be argued to have been more extreme. However, as platoons, professional soldiers do seem to demonstrate greater effectiveness and this efficacy seems most obviously attributable to the collective process of training which they undergo.

Interestingly, just as training has become increasingly central to the professional army, scholars have also emphasized the importance of training in explaining cohesion in battle than traditional concepts of comradeship. They have effectively prioritized a practically oriented concept of task cohesion above the 'pure cohesion' of comradeship. William Cockerham's article on US paratroopers might be taken as the starting point of this interest in training and task cohesion.[1] He accepted Moskos's arguments about the importance of latent ideology and comradeship to military performance but argued that 'theories of primary group relations and latent ideology are not in themselves all-inclusive explanations of combat motivation'.[2] Rather, he describes a 'high level of identification with immediate superiors' and above all the 'strong value of teamwork' as critical to performance.[3] For him, drawing on Peter Bourne's work,[4] Cockerham maintained that in combat, soldiers were united through drills: 'one of the most efficient techniques which allows soldiers generally to adjust to combat is to ignore the danger by interpreting combat not as a threat to life but as a sequence of requirements to be met by an effective technical performance'. Citing Fehrenbach's work on Korea,[5] he records 'only knowing almost from rote what to do, can men carry out their tasks come what may'.[6] More recently the theme of training has been taken up by a number of other scholars.[7]

Hew Strachan has emphasized the importance of battle drills and, of course, inseparable from his observations about battle drill is a recognition of the close connection between training and combat performance. However, Strachan is not the only historian to have begun to highlight the role of training in generating cohesion. Edward Coss has recently explored the performance of Wellington's army in the Peninsular War between 1808 and 1814 prioritizing training in comprehending cohesion. His work represents a vivid and important contribution to the analysis of training and the armed forces. Although historical, its central argument that training is critical to performance is immediately relevant to contemporary debates in the social sciences about cohesion and to understanding the performance of professional soldiers in the twenty-first century. Despite the very different tactics of the time, Wellington's army were, like today's western soldiers, all professionals. Indeed, in the course of the work Coss explicitly addresses the work of major contributors to the debate about cohesion including Janowitz and Shils, Moskos, Bartov, and Little. Coss seems to see his work quite deliberately as a historical intervention into current debates about cohesion in the professional army.

Coss's analysis is animated by his opposition to the common perception that British troops under Wellington were drawn from the lowest sections of British society with criminal or antisocial tendencies; Coss rejects Wellington's own claim that his soldiers were 'the scum of the earth'.[8] Coss attempts to show that troops were not always drawn from the bottom of British society but were typically labourers and weavers displaced by rural and industrial transformation. In addition, he demonstrates that their proclivity for larceny in Spain was not the product of innate criminal tendencies. On the contrary, with the exception of atrocities after sieges, which had their own special dynamics, the appropriation of local goods by British soldiers was typically a response to chronic and systematic failings in logistics; they simply did not have enough food.[9] 'Even when supplied, the provisions of the ranker fell almost 19 per cent below the absolute minimums [for calorie intake] needed for men not facing the stresses and exertions of campaign and battle.'[10] British soldiers stole to survive.

Coss establishes then that professional British soldiers in this era were neither 'scum' nor career criminals. He then plausibly suggests that their excellent battlefield performance would be incompatible with a force comprised of such individuals: 'the redcoats' exemplary conduct under fire was also not engendered by their base natures... Courage and related cooperative actions have no meaning for vagabonds, rogues and sociopaths disconnected from society on every level.'[11] Coss also quickly dismisses the importance of patriotism. He does not, however, deny the significance of discipline, which was enforced brutally, but, for him, the performance of British soldiers can be explained finally only by reference to the formation of small primary groups. Crucially, British professional soldiers lived and operated together in close proximity in small mess groups of about six individuals. Close associations were developed between these groups in barracks but, on operations, those associations became critical. An individual's survival relied upon their colleagues in their mess group. Specifically, it was only as part of a mess group that an individual soldier could hope to acquire sufficient rations to live. Each individual was dependent on his comrades to procure and share provisions with him; 'On campaign the ranker discovered that his mess group, the men with whom he lived and next to whom he stood in combat, played a significant role in determining his survival. When rations were short, the group scrounged, stole, and, shared, distributing an equal portion to each member of the band.'[12] The mess group provided a variety of other essential supports to its members, and to fail to support the mess group or to act selfishly was to risk ostracism. On operations, such ostracism substantially increased the chances of an individual's death. Fear of exclusion compelled soldiers to cooperate with each other.

Out of this mutual dependence, a deep sense of collective obligation emerged which, according to Coss, informed battlefield performance: 'the ranker's steadiness was based on an acquired self-confidence and the knowledge that, as part of a squad, he was not alone.'[13] For Coss, 'this ongoing mutual support on campaign and the close relationships among the men had a collateral effect: enhanced cohesion during combat'.[14] Soldiers became accustomed to relying on each other away from the battlefield and on expecting their support to be reciprocated. That trust seems to have been transferred to the battlefield. Indeed, while acts of selfishness might be routinely sanctioned by the mess group, warning the individual to comply to the group's needs, cowardice on the field of battle typically resulted in immediate and final ostracism. Here, the individual had deserted his comrades at the point of most risk; his selfishness had endangered their lives. Coss gives a number of useful illustrations. For instance, 'Long Tom' left his comrades during a skirmish at Redinha in 1811, supposedly to help the wounded. He was excluded from mess groups that evening. Indeed, Coss produces a number of examples where soldiers who had been disgraced and were facing ostracism engaged in extravagant acts of bravery in order to reaffirm their status with their comrades.[15] Coss draws heavily on the memoirs of William Lawrence, who participated in the storming of Badajoz in 1812, during which he was wounded while his two closest friends were killed.[16] However, precisely because Lawrence's comrades were so densely bound together in mutual commitments, they could not be seen to let each other down. They advanced into the breach together, favouring the high probability of death over the certainty of exclusion and dishonour.

Coss prioritizes small-group bonds, initially generated by mutual dependence for provisions and other needs. It is therefore understandable why he draws primarily upon Janowitz and Shils in his analytical discussions of cohesion. They too identify personality needs as the primary motivating factor in the formation of small groups. However, dense interpersonal bonding does not, according to Coss, fully explain the performance of Wellington's troops, and it is here that he becomes especially relevant to professional soldiers today. British soldiers were successful not only because of the dense bonds of obligations between them and their great sensitivity to their status in the eyes of their peers but also because they were sufficiently disciplined to be able to execute effective tactics. Wellington's infantry developed distinctive linear tactics which contrasted with the column tactics typically used by the French especially in the later period of the Napoloeonic Wars.[17] The British infantry's favoured combat formation was the two-rank line. This maximized the amount of firepower which units could generate, while simultaneously giving them the most frontage, allowing them to outflank French formations, with the obvious advantage which that provided. The favoured British tactic at this time was to await the French advance silently, holding fire until very close range: the average range at which the British line opened fire was 64 yards.[18] At this point, the British troops would fire a single mass volley, give three cheers and mount a bayonet charge, the importance of which Coss emphasizes over the typical assumption that firepower was critical.[19] Major Thomas Bugeaud recorded his experience of this tactic when:

The English line remained silent, still and immovable, even when we were only 300 yards distant, and it appeared to ignore the storm about to break. The contrast was striking; in our innermost thoughts we all felt the enemy was a long time in firing, and that this fire, reserved so long, would be very unpleasant when it came. Our ardour cooled. The moral power of steadiness which nothing can shake (even if it be only appearance), over disorder which stupefies itself with noise, overcame our minds. At this moment of intense excitement, the English wall shouldered arms: an indescribable feeling would root many of our men to the spot; they began to fire. The enemy's steady, concentrated volleys swept our ranks; decimated we turned round seeking to recover our equilibrium; then three deafening cheers broke the silence of our opponents; at the third they were on us, pushing our disorganized flight.[20]

At the battles of Vimerio, Talavera, Quatre Bras, and numerous smaller skirmishes, the British employed this technique successfully. For instance, one officer recorded at Quatre Bras in 1815: 'We gave them a beautiful volley and charged, but they ran faster than our troops.'[21]

Coss notes that these line tactics demanded high levels of training and professionalism from the troops: 'the spacing of a line formation, with approximately 18 inches between men, put a greater demand on each soldier to stand his ground. The flight of but a few men in a two-rank line could have catastrophic consequences.'[22] Moreover, in order to protect themselves from artillery fire, British infantry would often lie down until the French attack was close: 'Getting men to rise from the prone position and advance while under fire was a difficult manoeuvre that depended heavily on unit discipline and *esprit de corps*.'[23] Indeed, it might be suggested that the tactic of getting British soldiers to lie down potentially introduced the problem of the Marshall effect onto the battlefield. Theoretically, soldiers might refuse to stand up in the fire that they would have been able to feel passing over or near them; they could have feigned injury or the action of standing up might in fact induce a general panic in the moments of uncoordinated unit action as individuals lifted themselves to a standing position. There was a potential moment of uncertainty and a presumption during this transition that others would also stand up. Yet, Coss records no instance of British soldiers failing to stand and receive an assault or to press forward with the bayonet charge. However, Coss notes that 'it was not genetic or cultural superiority that made the redcoat into good soldier'. On the contrary, the British troops were successful because they had an effective doctrine and were extremely well trained in its execution.

Training is central to Coss's account. In 1788, Sir David Dundas published *Rules and Regulations for the Movements of His Majesty's Infantry*, which described the basic infantry evolutions and drills.[24] The development of a coherent doctrine was important; it codified practice, just as contemporary doctrine does. However, critically, British regiments in this period practised their manoeuvres and drills intensively. Drill was a rigorous daily occurrence. A new ensign with the 88th Foot recorded that he had 'been out at six o'clock in the morning for some time past—since I joined the regiment. We are drilled with the men exactly the same as the privates.'[25] Coss observes that, in fact, the historical archive does not record drill and training extensively: 'Soldier accounts (including Lawrence's) do not often mention close order drill, however, perhaps because it was such a standard part of daily life that it did not merit highlighting.'[26] However, Coss plausibly presumes that 'drill must have been fairly frequent' and indeed, Captain Thomas Browne records how the men learned 'manual exercise, platoon exercise, the evolutions, the firings and the manoeuvres' in his *Napoleonic War Journal*:[27] 'the required movements were assimilated through slow repetition at first, then executed more rapidly as the soldiers became more proficient.'[28] The training of the British was in stark contrast to the poor quality of French infantry tactics especially after 1796. The French had to resort to unsophisticated column tactics and three-rank line tactics as a result of 'lack of training time, coupled with the continued influx of raw recruits'.[29]

Coss is extremely critical of Wellington's descriptions of his troops as scum and cites Wellington's famous comment about the relevance of patriotism in encouraging soldiers to enlist: 'People talk of enlisting from their fine military feeling—no such thing. Some of our men enlist from having got bastard children—some for minor offenses—many more for drink; but you can hardly conceive such a set brought together, and it really is wonderful that we should have made them the fine fellows they are.'[30] For Coss, the passage is clear evidence of Wellington's misplaced contempt for his troops. Yet, Cope misses the significance of the final phrase. While patriotic ideals were more or less irrelevant to the recruitment of soldiers and their subsequent performance, Wellington fully recognized and praised the quality of his infantry—after they had been trained. Whatever their rationale for joining, the regiments united their recruits into companies of 'fine fellows'. As Wellington's notorious comments about corporal punishment reveal, the creation of such a body of men required harsh measures, in his view. Yet, the phrase 'fine fellows' implies not merely docile slaves or automatons; individuals subordinate to some centralizing authority. On the contrary, it seems that Wellington himself recognized that battlefield performance—the generation of cohesion—required some act of positive collective agency on the part of the troops themselves. They collectively committed themselves to dangerous and difficult endeavours. Training, drill, and collective expertise were a central part of this collective competence. Coss concludes: 'The ranker believed in his officers, his training and the British volley-and-charge tactical speciality almost as much as he trusted the men with whom he fought.'[31] The British soldiers whom Coss analyses fought a war which finished nearly 200 years ago and yet, underlying this divide, Coss has produced a work which has direct relevance to understanding cohesion and combat performance in a professional army today. For Coss, the cohesiveness of Wellington's troops was a product of their collective training, supported by their mutual dependence on each other as mess groups.

It is worth stressing the collective rather than the individual effects of training which is central to Coss's account because the social dimension of training has often been missed, although it is decisive. For instance, in his work on Special Operations Forces, Rune Henriksen adopts an individualist approach to training. He has claimed that training does not produce Special Forces soldiers but only reveals the warrior characteristics which were already inherent in individuals: 'The SOF approach to warrior selection implies that warriors are revealed and that they cannot be made.'[32] Although individuals need to possess certain physical and emotional capabilities if they are to become infantry soldiers and, especially, if they are to join the Special Forces, Coss's work suggests that Henriksen may misunderstand the role of training in the armed forces. Training certainly equips the individual with new skills which are vital in combat, but the central aim of military training is not only the individual but the group. It would be entirely conceivable to imagine a group of brilliantly trained individual warriors who could not or would not fight as a team as they all preferred to operate according to their own personal drills. Hans Delbrück's descriptions of mounted knights in the Middle Ages record a military force which comes close to this kind of individualized approach.[33] Training in a professional army, by contrast, aims not merely to teach individual skills but crucially to unite troops around shared collective drills. Indeed, as close-quarters battle drills indicate, one of the central aims of even purely individual training is to inculcate a single shared template of action on everyone. Philip Smith's analysis of military training usefully emphasizes its collective rather than individual nature. Citing the United States Marine Corps's training exercise 'the Crucible', Smith notes that 'the focus here is not on building fitness [an individual attribute] but on generating teamwork'.[34] He concludes: 'drill is not so much about individuating in the service of a new capillary power but rather generating a sense of fusion with the group that starts to occlude egoism'.[35] Training, for Smith, explicitly aims at suppressing the individual and prioritizing the group. Training is, then, as critical to today's professional soldiers as it was to paid infantrymen in Wellington's army; but even in a more tolerant, perhaps individualized society, training is not primarily about the individual. Through repeated training, professional soldiers develop a common set of drills which they are able to reproduce instinctively and without recourse to contemplation, discussion, and extensive coordination in combat. Through training, drills become automatic not just for the individual but for the platoon as a whole. Having trained for long periods together, shared drills become deeply ingrained; soldiers are united around common practice and are finely tuned to even small cues from their colleagues. The shared experience of past training induces common and instinctive responses from these troops when they are on operations. Just as in Wellington's army, training aims at generating cohesion; it aims to enhance combat performance of military units from the platoon of thirty soldiers up to the highest formation of tens of thousands.

THE IMPORTANCE OF TRAINING

One of the key differences between a professional and mass citizen force is precisely that all soldiers in a professional army are trained to a high level to perform a common set of drills. Indeed, contemporary military doctrine repeatedly highlights the importance of training: 'Collective performance is only achieved through an understanding of common doctrine combined with collective training and exercising to rehearse and sharpen the ability to apply it... There can be no compromise on this, for the ability to deploy fully prepared for combat is at the core of fighting power.'[36] In the previous chapter, the concept of professionalism was summarized as a collective attention to detail and, in a professional military force, training is the sphere in which that attention to minutiae becomes particularly obvious. Combat performance is self-consciously related by professional soldiers to this refinement of a myriad of military skills, each small, often apparently insignificant in themselves, but together generating a complex of battlefield competence. Bryan McCoy, the commander of 3/4 Marines in Iraq in 2003, has usefully summarized this requirement: 'Focus on the basics and become brilliant at them.'[37] He observed: 'Great units do the basics with a high degree of proficiency and as habit.'[38] There is, in a sense, no mystery to the cohesiveness displayed by professional troops. It is a product of training. Indeed, the higher level of training in a professional force may seem like a relatively small and mundane difference. Yet, the experience of training fundamentally alters the relations between soldiers in an infantry platoon and the kind of association which they generate with each other. Moreover, training does not simply inculcate individual skills, though it certainly does that, but fundamentally alters the way the members of the platoon are able to interact and to perform together. Intense training changes the very nature of the solidarity—and, therefore, the cohesiveness—of the platoon. It, therefore, fundamentally changes the collective performance of which combat soldiers are capable especially in the confusing environment of battle.

The dramatic effect of training on cohesion has been widely recognized by professional soldiers. Indeed, this instinctive knowledge of platoon battle drills has often been called 'muscle memory' by British soldiers. British soldiers and Marines widely regard the junior and senior command courses run by the British army at Brecon and the Royal Marines at Lympstone for corporals and sergeants as critical to the inculcation and preservation of this 'muscle memory'. These courses involve much more than section and platoon tactics since candidates are being trained to perform all the various functions required of a corporal or sergeant. However, the central element of these courses is tactics and, specifically, the section attack on the junior command course and the platoon attack on the senior command course. Candidates could not pass these courses if they had failed to demonstrate the ability to command a section or platoon attack competently. Accordingly, many days on these courses are assigned to the section and platoon attack which are repeated many times to allow each individual to be assessed as a commander. Most sergeants and corporals regard these courses as critical to the professional competence of the British forces, which they believe has been proven at the tactical level in Iraq and Afghanistan. Brecon—and Lympstone—institutionalize tactical drills; they instil a common 'muscle memory' in British troops.

Perhaps predictably, training was especially important to the Special Forces and this is very evident in the Canadian Special Forces unit Joint Task Force 2, the equivalent of the SAS or Delta Force. Like British non-commissioned officers, they recognized the direct correlation between training and performance, employing an epigram from sports science to eliminate complacency: 'In the gun-fight, you don't raise your game. You sink to the level of your training.'[39] Joint Task Force 2 comprises some of the most competent troops in Canada and it would be entirely expected that they would value training. Yet, their advocacy of training was affirmed in the regular infantry. Indeed, the centrality of training was underscored throughout the education and instruction of soldiers and officers in Canada. One female infantry sergeant, who had been on operations in Kandahar and involved in many fire-fights, impressed the importance of training on a group of very junior officers at the start of their training, whom she was instructing: 'If you train people properly, they won't be able to tell a drill from the real thing. If anything, the real thing will be easier.'[40] Her point was summarized in a widely used military aphorism: 'Train hard, fight easy.' Training has been widely and self-consciously identified by professional soldiers as essential to successful combat performance.

The centrality of training to operational performance has been particularly evident in the statements of soldiers who have been on the most recent operations in Afghanistan and Iraq. There are numerous examples where soldiers have described how they executed drills automatically as a result of their training.[41] However, 42 Commando Royal Marines' operations between October 2008 and March 2009 provide a particularly useful and well-documented example of the importance of training to battlefield performance. 42 Commando were deployed to Afghanistan in October 2008 for a six-month tour as Regional Command South Reserve Battle Group. They were stationed in Kandahar Airfield and used on a series of offensive operations in Helmand, Kandahar, and Oruzgan. Their experience of the campaign was very different therefore from NATO ground-holding battalions who were stationed in Forward Operating Bases (FOB) and tasked to secure particular districts.

42 Commando was used as a conventional strike force fighting insurgents. Usefully (in terms of the question of cohesion), because they typically inserted by air, surprising the Taliban, they did not have to adjust their tactics to counter the threat of IEDs. By contrast, many of their peer units were typically unable to carry out fire and movement tactics because the Taliban often concealed IEDs in positions which could be used as cover. Platoons and sections were not able to form normal assault formations with bases of fire supporting attacking sections bounding forward in fire teams. Platoons and sections formed snakes behind the lead soldier clearing the way with an electronic mine-detector to identify possible IEDs. Moreover, the Taliban would often initiate a contact from a firing point, hoping that NATO forces would mount a conventional platoon assault to clear the position which was laced with IEDs. The aim of the Taliban was merely to inflict casualties. Accordingly, for many units, Afghanistan represented a negation of normal platoon assault tactics; it was a form of siege warfare in which IED clearance was the priority. In response to enemy fire, western troops often preferred simply to stand their ground and trade fire where they stood, trusting their body armour and the poor marksmanship of the insurgents over obvious ditches nearby.[42] 42 Commando were in the enviable position in which its sub-units were able to conduct normal infantry tactics in line with established doctrine and crucially in line with their training.

Within 42 Commando, Lima Company, consisting of approximately 120 Marines, widely regarded as one of the best British sub-units in theatre at that time, engaged in many of the most ambitious and largest assaults the British had conducted up to that point, including the attack on Nad-E-Ali, as part of Operation Sond Chara in December 2008. The officer commanding Lima Company, a Royal Marine major, recorded his experiences during that tour in a distinctive manner. He emphasized, first, the importance of apparently very simple drills, like being able to move quietly at night to tactical success.

Moving quietly at night does not seem to be a complicated thing to do. However it is crucial to tactical effectiveness. Moreover, it is very difficult to move quietly at night as a company. The hours we all spent on Woodbury Common on exercises was crucial to our ability to do this and it was crucial to our success. I don't think a conscript force could have developed such a skill.[43]

Indeed, in Normandy in 1944, some of the Wehrmacht noted how noisy some US infantry divisions were at night, especially ones new to the theatre, facilitating the job of interdicting them very significantly. Clearly, packing equipment up and moving at night includes an individual dimension: a single noisy soldier would reveal the location of the unit. However, maintaining silence at night is also a collective skill. It required individuals to coordinate their movements without speaking, ensuring that they remained in contact with each other even though they might barely be able to see one another.

Of course, the major commanding Lima Company also described the importance of training to the execution of fire and movement tactics. During training and exercises, the officer had undergone significant instruction in infantry tactics, but one of the most useful periods of training, which he discovered in Helmand, was earning his range qualification. The British armed forces employ open ranges in which live ammunition is used and troops are given freedom to move and fire as the designated range controller judges best. Clearly, these open ranges require careful administration and control in terms of the position of targets, indicating objectives, and the exercising troops. The qualification takes six weeks in which time candidates practise setting up open ranges, but they also have to conduct numerous live attacks, so that the adequacy of other candidates on the course can be tested. 'In Young Officer training, one time out of ten, you are in charge but the other nine times, you are the rifleman and so you did section attacks time after time.'[44] This experience of repeated attacks, day after day, drilled the practice of section attacks into the officers until the choreography of the assault became instinctive; it became automatic to go left or right flanking and to establish points of fire to facilitate such a manoeuvre. On operations, in Helmand, the Royal Marine major found the experience of running open ranges central to his effectiveness. So deeply had the training ingrained the battle drill into him that he sought out flank attacks instinctively: 'There is no need for loads of orders or commands. It becomes instinctive. The enemy's there. You put your fire support there. I am going there.'[45]

On 27 February 2009, his company were deployed into the village of Khan Neshin on the so-called 'Fishook' on the Helmand River in southern Helmand to search and clear the area of any potential Taliban fighters as part of Operation Aabi Toorah 2B.[46] The operation provides a very good and well-documented example of the importance of training in generating combat performance. The company advanced during the night from the north, approaching quietly on foot, while K Company inserted from the south. A suicide bomber detonated himself ineffectively in front of K Company, but as L Company neared the settlement they began to sense the presence of the enemy who seemed to be hidden in and around the fort. In order to identify their firing positions and their strength, the company commander sent one of his platoons (8 Troop) across open ground to draw enemy fire, with the other platoons in support. 8 Troop, as expected, encountered the first serious resistance, coming under fire from a compound which wounded a Marine. In response to this enemy fire, the company commander ordered 8 Troop to mount a left-flanking attack on the compounds from which the fire had come. For this officer, intense repeated training was explicitly fundamental to operational performance. In an official account of the tour, Ewen Southby-Tailyour recorded this assault: '8 Troop moved into an engagement that included an exchange of grenades at very close quarters, Apaches firing

30mm cannon at danger-close range, and a command course-standard troop attack followed by a section conducting an individual fire and movement assault.'[47] Southby-Tailyour concluded: 'The clearance of the objective—a small, fortified enemy compound—had been done by the book.'[48]

The assault did, in the end, go according to plan and according to contemporary infantry doctrine. However, Southby-Tailyour ignores many of the problems which 8 Troops encountered. Consequently, he actually underestimates the significance of training to combat performance in general and to this action in particular. On the initial entry into the compound, the assault had met with ferocious resistance from insurgents inside and the assault section became badly confused and was initially forced to withdraw to re-organize itself. It then re-mounted its attack and during the final assault, the Royal Marines major recalled hearing the commands of the section commander, a physical training instructor, 'Delta fire team: final position', 'Charlie Fire team: suppressive fire' as he re-entered the compound. On this occasion, prepared for the level of resistance they would meet, the Marines breached the compound successfully and killed the Taliban fighters there, who refused to surrender. The company commander emphasized that during training the meaning of these commands and, therefore, the response to them had become almost instinctive. Indeed, the commands were not potent simply because their meaning had been repeatedly emphasized in training but the very intonation of the words was deeply evocative to the Marines in Lima Company. During training at the Commando Training Centre at Lympstone, the physical training instructors organize sessions in the gymnasium by the use of very distinctive clipped, high-pitched, single-word commands. The actual words of the commands are themselves often indistinct especially in the noise of a large gym with echoing acoustics. Marines are trained to respond automatically to the tone of these commands and, because they know the next exercise and what movement is expected from them, they do not need to hear the actual word. In Khan Neshin, the Marine corporal's commands, which echoed across the battlefield, were delivered in the same distinctive tones which all the Marines had heard throughout their training at Lympstone. The section under his command—and indeed the whole company—knew what they were being commanded to do by these simple and short instructions. Moreover, familiar with the tone of the command, each Marine, on the basis of their experiences in training, was confident that both they and their comrades would respond appropriately to these commands. For these Marines, muscle memory did not simply refer to the drills of the platoon attack taught repeatedly in training, but the commands which cued these drills had themselves become deeply resonant and meaningful. In order to execute tactics 'by the book', the Royal Marines in 8 troop relied not simply on rote learning one specific choreography but on a deeply ingrained shared professional culture.

The Royal Marines may be particularly advantaged in the quality of training they receive. They are an elite regiment in a highly professionalized military which has been engaged in intense operations for a decade. Indeed, even during the Cold War, British infantry, and especially the Royal Marines and Parachute Regiment, were subjected to intense and realistic training as a result of the Troubles in Northern Ireland and the Falklands/Malvinas War in 1982. It is, of course, necessary to consider other western forces to determine whether the patterns evident in Britain can be detected more generally. Significantly, while in Europe British infantry—especially elite infantry— might represent an extreme, even in the case of the Bundeswehr, which has been subjected to the most political control in the post-Cold War period and been proportionally less involved in intense military operations, a similar emphasis on training and professionalism is becoming increasingly evident. The Bundeswehr has only just abolished conscription fully, although it has been moving towards a professional model since the end of the Cold War and, especially, in the light of operations in Afghanistan. Consequently it represents an interesting case study since the process of professionalization and the move from a mass, citizen army is currently observable. The emergence of compatible training regimes in Germany would be empirically indicative. Interestingly, as they professionalize, German infantry soldiers are indeed emphasizing the importance of training ever more strongly. An experienced captain from one of the parachute brigades who had been on operations three times, once to the Congo and twice to Afghanistan (on the last of which tours he was wounded), described training as central to a professional approach.[49] For him, cohesion was the 'alpha and omega' of the armed forces and especially the infantry. He offered up his own company as an example of how cohesion can be generated in the infantry. His company was part of an infantry battalion from the 26th Airborne Brigade which was not only among the most experienced formations in the Bundeswehr but received specialist training. As a result of operations in Afghanistan and observation of allied militaries, the Bundeswehr noted that a gap had appeared in its military capabilities between Special Forces and conventional infantry. It identified the need to create a tier below the Special Forces which was more highly trained than line infantry so that it could operate in more demanding conditions and support the Special Forces; in the USA, the Rangers fulfil this role. The captain's company was selected as one of the first infantry sub-units to be turned into an *Erweiterte Ausbildung* (advanced training) company. On the basis of his experience and his observation of other units in the Bundeswehr, the captain emphasized that 'what we require is specialist training, not broad generic training'. Specifically, he regarded training at the section and platoon level as critical. They were in his view not the lowest element of the military but the most forward element and training should be focused on them. The captain had already emphasized that drills (which could be developed only through intensive training) were essential to the performance of small groups.

Another German captain from the paratroopers confirmed the point.[50] When he had been a company commander he prioritized training; 'not theoretical training but practical training'. He had studied sports science at university and drew from this discipline to inform the way he had trained his troops; specifically he emphasized the importance of practical, repetitive training. Typically, the paratroopers whom he commanded learnt to strip and clean their weapons standing up, laying the pieces on a chair in front of them. He trained his company so that they could strip weapons on the march or strip weapons they had not seen before. Moreover, he emphasized the efficiency and intensity of training; 'We have less time for training in the Bundeswehr now so we have to make every session count.' The captain remarked that he used to compete with the sergeant major about who could maximize training time most effectively. Thus, if the training topic was loading a weapon, he would test his troops by seeing how many times they could load the weapon in 45 minutes; he observed, 'the more the better: you learn from repetition'. There are signs of a growing interest in the latest forms of infantry drill—close-quarters battle—and the Bundeswehr has liaison officers at some foreign training establishments. Among the younger cohort of operationally experienced Bundeswehr officers, there is evidence that professionalism, generated by training, is becoming more important. In order to achieve a higher level of combat performance, these officers prioritize experience, stable periods of collective training, and drills.

Consequently, these field officers are deeply interested in their NATO partners and especially the USA and the UK, which are openly admired. One officer had worked with a British infantry battalion in Kosovo which had just come back from Iraq. He described how he was in 'awe' of them because they were so 'polite but so professional'.[51] The captain from the *Erweiterte Ausbildungs* company stressed the importance of these connections to the USA and the UK. He had observed the higher levels of experience and expertise displayed by forces from these countries and argued that training and operational experience were the central means of professionalizing the Bundeswehr: 'the US and the British have lots of experience, the Bundeswehr soldier stands in the middle between them. We have finally recognized that. We need to learn and to train. Our experience must be translated into training and into training with our partners.'[52] Indeed, since the Bundeswehr was dependent upon its allies for strategic military resources, deepening the army's multinationality was decisive. Crucially, in order to be able to collaborate with US and British partners, professionalism was institutionalized even more intensively. The officer noted that at the Special Operations Forces training centre in Pfullendorf, there were directing staff from Britain, the USA, and a number of other countries. More of these links were needed.

He suggested that officer exchanges might assist in this in order to overcome the fact that 'we are behind the US and the UK'. While these experienced German officers might disparage the Bundeswehr, they provide clear evidence of the emergence of a training ethos and regime which is compatible with their NATO allies.

BATTLE PREPARATION

Training is critical for effective combat performance because it is only through this process of collective instruction that sections and platoons are able to execute battle drills in combat together. Moreover, since combat always presents unforeseen circumstances, groups of soldiers must be able to improvise individually and collectively to the unseen threats on the battlefield. They must know their drills well enough to adapt quickly and coherently. However, in order to maximize the effectiveness of drills, armies have recognized the importance of preparing their soldiers for specific missions by generating a detailed collective understanding of what the military unit is trying to achieve. Battle preparation has been a critical means by which generic training and drills are tailored and refined for the context of a particular fight. Battle preparation represents a form of mission-specific training, therefore. It is a critical element of the passionate attention to detail which is central to professionalism. Indeed, this willingness to prepare exhaustively and to pay attention to every possible detail almost paradigmatically separates the professional from the citizen army. Indeed, professional soldiers seem to be self-consciously aware of this difference. Following the completion of the offensive phase of their Military Operations in Urban Terrain Exercise, one of the instructors from the Marines Infantry Officer Course declared in his debrief to his students: 'Amateurs train till they get it right. Professionals train until they cannot get it wrong.'[53] He referenced the intense process of preparation, planning, and rehearsal which had been necessary to conduct the final assault properly: 'You have now seen the amount of preparation and rehearsal which is necessary to conduct operations.'[54] His point is affirmed by numerous other professional soldiers. Lieutenant Harry Tunnell, commanding officer of 1st Battalion, 508th Infantry, 173rd Airborne, jumped into northern Iraq with his battalion at the beginning of the war and subsequently conducted combat operations for eight months before he was wounded. He, too, identified training as the prerequisite of combat success: 'Define standards, train people on what they are, and enforce them. It is not a standard until it is written and understood. Your unit will fight the way they have trained regardless of whether you want them to or not.'[55] British officers have been equally emphatic about the importance of training. An excellent company commander in 5 Scots recorded how he had trained attacks repeatedly when he was a platoon commander. He would take his platoon onto the training area and run through section and platoon attacks assaulting in different directions; 'You train two three four times. Once you have trained six or seven times it becomes a drill.'[56]

As specialist training, battle preparation typically involves three key elements; orders which specify the mission, models which illustrate that mission, and rehearsals in which troops physically practise their manoeuvres. As discussed in Chapter 6, all of these techniques were regularly practised especially by highly trained forces often on special missions from the First World War onwards. Yet, for the most part, battle preparation for the mass army was rudimentary and inadequate. Although commanders were given orders, line infantry typically received little information about the precise nature of their missions. They followed their commanders and their orders. In the professional forces of the twenty-first century, battle preparation is regarded as a critical and indispensable part of tactics. Rapid and unanticipated reactions to enemy action are common but it would be highly atypical for even fairly routine military operations to be conducted without the gathering of intelligence and a formal battle preparation process. Before a mission, professional soldiers are given a detailed set of orders in which their commander explains the plan and his troops' role in it. The importance of an orders process was recognized in the First World War, since which time western infantry doctrine has recommended standardized procedures for delivery in each country. During the Cold War, NATO established a standard method for giving orders which is now evident across all member states. Orders conform to a five-paragraph format in which the situation, mission, execution, logistics, and command and signals are communicated. The most important paragraph is, of course, the mission and, if everything else is forgotten (or incorrect), the orders process is structured so the mission will be clear. By institutionalizing a single method of giving orders, western forces have sought to obviate two potential problems: that commanders will omit some crucial piece of information or that troops receiving orders will be unclear about what they are to do. By standardizing the process, commanders are less likely to make mistakes, while troops (often tired and hungry) are assisted by having their tasks communicated to them in a familiar structure. Indeed, emphasizing the importance of standardization and predictability, the US Army and Marines have reduced the orders process to an acronym, SMEAC (situation, mission, execution, administration, and command and signals), in order to minimize the chance of mistakes and omissions.

The standardization of the orders process has been important in clarifying the mission. However, one of the most interesting and important aspects of the orders process is the use of models which have sociologically profound implications in terms of the question of social cohesion. At the highest level of command, professional model-builders are employed to create detailed dioramas of the areas in which the missions will be carried out, physically detailing the ground and the enemy positions. However, even at the lowest section and platoon levels in the infantry, models are always preferred even by the Special Forces who are adept at using maps and usually operate in small numbers.[57] The use of models has become all but universal among western forces. On the Canadian army's Advanced Reconnaissance Course, which is probably the most demanding infantry training course on offer in Canada (outside the Special Forces), preparing senior non-commissioned officers and officers to command and train battalion reconnaissance platoons, enormous emphasis is placed upon the ability of patrol commanders to deliver orders clearly and, therefore, on very accurate models. Reconnaissance patrols rely on stealth in order to achieve their mission. Accordingly, a precise understanding of the terrain in which the patrol is to operate is critical to success. The better understanding a patrol has of the ground, the better chance it has of concealing itself on its approach, in its observation posts, and during its withdrawal. In the regular infantry, the ground brief typically lasts two or three minutes, but for a reconnaissance patrol, this brief had to last at least ten minutes. It is important to go into far greater detail about the topography. Accordingly, great stress is placed on candidates on this course to produce very high-quality models, from which to receive their orders, and they were closely assessed on them. Without a good model, it is impossible to give an accurate and collectively comprehensible ground brief; 'a basic skill is model-making and in normal circumstances, a model has to be workable. But here, the model has to be perfect.'[58] As the officer running the course noted, 'the level of professionalism is a lot more demanding'.[59] During the course, which was run in October 2011 in the Combat Training Centre, Gagetown, patrols returned to their base every 36 hours to prepare themselves for their next mission and to receive orders. It was here, in this base area, that they built their models. The directing staff on the course were delighted with the standard of model making which was among the best that they had seen; the models were extraordinarily realistic and accurate. Candidates on the Reconnaissance Patrols Course had shown great creativity in building these models, using local twigs and lichen to build very impressive replica trees and bushes. They were provided with or had procured specialist model-making equipment, Sully's Reconnaissance Supplies, produced locally by an ex-Canadian soldier. The Reconnaissance Supplies, packed in a small plastic box, contained an extensive array of specially designed material for helping to mark models: string to mark grid squares, different coloured pipe-cleaners to mark roads, rivers, approach routes, coloured tags and tallies denoting Rendezvous Points and patrols.

The Canadian Reconnaissance Course is very advanced but the specific attention paid to model building—and prepared materials to do it—is replicated in other forces. The US Marine Corps's Infantry Officer Course has an extensive kit consisting of markers, blocks, and tape specifically for the building of models which students are expected to use before they give orders and which is brought out on exercises. Even without specialist kits, troops often carry with them coloured ribbons and tags to designate distinctive features and they use surprising creativity in sculpturing the earth inside the model pits to represent ridges and valleys and to improvise with stones, grass, twigs, and leaves to signify woods, streams, paths, and roads. In this way, the models facilitate the orders process by physically representing enemy positions and ground over which the action will take place. As they give their orders, commanders are able to illustrate the precise movements of their troops on the ground by reference to the models, highlighting specific features which will orient the action or which may pose problems for the advancing troops. The model is a graphic device aimed at facilitating communication, and the more detailed and accurate the model the more effective it is.

There are pragmatic reasons why models are employed by troops. It is impractical to use maps in a tactical situation. Commanders and troops must be able to identify central landmarks by sight as they advance under fire or as they lie in tactical concealment; in these situations it will be difficult for them to examine a map. Consequently, the models represent what the audience will see as they advance on their target, such as trees, hills, or rivers, preparing them with a series of visual cues. However, the models assist collective action not merely because they depict the ground as the troops themselves will see it. More importantly, models are used because this ensures that there is only a single representation of the terrain, to which the attention of all is directed. Maps, by contrast, display a multitude of extraneous and often irrelevant topographic features in extensive detail. If each soldier received orders by reference to their own individual map, individual misinterpretations could occur. Soldiers could easily interpret their maps and the features depicted on them differently; they could focus on alternative geographic points, or misread the lie of the land from the contours of the map. Misinterpretation is a major problem on operations, when soldiers are very tired. The captain running the Canadian Advanced Reconnaissance Patrol Course emphasized the way in which models minimized the chances of mistakes or forgetfulness: 'The point is when troops are very tired and they have to go out again and the orders last one hour and fifteen minutes, you might forget everything except the model and your movement on it. You carry that image of the model inside your head so that when you are inserted you know that after fifty metres there is a tree on your right and then a black trail [an unmetalled road] going off from there.'[60]

Models are deliberately designed to eliminate individual deviance, then. The soldiers who are tasked to construct the models work from a map but they deliberately ignore subsidiary and irrelevant data which could mislead. The model is not simply a large-scale map, although it is on a bigger scale and uses far more detail than a map. The model makers are concerned with minutiae but model makers also focus on those decisive points which soldiers have been trained to employ as orienting axes and reference points in tactical situations and which troops will confront on the operation. Thus the model makers and the commander using the model will emphasis wood-lines (which stand out clearly even at night), roads, rivers, and pylons which are unmistakable physical features. Above all, objectives, whether they are natural features like hills or artificial ones like buildings, are especially highlighted on models. The result is evident. All are knowingly oriented to a single collective representation which consists of clear and distinct symbols. As the Canadian Reconnaissance Course demonstrated, the more accurate the representation, the more effective the model was as a shared symbol coordinating the actions of the soldiers on the ground. In this way, the chances of misinterpretation and deviance are minimized. The models also serve another important purpose. Precisely because troops are oriented to common objectives, signified by the model, they are more able to improvise during the mission itself. Since troops all understand the primary objective, expressed in orders as the Commander's Intent, and know the ground on which they will be operating, the models are intended to allow the troops to pursue this objective in alternative ways, as the situation demands. The models have become a means of institutionalizing collective flexibility.

The model has become the standard technique of uniting professional military groups around the pursuit of dangerous and difficult goals. In the past, the orders process culminated in rehearsals which could in the case of large-scale or important operations consist of major exercises in which thousands of troops would practise their manoeuvres on prepared ground. Alternatively, a reconnaissance team or platoon might rehearse, walking through its mission conceptually near its bivouac. In Chapter 6, the origin of this practice in the First World War among assault battalions on both sides of the conflict was discussed. Rehearsals of this kind are typical in the professional army. However, there has also been an attempt to fuse the model and the rehearsal into a single, graphic Rehearsal of Concept (ROC) drill. In the ROC drill, the commanders and troops run through the sequence of the operation on a very large (often uncontoured) diorama of the operating area in which buildings, roads, and geographic features are represented on the ground. The commanders of each unit and sub-unit then stand in the position they are assigned at the beginning of the operation and, as the commanding officer narrates the sequence of actions, they move about the diorama to indicate their manoeuvres and their positions throughout the operation. The rest of the troops watch the ROC drill. The ROC drill seems to have been a US innovation. At the lowest level the ROC drill has become a standard procedure in the US Marines and is a central part of the training curriculum for officers (see Figures 9.1–9.3). For instance, during the Military Operations in Urban Terrain phase of the US Marines' Infantry Officer's Course's Final Exercise,

Figure 9.1. Infantry officer course ROC drill, Range 220, 4 June. The instructor inspects the model before the drill begins. Objective markers, model tanks, and vehicles visible in the foreground.

the officers and instructors conducted a ROC drill before the final clearance operation. The students prepared a model of the urban area which they would have to clear, using blocks of wood which the IOC used specially for the process. The students used the blocks to indicate the shape and height of buildings on the basis of their maps; a single block showed a single-storey building, two blocks two storeys, and so on. Buildings which had been identified as important objectives were marked with their objective name on a piece of card placed on top of them. The roads were marked with white mining tape, with three successive phase lines (including Phase Line Green, mentioned in the previous chapter) indicated. Tallies signifying the sections, tanks, Humvees, and Amtraks had all been prepared to indicate the lines of assault and the relative position of the troops to each other and their support

Figure 9.2. ROC drill. The student officer in the foreground holds his platoon tallies behind his back. They will be placed on the model to indicate the scheme of manoeuvre. The uppermost tally represents Section 1, Platoon 1.

vehicles. Before the ROC drill began, one of the instructors, a Marine captain, explained to the gathered students the purpose of the process: 'The ROC drill is used to clarify the plan and then to work out the friction points.'[61] The process is now all but universal in the armed forces.

The function of the model and ROC drills as a collective representation, enjoining coordinated action, is eminently observable on operations. Famously, before the Iraq invasion, the charismatic but somewhat idiosyncratic commander of the 1 Marine Division, Major General James Mattis, conducted a ROC drill for the whole division on a very large (non-relief) model depicting Iraq. He issued each of the commanders of his battalions with a rugby shirt with their unit designation printed on it. While he sat in a raised chair, the rugby-shirted battalion commanders moved about the model in accordance

Figure 9.3. British army ROC drill and model before an urban assault on the old village of Beauséjour, CENZUB.

with the plan. In this way, both Mattis and his subordinates could gain a collective understanding of the plan and identify possible friction points in it. Given the complexity and ambition of the operation—driving all the way to Baghdad—this process was seen as essential, although some battalion commanders did not appreciate having to wear Mattis's rugby shirts. In their film recording the tour of a platoon from the 173rd Airborne Brigade in Kunar Province, Tim Hetherington and Sebastian Junger include footage of the platoon receiving orders before their most demanding mission, Operation Rock Avalanche, when they would push into previously enemy held territory to establish a new combat outpost. The platoon had built a large-scale relief model about 5 metres square of the operating area, depicting the major geographic features in which they would be working, inside one of their bases. The model was constructed on flattened square sheets of cardboard on top of which earth had been placed and sculpted to match the contours of the valley; the grid lines were represented by cords stretched just above the model which not only aided orientation of the platoon as it received orders but helped the model builders to ensure that their model was accurate. Roads running along the valleys were marked by stones and the compounds and villages denoted by taped ration boxes. At the centre of the model, and somewhat incongruously, were four green plastic model soldiers and a red plastic cowboy on a horse, marking the location of the company headquarters. The lines of advance of each sub-unit were marked with small cardboard arrows to indicate their specific mission and their relations to each other. In one case the arrows led up to the top of a hill, denoted by a cardboard marker, with the words 'Objective Taylor/West' written onto it.

In the film, the platoon is gathered around the model, listening intently to the plan for the operation; prominent among them is Staff Sergeant Rougle who would subsequently be killed on the operation. In turn each of the section commanders ran through the movement of their squads during the operation, indicating their movements with pointers. In the filmed sequence, the location of the helicopter landing zones and the platoon's movement from them were being described; one staff sergeant clarified, 'We're going to move from HLZ Cub into HLZ Pope'. The anxious concentration of the members of the platoon was evident and seems to have been intensified by the fact that the soldiers were well aware of the dangers of this operation. In a number of post-tour interviews, the soldiers reported that Rock Avalanche was the low point of the tour for them. Indeed, one soldier noted that he 'saw a lot of professional tough guys go weak in the knees' at the prospect of this operation. For 2nd Platoon, the rehearsal of concept for Rock Avalanche represented an intense forum in which every member of the group was focused on the unit's collective task and their personal role in it.

It is not only the United States Marines or Army which use the ROC drill. During the Iraq invasion of 2003, the Royal Marines and specifically 42

Commando adopted the ROC drill in their planning and preparation process, having worked with the US Marines on that operation, so that by 2005, it had become a standardized part of the unit's battle procedure. In 2008–9, when the unit was on operations in Afghanistan, Lima Company, 42 Commando, used ROC drills before every significant operation. The officer commanding gave a formal set of company orders in the unit's briefing room to his platoon commanders, sergeants, and specialists which included maps, photographs, and satellite images so that the soldiers knew the ground. They would then relay these orders to their subordinates, but before the operation, the whole company group would go through a ROC drill. To this end, the company constructed large models using ammunition boxes to represent buildings and compounds, each numbered for easy reference, and dug out small ditches to represent rivers or wadis and used tape to indicate roads. 'There was no reason why we did not build a model but we just did a ROC drill instead. If there were obvious really high features, we would make them.'[62] 42 Commando's ROC models were, then, a hybrid of the traditional model and the more conventional flat ROC diorama. Members of the company were explicit about the collective effect of the ROC process. 'This oriented people to the ground. The whole company watched the ROC Drill and the O/C [Officer Commanding] narrated it with people walking over the model. For instance, the helos were these guys. The troops were represented by the troop commanders. As they moved over the model they would say "This is me walking out", "These are my arcs". Then the next troop commander would go through his manoeuvres until everyone was completely happy with the plan. It was fundamental for the lads because it allowed them to paint a picture in their own minds. I would have been a lot more nervous had we not done a ROC drill. I had it in my mind's eye. It was fundamental to us.'[63] Through the ROC drill, the company united itself around a common understanding of its objective and what it intended to do.

Although Junger's Battle Company were plainly concerned during their orders process, troops do not tend to display extreme emotion, even less the euphoria which Durkheim noted of aboriginal clans; but the orders process has resonances with the aboriginal ritual. It is deliberately designed to invoke unity and, in moving about the model, troops seek to generate the homogeneity of movement which Durkheim noted as essential for solidarity. Moreover, while the collective effervescence of the religious ceremony is perhaps absent in the orders process, the ROC drills enjoin powerful emotions among the troops, whose concentration is demanded. Sometimes, as the ROC drill for Operation Rock Avalanche shows, the orders process can stimulate a collective sense of dread and apprehension. That visceral fear was infused into the model as a collective representation, heightening each individual's attention to the group's goals in an effort to minimize the potential for failure and indeed disaster.

As part of their attempts to become more interoperable especially with the United States and Britain and to professionalize, the French army has recently adopted the ROC drill as a method of battle preparation. In the past, French units gave orders from a map or alternatively by reference to a sand box. The sand box is a piece of military equipment of long standing, initially used as a means of examining tactical command problems. French troops effectively used the sand box as an efficient means of creating a model. However, as a result of their experiences in Afghanistan, where they have worked closely with the Americans, from whom, like the British, they learnt the technique, they have sought to introduce the ROC drill into their battle procedures, although it is not mandatory at the unit level. Innovative training centres like CENZUB have been at the forefront of this innovation, insisting that French troops training at the facility employ ROC drills before major evolutions in their exercises.[64] Officers explicitly regarded the institution of the drill as an important and useful development since the sand box was too small for tactical manoeuvres to be demonstrated; because of its size it was difficult to attain the level of detail which was possible on a large ROC model. On a large and well-constructed ROC model, it is possible for soldiers to imagine their operational surroundings very well so that they can orient themselves on the ground. French trainers have also noted the added advantage of physically walking through a rehearsal; 'It is easier to understand with your feet than by hours of talking.'[65]

The introduction of the ROC drill into France represents an interesting development which is likely to improve combat performance but it is also noticeable that in comparison with anglophone troops, French forces may not yet have developed refined model-making techniques. The staff at CENZUB were impressed by the models constructed by a visiting British company which used the lichen from local trees to represent the woods around Beauséjour in a highly realistic fashion (see Figure 9.3). The French infantry company undergoing a validation exercise in early December 2011 conducted a certainly proficient ROC drill in which the central movements of their assault on the major apartment blocks in Jeoffrecourt were rehearsed.[66] The fire support platoon stood on a hill overlooking the town, while the three rifle platoon commanders walked through their attack on the housing blocks at the direction of the company commander. The scheme of manoeuvre for the assault, for observing French troops, was unmistakable. Nevertheless, the model was not particularly accurate. It represented the objective schematically rather than mapping it accurately. The apartment blocks themselves were not to scale, nor were the distances between them. Small blocks of concrete had been placed on the model in loose relation rather than in strict triangulation with each other. At the same time, important features such as a kindergarten with a significant play area surrounded by walls adjacent to one of the apartment buildings which was of potential tactical importance, since it could be used as cover by either opponents or the French troops themselves, was not on the model at all. Consequently, while the general company manoeuvre was adequately depicted, the model was not sufficiently detailed to facilitate platoon and section preparation. Although the difficulties into which the assault subsequently ran, as the platoon entered the first apartment blocks (to be heavily attritted by an opposing force), could not be put down entirely to the shortcomings of the model, its inadequacies are not likely to have helped the troops' preparations (see Figure 9.4).

Models seek to create a collective representation which generates a common understanding of the battlefield for troops. There is clearly a vital and purely intellectual element to this process, whereby common understanding is developed. The coordination of practice is extremely difficult if not impossible without at least a degree of shared understanding of the situation, and this is particularly the case on the battlefield when the typical problems of coordination are multiplied many times over. At one level, models simply depict the terrain as it is so that all the troops are able to recognize and visualize it; they have a common understanding of the ground and their mission. However, collective practice is not merely a theoretical problem. It is not enough for participants merely to define the situation in common and to know what to do. They must be prepared to act in concert as a result of their understanding. Although it is very easy to overlook the point because they seem to be such mundane objects, models have to engender moral obligation as they communicate information. At this point, models and ROC drills become deeply interesting and it is worth considering at some length how they are able to fuse knowledge and obligation. Consequently and perhaps surprisingly, the sociology of knowledge, and, above all, Durkheim's engagement with Kant's philosophy, can usefully illustrate the way in which moral obligations may be inscribed in the military model.

Durkheim certainly maintained that the act of comprehension necessarily involves a moral obligation. Indeed, his concept of the 'conscience collective' was explicitly intended to communicate the indivisibility of knowledge and morality. Unfortunately, in translation to the English 'collective consciousness',[67] the moral aspect of his argument can be easily overlooked. Yet, one of the less obvious objectives of the *Elementary Forms* was to relocate Kant's concepts of both Pure and Practical Reason in the human community as opposed to the individual mind; Durkheim sought to provide a sociological account of Kant's concept of Reason and the Categorical Imperative specifically to demonstrate this necessary fusion of human knowledge and morality. This aim is, naturally, most obvious in his analysis of the moral implications of belief. In his *Critique of Pure Reason*, Kant sought to identify the necessary structure of human reason in order to establish what humans can know;[68] Kant was concerned with ascertaining the transcendental conceptual conditions on the basis of which coherent experience was possible at all. He believed that it was possible to deduce the existence of synthetic a priori categories

Figure 9.4. French army ROC drill CENZUB, before an attack on the new town of Jeoffrecourt. The apartment blocks which the company was tasked to attack are represented by the concrete blocks behind which stand the three platoon commanders tasked to take them (see photographs of CENZUB, Figures 9.10a-d, for images of the actual ground).

which underpinned all human perception and understanding. On the basis of this epistemology of reason, he subsequently sought to erect a compatible system of morals so that humans were obliged to act in a certain way because their own reason dictated it. The Categorical Imperative followed, in Kant's view, logically and necessarily from the synthetic a priori categories of Reason. To act against the categorical imperative was to act against Reason (and, therefore, what humans know to be true); it was ultimately to be irrational, non-human, and, in fact, un-free. For Kant, what the human could know should also determine what they should do. To understand and experience the world according to the categories of reason and then to act in contradiction to those categories was irrational and wrong.

Durkheim rejected Kant's claims that human understanding and especially morality could be determined by individual reason alone. Yet he was deeply impressed by the idea that the way humans comprehended the world was very substantially predetermined for them; it was not a matter of individual opinion or personal experience. Moreover, he was convinced that morality had a force which quite transcended mere utility or individual preference.[69] Morality had a necessity, although Kant may have also underestimated the voluntary and desirable character of genuine morality. Humans actively want to fulfil their duty: 'morality must be not only obligatory but desirable.'[70] In his work on aboriginal ritual, Durkheim sought to show how intellectual comprehension and social solidarity—understanding and morality in a deontological Kantian sense—were, indeed, intimately connected. He wanted to demonstrate that the way in which a community understood its world necessarily involved moral imperatives about how group members should act in that world. Indeed, he was explicit that science and philosophy both have their origins in religion.[71]

At the roots of all our judgements there are a certain number of essential ideas which dominate all our intellectual life; they are what philosophers since Aristotle have called the categories of understanding: ideas of time, space, class, number, cause, substance, personality, etc. They correspond to the most universal properties. They are like the solid frame which encloses all thought... They are like the framework of the intelligence. Now when primitive religious beliefs are systematically analysed, the principal categories are naturally found.[72]

For Durkheim, the origin of Kant's synthetic a priori categories lies not in logic or the structure of reason itself but on the contrary in social experience and collective representations. Durkheim gives the examples of time and space:

It [time] is an abstract and impersonal frame which surrounds, not only our individual existence, but that of all humanity. It is like an endless chart, where all duration is spread out before the mind, and upon which all possible events can be located in relation to fixed and determined guide lines. It is not *my time* that is thus arranged; it is time in general, such as it is objectively thought of by everybody in a single civilization.

That alone is enough to give us a hint that such an arrangement ought to be collective.[73]

In his discussion of space, which he sees as 'the same thing', Durkheim explicitly opposes his own sociological position to Kant's rationalism: 'Space is not the vague and indetermined medium which Kant imagined; if purely and absolutely homogenous, it would be of no use, and could not be grasped by the mind.'[74] Durkheim asks: 'But whence come these divisions which are so essential? By themselves, there are neither right nor left, up nor down, nor north not south, etc.' Durkheim claims that since 'all the men of a single civilization represent space in the same way', they must derive from 'sympathetic values', and this fact 'almost necessarily implies that they be of social origin'.[75] Durkheim gives the example of societies 'in Australia and North America where space is conceived in the form of an immense circle, because the camp has a circular form'.[76]

The origin of purely intellectual beliefs and frameworks is for Durkheim collective and social. However, in *The Elementary Forms*, Durkheim also wanted to show how a moral imperative was attached to categories of understanding as a result of the social origin of belief. A social group was, for Durkheim, united not only intellectually but also morally through its shared beliefs. By worshipping their totem and publicly displaying their belief in this deity, the members of a tribe were not simply consenting to a common intellectual framework, although this was vital; they were also committing themselves to the practices which were represented by that god. Since, as Durkheim famously argued, the aboriginal deity in reality represented the tribe itself, the aborigines' collective confession of faith was simultaneously a demonstration of their solidarity with each other. By worshipping their god together, they were simultaneously reaffirming their bonds to each other. The belief in this particular totem concretely prescribed a code of conduct which the tribespeople were expected to follow, then. Indeed, the ecstasy which the aborigines felt in the course of their rites was not so much induced by their intellectual consensus about the nature of the world but by the intense feeling of community of what they should do in it. Their submission to a shared deity represented this affective (and intellectual) union. Shared belief especially when publicly displayed in worship in and of itself involved a shared moral commitment. Certainly, it was entirely conceivable that an aborigine might subsequently renege on this faith and betray the tribe but such recidivism was, after a religious profession of faith and solidarity, recognized as treachery. The traitor was in bad faith, contradicting what he or she had publicly confessed as the truth. To share a belief is, according to Durkheim, in and of itself to be morally committed to a set of practices implied by that belief and to expect others to be so obligated as well. To accept something as epistemologically true is also necessarily to accept that certain forms of action which follow from that belief are morally right. Durkheim's focus in the *Elementary Forms* was the Australian aborigine but it was quite clear within that text, where he draws a parallel between aborigines and scientists and his other writings on the professions, that this intellectual and moral union was critical to the generation of any community.

In the light of this Durkheimian fusion of the intellectual and the moral, the use of models and ROC drills may be deeply significant for professional troops. Battle preparation is, of course, primarily oriented to improving the collective understanding of the platoon; its purpose is purely intellectual and practical. As the captain commanding the Canadian Advanced Reconnaissance Course stressed, models helped soldiers to understand and remember the mission. Yet, in generating a shared understanding about how they were going to operate, the process of battle preparation also seems to engender a moral obligation among the troops receiving orders. By agreeing to a common understanding of the situation in the orders process, soldiers are not simply demonstrating their comprehension of the mission in theory; they are also committing themselves to the prescribed forms of collective action represented by the model or in the ROC drills. They are committing themselves morally to the mission and, should they fail to fulfil their tasks, it would be impossible for them to claim that they did not understand what was being asked of them. Once intellectual clarity has been achieved in orders, failure becomes a moral issue which can be excused only on the ground of dramatic unforeseen circumstances which genuinely obstructed the prescribed action. Failure on any other grounds must be dereliction of duty for, in the orders process, soldiers had demonstrated their understanding of what was required of them. In this way, the use of models and the ROC drills may be even more significant than it appears because it suggests that the motivation to perform on military operations for the professional soldier may not come primarily from extraneous sources, as it did for the citizen soldier. Where citizen soldiers committed themselves to perform by reference to masculinity or nationalism, professional soldiers may not be so worried about their reputation as a man or be so reliant on political motivation to sustain themselves. Rather, motivation may come from the mission and the act of performance itself. Once they have committed themselves to it in the preparatory phase, it becomes a matter of collective and individual pride that soldiers succeed.

The armed forces seem to be more than implicitly aware of the moral imperative in the process of battle preparation. For instance, all western forces stress that commanders must give their orders confidently and firmly in order to enthuse their soldiers. More specifically, US doctrine states that 'the platoon leader issues the order in person, looking into the eyes of all his Soldiers to ensure each leader and Soldier understands the mission and what his element must achieve'.[77] By looking into the eyes of their soldiers, platoon leaders are seeking to confirm that the orders have been understood; confusion or incomprehension are communicated very easily by the ways in which the eyes are focused. However, the recommendation that commanders look into their soldiers' eyes seems to be more than a pedagogical method to ensure understanding. It seems to be simultaneously an explicitly moral gesture which binds commander and soldiers together in a morally obliging contract. In showing that they have understood the mission by returning the commander's stare openly and honestly, the members of the platoon are also demonstrating their personal commitment to the plan. Decisively, in order to augment this sense of moral obligation, commanders actively try to fuse immediate tactical activity with wider regimental identity in the imagination of the troops. Thus, objectives on operations are typically given emotive names which invoke collective memories and heighten the sense of moral obligation between the troops immediately present and their commitment to their fellow troops both now and in the past. On operations in Iraq and Afghanistan, the Royal Marines have used names like Taunton, Arbroath, or Gibraltar referring respectively to towns in which commando units are based or to the Marines' regimental insignia. The British Parachute Regiment has used the name 'Arnhem', referring to its most famous battle in the Second World War. Junger's Airborne platoon, of course, named the vital combat outpost they built in the Korengal Valley after their medic, Doc Restrepo, killed earlier in the tour. In this way, models imbue military objectives with an intensely shared meaning. The objective comes to represent the platoon itself and to fail to perform in the manner agreed upon in orders is to renege personally upon one's fellow soldiers. The models, therefore, unite pure and practical reasoning, engendering unity among the troops and encouraging coordinated military activity. Totally consistent with Durkheim's analysis of religion, moral obligation is fused with shared knowledge through the institution of models. They connect a concrete series of mundane tactical practices in the here and now with a sacred shared idea of professionalism. Interestingly, of course, the increasing salience of models and ROC drills suggests that the appeals to general masculinity, nationalism, and ethnicity, so important for the citizen army, have become less important in the professional force. The professional army does not encourage participation by general appeals to manhood or patriotism but by mission-specific forms of preparation which fuse intellectual and moral imperatives in the immediate task. A self-referential ideal of professionalism motivates the troops. This theme will be discussed further in the next chapter.

The importance of proper battle preparation to combat performance in engendering intellectual and moral unity is perhaps best demonstrated by operations when levels of planning and preparation have been inadequate. There are a number of well-known cases when the battle preparation on recent operations has fallen below standards which professional soldiers might themselves regard as appropriate. One of the most obvious was the American operation in November 2001 against an Al Qaeda position in the Tora Bora mountains, Operation Anaconda.[78] The mountainous area in which the operation took place complicated the mission from the outset but American forces failed to prepare properly for the operation as a result of a fragmented chain of command and diversity of force elements.[79] Signally, in planning, rehearsing, and executing the operation, American commanders could not coordinate the airborne force from the 101st Airborne Division, the 'Rakkasans', acting as a block in the mountains and Special Forces with their Afghan mentees clearing the villages in the valley.[80] Indeed, there was unclarity in the headquarters and throughout the two forces which was the 'main effort' (i.e. playing the decisive role) even though the Special Forces element was formally assigned to it.[81] The result was confusion about which force had priority in terms of support during the battle with, ultimately, disastrous consequences. There are many lesser-known but equally instructive examples.

In September 2006, NATO's ISAF, having just taken over responsibility for the south of Afghanistan, launched its first major offensive operation, Operation Medusa,[82] against the Taliban in the Panjwai'i district of Kandahar approximately 30 kilometres to the south-west of Kandahar city. There was a large force of Taliban gathering in the area who, NATO believed, were about to cut off the city by seizing Highway 1, the main road which runs through the city from east to west. Fighters were heavily dug into prepared positions around the small village of Pashmul in the Argandhab River valley, just to the north of the peak Masum Ghar, concentrated in the area of a large white school house where they had prepared an ambush site. The 1st Battalion the Royal Canadian Regiment had taken the high point Masum Ghar and were preparing to clear this area designated Objective Rugby, after a planned and very heavy artillery strike, when, on 2 September, the Regional Command South commander, Brigadier General David Fraser, ordered them to mount a rapid assault on Pashmul from the south, across a major wadi in the Argandhab Valley, directly into the prepared Taliban positions. Brigadier Fraser had himself come under pressure from his ISAF commander, General David Richards, and the US commander of Operation Enduring Freedom, Major General Ben Freakley, to assault quickly. US troops under Freakley were concurrently conducting an anti-guerrilla offensive called 'Mountain Thrust' and wanted to capture or kill as many Taliban fighters in this period as possible; Richards recognized that it was important for NATO's credibility that it demonstrated its resolve and its ability to conduct intense combat operations. C Company 1 Royal Canadian Regiment attacked, as ordered across the wadi, into the Taliban killing zone. As their four light armoured vehicles approached the school house, they were subjected to intense small arms and rocket-propelled grenade fire. Four Canadian soldiers were killed and the attack failed badly, with C Company being forced to extract itself back across the wadi. In the end, Operation Medusa, which continued to

17 September, was successful; a combination of further attacks from the north, involving US and British Special Forces, and airpower inflicted some 200 casualties on the Taliban and, if the Taliban were indeed intending to advance *en masse* on Kandahar city, their attempt was categorically interdicted. However, the initial assault on Objective Rugby on 2 September was severely criticized by the Canadian troops involved in it. As one soldier noted: 'We were pushed across the wadi but then what? We weren't even sure what the objective was... we were going into the unknown.'[83] Master Warrant Officer Keith Olstad of C Company was deeply critical of the chain of command: 'There was a plan and you only vary from the plan if the enemy influences events... [the change in plan] was influenced by decisions not by the enemy.'[84] A staff officer confirmed the point: 'The intelligence was clear! The General was clearly informed! And clearly chose to ignore the intelligence.'[85] The first assault on Objective Rugby demonstrated that for political reasons, normal processes of planning and preparation—professionalism—were short-circuited by the requirement to achieve immediate results and to demonstrate aggression. The result was a poorly conceived and executed attack which allowed the Taliban to fight from a position of their choosing, resulting in the deaths of four Canadian soldiers for no tactical gain.

The Royal Marines, like all British troops at one time or another in Helmand, have been involved in a number of problematic operations which have parallels with Operations Anaconda and Medusa. These operations are particularly useful precisely because the Royal Marines are recognized to be among the best-trained and most professional infantry regiments in NATO. The difficulties they encountered, despite the high quality of the Marines, highlight the indispensability of good battle preparation even for professional soldiers; their performance relies upon it. Operation Sond Chara, in December 2008 was one of the most important operations which the British Task Force Helmand had conducted up to that point.[86] With Sond Chara 3 Commando Brigade aimed to clear the village of Zargan Kallay in Nad-e-Ali. For the first time, the British focused on the critical area in central Helmand around the provincial capital Lashkar Gar, and after Sond Chara, the British efforts became notably more coherent and, in counter-insurgency terms, more successful. Operationally, it was, therefore successful, even. However, although the operation represented an important moment in the campaign when the British Task Force began to exert their efforts in one contiguous area around Lashkar Gar, the tactical execution of the operation was deeply troubled. Juliet and Lima Companies 42 Commando were put under the command of a new ad hoc battle group which aimed to attack Zargan Kally from the south along a single road, a junction on which was to be secured by Juliet Company and was to act as a start line. In the event, the battle group became congested at this junction, partly because of a defensive ditch which the Taliban had dug. The delay gave the Taliban ample time to strengthen their defence of the village and resulted in an intense and very difficult three-day operation which was, at great cost, ultimately successful in purely tactical terms; the Taliban were temporarily driven from the village. Marines involved in the operation were deeply critical of the planning and preparation, which was inadequate, in their view. Zargan Kallay is built on a block system with a main road running through the middle of the settlement and it seemed obvious to many Marines that Juliet and Lima Companies, advancing on foot simultaneously under the cover of night, should have cleared the village on either side of this road together. Orders were given for Operation Sond Chara but Marines in Juliet Company were not involved in a ROC drill or in rehearsals. The lack of a ROC drill may have been particularly significant here as it might have indicated the flaw of advancing on a single road and using the junction as a starting point. Specifically, the ROC drill gives subordinate commanders the opportunity to express their concerns about the plan or to offer ways of improving it.[87] The fact that an important operation could have been conceived and planned in a way which junior non-commissioned officers were immediately able to recognize as flawed demonstrates that professional performance on the battlefield is an aspiration which has to be attained by training, planning, and rehearsal. Even among the most professional forces, it is a possibility not an inevitability.

SHARED DEFINITIONS

It has been argued that battle preparation and specifically the orders process fuses the intellectual understanding with moral imperative to generate a self-referential task-orientation among troops. Professional troops understand their mission completely and are therefore obliged to execute it. In this way, battle preparation and training aims to improve collective performance. However, practice and preparation does not merely improve individual and collective performance in combat, although there seems to be evidence that it does that. Nor does it only unite an intellectual understanding of the mission with a moral requirement to complete it. Perhaps most interestingly, training actually changes soldiers' shared experience and collective understanding of combat itself. As they undergo intensive training, soldiers begin to understand the battlefield differently not only individually but, crucially, collectively. They apply new shared definitions to this domain, enabling them to perform tactical drills together because the platoon is oriented to a common definition of the situation and the appropriate joint response to it. Specifically, this professionalized re-definition of the battlefield is facilitated by and intimately related to the Durkheimian process of giving orders and using models—which invest objectives with professional significance—but it goes substantially beyond it and leads to a whole regime of training and a dense configuration of practices and relations within an infantry unit which sustain their definition recurrently among its members.

It is useful to consider the role which shared definitions play in social action more deeply. Indeed, in his work on social reality,[88] John Searle has emphasized that 'an understanding of collective intentionality is essential to understanding social facts'.[89] Uniquely, and in sharp contrast to natural phenomena, social reality is constituted by the very definitions which social actors put upon it; Searle rightly notes this is the 'remarkable feature of social facts'.[90] Marriage ceremonies in which couples declare 'I do' to each other or meetings in which an individual is designated chairperson do not exist prior to speech acts which merely describe a reality which already exists.[91] Speech acts are constitutive of social reality itself. Searle reduces the speech act to a 'status function' which takes the basic constitutive rule of 'X counts for Y in [context] C'.[92] Searle uses money to illustrate the point. Money is objectively merely bits of printed paper, but because these pieces of paper are collectively defined as valuable in society, they come to count as having the status of—and, indeed, are—five or ten pounds or dollars. The act of collective definition makes money what it is—it gives it its status function—with prodigious social consequences.[93] Ultimately, Searle argues that social reality consists of a vast and complex web of these status functions, where speech acts in innumerable life-worlds give a variety of objects, people, or statements status functions on the basis of which participants coordinate their interactions with each other. The status functions articulated in speech acts define and constitute social reality, orienting actors to common goals.

John Searle's work has certainly been one of the most elegant and profound discussions of the constitutive role which collective definitions play in social life. However, the sociologists have long been aware of the centrality of shared definitions to social activity. It was a central element of Parsons's work, especially in *The Structure of Social Action*, while Weber's *Economy and Society*, subtitled 'An Outline of Interpretive Sociology', begins with a long discussion of the distinctive centrality of subjective understandings to social reality. The constitutive role of shared understandings in social life has always been central to sociology, then. However, some research programmes within the discipline have concentrated on the connection between collective definitions and social reality in a particularly suggestive manner. In particular, in the 1960s, Harold Garfinkel, a former student of Talcott Parsons, sought to develop a new research programme in sociology against the existing functionalist consensus which he termed 'ethnomethodology'. In understanding the role of collective definitions in combat, it is worth considering the ethnomethodological approach more fully. Ethnomethodology was an exploration of the actual ways by which participants sustained social interaction through creative use of shared understandings and expectations. In his development of ethnomethodology, Harold Garfinkel believed that he was 'working out Durkheim's aphorism' that 'the objective reality of social facts is sociology's fundamental principle/phenomenon'.[94] Garfinkel rejected Durkheim's notion of the social fact which seemed to assume that society reproduces itself by simply imposing itself on the individual. Sociology was not interested in individual action but in structural forces. For Garfinkel, such a position represented a facile determinism; social forces simply manifested themselves in individual action irrespective of the dynamics of the social situation itself. Yet Garfinkel maintained that these interactional dynamics as participants made their own world were decisive and he aimed to address this putative gap in sociological thinking. He forcefully argued that social interactions do not make themselves; they have to be actively constituted by the actors involved in them. In fact, Durkheim always fully recognized the potency and complexity of interactional dynamics; he was deeply sensitive to the way in which different forms of association could give rise to different patterns of individual practice. Nevertheless, despite his misrepresentation of Durkheim, Garfinkel has usefully emphasized the centrality of 'accountability' in human social interaction and the production of situations (which is consistent with a Durkheimian approach). Specifically, Garfinkel claimed that actors have to define their encounters collectively in order that the situation becomes what it is; they have to ascribe a shared meaning to it on the basis of which their actions towards each other are comprehensible: 'The activities whereby members produce and manage settings of ordinary everyday affairs are identical with members' procedures for making those settings account-able.'[95] These accounts 'normalize' the situation, not only by making them meaningful to the participants, allowing them to interact successfully, but in so doing by investing them with normative expectations. Once defined as a certain situation, Garfinkel's actors mutually expect each other and themselves to act in certain ways. Garfinkel's breaching experiments when investigators deliberately acted in opposition to expected collective accounts were intended to demonstrate the intellectual and moral importance of these shared definitions. Shared understandings define a situation for a group of participants and, therefore, identify practically effective and morally appropriate collective activity for that group. Appositely, Garfinkel stressed that these accounts (albeit often unacknowledged) were not supererogatory, social interactions naturally and successfully being executed without them. Without shared accounts, actors could not perform coherently; there could be no successful practice and ultimately no stable situation.

Garfinkel's concept of accountability applies with great force to combat because in this situation, humans are subjected to one of the most extreme and confusing environments. On the battlefield, soldiers are submitted to a terrifying experience. Indeed, if the cultural representations discussed in Chapter 1 are to be believed—and they are certainly compelling—the front is hellish. It is a grotesque confusion of sensations. If it is difficult to coordinate activity in normal social life, the Garfinkelian problem is multiplied in this chaos. Given the intense emotions which combat generates, it would be entirely conceivable that individual soldiers would fail to generate a collective account of their situation. Each would respond to their terrors alone, some fleeing, others freezing, a few becoming fighting mad. Soldiers could respond in a multiplicity of ways. However, through preparation and especially through training, orders, and rehearsals, troops are able to generate a common account of combat which they are able to sustain during the fighting itself. In the light of the careful construction of common definitions of the combat situation and their mission in it, soldiers are able to maintain their cohesion and to continue to cooperate with each other.

Junger's descriptions of the Operation Rock Avalanche and Gatigal ambush have already been used in this chapter and the last to show the importance of drills and training. The Gatigal ambush also illustrates how soldiers can begin to generate new collective definitions of the combat situation as a result of training in line with Garfinkel's concept of accountability. According to Junger, the US Army have been interested in the response of Giunta and his comrades since it was viewed as a remarkable and ideal response to an ambush. Giunta's explanation of his performance is deeply interesting:

I did what I did because that's what I was trained to do. There was a task that had to be done, and the part that I was gonna do was to link alpha and bravo teams. I didn't run through fire to save a buddy—I ran through fire to see what was going on with him and maybe we could hide behind the same rock and shoot together. I didn't run through fire to do anything heroic or brave. I did what I believe *anyone* would have done.[96]

It would be wrong merely to take Giunta's account as an expression of modesty, although it certainly is self-deprecating. Giunta's humility is itself an expression of a deeper concept of professionalism which provides a framework of meaning for his actions. Specifically, the instinctive reaction displayed by Giunta was the product of an intense training regime. Equally interestingly, Giunta did not see his actions as individualistic. His actions were defined by him in collective institutional terms. He was following a training drill in order to reconnect two tactical sub-units. Giunta's account of his actions is deeply significant. He defined the situation he confronted in professional military terms. He did not face a personal existential crisis of whether he was a coward or a hero, nor did he mobilize political ideology to define the situation and his actions in it. Nor, interestingly, was it simply about saving a comrade. On the contrary, the situation and his response were defined in technical terms: he faced a tactical conundrum which had to be resolved by his own actions in conjunction with those of his fellow soldiers (including those whom other accounts might suggest he saved). He drew automatically on his training and his knowledge of anti-ambush drills to solve that problem. Moreover, his definition of the situation and, therefore, his reaction to it was always collective. Within seconds of the fire-fight starting, he and two other colleagues had come to an implicit agreement about the definition of this situation and the optimal collective response which they must therefore enact in the face of it. In a sense, although undoubtedly a reflection of his personal modesty, Giunta is absolutely correct when he claims that *anyone* would have done what he did. His point is that any *professional* soldier with the same level of training which he and his colleagues had received would be able and expected to perform the anti-ambush drill which he had executed so successfully on the Gatigal. Once the situation has been collectively defined as an ambush, certain actions were necessarily expected of professional soldiers by their colleagues. Individual and collective skill, local circumstances, the opposition, and, of course, chance would finally determine whether the action demanded by the way these definitions had constituted the situation would be successful.

British forces have been engaged in intense combat in both Iraq and Afghanistan and there are many examples of this application of professional concepts to define the combat situation in which they have been involved. For instance, Corporal Bradley Malone, of 45 Commando Royal Marines, rescued his sergeant who had become separated from the rest of the troop during a fire-fight in Sangin. He ran out from cover to give his sergeant covering fire so that he could extract himself, for which action he was awarded the Conspicuous Gallantry Cross, one award below a Victoria Cross. Malone explained his actions in the technical manner which closely echoes Sal Giunta's own account: 'We were surrounded and were taking fire. I realized the Sergeant was stuck in the middle and was pinned down. I just went to give him some covering fire, to engage the enemy so he could get back. You don't think about anything at the time, no emotions, you just get on with it.'[97] He understood himself to have been presented with a problem which he employed standard battle drills to solve.

Interestingly, although the Bundeswehr are much less committed to combat operations and, in Afghanistan, have operated in the more benign north, German soldiers involved in fire-fights have reacted similarly to their experiences. Thus, on 13 May 2010, an official Bundeswehr social scientist, Philip Langer, was conducting research among the Quick Reaction Force (QRF) stationed with the German Provincial Reconstruction Team (PRT) in Kunduz. Typically this research involves reasonably uncontroversial surveys of individual attitudes towards the operation, the Bundeswehr, allies, and a variety of other matters. However, he was unexpectedly faced with 'an uncontrollable group'.[98] The QRF had just returned to the PRT after a successful two-hour fire-fight with the Taliban. Dr Langer found the intense scenes which followed, where soldiers engaged in an emotional collective discussion about what had just happened, almost impossible to record by traditional methods. As the group sought to attach a stable collective meaning to this extraordinary event and to stabilize their adrenalin-heightened emotions, soldiers cut across each other, talked over each other, and made a series of emotive but unclear interventions which sought to communicate the extremity of the experience of combat. One sergeant stammered: 'Yes, that was a new experience... To be shot at for the first time was interesting and yet somehow... Well, ok, it was all a bit unreal for me, to be honest. At first, I thought; mmm, is that all there is to the real thing? Like a Bad movie, but...'[99] So complex did the conversation become that the transcriber subsequently admitted defeat and informed Dr Langer: 'For at least 45 minutes everything gets completely confused, it's unclear, how many persons participate in the interview, definite identification [of interviewees] was no longer possible.'[100] However, as the interaction became more subdued, a coherent collective line began to emerge about the experience which the troops had just undergone. A number of professional analogies were forwarded but rejected. For instance, the suggestion of butchery was rejected because it did not match the violence which they had experienced. In the end the QRF decided that 'the fighting was like pre-deployment training on the exercise area'. Indeed, Dr Langer concluded that the 'message of the narration is obvious: we were prepared and if we could manage these situations, we can successfully handle future combat as well.'[101] Perhaps because this was their first combat experience, the German soldiers were more initially affected than Junger's paratroopers, but what is striking is how they eventually came to a very similar understanding of the event as Sal Giunta. The German QRF eventually understood the fire-fight in professional terms. They had executed the drills they had learnt in training and not only did this ensure their survival during this fight but it was to become established as the basis of the unity for the rest of their deployment. Their professional competence had proved itself as the most powerful and effective way of understanding what they had all just done and this competence would be the most potent way of ensuring their survival on future missions. As a result of their experience of actual combat, these professional definitions are also evident at higher levels of command; 'I think that we have learnt that we are good. To draw this conclusion, a soldier has to have the possibility to experience in combat what we have been trained for. Those who have performed in combat exude a completely different degree of calmness and confidence because they know "I have proven myself, I could do it again".'[102]

Professional definitions—inculcated through training regimes—have effectively changed the way in which professional soldiers understand combat and, consequently, how they respond to it. In effect, although combat may have been as individually frightening for Private Giunta or for German soldiers in Kunduz as it was for conscript soldiers in the twentieth century, the battlefield is a different place for the professional military which collectively defines it in technical and professional terms. Individual virtuosity and, therefore, the kinds of individual acts of bravery typical in the First and Second World Wars have become much less obvious and indeed necessary. Defining combat in common professional terms, soldiers today have enabled themselves to respond collectively. Training has been central to the ability of professional soldiers to execute these collective drills not simply because it has refined individual and collective skills, although it has certainly done that, but the embodiment of these skills has also simultaneously involved an internalization of a series of common definitions which unify soldiers intellectually and morally. They have begun to understand the battlefield in common and, therefore, expect to act on it together. Drills and training have not only altered the capability and preparedness of the infantry platoon, they have fundamentally changed the experience of combat for the soldier. Combat has become a domain of collective professional activity.

NEW URBAN TRAINING FACILITIES

In the last chapter, new urban combat techniques were discussed to illustrate the refinement of existing battle drills. In this chapter, an attempt has been made to demonstrate the importance of training to the professional army in inculcating drills. Especially since its drills are unusual and complex, close-quarters battle has also been an important catalyst in improving training methods among western forces with a view to increasing combat performance. One of the most obvious and noticeable developments in the last decade, consistent with the dissemination of close-quarters battle, has been the appearance of elaborate new training facilities in which troops have been able to practise their urban warfare techniques. Thus, as they have developed new doctrines, to expedite this process of training, the US Army and Marines began to construct new ranges at which close-quarters battle techniques could be practised. To that end, in 2007, the US Marine Commandant General, James Mattis, ordered the creation of a dedicated infantry training centre. Widely recognized as an unusual but very talented general, Mattis had noticed that air force and Marine pilots undertook hours of training on simulators before they undertook genuine combat missions. In the light of the importance and complexity of infantry operations, Mattis wondered whether Marines could not benefit from a similar process of simulatory training, and, to this end, the US Marine Corps built the 'Infantry Immersion Trainer' at Camp Pendleton in a disused tomato packing plant beside the Interstate 5 and a short distance from the Pacific Coast. The old packing plant, some 40 metres wide and 60 metres long, was renovated by Hollywood set designers who were brought in from nearby Los Angeles to design and build the simulation trainer. Within this structure, they constructed a simulation Arabic village, consisting of tiny alleyways, numerous buildings and rooms, a small mosque, an open courtyard, and a market place with stands (with automated stall-owners capable of moving and talking) and a well. The set was designed to look, sound, feel, and smell like the kind of urban area in which the Marines would be operating in Iraq and Afghanistan. In order to add realism, the Infantry Immersion Trainer has hired approximately forty Afghan role players on a regular or permanent basis to play a variety of interactional roles, while (ex-Marine) contractors, who run the centre, play the role of insurgents. The set is also laced with IEDs and rocket-propelled grenades to simulate attacks. Simunitions and blank rounds are used to practise infantry drills. The Immersion Trainer now features as an institutionalized part of all Marine training, with every platoon undergoing a day's training at the centre as the first part of their pre-deployment training. Typically companies are assigned to the Trainer before going onto their month-long training programme in Marines Ground Air Combat Centre in 29 Palms, approximately 100 miles to the east in the Mojave Desert. The Trainer was expanded in November 2010 with the extension of a large external set approximately 400 metres by 400 metres which is based on Afghanistan and specifically on Now Zad, where the Marines have been operating since 2007. The larger area consists of schools, government buildings, mosques, and stalls, giving multiple opportunities for interaction with role players as well as IED, rocket, and sniper assaults (see Figure 9.5).

Figure 9.5. The infantry immersion trainer, Camp Pendleton, 7 June 2011: the external area with training in progress.

At a similar time, the Marines have invested heavily in urban combat trainers at the Marine Corps Air Ground Air Centre (MCAGCC) at 29 Palms. This range has included a large live-fire urban range, consisting of specialized concrete which allows live rounds to be fired inside buildings, in which company operations (of approximately 200) can be conducted. Two years ago, the Marines also constructed the modestly named Range 220 some 10 miles east of the main 29 Palms base beneath Grizzly Mountain. Range 220 is modelled on an Iraqi city (Fullujah or Ramadi) and consists of over 2,000 buildings constructed from iso-containers, which have been designed and laid out to look like Iraqi houses, markets, mosques, sports stadia, and shops; there is also a section of the town consisting of partly destroyed houses made from breeze blocks. The range covers an area of approximately one square mile and, consequently, replicates the size and complexity of a real town. On the pre-deployment Mojave Viper, the US Marines employ 500 Afghan role players to populate the town as a full Marine battalion exercises for ten days in this environment. The sheer scale and intricacy of the environment of Range 220 imitates reality as closely as possible and tests Marines and their commanders in a way which smaller facilities simply cannot (see Figure 9.6).

The US Army has similarly invested in new urban training facilities at the National Training Centre (NTC), Fort Irwin, which, according to some individuals in the Marines, dwarf the Immersion Trainer and Range 220. The US Army has been training at Fort Irwin in the Mojave Desert since the 1940s and successes in the Gulf War in 1991 were substantially attributed to this area. In the late 1990s, Fort Irwin was identified as a key site for army 'Transformation'; it would provide a training and exercise area in which the digitalization of the army could be implemented. Fort Irwin subsequently continued to be critical in reorienting the army from its thirty-year focus on conventional manoeuvre warfare to new counter-insurgency operations in Iraq and Afghanistan, a significant element of which involved urban missions. Fort Irwin describes itself as 'the world's premier training centre for the world's finest military'. It is an immodest but not un-evidenced claim: 'NTC trains the transformed Army by conducting force-on-force and live-fire training for ground and aviation brigades in a joint scenario across the spectrum of conflict, using a live-virtual-constructive training model, as portrayed by a highly lethal and capable Opposing Force and controlled by an expert and experienced Operations Group.'[103] Specifically, at the National Training Centre at Fort Irwin, the US Army has built a mock Iraqi town which is, like Range 220, populated not only by Afghan role players but by authentic farm animals, such as sheep, goats, and donkeys. The NTC has also employed iso-containers, like Range 220, but has painted them realistically so that they look startlingly like a run-down Iraqi—or Afghan—town. In addition to role players, impressive ballistic simulations are employed to maximize realism.

Figure 9.6. Range 220, 29 Palms, MCAGCC.

British developments have necessarily been on a smaller scale than the impressive construction of urban ranges by the US Army and Marines. However, especially in proportional terms, the construction or development of new ranges has been very significant. Despite their small size, the Royal Marines have been an important innovator of close-quarters battle, constructing one of the first new ranges in Britain. The Commando Training Centre (and its platoon weapons branch) has become a critical site of innovation here and the Royal Marines have invested £300,000 in 2009 to 2010 in a close-quarter battle facility in which troops can be trained in these latest techniques. The facility, located at Lympstone and drawing on the US Department of Energy's live-fire shoot house, consists of a series of rooms and corridors, constructed from breeze blocks, over which there is a viewing gantry (for instructional purposes) and a roof, so that the whole facility looks like a large barn (see Figures 9.7a–c). Live rounds are not used in the building but the Royal Marines employ 'simunition'.

The British army has also sought to invest significantly in training infrastructure. The British army has never been over-provided with urban combat facilities, due primarily to limited training estates and the Cold War army's orientation to open warfare in the countryside (primarily with tanks). Sennybridge and Salisbury Plain Training Areas both possess two relatively small villages which provide challenging environments up to company size but, until

Figure 9.7a. The compound, Commando Training Centre.

Figure 9.7b. The compound, Commando Training Centre.

2009, the British military had no new urban training environment nor any which imitated the environments in which they might be operating. At that point, General Sir David Richards was appointed Chief of the General Staff (i.e. head of the army). One of his first actions was to launch 'Operation Entirety' which put the British army on a war footing for the first time since the 9/11 attacks. Specifically, Operation Entirety prioritized operations in Afghanistan—and the preparation for them; many commanders, including Richards himself, had noted the inadequacies in training which had been a feature of Iraq and the early years in Helmand. As a result, the British army made a major investment in training infrastructure, building a mock Afghan village on the Stanford Training Area in Norfolk (Shin Kallay), at a cost of £13 million, and renovating a second village with a series of compounds (see Figures 9.8 and 9.9). In July 2011, the army opened its newest and most

Figure 9.7c. The compound, Commando Training Centre.

advanced close-quarters battle Range, a live killing house consisting of seventeen rooms, at Lydd in Kent at a cost of £3.75 million.[104]

The French army has identified the increasing importance of urban operations after its own experiences in the Balkans and especially in Sarajevo, where French troops had to clear sniper alley when IFOR was deployed, and its observations of Russian operations in Grozny and the USA in Mogadishu. Operations in Iraq and Afghanistan have only affirmed the importance of urban fighting to the French. There have been internal institutional reasons for the prioritization of urban operations and the continuing investment in it. Having professionalized in 2002, the French army has assessed that it is still behind the USA and UK. Partly as a result of the pressure to be interoperable with these allies and a desire to take the lead in initiating a European pillar within NATO, it announced its commitment to becoming the lead nation for urban combat in NATO in 2010. The French army decided to create a

Figure 9.8. Stanford Training Area, mock Afghan compounds. Compound clearing course, September 2010.

Figure 9.9. Shin Kallay, Stanford Training Area, compound clearing course: September 2010.

unified area of urban training in 1999 and eventually selected Camp Sissonne in Picardy in 2003. Centre d'Entraînement aux Actions en Zone Urbaine (CENZUB) started to be constructed in 2004 and was officially opened in 2006; it initially consists of an old village of sixty-three buildings (Beauséjour) and a shanty-town area which cost 80 million Euros to construct. The facility also now includes structures with gantries so that trainers can observe the execution of tactics inside buildings themselves. In 2011, CENZUB was expanded with the construction of a modern town (Jeoffrecourt) with a population of potentially 5,000 people, allowing for the training of brigades including engineer, armour, artillery, and air support. The new town is the largest and most advanced urban operations facility in Europe, consisting of four major areas; the modern town (dominated by large four- to six-storey apartment blocks), the historic area, consisting of a town hall, church or mosque, and some other large buildings, a suburban area of a series of smaller domestic houses, and an industrial area of factories and warehouse. An artificial river some 10 to 12 feet deep runs through Jeoffrecourt to create a realistic tactical problem for exercising troops. The training area also includes a major live urban firing range. Indeed, the facility is so impressive that the commander of the British army's new Force Development and Training Command has expressed concerns that in comparison, the British army has now fallen behind its French ally. With this facility, it is possible that the French will be able to take the lead in urban combat training (see Figures 9.10a–9.10d).

Like the French, the Canadian army had no dedicated urban training facility until very recently. However, in the light of operations in Afghanistan, a section-level and a larger company-level urban facility consisting of specially constructed buildings and a number of improvised iso-containers (known as C-Can Village) have been erected on the training area at Gagetown (see Figures 9.11a–9.11b). The German army's principal urban training area remains Bonnland, which is the largest and oldest facility of its type in western Europe. It was originally requisitioned for the Prussian army in the late nineteenth century and has remained as a training facility since that time.

The development of these facilities represents a significant investment in urban combat and in improving relevant training for the infantry soldier. In order to operate in the non-permissive and complex urban environment, it is necessary to train for it, and this training is possible only if there are sufficiently large and realistic facilities to provide such an environment.

Figure 9.10a. CENZUB, Jeoffrecourt. The modern town. The apartment blocks, represented by the ROC drill in Figure 9.4. The attack is under way with troops in the building to the left, behind which armoured vehicles are parked. The platoons are inside the building. The exercise operations centre is the building on the hill in the background (December 2011).

Figure 9.10b. CENZUB, Jeoffrecourt. The old town.

THE REFINEMENT OF TRAINING: CLOSE-QUARTERS BATTLE

Western armies have invested in new urban training infrastructure in which their soldiers can practise their close-quarters battle drills. However, training facilities alone, while essential to performance, cannot generate expertise independently. Physically, the facilities which have appeared in the last decade are extremely impressive both in terms of size and often in terms of detail. Most rooms in Range 220 are dressed so that Marines become familiar with operating in buildings which are furnished just as they will be on operations. At CENZUB, there are a variety of building structures and some of the apartment blocks consist of false stairwells, imitating barricades. Yet, however sophisticated, these facilities are ultimately only uninhabited buildings. They cannot teach close-quarters battle themselves although they are indispensable to its instruction. For these facilities to be useful, western armies need to teach their troops new skills. Accordingly, at the same time as these new structures have appeared, new training techniques have been developed in order to inculcate complex urban skills. Effectively, the higher level of performance demanded by close-quarters battle has required an improvement in training methods. A new pedagogy is appearing—and being invested in—which matches the transformed physical environment in which western troops now train.

Figure 9.10c. CENZUB, Jeoffrecourt. The suburbs. Assault in progress.

Figure 9.10d. CENZUB, Jeoffrecourt. Attack on new town.

Figure 9.11a. Groningen urban combat facility, Combat Training Centre, Gagetown.

Figure 9.11b. C-Can Village, Combat Training Centre, Gagetown.

One of the most obvious areas in which these new methods of training have appeared is shooting. Close-quarters battle relies upon improved marksmanship and this in turn requires a distinctive instructional approach. The development of this new method of teaching is clearly evident in the Royal Marines. The Royal Marines have been long regarded as one of the best and most proficient infantry regiments in NATO. Observing their performance in Kosovo in 1999, Christopher Bellamy suggested that they were among the most professional troops he had seen.[105] It is interesting to note then that even in the Royal Marines, regarded as advanced in the 1990s, there has been a requirement to reform training techniques, especially in marksmanship. Marine close-quarters battle instructors contrast the training which they now deliver in shooting to that which they initially received as recruits a decade ago in the 1990s: 'When I was training, the coaching [for shooting] was terrible. Basically, it was remedial phys [physical instruction]. If you shot badly, you were made to run.'[106] Poor shooting was seen as some sort of moral or disciplinary weakness which could be best remedied by physical abuse. This reflected the ethos of Royal Marine training at that time which sometimes tended towards the crudely aggressive. However, the Royal Marines were far from alone in this. At the moment, the British army is deeply concerned with its combat marksmanship, which, as noted in the previous chapter, was very inaccurate. One of the problems which the army has identified in training is the inappropriate teaching methods and, specifically, an overly aggressive approach by instructors. The British army still seems to see poor shooting as a moral or physical defect to be rectified by being shouted at. Although Royal Marine training is certainly still robust, the regiment has sought to train instructors to trace poor marksmanship back to inappropriate body positioning or movement at the moment of firing. The Marines' poor technique is to be improved by precise remedial training. A more careful pedagogy is evident in normal training.

On the Royal Marines close-quarters battle course, this pedagogy is highly developed. Indeed, close-quarters battle instructors compare their careful but relaxed approach to close-quarters battle training with the US Marines, who still tend to shout at and abuse candidates when they make a mistake. For them, this aggression cannot help in teaching 'an acquired skill'. The Royal Marines' coaching is aimed to be progressive, building skills slowly: 'It is necessary to slow down to learn the basics and then to get quicker.'[107] As a result of the new techniques of precision firing, the accuracy of trained professional soldiers who have undergone close-quarters battle training has improved. This is especially obvious in pistol shooting, a weapon with which few Marines or soldiers are familiar in contrast to the rifle. Consequently, on close-quarters battle courses like the Royal Marines', substantial time is required to develop the appropriate levels of skill with a pistol and it is here that the new pedagogy is particularly apparent. For instance, in May 2011,

Royal Marine corporals participating in the Commando Training Centre's close-quarters battle course went through extensive instruction and practice with pistol shooting. Demonstrating the importance of proper weapon manipulation, one member of the course failed, recording scores well below the required standard. There was some concern about his personal circumstances, which had caused him to miss a day's training on the pistol, but, more specifically, instructors noted that, unused to pistol shooting (although he had served in Iraq and Helmand where he had been involved in numerous fire-fights), he held the weapon incorrectly.[108] Specifically, his thumbs pointed apart from each other and up in the air when he fired. The thumbs should, in fact, be aligned along the side of the pistol, with the right thumb overlaying the knuckle of the left thumb, with both nails pointing upward. These details may seem trivial but they are vital to accurate fire, especially when in the dynamic context of close-quarters battle. By pointing his thumbs upwards, the corporal loosened his grip on the pistol just as it fired, causing his shots to become inaccurate, as the weapon recoiled in his hand. In an additional individual training session, when he and an instructor conducted a pistol shoot alone on the range together, the instructor repeatedly stressed his thumb position. His shooting improved dramatically as a result of this altered thumb position and he was able to pass the required test which qualified him for the tactical phase of the course. There is a clear precedence to this kind of instruction in elite sporting performance. In his work on martial arts, George Girton showed how Kung Fu fighters developed their skills through tiny, apparently trivial, adjustments to their physical repertoires. One fighter held his thumb incorrectly: 'I found out today that I had been doing certain movements wrong, for at least a long time: say a year for some, several months for others. One of them was that I was holding my thumb wrong on my hand.'[109] Merely by altering the position of the thumb, the fighter's entire sequence of moves became 'more integrated and felt more powerful and flowing'. He concluded: 'So the amazement is the large difference the way you hold your thumb can make in what you do.'[110] The Royal Marine corporal had a similar experience but he was only able to improve his shooting because of the instructors' careful observation of his shooting technique and its failings.

Training seeks to generate a level of familiarity with drills so that they become instinctive. Indeed, trainers explicitly aim to improve response times by moving the actions from conscious to subconscious responses which take a quarter of the reaction time.[111] In some cases, there has been difficulty in overcoming established practices. For instance, in current army doctrine, soldiers are taught to stop and clear their rifles if their weapons jam. In close-quarters battle, there is usually not enough time to clear a weapon. Instead, soldiers train to transition instantly to their pistol and to keep firing, but this transition requires extensive relearning and familiarization. A precisely defined series of practices need to be identified and learnt until they become instinctive reactions rather than conscious actions. An important element of this individual training process has to be conducted independently and voluntarily by the individuals themselves. Instructors emphasize the importance of 'willingness to train at your own sacrifice'.[112] In order to develop the appropriate level of skill to be able to shoot instinctively, extensive 'dry training' is necessary when soldiers run through their weapons handling procedures and movements on their own without firing their weapon or even without a weapon at all: rehearsing the movements. One instructor described it as the equivalent of 'professional footballers trying out new techniques at the end of training sessions'. In this instructor's experience, 'once you had done an action a hundred times, it became a muscle memory' and could be reproduced instinctively without recourse to self-conscious thought.[113] The concept of the dry run had been institutionalized on the close-quarters battle course at Lympstone. Before shoots at the range, the students on the course spent fifteen to twenty minutes warming up. On 31 May 2011, the students found an isolated position on the range, as the instructors set up the range equipment, and proceeded to run through their transitioning drills, with unloaded weapons. Initially, they drew their pistol over and over again, ensuring that they could seize the grip accurately and quickly, and then aimed at an imaginary target with their rifle before releasing it to simulate a stoppage and drawing their pistol. At the end of the shooting phase of the course, before going into the tactical phase, students were tested on the accuracy and speed of their shooting which included having to transition to and fire the pistol. Since it was mandatory to pass these tests, the dry runs were treated seriously by the students as essential to their progress on the course. Of course, while dry training is useful, live firing is essential for improving marksmanship. The close-quarters battle course lasts for twenty-seven days, during which the twenty-four candidates on the course expend 38,000 live rounds of rifle and pistol ammunition; each candidate fires over 1,500 rounds. Instructors on the course stressed that it was impossible to become a proficient close-quarters marksman without this extensive practice in live firing. Nevertheless, without careful instruction, it is doubtful whether so many troops would reach the high standards required.

The Canadian army's 'Gunfighter' programme follows very similar training precepts to the Royal Marines' course. There, the directing staff similarly aim to coach through demonstration and careful instruction, paying great attention to small mistakes. Repetition was also central to this course with a huge amount of rounds being expended during range days which lasted from 8 a.m. to 11 p.m. The course, divided into two halves, one English-speaking and one French-speaking, consisted of some sixty students who, in the duration of the six-week course, would fire some 150,000 rounds of ammunition, approximately 2,500 per soldier (see Figures 9.12 and 9.13).

Figure 9.12a. Royal Marines practise pivoting observed closely by instructor, Straight Point Range, Devon, May 2011.

Figure 9.12b. Pivoting. The manoeuvre was repeated many times at different ranges, with Marines practising left, right, and 180 degree pivots.

Figure 9.13a. Canadian soldiers practise the pivot on the 'Gunfighter' course at the Combat Training Centre, Gagetown, New Brunswick

Figure 9.13b. Pivot.

Individual marksmanship is critical to close-quarters battle but, in order to prevail in the urban environment, teamwork is essential. To this end, a suite of established drills including the five-step entry has been developed by western forces. However, paramount for team coordination is that the assault team develop a repertoire of precise commands which cue a completely uniform and predictable response from all members of the team. Verbal commands were central to Royal Marine close-quarters battle training. Thus, the Royal Marines' Instructor's Manual stresses the importance of coordinating commands which should 'be answered and repeated by other firers and team members' to ensure 'situational awareness'.[114] The manual records a series of endorsed commands such as 'stack on me', instructing other members of the team to form an assault line behind the caller, 'clear', indicating that the room is clear and that the team is authorized to enter, 'support', denoting that assistance is required, 'coming in', 'coming out', signalling the movement of an individual in and out of a room, and 'last out', confirming that the entire team has exited a cleared room.[115] Throughout the close-quarters battle course, the instructors emphasized the need for clear, precise use of these officially recognized commands. The sergeant running the course in July 2011 repeatedly emphasized that there should be 'no gobbing off' (i.e. irrelevant chatter) and he amusingly satirized poor communication, which needed to be eliminated, with a rambling and unclear monologue vaguely expressing the impressions of an individual. He contrasted this with the ideal: 'Obstruction right and left, back to back: Move.' He stressed the importance of 'orchestration'.[116] Another sergeant reiterated the point: 'You've got to make sure other people understand what you are doing', and, to confirm the point, one of the students, irritated by how badly a clearance had gone, emphasized that they needed to use the correct words, not just making personal commands up or saying 'Move' to yourself (and therefore failing to confirm that others had understood the command and were ready to execute it).[117]

The connection between clear and decisive commands and effective execution was demonstrated during the course of the training. The best teams used sharp, short commands. They had identified the key threats, prioritized them, and rapidly decided how they were going to address them; they had in effect resolved the collective action presented by the room clearance. Weaker groups failed to communicate so well. In one group, it was particularly noticeable that a senior NCO acted as leader but, perhaps habituated to standard infantry tactics, he struggled to identify the threats and to recommend appropriate actions. Consequently, he would stumble over his commands, correct them, or use long overly descriptive commands to identify the threats in the room. Neither he nor his colleagues knew which threat was primary, nor what drill to conduct, nor when precisely to execute it. Team members were not certain that others would move simultaneously with them and consequently hesitancy and mistrust undermined the coordination of the group. By contrast, in an effective team, the actions were initiated or concluded by sharp, clear commands typically of one or two words, followed by a decisive and smooth execution by the assault team.[118] When commands are accurate a room clearance by an assault team takes on a highly choreographed form, punctuated by a series of short commands which initiate appropriate drills from designated team members. In their training, close-quarters battle instructors emphasized the importance of these clear command words in generating coherent action.

Close-quarters battle is a physical act and there is no substitute for repetitive training, as the intensity of the courses in Canada and the UK demonstrate. Nevertheless, close-quarters battle training in western armies has increasingly focused not only on the indispensable physical performance of soldiers but also on their mental states. For instance, Royal Marine instructors at Lympstone have not only focused on improving marksmanship coaching but they have increasingly recognized the psychological dimension to military performance. Combat is self-evidently brutal and disorienting even for trained troops who have not experienced it before. However, this is even more so in terms of close-quarters combat, where noises, sights, and smells are intensified by the extreme close range at which violence is prosecuted and the enclosed environment in which the fighting typically takes place. In order to sustain performance in this intense arena, the Royal Marines have innovated a method of mental preparation for close-quarters combatants. They have sought to generate the optimal 'mindset' for these operations. Physical training is critical to developing the right mindset but, once the skills have been learnt, performance can be improved by purely mental techniques of visualization and attitudinal conditioning. Visualization of combat techniques improves execution by sensitizing individual response to external cues and ingraining reactions physiologically; it has been proven to increase sporting performance. Visualization might perhaps be seen as the individual equivalent of the rehearsal process in which individual soldiers run through a sequence of actions in their own minds in order to expedite their subsequent performance of these actions.

The attitudinal conditioning also involves developing the correct motivation for combat; 'the combat mindset' as the Marine close-quarters battle instructors call it. They define this mindset in a distinctive way. Typically, in combat, the natural human reaction is to seek merely to survive and to act in a way which maximizes survival. Indeed, even aggressive weapon use is sometimes motivated by fear and the desire to survive; it is an action of panic. In the twentieth century, conscript armies relied on extreme aggression to encourage this motivation to prevail and troops would sometimes be de-sensitized through exposure to slaughterhouses or to carcasses in training. Aggression is not irrelevant to professional forces but a rather different approach is adopted, especially in close-quarters battle training. In close-quarters battle training, the instructors seek to inculcate a requirement not just to survive but to prevail. This requires a different mindset—or attitude. Despite the fearfulness of their situation, a heightened but calm mental state is idealized in which combatants are highly attuned to their senses but in control.

The Marine instructors understand the 'combat mindset' as psychological. It is an individual phenomenon which exists only within the individual and is characterized as such. Thus, drawing on a schema developed by the former US Marine John Dean Cooper and then developed by the US Marine Corps, the Royal Marines close-quarters battle instructors have coded mental states on a spectrum, black, red, orange, yellow, and white: the 'Cooper Colour Code'. White refers to a state of disengagement, relaxation, or boredom when an individual is not focusing on anything. Yellow refers to initial arousal, orange to significant levels of concentration and engagement, while red represents the optimal performance state; the individual is completely focused on the action in hand but is also sufficiently relaxed and attuned to the performance that they retain control over their thoughts and actions. The Canadian army has employed exactly the same schema in their close-quarters battle training.[119] In both British and Canadian cases, Csikszentmihalyi's concept of 'flow',[120] where individuals are balanced between over-arousal and boredom, seems to be a close equivalent of this red state, pointing to those periods when an individual is balanced between boredom and over-exertion. Black, by contrast, refers to a state of panic or rage when individuals are overwhelmed by the situation which confronts them and are dominated by adrenalin-induced physical responses for flight or fight. The citizen army seemed to see this state of excessive rage—now coded black and seen as inappropriate—as the optimal emotional state for the warrior. With its requirement for individual precision and collective coordination, a state of heightened arousal, coded red, is viewed as optimal by close-quarters battle instructors.

The institution of such a psychological coding mechanism is interesting in itself. The mental states of individuals are likely to be significant to their performance. However, it is important to recognize the collective effect of these psychological techniques. This seems to be very clear in the way that the Royal Marines use the Cooper Colour Code. Specifically, the timing of its introduction into the Royal Marine close-quarters battle course implies that this method not only has an individual but also a self-consciously collective reference point. The Royal Marine close-quarters battle instructors introduce the concept of mental states just as the course moves from the individual weapon handling phase on the range to collective training in Modern Urban Combat: the tactics phase. At this point, the candidates gather for a lecture on the combat mindset. The lecture is self-consciously situated at this point as a signifier, denoting a rite of passage from marksmanship training to collective close-quarters battle training. Overtly, the intended reference point for the concept of mental states forwarded in the lecture is the individual and individual psychology, but the main effect of this lecture and of the concept of the mindset seems to be deliberately collective. It is explicitly intended to prepare students for working in teams. Indeed, one of the purposes of the lecture seems to be to re-attune the very experienced soldiers and Marines who have been firing individually on the range and thinking only about themselves to the collective demands of close-quarters battle. By publicly defining the approach state of arousal as 'measured awareness' and colour-coding it red, the Royal Marine instructors try to establish a collective reference point. Close-quarter combatants have a common understanding of the optimal individual emotional state to which they can adjust their own behaviour and influence the behaviour of their team-mates. The emotional state of the individual has become a domain of interest for the group and, in that way, it has become possible to align the attitudes of the individuals, who are now self-consciously and collectively aware of the importance of this dimension of their practice.

Significantly, following the combat mindset lecture, the colour codes became a collective resource for the course as they conducted their tactical training phase. Instructors periodically drew on the colour code to illustrate a training point and to improve the performance of the course. Thus, on the first day in the compound as the course began the tactical phase, the students had to change the firing mechanisms on their rifles and pistols so that they could fire 'simunition' rounds. Inserting the new simunition firing mechanism can be intricate and one of the students started to have great difficulty with the procedure. He began to get visibly irritated with his weapon and unsuccessfully started to force the mechanism into the rifle. At this point, the sergeant running the course stepped forward to help him, joking, 'You're going into the black already', at which comment both the struggling Marine and the rest of the course laughed. The concept of 'going into the black' became a leitmotif over the next few days' training in the compound, and at various points the same sergeant, noting panic and confusion on the faces of some of the students as they entered complex rooms, would shout at them not 'to go into the black' or, in debriefs, note that they had 'gone into the black'. He periodically re-emphasized that 'you don't want to go into the black', until it became an amusing joke for the group.[121] Yet the comic element of the remark only served to show that the mental state of each soldier was relevant to the team. In this way, the concern about mental attitude intensified the power of the collective over the individual. It represents a heightened form of professionalism where the attention to detail invades the most apparently private and individual space of the emotions and attitudinal states themselves.

The Royal Marines are by no means alone in this increasing concern not only with the obvious physical demands of combat but in addressing the mental dimensions of soldiering. The US Marines have also increasingly sought to improve performance on the battlefield through mental preparation.

In his analysis of combat, Bryan McCoy regarded the battle drill and training as fundamental. Certainly, no military unit could be successful without intense realistic training, and it is clear from his account of the actions of 3/4 Marines that they underwent very arduous training in the Mojave Desert before their deployment in 2003. Yet, this practical training was augmented by a reflexive programme of mental training: 'imaging' as McCoy called it.

> Battle drills—the physical acts of responding to a situation, be it 'contact right', call for fire, a hip shoot, putting a rocket shot or grenade into an enemy bunker—were carefully linked with mental imaging. These images were not foolish delusions of grandeur; we did our best to conjure up the fear and confusion of the battlefield, and mentally associated the emotions of the situation calling for a particular battle drill.[122]

It is important to note here not only the individual psychological role which this imaging may have performed in preparing the Marines for battle but rather the explicitly collective role of imaging. McCoy did not allow his Marines simply to 'image' alone although he undoubtedly wanted them to be conducting this mental training independently. Rather, even psychological training was a collective activity for his unit. The images which he wanted to establish were shared ones which programmed a common pattern of action into every single Marine. It is noticeable that in his description of imaging the first person plural 'We' is the preferred pronoun: 'We dealt with the emotions of killing... We imaged riding buttoned up in AAVs... We imaged the sound of enemy fire.'[123] Individual visualization techniques might well have aided his Marines to overcome their fear or to improve their individual performance but such imaging would not have aided the collective performance of this unit. By imaging together, McCoy self-consciously sought to impose a common suite of battle drills onto his Marines which they had practised physically and rehearsed mentally together on numerous occasions and, crucially, in the case of imaging, knew that they had all conducted this preparation, even though it was an internal process. For McCoy, the process of mental rehearsal tied to intense training explained the ability of his unit to survive the ambush at Al Kut: 'The battle drill of "contact right" used in the AL Kut firefight had been rehearsed and imaged dozens of times before Al Kut became part of our lexicon and unit history.'[124]

In the light of operations in Kandahar, the Canadian forces have become increasingly concerned about the mental welfare of their returning 'members'. They are concerned to minimize the incidents of psychological breakdown and post-traumatic stress disorder. They have been influenced by the concept of 'resilience' which was initially developed by the US Army, in response to extreme operational pressures. Although the concept of resilience was originally developed as a means of highlighting the problem of psychological injury and sensitizing officers and soldiers to its symptoms in an attempt to reduce the stigma often associated with breakdown in the military, the concept of resilience has involved an important innovation in training. Specifically, resilience training, which Canadian forces all receive from the early stages in their career, aims at improving the performance of Canadian troops by identifying the importance of mental preparation. Four specific tools, derived from elite sports and the training of snipers, have now been disseminated to all Canadian forces in the form of an aide-mémoire, 'Road to Mental Readiness', training lectures, and a video, *Resilience: the warrior's edge* (produced in September 2011), to which all troops are exposed. These tools, called 'the Big Four', involve 'goal-setting', 'mental rehearsal/visualization', 'self-talk', and 'arousal reduction' (i.e. controlled breathing). These mental techniques are intended to augment the standard cycles of training and exercises which precede deployment to improve the performance and determination of Canadian troops on operations. The instruction process emphasizes that these techniques need to be not merely understood but practised regularly. If they are to be effective, the Canadian forces recommend that the four psychological techniques must be repeated until they themselves become 'muscle memories' in the brain.[125]

In order to illustrate how these techniques overcome the natural reaction to panic in combat, the lectures and video on resilience describe the structure and function of the brain in considerable detail. Effectively, the instruction claims that by conducting sufficient mental preparation, the frontal lobes and the cortex can ultimately control the instinctive responses in the brain stem and the amygdala from which primitive flight or fight responses are generated. Clearly, as with the Royal Marines' concept of the combat mindset, the individual mental state is the ostensible focus. Indeed, in instruction lectures and the Canadian forces video itself, images of an individual brain and spinal column feature centrally, showing the structure of the brain in some detail. The aim is to strengthen the psychological preparedness of each individual soldier to the traumas of combat by improving everyone's understanding of their own cerebral functioning. However, although it seems likely that neurological pathways are developed by repetition so that it is not absurd to talk about physiological memories, it is important to recognize the sociological dimensions of these apparently private and individual methods of arousal control. Decisively, the concept of resilience and the institutionalization of mental training operate at and have had a manifest collective effect on Canadian troops. Specifically, while designed to aid the individual, resilience training has actually involved a colonization of the individual psyche by the Canadian forces. The internal workings of the individual brain are no longer the private possession of each soldier, unknown and irrelevant to the army, but they have become a domain of concern and control. While the mental techniques are intended to benefit the individual, protecting them from psychological harm, the institutionalization of resilience training is intended to ensure that psychological preparation is now required of the soldier. By identifying the brain as an object of collective concern, Canadian troops are obliged to include mental preparation in their formal military training. Indeed, as part of the resilience programme, the Canadian forces introduced, again derived from the US Army, the concept of the Warrior's Ethos. This ethos is depicted as a three-columned structure, supporting a pediment defined as 'Preparedness'. The three columns are the intellectual, the physical, and resilience or the mental. Significantly, together these three attributes, intellectual, physical, and mental (resilience), generate the key characteristics which are displayed by a warrior which notably include 'professional bearing' or 'professionalisme' (in French). In order to be a professional soldier in the Canadian army, soldiers have to comply not just with physical standards but with an institutionalized concept of mental standards. Moreover, this psychic training is now standardized. All soldiers understand themselves and each other to respond to the trauma of combat in the same way, and resilience training equips them not only to respond to this pressure in a predictable way but to expect their colleagues to do so as well. Indeed, a number of Canadian soldiers identified a potential paradox of resilience training. This training was partly intended to reduce the risk of psychological breakdown and, crucially, the stigma around it. However, some Canadian officers were concerned that resilience training would actually re-inscribe the stigma of breakdown as a form of weakness because since everyone had received mental training, those who collapsed could no longer claim that they had been unprepared for operations. Mental preparation has become a collective obligation and mental breakdown may, some officers worry, become evidence of a lack of professionalism and, therefore, individual weakness. It is not clear whether these fears will be realized but the concept of resilience seems to indicate that the professionalization of the Canadian forces has penetrated into the very psyche of the individual soldier.[126] It has standardized and collectivized even the emotional and psychological responses to combat and identified these once private domains as public concerns.

Perhaps unsurprisingly, the institutionalization of psychological training has been most pronounced in the Special Forces community, and Canada's Joint Task Force 2 based at Dwyer Hill provides an interesting example of this process. Like the US Navy Seals, the Canadian Special Forces started to become worried from about 2005 about the number of candidates failing their selection course. As a small military already, the numbers of failures was potentially very serious for Joint Task Force 2 especially in the light of intense operations in Afghanistan. While quality is typically emphasized by the Special Forces in all countries, if the numbers of qualified operatives fall below a certain level, the organization becomes ineffective no matter how competent each individual. Accordingly, Canada's Joint Task Force 2 began (like the Navy Seals before it) to explore whether sports psychology, which had been successful in raising pass rates in the US Navy Seals from one-quarter to two-thirds, might alleviate the problem. To that end, and on an informal basis, the Training Squadron at Dwyer Hill hired the services of a well-known sports psychologist who had been part of Canada's Olympic team for a number of years. A clear parallel was seen between elite sporting and elite military performance. In both cases, highly selected individuals, trained to the highest level, were being asked to perform extraordinary tasks in difficult, competitive circumstances in which the outcome was uncertain. The sports psychologist institutionalized many of the standard mental preparation techniques in Joint Task Force 2, such as goal identification and visualization, which are central to professional sport today (and indeed which the Canadian forces have themselves subsequently adopted). In each case, the focus was individual but, by institutionalizing these techniques of mental awareness, the Special Forces had effectively made once private mental states public, regimental property. The very mental state of an individual trooper was an aspect of their professional performance and therefore a concern to commanders, trainers, and fellow soldiers at Dwyer Hill. The Canadian forces, especially those involved in the development of resilience training, are deeply interested in Joint Task Force 2, which they regard as optimal in terms of its mental preparedness.[127]

The sports psychologist also introduced other novel techniques, the most interesting of which was the concept of the 'Banquet Speech'. The Banquet Speech required each Special Forces candidate (and badged members of Joint Task Force 2) to articulate, preferably in writing, how they would like to be remembered at the end of their careers. The Banquet Speech is the oration which is given to Canadian soldiers at the formal dinner for all their comrades at the end of their service. It represents the pinnacle of their military career in which they are publicly honoured by all those with whom they have served most closely and whom they value the most. This imagined Banquet Speech would then be read to the rest of the selection course, to other members of the Special Forces team, or to the directing staff responsible for training. The sports psychologist rightly described the Banquet Speech as 'incredibly powerful' and both he and members of the Canadian Special Forces confirmed the motivational effects which it has had on members of the organization. The Banquet Speech seems to be a powerful mental and motivational tool in improving performance for two reasons. It clearly has an individual psychological effect, self-consciously identifying, perhaps for the first time, precisely what individual soldiers want from their careers. It is a means of clarifying long-term goals on the basis of which immediate and medium-term actions can be plotted. As a result, 'they become the person they want to be remembered as'; soldiers have a clear idea of how they must act in the present to attain future objectives. It is effectively a statement of personal conscience. However, and more importantly, the Banquet Speech has a collective function which generates its moral force; indeed, the individual and psychological impact of the Speech comes primarily from these external social relations with fellow group members. It explicitly collectivizes individual action, establishing a soldier's comrades as the tribunal of all his actions. By publicly announcing how they would like to be remembered, Special Forces soldiers can be called to account by their colleagues when they fall short of the standards which they have set for themselves. Soldiers can point out when a comrade is behaving in a way which is inconsistent with his own standards. The Banquet Speech, consequently, deliberately exposes every aspect and every act of each soldier to public accountability. Each soldier in the Special Forces has a moral authority over his comrades, who similarly have moral authority over them. In this way, the Banquet Speech by establishing collective standards 'keeps guys on the path' and ensures that Special Forces soldiers are mutually able to demand high levels of performance from each other.[128] The intense social occasion of the Banquet Speech, when soldiers profess their own ambitions, substantially aids this process of collectivization and ensures that the mutual and personal commitments are invested with great potency; the declarations become, in effect, solemn oaths of group loyalty. It is unlikely that an invasive technique like the Banquet Speech, which exposes the soldier so completely to his comrades, could be institutionalized among normal troops. The Special Forces operate in very small teams of highly specialized operatives. Consequently, the level of trust and confidentiality between Special Forces troopers is very high. It is unlikely that an individual would expose him- or herself so completely to a larger and less integrated social group. The Special Forces may represent an extreme but, precisely because of this fact, they illustrate the way in which professionalization has explicitly colonized the individual psyche submitting individual emotions and mental states to public accountability and seeking to strengthen resolve through the institution of mental techniques as established as physical drills like the section or platoon attack.

FAILURES OF PROFESSIONALISM

Training seems to have genuinely transformed the capabilities and collective understandings displayed by the professional infantry. Nevertheless, it would be wrong to idealize the performance of professional soldiers. Their performance levels may often exceed that of citizen soldiers, yet, precisely because infantry tactics on the modern battlefield are complicated by the terrain and the enemy, coordination or even basic procedures can be poorly executed on operations. This was very evident on the Royal Marines Junior Command course in September 2009. This course trains experienced Marines or lance corporals who have been selected for promotion to corporal. On current courses in Britain, the level of experience among the students is significant; almost all have served in Helmand, often on numerous occasions, and some are from the Special Forces. They have been involved in numerous operations and fire-fights. Yet, training was still necessary and mistakes were still made during the exercises. For instance, the final exercise of the course on 22 September 2009 involved an assault on the purpose built 'FIBUA' (Fighting in Built Up Areas) village at the Sennybridge Training Area. The course, consisting of approximately forty individuals in two platoons, approached the village at night and organized themselves on the line of departure just before dawn in a small defile about 800 metres from the target. As dawn broke, they began their assault which involved fighting through the village with one platoon clearing the buildings on the right-hand side of the main street, the other clearing the left-hand side, while they mutually supported each other with rifle and, especially, machine-gun fire. The initial assault was adequate but as the Marines entered the first buildings, the coordination of fire and movement between the machine-gun section and the assaulting section broke down. The assault section repeatedly advanced without support fire. At this point, a colour sergeant, one of the instructors on the course, a decorated Marine of formidable appearance and reputation, became incensed. After the first error, he had warned the Marines to coordinate their fire and movement: 'There was only one GPMG covering that attack.'[129] When the mistake was repeated with one of the troops going forward to the next house without covering fire and, indeed, without a breaching charge, he became visibly agitated at first to himself, cursing and whipping a large stick he was holding against the ground then shouting furiously at the members of the platoon, 'If you move without covering fire, I swear I will fucking kill you.'[130] Eventually, coordination returned and the final assault on the last building in the village was particularly well executed. A heavy weight of covering fire was laid down and then a number of smoke grenades were thrown towards the building to obscure the defenders' vision. By the time it had cleared members of the assaulting section had crossed the street and had already made an entry into the building. In order to finish the course positively, the colour sergeant deliberately did not dwell on the initial mal-coordination of one of the assaulting platoons but emphasized the last assault which had been well executed. The point is that for all the refinement of doctrine and training, professional troops are still capable of poor performances, demonstrating the difficulty of modern infantry tactics. Combat performance—cohesion—is a complex and arduous achievement which requires prodigious levels of individual and collective training. The failure to execute collective performance, that is, the collapse of cohesion, is always an eminent possibility and it is only the greatest attention to individual and collective detail which may prevent it.

These problems are replicated in other western forces. In the now professionalized Bundeswehr, training is regarded as paramount. The importance of training together in generating levels of cohesion adequate to current operations was demonstrated very clearly on an exercise which this captain was involved in organizing on the Hammelburg Training Area for the Infantry School. The exercise was an early pre-deployment training for two companies from a Jäger battalion from the 1st Air Mobile Brigade based in Frizlar. The exercise scenario centred on the eighteenth-century village of Bonnland which is one of the largest and oldest urban combat facilities in western Europe. The village was held by an insurgent force which had to be neutralized. The Jäger companies invested and reconnoitred the village on the first two days of the exercise; on the third day, 11 December 2010, they assaulted the village soon after first light. The main assault was conducted by a single company from the south. The assault was extremely challenging for the company. Not only is Bonnland large but the micro-geography of the village is extremely complex with alleyways, courtyards, and large buildings offering numerous access and entry points for the individuals playing the role of hostile forces. In addition, the exercise was conducted with electronic receptor devices fixed onto the soldiers' weapons and equipment so that if a soldier was hit by a laser, his equipment started flashing and he would be eliminated from the exercise as a casualty. Adding to the problems for the assaulting company, the soldiers playing the opposing force were all experienced soldiers who had recently been deployed together to Afghanistan as part of an infantry company. They demonstrated great skill in manoeuvring around the village to inconvenience the attacking Jäger company. As a result of these difficulties, the assault did not go well. Indeed, the directing staff were visibly annoyed with some of the most obvious mistakes which were being made.[131]

The initial entry into the village was not properly supported and several soldiers were lasered before entering the village (see Figure 9.14). More soldiers were lost crossing the first courtyard because fire and the use of smoke was not fused with the movement of the troops. Finally, in the face of some clever resistance, the assault lost all momentum. Interestingly, the assault took on the appearance of the kind of tactics which is often found in description of the conscript armies of the First and Second World Wars. In the absence of a high level of collective competence among the sections and fire teams, the captain commanding the company had to lead the assault physically; he did not coordinate the interlocking but independent actions of his manipular sub-units but was simply followed by the mass of his company from building to building, with platoons and sections intermixed. The enemy force was withdrawn at this point by the directing staff because the company was struggling so badly to make headway through the village.[132] The difficulties experienced by this Jäger company during this single exercise should not be over-emphasized, nor should excessive conclusions about either the company or the Bundeswehr be drawn from them. The exercise was interesting because of the explanation which directing staff put upon the suboptimal performance by this company. For the directing staff, who had a clear idea of how this village should have been taken, the company's performance was undermined

Figure 9.14. Jäger Regiment training at Bonnland. The company's attack begins to stall on entry to the second building and the assault platoon begins to bunch.

by the inexperience of the soldiers. Specifically, although the company consisted almost entirely of professional soldiers (minus three or four conscripts), the majority of them were new to the company and were young soldiers with little operational experience. Consequently, the kind of cohesion demonstrated by the *Erweiterte Ausbildungs* company (discussed in the previous chapter by the German paratroop captain), which had trained and operated together for some four years, was impossible. Both the directing staff and the company commander himself emphasized that collective performance (cohesion) in an infantry company relied on stable membership and therefore long-term collective participation in a regime of training and exercising. As in the British army, professionalism is an ideal to be continually striven for rather than a constant state. Professional performance is the fragile and elusive product of intensive training and preparation.

Similar difficulties are recurrent in the United States Marine Corps. The US Marine Corps has been at the forefront of many developments in infantry tactics, including close-quarters battle. The US Marines run a Basic Urban Skills Training Course as well as more advanced close-quarters battle courses which have been crucial reference points for the Canadian and British forces. Many Marines, especially long-serving ones, have been through these courses and demonstrate heightened levels of combat skill. However, the standard

Marine infantry platoon is not always the focus of this intensive training. The very high turnover rates in the US Marines, where 65 per cent of personnel serve for only four years, accentuates this problem. The US Marines already privilege large-scale, joint combat operations, and with this scale of personnel turbulence there is not always enough time to train troops at the platoon level up to the highest condition. Consequently, the standard of infantry tactics in the US Marine Corps at the squad and platoon level is not always particularly high. These troops are well equipped and are always tough and determined with a very high morale. However, individual skills and collective choreographies are not always perfected. In his memoir of the Iraq War, Tyler Boudreau describes the relatively low level of training which his battalion received in 1999 when stationed on Okinawa: 'We'd received the vast majority of our Marines only two months prior to going through the readiness evaluation.'[133] Boudreau found it troubling that his unit passed its Marine Corps Combat Readiness Evaluation even though many of its Marines were not of the requisite standard with the limited training they had received; his own company made some glaring errors, like attacking its own machine-guns.[134] Boudreau's account of the Marine Corps contrasts markedly with Bryan McCoy's and this difference might be partially explained by the disillusionment which Boudreau experienced during the war. Yet, his claim that riflemen in the Marines were not particularly well trained seems to be broadly sustainable. Indeed, four months before the invasion of Iraq, 'two-thirds of the Marines in the battalion were brand new to the Corps', having trained with Boudreau's company for two months and been in the military for less than six months.[135] Boudreau emphasized to them that, 'whether they trained hard or they didn't',[136] they would be deployed, because ultimately, in his view, a Marine was dispensable. Nevertheless, despite Boudreau's observations and the higher formation focus of senior Marine commanders, there is little doubt that even the least experienced Marine platoon demonstrates a level of performance significantly above those of the First or Second World War platoon. They have received far more tactical training than their forebears and, although extremely robust, the training is rather more sophisticated than the almost sadistic regime which Eisenhardt recalled at the time of Vietnam. Indeed, Boudreau himself draws the contrast, noting that some US Marines were deployed on operations in the Second World War without firing their rifles.

Recent scholarship on cohesion has prioritized training. Evidence from professional forces seems to affirm this interpretation. Professional armies are distinguished from their citizen predecessors not only by the fact that all their soldiers undergo a rigorous training regime but training and battle preparation is becoming increasingly sophisticated and refined. New techniques of training and battle preparation have been introduced. In line with the concept of professionalism as attention to detail, training and preparation now focus on the apparently trivial but in fact critical minutiae of individual and collective performance. Indeed, training now focuses not only on physical performance but identifies attitude and mental state as an area of collective concern. Training has colonized every aspect of combat performance in an attempt to increase individual competence and group coordination. Even when focused on the 'mindset' of the soldier, training is ostensibly primarily concerned with physical performance and, certainly, the practical effects of training are evident. As a result of practice, individual soldiers and platoons can perform drills which they could not initially execute or their performance improves, often dramatically. However, it is important not to ignore the intellectual effects of training especially at the collective level. Crucially, as it inculcates competence, training simultaneously unites sections and platoons around a common set of understandings. Training engenders a collective consciousness among the platoon not only of its own unity but crucially a common understanding of the battlefield itself. Soldiers in a trained platoon begin to define the combat environment in the same way and, consequently, coordinated, unified social practice becomes all but instinctive for them. Since all the soldiers understand their situation in the same way—they all define it by reference to the same professional terms—they necessarily respond to it uniformly. Training normalizes the battlefield, not simply reducing its terrors but investing it with shared professional and technical significance. This intellectual unification would be impossible without the physical practice of training. Without physically enacting drills, the meaning of the professional concepts would remain abstract. Indeed, without physical demonstration, soldiers might unwittingly entertain quite different ideas about what professional concepts mean in combat. Nevertheless, although the physical dimension of training is a prerequisite, the collective intellectual effect of practice on performance, uniting the platoon around a common consciousness, seems to be equally critical. Training is critical to professionalism not only because it imbues soldiers with appropriate competences but because it also induces a common definition of combat and what constitutes combat performance. Crucially, this shared understanding of combat tactics is not simply a practical skill; it involves a moral dimension. To have undergone training in tactics or to have gone through the specific process of battle preparation is not merely to signal that soldiers understand their role, it is to hold them morally accountable to perform it. In this way training generates solidarity in and of itself because it unites technical competences with a moral imperative to utilize them, even at personal risk. By uniting competence and morality (skill and morale), training is critical to combat performance and to the generation of cohesion.

10

Professionalism

PROFESSIONAL ETHOS

In the last two chapters, the analysis has focused on expertise in the form of practical military skills. In effect, they explored Huntington's first two definitions of professionalism, expertise and responsibility, in relation to the infantry platoon. However, professionalism—substantially engendered through training—does not merely improve the practical performance of soldiers. It fundamentally alters the social relations between them; it transforms the nature of the associations between them. Professionalism generates a solidarity whose distinctiveness is often overlooked. Indeed, although the main focus up to this point has been expertise, the question of morality and moral obligation has inevitably begun to be addressed. Training and battle preparation did not merely inculcate the appropriate drills but, also, simultaneously imposed a moral obligation upon soldiers who were expected to perform. With tactical competence and public acknowledgement that soldiers had understood the mission came the expectation that soldiers were obliged to fulfil their duties on it. Mutual obligation was already implicit within the purely technical competences of the professional soldier, therefore. However, this solidarity is so different from that demonstrated by the citizen soldier and so important to combat performance that it is worth exploring its characteristics in far greater detail.

In his definition of status groups, Max Weber emphasized, as noted in Chapter 8, the importance of status honour in uniting would-be members so that they could easily identify and exclude outsiders. Status honour referred to the special respect which status group members accorded each other and the expectations which they imposed on each other; status group members were supposed to contribute to the collective good of the group and their reputation reflected the perception of that contribution by their fellow members. Status honour was defined both as the privileges of the group (as a result of the monopoly it enjoyed) and also as the expectation which members placed upon each other. Members of a status group expected their fellows to comport themselves in a manner befitting their distinctive position.

Honourable members of a status group dressed or spoke correctly and did not fraternize with non-status group members. As Weber's discussion of the importance of training to the establishment of a status group demonstrated, he was intimately aware that expertise could be a form of status honour. Members of a professional group were expected to work properly in order to maintain the reputation of the specialism. Nevertheless, while Weber did not provide a systematic account of how professionalism itself might become a status honour, this theme was central to the work of Weber's great French contemporary, Émile Durkheim.

Although Durkheim wrote about a diversity of topics and there is a dispute about whether his early pre-1896 work can be fundamentally divided from his later research on religion, his oeuvre might be understood as addressing two central questions. First, at a philosophical and universal level, Durkheim was fascinated by the problem of social solidarity. All his work is oriented to the question of how human groups form—how social solidarity is possible—and once formed how the dynamics of association influence and even, perhaps, determine individual and collective behaviour. This theme is as central to his first monograph, *The Division of Labour*, as it is to his last and greatest opus, *The Elementary Forms of the Religious Life*. Secondly, he remained concerned throughout his life about the special character of social solidarity in modern industrial society and its potential pathologies.

As his work on suicide demonstrated, Durkheim was deeply concerned about the problem of anomie. By anomie, Durkheim did not mean alienation in the Marxian sense, though the condition may be related to it. Rather, anomie described a condition of extreme individualism in which the greatest problem for the individual was not isolation (or alienation) but the enormous freedoms which society now granted its members. The individual, recognized by law and operating in a free market, was free to choose what to do and be in a manner quite unprecedented in human history. This liberation represented for Durkheim a great danger both for the individual and for society. It threatened the cohesiveness of the social order since individuals were no longer obligated to their society or to each other, while simultaneously undermining the possibility of happiness and fulfilment for the individual. The possibility of choosing any life-path undermined the intrinsic meaning of the one which an individual actually trod; there was no need for an individual to follow this course since it was no better than any other possible career which might equally well be chosen. Moreover, under the Cult of Individualism in which each individual was autonomous, the possibility of collective recognition and approbation also declined. For Durkheim, as he made clear in *The Division of Labour*, individual health and happiness actually required communal bonds. In that work, Durkheim identified two basic forms of social solidarity, mechanical and organic, which will be discussed more fully later. Mechanical solidarity referred to the solidarity of small tribal groups whose members were all alike. Solidarity was more or less automatic—or mechanical—as a result. Organic solidarity referred to a more complex and higher level of association where members were united by the very fact that they were different and, therefore, interdependent upon one another; the analogy which Durkheim seems to have had in mind was the biological organ whose parts mutually supported each other. According to Durkheim, properly developed organic solidarity, based on the intimate interdependence of individuals, might provide the context (for perhaps the first time in human history) for the development of a socially grounded individualism.

However, while modern industrial society offered the possibility of genuine individual liberation through new forms of community, Durkheim's observations at the end of the nineteenth and early twentieth centuries did not inspire him with optimism. On the contrary, instead of a healthy organic solidarity, he implied that anomie was the more typical condition. There seemed to be no obvious way to counter anomie and the corrosive effects of the Cult of the Individual which predominated in industrial society. The modern state, expanding dramatically in power and authority, might have been thought to be able to limit and even halt the dissolution threatened by anomie but Durkheim was deeply sceptical about the powers of the state to unite society into a community: 'While the State becomes inflated and hypertrophied in order to obtain a firm enough grip upon individuals, but without succeeding, the latter, without mutual relationships, tumble over one another like so many liquid molecules, encountering no central energy to retain, fix or organize them.'[1] The state was a huge administrative mechanism which organized and ruled individuals, as independent citizens, but precisely because all were equal and separate before it, the state could not unite them together in a common community.

For Durkheim, the solution to the problem of anomic dislocation was evident, if not yet fully realized, in modern society. The rise of professional status groups represented precisely the dense forms of social solidarity which he regarded as essential for individual fulfilment and social cohesion overall. Durkheim advocated that professional guilds and associations could become secular churches uniting the members of modern society around common endeavours: 'The facts related show that the professional group is by no means incapable of being in itself a moral sphere, since this was its character in the past.'[2] Durkheim continues:

Within any political society, we get a number of individuals who share the same ideas and interests, sentiments and occupations, in which the rest of the population have no part. When that occurs, it is inevitable that these individuals are carried along by the current of their similarities, as if under impulsion; they feel mutual attraction, they seek out one another, they enter into relations with one another and form compacts and so, by degree, become a limited group with recognizable features, within general society. Now once the group is formed, nothing can hinder an appropriate moral life from evolving, a life that will carry the mark of the special conditions that brought it into being... It is at this point we have a corpus of moral rules already well on their way to being founded.[3]

Crucially, Durkheim did not see the professional group simply in terms of practical expertise, although that was plainly crucial. For Durkheim, the professional group might have the capability of becoming a moral community, of developing a 'moral life', on the basis of its distinctive skills. Here, the professional group does not merely train its members to perform a specialist function but enforces a distinctive 'status honour', as Weber would call it, upon them. On the basis of its skills, the professional group imposes expectations about the behaviour and conduct of the individual professional. In the first instance, Durkheim seems to be referring to the moral rules of professional conduct; doctors are bound by the Hippocratic oath not to harm their patients, lawyers to represent their clients, and academics to the truth. For Durkheim, moral rules are different from mere regulations, however, enforced externally on the individual. They are mutually imposed on group members together and thus, individuals enforce these rules not only on each other but on themselves. Specifically, not only do these rules direct members to the collective goods of the group from which all benefit but adherence to the group's rules also earns individuals the respect and admiration of their peers. Consequently, the imposition of rules is not regarded negatively. On the contrary, individual self-fulfilment and happiness is substantially dependent on their successful embodiment of the moral life of their professional group. The professional vocation provides a firm moral framework for their life, providing discipline and coherence, and adherence to this discipline ensures recognition from other group members and therefore a sense of self-worth.

It does not seem implausible to suggest that Durkheim conceived of the 'moral rules' of a professional as binding well beyond the confines of professional performance. The profession necessarily involves a lifestyle (partly dependent upon the income a profession receives for its services) that incorporates both public professional performance and private behaviour. In particular, the moral life of the professional is demonstrated and enforced by socializing between members of the profession away from their work. Doctors and lawyers interact with each other at dinners and parties and their performance in those domains is a subordinate but important index of their membership of the group. Indeed, it might be inferred that this moral life extends into the private and domestic lives of professionals who are expected to conduct themselves in their relations towards their partners and children in an appropriate manner. In the absence of a binding national identity, Durkheim advocates the professional group as the basis for community in a modern industrial society. Individuals are bound to these groups through which they earn their livings and in which they spend their entire public and private lives. In a profession, moral and economic imperatives are fused to unite individuals into coherent communities, thereby eliminating anomie. Although he does not explicitly connect his writing on professions with his theory of the division of labour, he seems to imply that genuine organic solidarity might be developed when dense and coherent professional groups, each specializing in a particular activity and each interdependent with the others, have emerged. Each has its own distinctive moral life.

Durkheim's writings on professionalism are deeply suggestive and certainly many other social scientists have pointed to the importance of the profession in modern society.[4] However, the first decades of the twentieth century seemed to partly disprove Durkheim's theory. With the First World War and the rise of right- and left-wing authoritarianism throughout most of the twentieth century, almost as the preferred form of governance in industrial or industrializing nations, the state does seem to have been substantially able, against Durkheim's claim, to have united its citizens typically by reference to concepts of nationalism. Ironically, Durkheim's comments about professionalism may be more relevant now in the twenty-first century as state power is widely regarded to be receding as a result of globalization than at the time he wrote just before the First World War. Specifically, Durkheim's promotion of professional associations may be directly relevant to the question of cohesion in the professional infantry. In line with Durkheim's intuition, the professional infantry may be increasingly integrated not by concepts of national identity, civic duty, comradeship, or by personal bonds but rather by a professional ethos. Military professionalism may provide the new 'moral sphere' which unites soldiers primarily in their public life but also in their private existences. Professionalism may not simply be a series of practical skills for soldiers today, though that expertise is vital, but also a morality which obligates soldiers to perform their role properly and, indeed, comport themselves generally in a manner which their professional status and colleagues demand.

Indeed, both Huntington and Janowitz recognized that professionalism might include not simply practical expertise but also a moral dimension. Huntington defined professionalism as expertise, responsibility, and corporateness. Expertise, as discussed, referred to the specialist skills of the profession and responsibility to the service it provided society. Corporateness referred, by contrast, to the relations between the members of the profession; their sense of community, commitment, and common obligation to other professionals: 'the members of a profession share a sense of organic unity and consciousness as a group apart from laymen. This collective sense has its origins in the lengthy discipline and training necessary for professional competence, the common bond of work, and the sharing of a unique social responsibility.'[5] Because of the uncertainty and danger of military operations, Huntington regarded corporateness as especially critical to the armed forces.

However, Huntington distinguished between the corporateness of the officer corps and that of the enlisted soldier. 'Officership is a public bureaucratized profession' in which 'the commission is to the officer what his licence is to a doctor'.[6] As a bureaucratic profession, the officer corps is divided from lay society by its symbols and insignia of rank. By contrast, 'the enlisted personnel have neither the intellectual skills nor the professional responsibility of the officer. They are specialists in the application of violence not the management of violence. Their vocation is a trade not a profession.'[7] Consequently, Huntington insists that there is a 'sharp line' between the officer corps and enlisted men 'in all the military forces in the world'.[8]

Huntington's division between the officer corps and the enlisted ranks has validity, especially at the level of general officers and their staffs. Indeed, there is a great distinction between officers and enlisted ranks in terms of the work performed by officers in headquarters, their skills, and their career structures. Staff officers are not so very different from civilian managers and executives in many respects. For Huntington, only officers are able to display a genuinely corporate identity and can, therefore, be truly described as professionals; he regarded enlisted men as the equivalent of technicians or artisans. It is understandable why Huntington made this division. It is true that enlisted soldiers, for the most part, perform manual or technical functions, operating or fixing weaponry or organizing personnel who do. Although often very skilful, their work does not usually involve an intellectual, analytical, or critical dimension which typifies a profession. However, it is important not to overstate this divide especially at the level of the infantry. At the battalion and especially at the company and platoon level, the divide between the enlisted soldiers (especially the senior non-commissioned officers) and the field officers, namely lieutenants, captains, and majors, is not nearly so clear. Field officers manage violence directly but they, like their subordinates, may be called upon to engage the enemy with their weapons; they also apply violence. Similarly, corporals and sergeants do not simply apply violence but also play an important role in managing it; they command their soldiers, instructing them how and when to fire their weapons. In the case of an incompetent platoon commander, sergeants and corporals will often take over the supposedly distinct function of the officer. At this level, the divide between technical and professional expertise is not clear. Moreover, the platoon's shared mission and the risks which go with it seem to generate a sense of professional solidarity which transcends rank boundaries. The careers of officers and enlisted soldiers may indeed be very different, as Huntington suggests, but at the level of the infantry platoon, the vocations are united in a single endeavour: fighting a close enemy.

Indeed, the corporate identity found in the infantry platoon does not seem to be limited to this small unit but seems to suffuse the entire army and is critical to its organizational coherence. Although many generals and staff officers may never have commanded an infantry platoon, a substantial part of their legitimacy is derived from their experiences at the lowest tactical level, especially in combat. It is noticeable that many generals deliberately seek to communicate their combat status to their troops. In the Second World War, General Patton famously carried pearl-handled pistols, General Montgomery continued to wear his black Royal Tank Regiment beret and a bomber jacket, and Lieutenant General Truscott wore a silk scarf, faded russet leather jacket, jodhpurs, and well-worn cavalry boots.[9] Even today, when the general's role has become more technical and political, most generals actively emphasize their combat status and their past experiences at the front. This warrior ethos remains particularly noticeable among US officers. Thus, the most prominent American generals in Iraq and Afghanistan, Generals Petraeus, McChrystal, and Mattis, all display the ascetic lifestyle of the field officer; they wake up very early, exercise, and eat modest amounts of food. The result is that, while every luxury is available to them, they in fact have the lean appearance of 50-year-old platoon commanders. General Petraeus, to the dismay of his close protection team, often toured Iraq, engaging with locals, without wearing body armour and helmet, while Stanley McChrystal regularly accompanied his Special Forces on night-time assault missions. British officers often display a rather more aristocratic habitus than their American peers but the combat credentials of senior generals remain important to their reputation. For instance, Major General Nick Carter commanded NATO's Regional Command South in Kandahar between October 2009 and November 2010. The central element of his campaign was focused on Kandahar City and involved engaging in major military operations against Taliban strongholds in Zhari and Panjw'ai. Despite the risks and the fact that he could have plausibly remained in the safety of his headquarters, he was to be found on the ground at the front line supporting his subordinate commanders to the very end of his tour, where he was at the very beginning of his career as a platoon commander.

The officer corps may indeed be primarily a bureaucratic profession dedicated to the management of violence but the competent management of violence seems to be indivisible from the experience of applying it both practically and morally. Indeed although Janowitz was also only interested in the professional officer and he claims that, in the post-war period, the military officer was becoming increasingly civilianized, he does not draw a division between the officer corps and enlisted soldiers so definitively as Huntington. The armed forces as a whole are united by their specialism in the prosecution of violence—rather than divided by those who manage and who apply this violence; 'the military profession' (not just the officer corps) are 'managers of the instruments of violence'.[10] Notwithstanding the differences between officers and enlisted soldiers, it seems plausible to claim, against Huntington, that professional soldiers across the ranks substantially share a corporate identity. They understand themselves as professionals in the application of violence and this corporate identity seems to generate a form of solidarity which is not only quite distinct from civilian professions but also from the bonds of association typically found in the citizen armies of the twentieth century.

PROFESSIONAL COMRADESHIP

Since at least Janowitz and Shils's famous article on cohesion in the Wehrmacht, it has been widely presumed in the social sciences that the primary group in the armed forces is based on friendship. Primary groups endure the rigours and dangers of combat because their members are obligated to each other as friends and comrades. Implicit within Janowitz and Shils's argument is the suggestion that primary group cohesion is based on social 'likeness'. Because the members share common social, ethnic, and gender backgrounds, it is possible for them to form the personal bonds of comradeship which constitute solidarity. The bonds of friendship between homogeneous groups of soldiers do seem to have played an important historical role in enjoining combat performance, especially in citizen armies. Chapter 4 explored this dynamic at length. However, it is necessary to be very careful about the definition of comradeship and, in particular, when analysing the professional army, to avoid the common presumption that the friendship between career soldiers is self-evidently based on prior social likeness or, indeed, even upon individual personality. The bonds to which social scientists and soldiers themselves often appeal in explaining combat performance may not be nearly so personal as is often presumed in a professional force. On the contrary, especially in a professional army, the nature of comradeship may be surprisingly and ironically impersonal.

In their intervention into the debates about cohesion, Robert MacCoun and Elizabeth Kier cite substantial evidence to suggest, perhaps somewhat surprisingly, that the members of successful military groups do not necessarily need to like each other in order to conduct operations together. Certainly, Kier and MacCoun do not deny that soldiers may become very close friends as a result of their shared experiences but 'likeness' is not a requisite for combat performance. In order for the platoon to operate on the battlefield, they claim that its soldiers need to be trained to execute their drills competently. In the previous chapter, the training process was examined at length and it was argued that training fundamentally altered the collective competences and combat performance of infantry soldiers. Training does not just inculcate a series of complex individual and collective skills, however, although these factors are clearly crucial. It also generates social solidarity in and of itself. In his work on interaction ritual chains, Randall Collins has noted the powerful but sometimes overlooked social effects of training; '"Training" is not simply a matter of learning; it is above all establishing identity with the group who carry out their skills collectively.'[11] Training generates a sense of social unity which is essential to collective performance; training does not simply inculcate individual skills, it unites the group so that it can complete the tasks it is set. In his study of rhythm, William McNeill affirmed the importance of training when he recalled his wartime experiences as a conscript in the Second World War when he and his fellow recruits were frequently ordered to march 'whenever our officers ran out of training films'; 'A more useless exercise would be hard to imagine.'[12] Nevertheless, McNeill describes that in the course of those long marches 'on a dusty, gravelled patch of the Texas plain', his recruit cadre experienced 'a swelling out'; 'words are inadequate to describe the emotion aroused by the prolonged movement in unison'.[13] For McNeill, 'something visceral was at work' which he tried to capture with the concept of 'muscular bonding'.[14] Training seems to alter the very associations between soldiers, redefining their relations with each other and generating new forms of solidarity among them. If McNeill's description is believed, this seems to be true of citizen soldiers but, in the light of the increasingly refined and intensified training regimes of today's professional armies, it would seem plausible to suggest that the solidarity which they display is likely to be particular and specialized. Concepts of pure cohesion may be inadequate and inaccurate here since it would seem unlikely that the cohesion of the professional infantry is reducible merely to social homogeneity or to personal friendships. The solidarity of the professional army seems to be far more interesting than any association of likeness.

As already noted, the British army was one of the first western forces to professionalize and, indeed, it was the first all-volunteer military to be engaged in combat operations after the Second World War; the Canadian armed forces acted as peace-keepers until the end of the Cold War. The professionalized British army (and Royal Marines) were involved in campaigns in Borneo, Aden, Oman, Northern Ireland, and the Falklands between 1960 and 1990. They represent a potentially useful pre-emptive example of the special cohesion which attends professionalization in the armed forces. In the First and Second World Wars, it has been argued that the mass British infantry, like their western peers, was primarily united by appeals to masculinity and patriotism. At this point, their cohesion accorded with much of the now classic literature on the subject. Masculinity certainly remained central to the British forces from the 1960s to the 1980s, and many of the descriptions of training have some resonances with the treatment which Eisenhart received from the US Marines in the 1960s. However, even in those units, like the Parachute Regiment or Royal Marines, where an aggressive masculinity was most highly prized, there is substantial evidence that cohesion was based on professional competence. Indeed, professionalism was perhaps most pronounced among the robust culture of British paratroopers and Marines especially since they played the major role in the Falklands War of 1982. For instance in his account of that conflict, Vince Bramley, a lance corporal in the Support Company of the 3rd Battalion, The Parachute Regiment, described how his friendship with a private in his section was undermined by the latter's professional incompetence. Bramley described how the private had stood up to put on waterproof trousers during the Battle of Longdon, thereby compromising the section's position: 'I felt embarrassed by him. Twice he had acted like a week-one recruit. Things were never to be the same between us.'[15] Robin Horsfall, a paratrooper in the 1970s who would go on to serve in the SAS including the storming of the Iranian Embassy, confirmed the point:

In the Paras the biggest insult that could be levelled against anyone was to accuse them of being a bad soldier. Every man prided himself on being one of the elite, and tried never to leave himself open to such a comment. He could be wet, queer, thick, an idiot or any number of other things but he could not allow himself to be called a bad soldier.[16]

In an account of his experiences in the Royal Marines at a similar time, Steven Preece has affirmed the point: 'Amongst the Marines there was always the same code of practice: don't gob off [boast] unless you can back it up with your fists, and always maintain the high soldiering standards required of a Marine or expect to be beaten up.'[17] The critical criterion of acceptance in the Royal Marines was the ability to adhere to collective military drills. Comradeship within the Royal Marines and Parachute Regiment was finally awarded only to those who were professionally competent. Clearly, some care needs to be taken with this evidence as it is brief and fragmentary but, although statements of this kind have often been overlooked, they begin to suggest that cohesion cannot be presumed to be a product merely of friendship. On the contrary, these statements suggest that in the professional military the collective sentiment of mutual affection or respect, so central to much of the scholarship on cohesion, actually follows effective performance.

Significantly, the kind of cohesion suggested by British paratroopers and Marines in the 1980s seems to be increasingly evidenced on current operations. There, too, traditional notions of cohesion seem inadequate. Although a piece of journalism, not social science, Sebastian Junger's recent work is a useful piece of evidence in assessing the nature of cohesion in a professional force today. He records the intense solidarity which was evident among the 2nd Platoon, Battle Company. His entire work might be read as a discourse on 'pure cohesion', as David Segal and Meyer Kestnbaum have felicitously called it, especially since the final part of Junger's book is entitled 'Love'. In order to explain the cohesion of this infantry platoon, Junger follows some of the classical interpretations of cohesion. For Junger, the 2nd Platoon were able to endure the trials of the Korengal as a result of their personal bonds with each other: 'Loyalty to the group drove men back into combat—and occasionally to their deaths—but the group also provided the only psychological refuge from the horror of what was going on.'[18] Accordingly, Junger reports a number of interviews in which the members of 2nd Platoon articulated their love for each other. For instance, recalling the death of the popular team medic 'Doc' Restrepo, one soldier, Cortez, observed: 'His death was a bit hard on us. We loved him like a brother. I actually saw him as an older brother, and after he went down, there was a time I didn't care about anything. I didn't care about getting shot or if I died over there.'[19] Junger asked the same soldier whether he would risk himself for others in the platoon: '"I'd actually throw myself on the hand grenade for them", he said. I asked him why. "Because I actually love my brothers" he said. "I mean it's a brotherhood. Being able to save their life so they can live, I think is rewarding. Any of them would do it for me."'[20]

For Junger, then, 2nd Platoon Battle Company displayed the same dense solidarity which is a historically recurrent feature of all combat units. Close personal bonds of friendship were intensified by the harsh and dangerous conditions of the operation to produce feelings of genuine comradeship. There is no doubting these sentiments, and, indeed, the intense grief which followed the deaths of Restrepo and Rougle evinced the emotional attachment which developed between these soldiers. Nevertheless, Junger ignores important pieces of evidence which he himself presents about the nature of the solidarity in 2nd Platoon and perhaps by extension in all combat units in the all-volunteer American forces. The putative love which the platoon's members displayed towards each other was strange especially as a demonstration of pure cohesion. Junger describes how the members of the platoon would regularly fight each other. New soldiers would often be made to fight each other or existing members of the platoon. Junger saw new officers being attacked by the platoon together which seemed to have been both a mark of acceptance and an expression of collective resistance against and even a warning to this new authority. The aggressive masculinity of the group clearly informed this recourse to violence. In the film *Restrepo* which Junger made simultaneously with the book with Tim Hetherington, a new soldier is forced to fight a member of the platoon. At the end of the wrestling match, one of the platoon explains the fight: 'We're making him into a man.' Junger observed the way violence was often used to mark a status passage; 'you got beat on your birthday, you got beat before you left the platoon—on leave, say—and you got beat when you came back.'[21] These soldiers demonstrated their love for each other in an unusual way, then; by beating each other up.

Yet, their use of violence may not have denied their affection for each other. In a highly physical culture, violence might have been seen as the appropriate way to express (or repress) male emotions but the patterns of violence were interesting. The violence was neither random nor universal as might be expected if the love between the members of this group was genuinely communal. If violence was a sublimated expression of love, every member of the platoon should have fought with all the others at various moments to articulate their mutual but sublimated affection for each other. Yet, in fact, violence was structured by sub-unit membership. Violence would be used by members of the same squad or platoon to mark a change of status but it emerged most dangerously along the lines of distinction between the functional sub-units within the platoon. Squads would protect their own members or band together to assault the members of other squads to initiate a round of retaliatory inter-squad violence: 'Jumping someone was risky because everyone was bound by affiliations that broke down by platoon, by squad, and finally by team. If a man in your squad got jumped by more than one guy you were honour-bound to help out.'[22] Cortez claimed that he loved everyone in the platoon and there is no reason to doubt him. Yet, Junger's evidence suggests that the claim of universal love requires qualification. Soldiers in 2nd Platoon were primarily bonded with their tactical sub-unit, their squad or fire team. Specifically, the apparently personal bonds of emotion were actually attached to the quite arbitrary assignment of individuals to particular functions. Whatever their personal characteristics and their prior compatibility, the individuals in the four-man fire teams became the most densely bonded sub-unit who would fight all outsiders precisely because they worked together most closely. The bonds of 'love' between the men of 2nd Platoon actually followed their membership of groups to which they had been assigned for purely military reasons; they were ordered to fulfil a particular role. Ironically, then, the most apparently personal and intimate form of relationship was generated among 2nd Platoon not because of their personal likeness but purely as a result of their performance of their institutionalized and impersonal roles. The soldiers formed dense associations on the basis of the collective execution of tactical drills which they had been trained to perform. The groups which trained the most together and were tasked to perform collective drills with each other in combat demonstrated the most 'love', as Junger called it. Love, somewhat surprisingly, was a function of role.

Indeed, it is even questionable whether the members of 2nd Platoon did love each other in the unconditional way which Junger sometimes claims. Thus, Sergeant O'Byrne, one of the central figures in Junger's account, made a surprising omission: 'There are guys in the platoon who straight up *hate* each other.'[23] Against Junger, O'Byrne recognizes that the solidarity displayed by his soldiers on this tour did not necessarily depend upon personal affection. On the contrary, at a personal level, many soldiers actively disliked each other within the platoon. However, he also noted a paradox: 'But they would also die for each other. So you kind of have to ask, "How much could I really hate the guy?"'[24] Although many of the soldiers in this platoon detested each other, they simultaneously 'loved' each other enough to be prepared to sacrifice themselves, if necessary. It is a very striking and apparently contradictory observation; these soldiers could simultaneously genuinely hate and 'love' each other. Junger fails to acknowledge the significance of this paradox. He sustains the simple model of pure cohesion, asserting, against the evidence, that these paratroopers did indeed (despite all the interpersonal aggression and animosity) love each other unconditionally. However, the paradox can be resolved more satisfactorily if it is recognized that among these professional soldiers, 'love' or cohesion is not necessarily based on personal affection. It is based on performance. Specifically, the members of the platoon unite around their training, their drills, and the execution of these collective practices eliminates potential differences between the men. It unites them concretely about a set of procedures which they must perform together if they are to prevail in combat.

For all his discussion of 'pure' cohesion and love, Junger is eventually eloquent about this point.

It's such a pure, clean standard, that men can completely remake themselves in war. You could be anything back home—shy, ugly, rich, poor, unpopular—and it won't matter because it's of no consequence in a firefight, and therefore of no consequence, period. The only thing that matters is your level of dedication to the rest of the group, and that is almost impossible to fake.[25]

Indeed, Junger records the surprising friendship between a black soldier and a white Southern soldier, who is explicitly racist. Yet, this racist would not allow anyone to slur his black friend racially on this tour. Dedication to the rest of the group is not merely a willingness to be collegial in a generic civilian sense, then. On the contrary, dedication to the group is demonstrated by the performance of a series of concrete practices; practices which are essential to combat performance and therefore the protection of the group. Each soldier is under constant and mutual surveillance to ensure that everyone fulfils their duties competently.

Margins were so small and errors potentially so catastrophic that every soldier had a kind of de facto authority to reprimand others—in some cases even officers. And because combat can hinge on the most absurd details, there was virtually nothing in a soldier's daily routine that fell outside the group's purview. Whether you tied your shoes or cleaned your weapon or drank enough water or secured your night vision gear were all matters of public concern and so were open to public scrutiny. Once I watched a private accost another private whose bootlaces were trailing on the ground. Not that he cared what it looked like, but if something happened suddenly—and out there, everything happened suddenly—the guy with the loose laces couldn't be counted on to keep his feet at a crucial moment. It was the *other* man's life he was risking, not just his own.[26]

The last chapter discussed the way in which even mental states have become an area of collective concern for professional soldiers. Here, private functions like urinating became public property because they indicated whether a man was hydrated or not (and therefore whether he might collapse in combat).[27] Junger notes that he himself as a journalist was brought into this intense circle of professional collective self-defence. He temporarily thought he had left an army-issue shirt at a shura he had attended. He was panicked that an Afghan might have picked it up and it would then be used by an enemy fighter 'to pass himself off as an American soldier'.[28] 'Eventually I found the shirt, but it was clear from the looks I was getting that I'd fucked up pretty badly and that it had better not happen again.'[29] Indeed, even emotions were a public concern in this group of soldiers. On his last operation, Junger betrayed fear as he packed his equipment.[30] He was reprimanded by a sergeant who had no problem with Junger being afraid—everyone was—but showing his fear was dangerous; it potentially de-stabilized the group by undermining their collective confidence in themselves.

The importance of performance to interpersonal affection explains the strange patterns of violence which were displayed by 2nd Platoon. Sub-units within the platoon were more densely unified around specific tasks than others and had therefore become more militarily interdependent on each other than on the other members of the platoon. They worked with these individuals more and, consequently, 'loved' them more. This is not to deny the intensity of the emotions these soldiers felt for each other or that they would not die for each other. Rather, it is to emphasize only that apparently deep personal affection was in this case, and perhaps typically, the product of the formal roles which these individuals happened to be assigned to perform together. Their sense of comradeship emerged out of their professionalism; it did not precede it.

The Bundeswehr represents an interesting example of the emergence of a compatible form of professional solidarity. Although a conscript force until 2011, there seems to be clear evidence that the kind of professional solidarities evident for some time in British and American forces are increasingly apparent in the Bundeswehr, especially in the light of overseas operations in the Balkans, Africa, but above all in Afghanistan. Members of the Bundeswehr self-consciously emphasize the importance of professionalism: 'The Bundewehr has undergone a big change over the past two to three years. It is unimaginable in comparison with the past. But we must learn and we must learn quicker especially from our partners. Professionalisation is the way to do that.'[31] An experienced captain from the infantry (discussed in Chapter 9) offered up his own company as an example of how cohesion can be generated in the infantry. His company was part of an infantry battalion from the 26th Airborne Brigade which was among the most experienced formations in the Bundeswehr. Consequently, his company had had a long time to unite itself. They had spent four years together before deploying to Kunduz in 2009, in which time they had undergone basic and advanced training (*Erweiterte Ausbildung*) and then a cycle of increasingly demanding operations: 'My soldiers all knew each other, they had trained together. They knew the SOPs because they had trained as a company.'[32] Another officer affirmed the point: 'For deployment the German army now—but not in the past—stabilises the companies in order to build up cohesion. They train intensively together so that everyone knows everyone. They know if someone is in a bad mood if they hold a coffee cup in the left or right hand. And the attached specialists—medic, engineers etc—also train together.'[33] Among this group of officers, interpersonal solidarity was essential to performance. Indeed, the captain from the 26 Airborne Brigade noted that another company had not had the opportunity to train so intensively nor over such a long period before deployment. As a result, not only did he (with some evidence) believe that his company was more effective but they had endured fewer casualties. The follow-on company had suffered three deaths very shortly after deploying to Kunduz. However, the solidarity, while intimate, was distinctive. It was, according to this officer, not based on mere friendship but on professional trust and competence. Friendship follows function, comradeship performance. Although it potentially disturbs western notions of friendship as authentic lasting relationships, the evidence seems to suggest that, in the all-volunteer force, comradeship is not some everlasting union of souls but, on the contrary, a contingent product of institutional assignment and function. The better the training and the higher the levels of performance, the more intense this feeling of solidarity becomes.

Indeed, scholars have begun to note the changing basis of cohesion. Ben-Shalom et al.'s recent work on the Israeli Defence Force was one of the earliest pieces to suggest an alternative model of cohesion which seems to reflect the nature of solidarity in a western professional army.[34] Clearly, some explanation is required here as the Israeli Defence Force is a predominantly conscript army. However, because of its strategic position, it is a highly distinctive conscript force. Israel has effectively been at war since 1967 and accordingly not only must its conscript soldiers serve much longer than was typical of conscripts in western countries in the twentieth century but they became highly experienced in military operations. Indeed, the intense military activity in which the IDF has been involved for the past four decades has engendered a high level of competence among its personnel. Significantly, Ben-Shalom et al. have identified military drills—rather than affective bonds—as the critical factor in explaining military performance. They have recently shown how members of the Israeli Defence Force (IDF) are able to engender 'swift trust' among themselves through adherence to common professional practices.[35] Members of the IDF are highly trained in established military drills which are disseminated across the force in training and subsequent operations. These collective drills have become firmly established within the IDF as shared reference points for all Israeli soldiers whether they know each other or not. Consequently, when unknown members of the organization are tasked to perform a role, they can coordinate their actions very effectively and very quickly through appeal to commonly recognized drills. They do not need to be friends and, indeed, they barely need to know each other. The basis of solidarity is common allegiance to impersonal but shared professional practices. In many cases, these professional practices were precisely those about which the professional infantries of the USA and UK united from the 1970s.

Interestingly, the equivalent of Ben-Shalom's quick trust seems to have become increasingly prevalent especially in the UK and USA as they have felt the pressures of recent operations. From the 1970s to the end of the twentieth century, western militaries were involved in few episodes of combat, the Falklands, Grenada, the Gulf War, and a series of smaller operations in central and southern America, Northern Ireland, the Middle East, and Africa. Consequently, the Canadian, British, American, and Australian armed forces, which had professionalized in 1945, 1960, and 1973 respectively, were able to sustain relatively high levels of personnel stability in infantry units; sections and platoons were often able to work together for months and years, with personnel trickling in and out through a process of promotion and replacement. Consequently, small groups were able to fuse during the cycle of training, exercising, and deployment. Soldiers were able to get to know each other very well both in the formal and informal domains and often became friends away from the military sphere. This stability has been very substantially dislocated in the past decade as western forces, especially in Britain, America, and Canada, have been engaged in intense combat operations in Iraq and Afghanistan, which have stretched personnel resources and inflicted significant casualties on infantry units. In addition, in order to conduct operations in these theatres, units and formations have often been assembled from or augmented with numerous specialists or troops from other formations. Moreover, once in theatre, the special demands of counter-insurgency operations in these highly challenging environments have forced military units to restructure themselves on a task basis so that even within the course of a deployment sub-units are being recurrently re-formed and sent on missions with different personnel, selected because they have the precise expertise for that particular operation. It is clearly important not to exaggerate here. Specialist augmentees are typically attached to existing sections, platoons, or companies whose personnel have trained together for a substantial time and who have formed bonds of professional solidarity. Yet, in the light of this personnel turbulence, the phenomenon of professionalized comradeship seems to have been accentuated in the twenty-first century. Relations between soldiers in these countries seem to have become potentially even more transient and impersonalized. The 'swift trust' which Ben Shalom et al. noted in the IDF may be increasingly pronounced in western forces.

Indeed, the shift towards a task-based 'quick' cohesion has been noticed widely by soldiers who have operated in Afghanistan and Iraq. In preparing for the Helmand mission, British troops undergo a long process of pre-deployment training provided by the Operational Training and Advisory Group (OPTAG), which culminates in a period of training in Camp Bastion, Helmand, provided by the forward staff before troops are deployed into the field. This staff consists primarily of senior NCOs who have extensive experience of the operating environment in Helmand, with at least one full tour of the province. Having served in the army for fifteen to twenty years, they have a privileged insight not only into current operating practices but how they differ from previous techniques. This group has observed a notable shift in the way in which the British army operates in Helmand and how it generates cohesion among platoons and companies.[36] The basis of infantry training in the British army in the past was the battle courses at Brecon which taught conventional infantry tactics. As a result, a high level of solidarity was generated within sections and platoons, the members of which had worked and trained together for a long period. In the current era, they note a decisive change in the manner of training. While collective tactical training remains valid, in the current era, pre-deployment mission-specific training focuses far more on individual skills. In order to survive in an environment like Helmand, it is essential that technical skills are disseminated down to the level of the individual: 'You are asking individuals to build up their knowledge base.'

> There is a lot more emphasis on the individual now. Before the emphasis was on the corporal or the section commander but, look at the Barma man [the IED-detector] now; everyone has responsibility... Everyone from the front Vallon man [IED-detector] to the last man needs to know his skills and drills. All training before the start of pre-deployment training is about core soldiering skills and teamwork but as soon as Pre-Deployment training starts the focus switches to theatre-specific individual skills. You focus on individual patrol skills, the Vallon man, the medic. That is quite different to the past. The soldier must be more self-reliant as you may not be in a position to rely on your mates. The dynamic is harder. Gone are the days when you had only one team medic and everyone else lacked that level of training; everyone has a high level of first aid training and the team is better for it.[37]

The need for new forms of specialism has increased the need for individual skills in favour of generic professional solidarity around common forms of military practice like the section attack in which riflemen all performed similar functions.

In Helmand, the typical formation for normal operations is a company group consisting of two platoons (of about thirty soldiers each), a fire-support group (mounted on vehicles with heavy weapons), and numerous specialist attachments. These companies are typically stationed in a network of Forward Operating Bases dispersed across the province from which they patrol assigned areas. The operating environment and this tactical grouping has demanded an increased level of skill among even junior soldiers. The requirement for specialist teams in Afghanistan has also resulted in a novel form of cohesion in these company groups in Helmand which has been termed 'FOB cohesion'. However, the new forms of solidarity have not been generated by the attachment of specialists to existing sections and platoons alone. The OPTAG team—and other soldiers who have served in Helmand—noted that even in the platoons, sections, and fire teams an increasing individualism was observable. The result is a striking accentuation of professional solidarity.

> You go out at section and multiple level but the individuals in the section are potentially new. That would have happened in the past when people were injured for instance in World War II or the Falklands. Those jobs were filled by generalist replacements. But guys are now trained to a level where they can just slot in to a particular role and crack on. This is now different even from Iraq, when manpower was tighter. There you patrolled more often as a multiple[38] rather than a troop or platoon and you kept multiple integrity in Iraq. But here you can't do that—you need greater numbers or you would get smashed. You must pick and choose to produce the right skill sets. Each patrol must have the right pieces. You breakdown established forms of cohesion to build new cohesive units in order to develop those capabilities that you need and, against expectations, it does not affect your capability. This is because the lads are not at the same level as in the past. The lads are a lot more intelligent and adaptable. It is no longer a question of: 'here is a 100lb Bergen: now walk'. They are taking on responsibility. The lads want responsibility and when they are maxed out, they relish it.[39]

The mixing and matching has not been a problem because this campaign has intensified the professionalism of the troops. There is no longer the need for section level cohesion. You go out with a platoon consisting of various elements; there is Patrol Based cohesion. After five or six tours out here, there are artillery guys out there with you whom you may not know working as part of your patrol. But, in my experience you might go for a drink with all the guys of Herrick 8 whatever their unit. There is FOB cohesion. From a psychological perspective, friendship is developed by professionalism not because someone is in your section.[40]

Specifically, in these specialized, individuated groups, cohesion arises through the acknowledgement of mutual professional interdependence. For instance, in the past, it was typical for infantry platoons and sections to generate a feeling of independence. A feeling of cohesion emerged out of this extended experience of communal activity in which a section and platoon did everything together. Because they were rarely under intense military pressure in the past, their dependence upon other platoons and arms was only rarely demonstrated. The sense of section or platoon solidarity was enhanced by a (false) belief in its autonomy as a small unit. Consequently, on exercises in the past, the support company with its heavy weapons was often viewed only as 'noise at the side of an attack' which would be instructed by the company commander to just 'go up on that hill'.[41] Under the pressures of operations in Helmand, the infantry has learnt the critical role of the fire support group on which they have come to rely and whose professionalism they respect. This recognition of interdependence extends to all the members of the FOB, each specialist of which provides an essential skill. Especially in the light of the personnel turnover on patrols and in FOBs, cohesion has been built upon an often impersonal recognition of the requirement for certain professional skills. Soldiers who do not necessarily know each other trust each other on the basis of presumed (and proven) professional expertise. The OPTAG team's concept of FOB cohesion emphasizes that small units now require a diversity of skills and that these skills—not individual bonds or general collective sentiments—unify patrols in Helmand. Nevertheless, they also stress that due to rotations and casualties even in infantry sections and platoons themselves whose members have the same skills, the basis of cohesion has become more competence-based and impersonal than by personal familiarity.

These statements seemed to be very important, documenting a considerable change in the way in which cohesion is generated in the armed forces. Indeed, the solidarity described here seems to accord very closely with Durkheim's own remarks on the way in which the division of labour transforms the kinds of associations of which a society is comprised. In his early work on solidarity,[42] Durkheim drew an important distinction between the alternative forms of cohesion generated in simple and modern societies. Primitive tribal societies demonstrated 'mechanical' solidarity where the homogeneity of the society's members generated an all-encompassing sense of community. Individuals were united by their 'similarities', which were total. Indeed, in this form of solidarity, individual differences are all but eliminated: 'The solidarity that derives from similarities is at its *maximum* when the collective consciousness completely envelops our total consciousness, coinciding with it at every point. At that moment our individuality is zero.'[43] On this basis Durkheim explains his selection of the term 'mechanical solidarity': 'The social molecules that can only cohere in this one manner cannot therefore move as a unit save in so far as they lack any movement of their own, as do the molecules of inorganic bodies. This is why we suggest that this kind of solidarity should be called mechanical.'[44] The solidarity of the primitive tribe is like the unity of matter whose molecules have no independent agency; they change together in response to environmental changes in accordance with chemical laws. 'The situation is entirely different in the case of solidarity that brings about the division of labour.'[45] Here, individuals 'are different from one another'; 'the collective consciousness leaves uncovered a part of the individual consciousness.'[46]

Society becomes more effective in moving in concert, at the same time as each of its elements has more movements that are peculiarly its own. This solidarity resembles that observed in the higher animals. In fact each organ has its own special characteristics and autonomy, yet the greater the unity of the organism, the more marked the individualisation of the parts.[47]

Durkheim clearly sees organic solidarity as a superior form of association since both the capacities of the society and the individual are augmented by it. However, the division of labour clearly represents a danger, potentially fragmenting the social order and eliminating all communal sentiments. Individualism becomes a danger. Accordingly, with a division of labour, modern society could only cohere if it generated 'organic solidarity' when members committed themselves to each other and recognized their collective interests precisely because of their differences. Individuals in a complex division of labour could be united only by recognizing their cooperative interdependence. Organic solidarity refers to a more impersonal, practically oriented association; the members of such a society have a dense feeling of mutual allegiance because they rely on each other's specialisms but they cannot feel the intimate sense of communal love generated in a small mechanical group where all are alike. The concepts of mechanical and organic solidarity have potential applicability to the citizen and professional army. Specifically, as a result of current military operations and the increased levels of expertise which they have demanded of the armed forces, an accentuated professional ethos seems to have appeared which has some echoes of Durkheim's concept of organic solidarity. Here, in contrast to the mechanical solidarity of the infantry platoon in the citizen army unified by its common social background, the professional platoon is united by its professional ethos and its functional interdependence.

Perhaps more strikingly, there is evidence that, in today's increasingly professionalized military, not only is there no requirement for soldiers to share deep personal bonds or to like each other but they can actually actively dislike each other and still perform their military functions effectively. Junger's Sergeant Byrne already intimated this possibility in his observation about the platoon up at Combat Outpost Restrepo but there are cases where antipathy between the members of a platoon has been explicit. In 2008–9, 45 Commando Royal Marines were stationed in the Sangin Valley, one of the most difficult and dangerous locations in the whole of Afghanistan. Even within this area, Patrol Base Wishtun to the east of the town was regarded as particularly demanding; it was surrounded by IEDs and it was impossible to patrol any distance from the base without coming under Taliban fire. During this tour, the troop[48] deployed to Patrol Base Wishtun contained an individual, Marine Smith (name changed to maintain anonymity), who was widely disliked by his fellow troop members; he was regarded as a difficult and awkward individual who was not a good Marine. The troop was subsequently caught in a very intense ambush in the alleys around their base in which Marine Smith was wounded. He was hit in an exposed position in a Taliban killing zone and any attempt to retrieve him was going to be dangerous. However, taking cover in the relative shelter provided by a wall, one of the Marines, who was known to dislike Marine Smith (and who Marine Smith knew disliked him), ran out and, under enemy fire, picked Smith up, carrying him back approximately 250 metres into cover at great personal risk to himself. Subsequently, while Smith was recovering from his significant but not serious gunshot wound in the hospital at Camp Bastion, the Marine who had rescued him and some of the other members of his section visited the hospital to check on his progress and to be thanked by Marine Smith. The exchange was awkward for all involved because Marine Smith had to thank his fellow Marines for saving his life, even though he knew that none of them liked him. They had rescued him out of an impersonal sense of professional duty, out of status honour, not out of a sense of personal comradeship.[49] On a personal basis, it might even be said that the Marines did not care whether Smith lived or died but their sentiments were entirely irrelevant in the operational context; in Helmand, they were professionals. The rescue of Marine Smith was executed as bravely and skilfully as if he had been a close friend. Yet, and this is precisely what was so embarrassing for all involved, his rescue was explicitly not personal.

Friendship is certainly not irrelevant to even a professional army but it does not seem to be as vital to successful combat performance as has often been presumed—or as it may have been in the citizen army. Moreover, even when comradeship is present and does inform action on the battlefield, comradeship is heavily inscribed by professionalism. Ultimately, only those soldiers who have proved their value as professional soldiers are worthy of association in the private sphere. One of the problems with Marine Smith was precisely that he was not seen to be a very good Marine. Friendship was impossible with him because he simply was not very professionally competent.

As numerous studies and the discussion in Chapter 4 have shown, the basis of solidarity in a citizen army was typically quite different. Then comradeship was typically based on 'likeness'—social homogeneity and personal affection. The professional solidarity displayed by today's soldiers may seem harsh and cold by civilian standards. Yet, in stark contrast to the rescue of Marine Smith, comradeship based on likeness and friendship could involve a darker and, indeed, cynical element which is often ignored. Against Bartov's argument that cohesion was impossible on the Eastern Front because of the attrition rates which the Wehrmacht suffered, evidence presented in Chapter 4 suggests that in fact citizen armies typically suffered two attrition rates in combat: a very high rate among replacement soldiers and a slow rate among the primary group of veterans who comprised the core of any functioning military unit. For the most part, these differential attrition rates can be satisfactorily explained by the inexperience of the replacements. They simply did not have the skill to survive in combat and, unlike the veterans, did not survive long enough to learn it. However, not only did the individual inexperience of replacement troops reduce their life-expectancy but group dynamics within the infantry platoon seem to have further exposed them. In particular, while veteran primary groups knew how to operate under fire, there seems to be evidence that new troops would be treated differently from established members of the primary group. Typically, often because they were killed or wounded so quickly, they were never integrated into the veteran primary groups: 'We pitied the scared, shy eager youngsters who were awe-struck around us old boys... In the first battle they usually died in heaps.'[50] One of the original members of E company, 506th Parachute Infantry Regiment, 101st Airborne described how he deliberately did not get to know new replacements because they were killed so quickly: 'I think maybe they were trying to impress the older guys; people like me or Shifty. I don't know why? But I got to a point where I didn't want to be friendly with replacements coming in because I didn't like to see them getting killed. It just tore me up.'[51] His indifference towards them shielded him from the sorrow at their loss. Yet, it is difficult not to infer from these pieces of evidence that the distance which veterans maintained from replacements may also have made it easy for them to be expended in ways which were of benefit to the primary group. Indeed, in the case of the E Company veteran, it seems likely that the very indifference displayed towards replacements in a socially—and psychologically—vulnerable position may have actually encouraged them to try to 'impress' their seniors in order to earn recognition from them. In this way, the indifference of veterans towards new soldiers may have actually encouraged new soldiers to risk themselves for the sake of the group. The replacements were willing, and indeed had little choice but to do what they were told or what seemed to be expected of them to prove themselves worthy of comradeship, especially since they did not know any better, significantly increasing their chances of becoming casualties. Indeed, in his work on D-Day, Anthony Beevor explicitly notes that veterans would use replacements to conduct more dangerous operations, in which there was a higher chance of them being killed: 'Replacements joined their platoon usually at night, having no idea where they were. The old hands shunned them, partly because their arrival came just after they had lost buddies and they would not open to newcomers. Also everyone knew that they would be the first to be killed and doomed men were seen as somehow contagious. It became a self-fulfilling prophecy, because replacements were often given the most dangerous tasks. A platoon did not want to waste experienced men.'[52]

Mateship has been identified as central to the performance of the Australian army in both world wars. Mateship seems to have been a factor in motivating Australian troops and has been seen as critical to their performance especially in the Great War, yet the informal bonds of comradeship also had a negative aspect. The reception for replacements, denigrated as 'Reos' in the Second World War, was minimally 'unfriendly':[53] 'The most helpless men in any front-line unit were, as a rule, the untried reinforcements. The distinction between them and the more experienced soldiers is one of the major status differences that adversely affected relations within front-line ranks.'[54]

Replacements recorded their alienation: 'We reinforcements had to take a very lowly place, and were hardly spoken to for awhile,' or they were 'paraded, tiraded, promulgated, investigated and castigated, inspected like cattle and then tolerated only until they could "prove" themselves'.[55] Veterans explained their denial of mateship to replacements on the grounds of their incompetence. At Buna, they were accused of causing the deaths of experienced soldiers: 'The Reo's were giving us a lot of trouble by hanging back and that is one of the reasons why so many of the old hands are getting bumped off.'[56] Replacements noted that they were 'just a reinforcement' or 'a flaming reo'. Johnston notes that the replacements' lack of training was central to the problem of their incorporation into the institution of mateship. However, instructively, he observes that while individual replacements were not to blame for their tactical failings 'the burden of that insufficiency, and of the consequent lack of coordination between veterans and "new chums", seems generally not to have fallen on the old hands, as in the Buna example, but on the reinforcements. Again and again, newcomers were killed or wounded in the first actions.'[57] Veterans made no attempt to help reinforcements and, therefore, at least passively colluded in their attrition. Yet, given the extraordinary influence which the veterans exerted over the replacements, it is difficult not to infer that in many cases, reinforcements chose to run excessive risks in combat in order to prove themselves worthy of mateship. Johnston records that one replacement officer in 2/6 Battalion, wounded in May 1945, was able to declare as a result of his injury, 'Now they won't be able to call me a reo.'[58] It seems likely that the reputation of the citizen armies of the two world wars in national traditions has obscured the darker sides of comradeship in those forces. Nevertheless, the comradeship sometimes displayed by the citizen soldiers may have been highly exclusionary to the extent that those who were denied membership to these combat fraternities had a much higher chance of being killed or wounded.

Perhaps expectedly, in Vietnam, disdain towards the replacement was institutionalized in the term 'FNG': 'Fucking New Guy'. While ostensibly a criticism of the replacement's inevitable combat incompetence (and the potential danger that posed for the primary group), the disparagement of the replacement soldier allowed veterans to take less care of him and to be willing to risk him. The neglect of replacements is perhaps most striking not among US troops with collapsing morale but among the Australian army in Vietnam whose discipline remained relatively high throughout the campaign. As discussed in Chapter 6, the Australian Task Force in Vietnam was generally better trained than its US Army allies and benefited from the generous ratio of regulars to national service personnel. Moreover, personnel turnover was minimized so that national servicemen often served in the same company for two years: as junior soldiers, their experiences were not very different from professional privates. However, as D Company 6 RAR found in 1966 after the Battle of Long Tan, national servicemen at the end of their service would be individually withdrawn and replaced by a new conscript who would often find it difficult to integrate at least initially. As one national serviceman who joined D Company noted, 'they accepted us but they were wary of us'.[59] The disdain for replacements seems to have been less in the Australian army. Nevertheless, inexperienced reinforcements suffered higher casualty rates than the regulars and, even here, the actions of regulars could actively endanger replacements. For instance, in 1971, there was an incident in D Company 4 RAR when, unfamiliar with the company's contact drills, a replacement had failed to lie down when coming under enemy fire, 'he just dropped to one knee or stood there and got hit'.[60] He was shot and killed by his veteran fellow soldiers. There is no suggestion, of course, that the killing was deliberate but it is perhaps surprising that the replacement was not instructed about so important a drill and that the experienced members of his company shot so wildly. Whether they would have fired as indiscriminately if there had been established comrades in front of them is unclear but, perhaps most significantly, the description of this incident by a regular in the company placed the blame not on the undisciplined firing of experienced combat veterans but solely and exclusively on the replacement. This seems to demonstrate that they were invested with less importance than established comrades to the point where they may have been viewed as expendable. Minimally, it is difficult to envisage a primary group of veterans being as willing to take such great risks to rescue a new guy, whom they actively disliked, as those Marines in Sangin took to save Marine Smith.

To suggest that the interpersonal bonds between soldiers are not essential to military performance is not to say that the general pastoral care of professional soldiers is not important. The company commander of A Company 5 Scots, a sub-unit which had performed well on operations in Afghanistan in 2010–11 and during training for the Airborne Task Force, emphasized that in order to maintain motivation and standards, he and his command team invested considerable effort in sustaining the morale of his soldiers. As a proficient officer, this involved extensive administration of the financial and often domestic situation of his soldiers as well as the organization of company social events. The latter were very important. While training at CENZUB in December 2011, the company commander took great trouble to organize a visit to Paris, which included an educational visit to Les Invalides and a night's drinking in the Châtelet area, despite the evident risks of a major incident, involving over 100 British men, such an expedition involved. Not only was the enjoyment of the young Scots soldiers evident during that evening's carousing, which included the accompaniment of the company's well-refreshed piper, much to the enjoyment of locals and tourists alike, but in the course of it, A company's soldiers openly expressed their admiration for their company commander, whom they respected as an individual and as a professional. They were grateful for his efforts in organizing the visit to Paris. The general treatment of professional soldiers by their superiors is important and this pastoral care for each individual can generate a sense of community and well-being which reduces discipline problems and increases performance. Nevertheless, even in the case of A Company 5 Scots, the company commander was finally valued by his soldiers primarily because he was already regarded as a very competent officer. It is possible that in a British Royal Marine or Parachute Regiment company, whose soldiers have been specially selected and have, therefore, already demonstrated a very high level of motivation, the care displayed by an officer might be equally appreciated but its importance in generating performance might be less critical. Indeed, in the recent past, there have been cases where commanders of these elite battalions have been deeply unpopular and the personnel within them have themselves been very unhappy. Yet, on operations, the performance of these same battalions has remained very high. Professional standards remained the definitive reference point. By contrast, the often unacknowledged weakness of cohesion based on affection is that when friendship does not exist, combat performance is significantly impaired. The professional solidarity which seems to be appearing among all-volunteer forces may seem heartless and impersonal in comparison with the warm fellowship of the citizen army. Indeed, the notion that comradeship is ultimately dependent on role and performance may disturb comforting ideas about friendship. Yet, not only does this professional solidarity seem to be different from the 'pure cohesion' of the citizen army but the functionally oriented association of professionals seems to have generated a higher level of performance than that which the mass army was typically capable of executing. As the key criterion of group membership has become an impersonal assessment of performance, the platoon seems to have become ever more cohesive.

DISCIPLINE

Chapter 4 examined the issue of combat motivation in the citizen army, highlighting the importance of masculine bonds of comradeship and political incentives as central to battlefield performance in the twentieth century. It seems plausible to suggest that in many cases these central forms of motivation were supported or augmented by recourse to military discipline. Effectively, and notwithstanding Wesbrook's claim that on a modern battlefield soldiers have to be normatively committed to the fight, coercion does not seem to have been irrelevant, although its effects are certainly differential and never easy to quantify. However, historians and social scientists have long been interested in the question of military discipline and how it affects combat performance. It is impossible to adjudicate finally on that debate here. However, the different systems of discipline which are evident in the citizen versus the professional army usefully illustrate the difference between the solidarities displayed by these two military types. Indeed, the alternative disciplinary systems seem to be one of the best ways of illustrating the very different way in which cohesion is engendered in a professional force. The central argument of this chapter has been that professionalism does not only involve expertise but also necessarily includes a moral dimension which obligates soldiers to perform. Professional soldiers mutually expect each other and themselves to perform precisely because they are professionals. Professionalism has become the status honour of the armed forces today; it is the normative standard by which members of this group judge themselves and others and enforce appropriate conduct on each other. The institutionalization of professionalism as the status honour of the armed forces seems to have had a significant effect on the role of discipline and especially its use to compel combat performance.

The citizen armies of the twentieth century all employed capital discipline to encourage combat performance. Cowardice in the face of the enemy or desertion were normally punishable by death. In the First World War, the British army tried several thousand for cowardice and eventually executed 340 soldiers. The French army shot approximately 600. After the mutinies of 1917 which involved some 30,000 troops, 554 soldiers were sentenced to death but only 49 were actually executed; there seem to have been three summary executions.[61] The US Army shot a few soldiers but their combat involvement in that conflict was limited to six months between March and November 1918. Distinctively as one of the major combatants, the German army shot only 46 soldiers in the First World War. The situation was reversed in the Second World War when the western Allies generally eschewed corporal and capital punishment. The British executed no one during the conflict for cowardice, although there was at least as much desertion as in the Great War, while the United States executed a very small number of soldiers, including the notorious case of Private Eddie Slovik.[62]

Historians and social scientists have sometimes tried to determine the effect of discipline—and especially the threat of capital punishment—on the citizen soldier. It is very difficult to establish the degree to which sanctions contributed to combat performance. Some scholars, noting that the immediate terror of combat must overwhelm all other considerations but the drive for immediate self-preservation, have suggested that discipline can have played little role. It is not an implausible argument and, certainly, among those forces where the sanction of death for non-combat performance was weak, it seems likely that formal discipline probably had little motivating effect on combat soldiers. For all the recent controversies about the executions conducted by the British and French military authorities in the First World War, the numbers executed were very small in comparison with the size of the armies and, especially in the

French case, with the scale of mutiny which they faced. Moreover, although the military justice which prevailed in these cases might fall some way short of standards of civilian justice, soldiers were not shot summarily. Consequently, it might be suggested that for all the combatants in the First World War on the Western Front, discipline was not a major motivating factor in enjoining cohesion in combat; it did not contribute markedly to performance.

This was quite different for the Italians fighting in the Dolomites and on the Isonzo, however. Perhaps because of the stereotype which coagulated around the Italian soldier in the Second World War, the extreme discipline exercised by the Italian army in the First World War is often forgotten. Mark Thompson has usefully illuminated the draconian discipline which was enforced by General Cadorno throughout the war until his removal after Caporetto. General Cadorno and his subordinates employed three disciplinary systems. First, they used official forms of justice as a means of deterrent. As Thompson clarifies, official military justice in Italy during the First World War was independent of political control and, consequently, was so extreme as to border on summary justice. Army commanders regarded desertion as the greatest danger because in a mass army not only was desertion easy but if enough individuals decided to withdraw, the army would collapse very quickly. Accordingly, the definition of acts punishable as desertion was extraordinarily wide. A soldier who returned to his unit at the front more than three days late after leave was considered a deserter and liable to execution.[63] In May 1917 alone, fifty-four men were executed and, in July 1917, some soldiers of the Catanzaro Brigade, exhausted from months of fighting, revolted at the prospect of being returned to the front after only a few days' rest. Twenty-eight men were charged with rebellion and executed on the spot.[64] In all 729 executions for desertion alone were carried out in an army half the size of the French army and about the same size as the British army.[65] In addition to formal justice, summary justice was used liberally by the Italian army. Three hundred cases have been corroborated but it is likely that several thousand soldiers were executed in this way.[66] Most disturbingly, the method of 'decimation' was introduced and employed on a number of occasions to discipline all units collectively. Thompson identifies the experience of the Ravenna Brigade in 1917 as one of the worst cases of the use of summary decimation in the entire war. Following the cancellation of leave, there was a minor protest by some members of the Brigade who were, according to their own commander, 'a bit annoyed, tired and almost all in a dreadful physical condition'. The men were 'amenable to reason, but a few shots were fired in the air'. At this point, the incident escalated and the corps ordered twenty soldiers to be picked from the most rebellious company. Five of these men were selected to be executed by a firing squad made up of their comrades. Two weeks later, nine further soldiers were tried, four of whom were executed, followed by a further eighteen.[67] Finally, the Italian army employed carabinieri behind the front line to ensure participation in assaults. Italian soldiers were reputed to know that a major assault was imminent when the military police mounted machine-guns behind their trenches, shooting any soldier who was reluctant to join the assault.[68] After one minor attack in the Dolomites, a doctor recorded 25 buttock wounds inflicted by the carabinieri.

Italian generals were clearly convinced of the necessity of harsh and summary justice in controlling a generally very poorly trained army, many of whose soldiers were illiterate peasants with potentially little commitment to Italy or to the war. They were also concerned about the infiltration of socialism into the ranks. Some Italian scholars have argued that this disciplinary regime suppressed potential rebellion and, without it, 'mutinies on the French scale would have erupted'.[69] Giorgio Rochat suggests that so extreme was the discipline that 'disobedience lay beyond the conception of men'.[70] Certainly, it would seem to be difficult for troops to refuse to take part in an attack with the immediate presence of armed carabinieri behind them. Thompson perceptively observes that summary justice had 'disastrous side-effects'; he claims that 'evidence that soldiers' morale was harmed by the almost arbitrary killing of their comrades is strong and ample'.[71] Yet, Thompson also admits that Cadorna's methods 'reinforced the obedience which almost all soldiers already displayed'.[72] Certainly, Italian soldiers cannot have enjoyed the regime and their morale, by which Thompson seems to mean their individual and collective happiness, was doubtless undermined by it, yet, as Thompson himself acknowledges, harsh justice does seem to have engendered discipline and ultimately enforced a fighting spirit on potentially reluctant Italian soldiers. For all its barbarity, summary justice does seem to have contributed to the morale of the Italian army during the First World War when it confronted uniquely appalling conditions. It was widely agreed that the Italian army, despite its poor equipment, training, and lack of educated troops, fought very hard and ultimately successfully on this front, albeit with assistance from the Allies following the disaster at Caporetto.[73] Unpalatable as it is, the system of summary justice does not seem to have been irrelevant to this performance.

For the Allied forces in the Second World War, sanctions for cowardice became even weaker and it would be very difficult to argue that the threat of discipline itself could have had any serious effect on Allied combat soldiers then. Indeed, discipline in general was not always enforced. Lieutenant General Lucian Truscott recorded for instance that he had issued two standing orders to the 3rd Army: helmets were to be worn in combat and only the specified number of individuals were to be carried in vehicles.[74] He noted that in each case his standing order was 'flagrantly violated'.[75] Indeed, the problem became so bad that he personally fined individuals he encountered without a helmet. He never told those soldiers that the fines would never be collected.[76] If a general was unable to enforce a minor but important dress regulation on his troops (for their own benefit), it is unlikely that the sanction of a distant and theoretical punishment had significant motivating effect upon US soldiers in combat. In the Second World War, Allied armies seemed to be genuinely imbued with a democratic and liberal spirit which reflected the polities which they defended. Armies may have been strict by civilian standards, and Truscott noted the discipline and tidiness of the British army, but they were not authoritarian. In fact, the British army treated not only its psychological casualties but even the under-performing and sometimes near mutinous 7th and 51st Divisions with great leniency. As Stephen Wesbrook suggested, western armies could rarely compel their soldiers to fight. It is of course possible that the threat of disgrace at being subject to a court-martial for cowardice did have an effect on Allied soldiers. In Chapter 4, social humiliation was used by the army and seemed to be mobilized by soldiers themselves to motivate themselves and make sense of their situation. Discipline was not severe and, probably as a consequence, not very effective in influencing combat performance among the democratic powers. It is perhaps because of this disciplinary latitude that the Allied infantry performance was so poor in a challenging campaign like Normandy where the defender held many advantages at the level of close combat. In the Second World War, the Italian army, fighting under quite different political and operational conditions, instituted less draconian discipline which was more compatible with its democratic allies in the First World War; ninety-two soldiers were condemned to death between 1940 and 1943.[77] It does not seem implausible to suggest that its generally weak performance may have been a product of this lax disciplinary regime and, if that was a factor, then the Italian army's experience would seem to support the contention that severe discipline is potentially very important to the citizen army.

The Wehrmacht and SS were, of course, quite different and, indeed, seem to illustrate precisely the importance of harsh discipline in the mass army. The German army executed 15,000 in the Second World War.[78] Under the authoritarian Nazi regime, discipline became almost fanatical, especially on the Eastern Front, where thousands of soldiers were punished or executed for dereliction of duty and cowardice. Hein Severloh, who was stationed at Widerstand Nest 62 on Easy Red Sector of Omaha Beach on D-Day and, armed with an MG 42, personally played a significant role in disrupting the landing, might never have reached Normandy as a result of this draconian discipline. He had been deployed to Russia in 1942 where he contracted frostbite in his feet. He was going to present himself to the hospital but was advised against it by the regimental doctor who warned him that the condition would be considered a case of self-wounding, for which the punishment was death.[79] Clearly, the doctor may have been exaggerating in order to reduce his work or to avoid enquiries about his judgement from the rear but the fact that Severloh believed him and returned to the front demonstrates that such punishments were regular. Indeed, the Wehrmacht actively publicized the sanctions for desertion and cowardice. Hans Kuss, an officer who initially served with the 94 Infantry Regiment on the Eastern Front but was reassigned throughout the war as his units were decimated, photographed a prominent warning sign on a road in east Poland which recorded the shameful execution of two soldiers, found guilty of desertion; 'Notice: On 21 February 1945, Grenadier G and Corporal de Lw. and, on 26 February 1945, Lance Corporal B and Engineer H were shot in Kahlberg following a courts martial.' The sign explicitly communicated the options for German soldiers by that stage in the war: 'The condemned came over the back-waters[80] as deserters and draft-dodgers. They shirked a hero's death before the enemy and died a death of disgrace and dishonour.'[81] It is not actually clear whether Kuss's sign referred to a genuine execution or whether it was merely a deterrent but sentences of this kind were frequently carried out, many of them arbitrary and unjustified. Moreover, many thousands were also summarily assigned to penal battalions or to penal roles in which chances of survival were small.[82]

Even in the apparently more civilized western campaign, discipline was harsh, especially as the Wehrmacht began to be defeated. During the Normandy campaign, after Operation Goodwood to the south of Caen, Feldgendarmerie detachments seized stragglers and deserters at bridges, executing many summarily, some of whom were hung from trees as a deterrent to others.[83] Of course, it is possible that faced with the prospect of immediate extermination by Allied and Russian firepower in combat, German soldiers forgot about the discipline which waited for them a few miles to the rear should they fail to perform. However, so severe were the penalties for even minor infractions and so common was recourse to summary justice that it seems at least likely that discipline did have some effect on the German soldier. Indeed, it might be suggested that the further forward and the more extreme the discipline, the more effect it was likely to have in motivating troops to fight. Clearly, the authoritarian culture of Nazism and the desperate strategic situation in which it found itself by late 1942 encouraged extreme discipline, but the fact that the German army had to resort to such extreme measures does seem to belie the notion that their troops were always willing combatants or that their units were highly trained and competent. The German experiences seem to show that, in a citizen army, good performance in combat does not seem to be entirely dissociated from harsh discipline. Perhaps the willingness of German commanders to sanction their troops severely in Normandy might explain their generally superior performance over the Allies, rather than any great military competence. Discipline and, especially, capital punishment was a coherent if brutal response to the problem of combat inertia and straggling in the citizen army. Although it might not have had a major effect on individual deserters, especially when discipline was enforced through a slow judicial system as it was by Allied armies, the possibility of capital punishment for cowardice and desertion is likely to have provided some additional motivation to soldiers in combat. Where discipline was extreme and immediate, it seems probable that it had an effect on frontline troops. Minimally, discipline and especially capital punishment was important to the citizen army.

Discipline has a quite different place in the professional army, indicating the way in which professional status honour now performs the role once served by the threat of judicial sanction. There are certainly cases of 'combat refusals' among today's professional forces. They are often kept quiet but there have been a number of refusals in both Iraq and Afghanistan. Indeed, there was reputedly a near mutiny in the regiment relieving the Princess of Wales Royal Regiment in Al-Amara in 2005, so extreme were the conditions with which they had been left to deal after a—possibly unnecessarily—bloody tour by their predecessors. There have also been cases of 'cowardice', as the military would define it. In the Falklands War, a Welsh Guardsman, Philip Williams, was widely regarded to have deserted on Sapper Hill and was found weeks later. In a lesser known but more striking case, a Parachute Regiment sergeant, apparently the 'victim of battleshock',[84] hid himself in a cave before the Battle of Mount Longdon being found only after the cease-fire: 'the soldier reappeared and the realization grew amongst his comrades that he had panicked, taken a load of blankets from an incoming medevac chopper, and built himself a nest amongst the nearby crags in which to hide, curled up asleep in foetal security.'[85] Professional soldiers demonstrate the same fears as citizen soldiers and, in some cases, they are prone to the same reactions; individuals panic and desert. However, there are evident differences with the professional forces. First, as a result of their training and socialization for combat, combat refusal and especially cowardice and desertion are rare. Collapses and panics, which were very common in citizen armies, do not seem to have happened in either Iraq or Afghanistan even though some of the fighting has been desperate at platoon and company level. Secondly, precisely because all the other soldiers are highly trained, the cowardice of one soldier does not tend to have the catastrophic effect which even small panics recurrently caused in the twentieth century. The cowardice or non-performance of a single soldier is simply that. The other soldiers are so tightly bound to the performance of their drills by their training that they are far less prone to panic as a group.

Crucially, however, discipline has become less necessary precisely because soldiers are imbued with a sense of professional honour. The prime motivating and disciplining factor in combat is that they will be respected by their peers for upholding professional standards. During the reconnaissance phase for Operation Sond Chara in Nad-e-Ali in November 2007, Juliet Company, 42 Commando were patrolling an area outside their base at night. The Fire Support Group, mounted on WMIKs [Land Rovers], was advancing along one road, while 2 Troop marched parallel to them on a second road, the town being organized on a grid system having been built in the 1950s by US developers. A French Mirage reported a number of insurgents on the roof of a compound to their north-west which the Fire Support Group decided to investigate. One of the corporals (later promoted to sergeant) led his section to the door of the compound but before initiating an entry he turned to his section and asked them, 'Does anyone not want to go through this door?' He explained his rationale subsequently: 'I would feel like shit, if something happened to one of them and they hadn't wanted to have gone in.'[86] In the event, no one refused and they entered to find that the fighters had disappeared leaving only families inside the compound. The sergeant's point was that only professional pride ensured that none of his Marines voluntarily withdrew from this operation, even though they had every opportunity to do so.

There are other examples of precisely this process of appealing to the professional pride of individual soldiers in order to demand performance from them. At Khan Neshin in February 2009 (see Chapter 8), the major commanding Lima Company wanted to reach the fort but knew he had to draw enemy fire in order to do it and, therefore, asked the commander of 8 troop to advance more or less across open desert in view of some compounds from which he would almost certainly be ambushed. The company commander subsequently described how he encouraged the troop commander to take on this unpalatable mission in which he would almost certainly come under enemy fire:

He [the troop commander] was circumspect but quickly got on with it—he was engaged from the flank as he moved, and the game was on; just as I had wanted. So what? My take on this is that this acceptance of a 'tricky' task was enabled through mutual trust (the key pillar of mission command). I literally briefed the new troop commanders on arrival (they had turned up halfway through the tour) that it was likely I would have to ask them to do seemingly ridiculous things but that I would do so in the knowledge that we could discuss it. IE: my hope was that by giving them the opportunity to voice a 'no', they would actually feel happy to always say 'yes'![87]

While individual soldiers might feel they have personally betrayed themselves if they refused these requests to perform, it should be noted that the main tribunal here is not psychological or internal but collective and public; it is the judgement of immediate colleagues. Professional soldiers are susceptible to these requests precisely because they actively want to sustain their reputation in the eyes of their comrades.

At the same time, the positive motivation of professional pride is supported by an unignorable sanction; ridicule and ostracism by peers. While the corporal in Juliet Company's Fire Support Group gave his Marines the opportunity to withdraw without apparent loss of face, inadequate performance is, in fact, subjected to cruel and merciless criticism. The Parachute Regiment sergeant who panicked on Longdon in 1982 was threatened with execution by a fellow

NCO on his discovery the following morning,[88] and his professional reputation was destroyed.[89] His treatment was typical. For instance, one Royal Marine lance corporal was publicly ridiculed in a report on Helmand in the regimental journal in 2007 when he cried for personal 'all round defence' because he was married with children; his rank and name, Terry, was changed for satirical effect to 'Supreme Commander Allied Forces (LCpl) Terry "fied"'.[90] His professional reputation was destroyed publicly and permanently in the official regimental account. There have been more extreme examples of this. In 2007, Britain's Task Force Helmand, at that point consisting of the Royal Marines' 3 Commando Brigade, organized a raid against a stronghold in the southern part of the province which was an important base for Taliban fighters in the area. Not only did the fort itself represent a formidable defensive position but access to it was impeded by the Helmand River which was difficult to ford in that area. Despite receiving warnings about the difficulty of the approach from their reconnaissance troops, an assault by Royal Marines was executed. The results were predictable. The company, mainly mounted in Viking vehicles, crossed the river with difficulty to be caught in a Taliban killing zone from which the terrain alone impeded extraction. The company was subjected to intense fire from numerous sources, causing a number of casualties. In the confusion, one member of the company was shot and killed. The company demonstrated evident competence and bravery during this fire-fight, returning fire and eventually disengaging from the contact. However, members of the company were scathing about the operation and, especially, the preparation for it. They believed that, in contravention of the professional standards they had come to expect, the operation was planned carelessly on the back of a 'fag packet'. Consequently, its objectives, methods, and overall coherence were inadequately developed and it was insufficiently tested by means of ROC drills and rehearsals. The officer who commanded the failed assault was vilified by his subordinates for whom his professional inadequacies and indeed, in their view, cowardice were beyond redemption. He was removed from his command immediately.[91]

Examples of shaming and exclusion are, naturally, equally evident in other western forces. One of the most notorious Canadian cases involved Brigadier General Menard who was removed from his operational post as Commander Task Force Kandahar in May 2010 and court-martialled following revelations of an affair with his female clerk, Master Corporal Bianka Langlois. Claiming that his career had been ruined by the case, Menard resigned from the military. The affair was in contravention of rulings about fraternization and was certainly dimly viewed by Menard's superiors. Yet, perhaps more damaging for Menard's reputation was the fact that he had had a serious negligent discharge on Kandahar Airfield while boarding a helicopter with the Canadian Chief of the Defence Staff, General Walt Natynczyk, in the same month.[92] Three rounds from his weapon passed between the Chief of the Defence Staff and his bodyguards. Although Menard was given a standard fine for the weapons handling misdemeanour, members of the Canadian forces suggest that it is possible that his subsequent sacking may have been more to do with this nearly lethal negligent discharge than his affair. His credibility had already been irreparably damaged by the negligent discharge—not only within the Canadian forces but with NATO allies. Yet, while roundly loathed, soldiers and officers who have been overawed, incompetent, or even cowardly on operations are not typically subject to formal discipline. They are an embarrassment rather than criminal. At worst, they are quickly removed from their unit or from command.

In his work on leadership, Bryan McCoy has perhaps surprisingly emphasized the importance of shame to maintaining and enforcing professional standards, which he calls 'the virtue of shame'.[93] McCoy outlines the general dynamics of shame and notes the great influence which the fear of shame exerts over soldiers: 'It is shame that prevents us from doing shameful things; it is shame that propels us forward despite fear, hunger, and sleeplessness. Shame is the knowledge that one's behavior is less than what is expected by the group. Shame in the eyes of our brothers is a powerful motivator. No one wants to be known or remembered for coming up short when most needed.'[94] As American citizens and US Marines, McCoy and his battalion were probably sensitive to the international reputation of the United States of which they were presumably proud. Yet, the shame which McCoy discusses does not refer to patriotism or national pride. On the contrary, the shame is entirely professional: 'Our greatest fear in battle was not death or maiming. It was the fear of being a coward in the eyes of our comrades.'[95] Professional failure was for McCoy and his subordinates the ultimate shame. Critically, failure in combat would mean a betrayal of the group, in this case the members of 3/4 Marines, and necessary exclusion from the bonds of fraternity within it. Indeed, in his pre-invasion speech on 19 March 2003, McCoy did not refer to the strategic rationale of the operation the Marines were about to conduct; 'I did not give a rousing "Sands of Iwo Jima" speech, but I spoke firm and soberly.'[96] The content of the speech was illuminating. He tied the performance of his Marines to the past performance of the Marine Corps and its survival in the future:

I reminded them that we, as a Corps, exist for one reason—to win on the battlefield. The day the Marine Corps fails to win in combat is the day we will cease to exist. I reminded them that our ancestors from Tripoli, Belleau Wood, Guadalcanal, Iwo Jima, Inchon, Chosin and Vietnam would be watching us. It was now our turn, our duty to take our place in the line of history and fight well. I appealed to the pride of being a Marine. Up till now they had reaped all the benefits of wearing the uniform, the prestige and honour of belonging to such a storied fighting unit.[97]

McCoy's speech is important. It ties shame to the profession of soldiering itself. Moreover, it highlights how this mechanism of shame operates. Marines are not only personally proud of their status as Marines but the Corps provides them with numerous material and ideational goods which define their lives; there are numerous benefits of 'wearing the uniform'. As members of the Corps, Marines had reaped very significant advantages from the collective institutional goods which the Corps possessed. However, to remain as members of this military organization, it was essential that McCoy's Marines contributed to its battlefield performance and its reputation even at the cost of their lives. Their honour as Marines relied upon it, for if they failed to perform, the collective goods generated by the Corps and on which all Marines relied would dwindle. Consequently, shame was the public means by which Marines, as professional soldiers, impelled each other to contribute to the maintenance of the Corps and therefore all the collective benefits it afforded its members. Professional shame was for McCoy the central motivating factor which drove him and his Marines into battle and sustained them in combat. At Al Kut, one Marine was killed and several others wounded. They demonstrated the potent influence of 'mere' reputation for the professional soldier. They were hit as they conducted the drills they had been trained to perform and which both they and their peers expected them to perform, whatever the costs.

The US Marines have been far from alone in institutionalizing a concept of professional shame in order to encourage combat performance. Canada's Joint Task Force 2's Banquet Speech harnesses the power of shame to encourage performance in a way which is closely compatible with Bryan McCoy's account. The Bundeswehr has only recently fully professionalized and has been less involved in combat operations than the French and anglophone western forces in Afghanistan. Nevertheless, although the concept of professional shame may not have reached such a high and self-conscious level of elaboration as found among the US Marines or Canadian Special Forces, there is evidence that German soldiers are increasingly susceptible to criticisms of their competence and motivation. In an interesting autobiography by a German paratrooper, Robert Eckhold records his emergency deployment to Kunduz in September 2008 following the death of a German soldier and the wounding of several others in an IED strike. He and some fellow paratroopers from 26/3[98] Battalion stationed in Zweibrücken volunteered to form an additional platoon to augment the company which had suffered the casualties, already deployed in Kunduz. He describes how they would always wear their maroon berets in camp as a sign of professional distinction but, even more important, he notes the critical basis of their professionalism. Before each operation, there were detailed orders in which in normal format, the situation, mission, execution, combat service support, and command and signals were communicated. As a result of this self-conscious competence, Eckhold's platoon were sensitive to appeals to their professionalism. Soon after their arrival, their company commander talked to each of the sections personally and motivated them with phrases like 'You are the professionals and the best for this job.'[99] Although these phrases might be taken as mere flattery, Eckhold, in fact, records that they 'did not fail to have an effect' on either him or his colleagues.[100] It is noticeable that when they handed over to their replacements, a company of Gebirgsjäger, Eckhold was disparaging of their lack of professionalism in comparison with his paratroop unit. The Gebirgsjäger had a number of negligent discharges and even their commanding officer admitted that there were things, like night operations, which the Gebirgsjäger could not do.[101] Professional pride is central to Eckhold's understanding of himself, his actions, and his motivation. Eckhold and his colleagues had volunteered for this mission out of a sense of duty (recognizing that in today's Bundeswehr status is earned by taking part in operations[102]) and were motivated to perform in Kunduz out of sensitivity to their professional reputation. In the end, Eckhold became deeply disillusioned with the mission in Afghanistan and his professional pride was not in itself enough to sustain him. Operations were cancelled as a result of suspected collusion by Afghan National Security Forces and interpreters with the Taliban and, Eckhold implies, the unwillingness of German commanders to initiate contact with the enemy.[103] Any sense of purpose is finally destroyed by the poor conduct of a major cordon mission, Operation Roosgaard, in Chahar Dara district in which two of Eckhold's comrades are killed by a suicide bomber riding into a check point on a motorcycle.[104] Nevertheless, despite all this, Eckhold concludes at the end of his account: 'As a soldier you should experience in first instance the support of your comrades—a comradeship which is not just understood in terms of military discipline but for that which it should be allowed to become and for which it really stands: fellowship, solidarity and unity.'[105] Here Eckhold demonstrates the importance of professional pride—and shame—to the new soldiers of the Bundeswehr but also illustrates how they may still be inflected with an understanding of themselves as representing democracy which reflects Germany's distinctive post-war history.

While shame has been used as the sanction for cowardice and incompetence, courts-martial have been primarily reserved for abuse cases, of which the best known are Abu Ghraib and Haditha. However, there have been a number of other atrocities in both Iraq and Afghanistan. In September 2003, a local Iraqi, Baha Mousa, died as a result of mistreatment while being held in custody by the 1st Queens Lancashire Regiment in Basra; although only one soldier was eventually found guilty of inhumane treatment, a public inquiry subsequently found that nineteen soldiers were directly involved in the abuses. On 12 March 2006, four members of 1st Platoon, Bravo Company, 1st Battalion, 502nd Regiment, 101st Airborne entered a house in Yusufihay, south of Baghdad, and murdered an entire family, whose daughter they also tortured and raped.[106] In Maiwand District Kandahar in January and

February 2010, five soldiers from 2nd Battalion, 1st Regiment, 5 Stryker Brigade murdered three Afghan civilians in separate incidents, allegedly collecting body parts as trophies. In each of these cases, soldiers and non-commissioned officers have been found guilty and imprisoned for their crimes which have included murder. However, courts-martial have not been staged for cowardice even though some individuals might be guilty of acts close to the military definition of it. Here, the professional reputation of the armed forces is at stake not the individual reputation of the professional soldier. However, those colleagues who fail professionally are subject to a perhaps worse sanction. Their professional reputation is ruined and they are dismissed and ostracized by former colleagues. They have no prospects of promotion and ultimately they are no longer part of the military profession even though they continue to be paid by defence ministries. They are excluded from their professional group having offended its central status honour: the competent execution of military duties. In this context, discipline is reduced to relatively minor infractions and illegal acts such as abuse. The sanction for combat failure is ostracism; it is professional rather than actual death.

It is a long-standing presumption that soldiers fight for their friends. Indeed, in the citizen armies of the twentieth century, there is significant evidence that this was the case. The Pals battalions on the Somme and ANZAC 'mates' were united by their personal and prior friendships which typically made up for their lack of military training. The result was that their personal bonds motivated them to fight but did not endow them with the collective skills to be able to survive and perform on the battlefield. Friendship was critical to their morale but did not, perhaps, ironically improve their cohesion; they were committed to the fight but in many cases did not have the competence to perform in combat. The emergence of professionalism primarily involves the institution of new training regimes in which soldiers are trained individually and collectively to a higher level. Training consequently seems to have transformed the ability of the professional platoon to perform. However, in and of itself, training also generates a different kind of solidarity among the professional infantry. It creates a new moral life for the platoon. In the professional army, soldiers unite with each other not on the basis of personal friendship but on the basis of their common adherence to established practices. They then assess and value each other substantially on the basis of their performance of these drills. Comradeship is now primarily based on competence, not on likeness. Soldiers are valued more for their knowledge and expertise, rather than so much for their personalities. As a result, soldiers increasingly seem to be able to cooperate with each other and form cohesive social groups on a recurrent basis not because they know each other or have a long history of shared experiences but because each has been intensely individually trained in common, compatible, or mutually supporting professional skills. Independently of their individual acquaintance, soldiers are increasingly united by their individual commitment to professional competence. Professional soldiers especially in an infantry platoon may often be united around the same broad suite of technical skills—they may, therefore, become very alike—and yet relationships between them have become de-personalized, focusing not on the personal attributes of each soldier as an individual but their abilities as a member of their section. Even a small, apparently mechanical group like a platoon may in the era of professionalization have become organic. Even here, where face-to-face interaction is very frequent and sometimes very intense, comradeship may have become a function of professionalism; cohesion, as a collective sentiment, may be a product, not a prerequisite of task performance.

11

The Female Soldier

WOMEN AND THE ARMED FORCES

Primarily as a result of its training, the infantry platoon in the professional army demonstrates a distinctive form of cohesion in comparison with its citizen forebear. The rise of professional solidarity seems to have improved the performance of combat soldiers but it may have even more radical implications in terms of who could be a member of an infantry platoon in the future. Specifically, professionalism opens up the possibility that women may be able to serve as infantry soldiers, performing a role which has often been regarded as essentially and exclusively male. Three quotations appear on the opening page of one of the Canadian army's 1998 publications on gender integration into the military, which the reader initially presumes refer to women: 'Integrating them into units will result in a complete breakdown of discipline and unit cohesion', 'They are only effective in service support roles', and 'they are only comfortable with their kind'.[1] In fact, the quotations were taken from the US Army in 1944 to justify the continuing segregation of black and white troops. As discussed in Chapter 4, citizen armies especially in the first half of the twentieth century employed ethnicity and race to engender group cohesion and performance on the battlefield among white soldiers, although systematic discrimination was clearly manifest as late as Vietnam. As a result of this recourse to ethnicity and race, various now plainly specious individual deficiencies were attributed to black American and, indeed, in the British and French armies, to African or Asian troops. Yet, in the US Army, once black soldiers were integrated into white units, which began to occur in the European theatre in 1945 as a result of personnel shortages and was made policy in Korea, and were trained, supported, and commanded in the same way as white soldiers, their performance was indistinguishable from their erstwhile racial superiors. The putative inferiority of black troops had nothing to do with individual traits or genetics, as assumed, but was a collective response to organizational and cultural factors within the armed forces and civil society; black units were treated unfavourably in comparison with their white counterparts and, therefore, sometimes performed worse. In the twentieth century, arbitrary social criteria were employed to exclude individuals from the armed forces and specifically from the infantry which have now been completely rejected. The presumptions underpinning these criteria have been shown to be false. The overwhelming majority of scholars and practitioners would regard it not only as unscientific but also disreputable even to suggest that black or Asian soldiers could not be in the infantry.

A similar process of gradual inclusion has occurred in relation to male homosexuals. Homosexuals were formally excluded from the citizen armies of the twentieth century, though, of course, in practice there were many homosexuals serving in the forces throughout this period. However, in the light of pressure from the gay movement, changing social perceptions, employment regulation, and personnel requirements, western armed forces began to alter their position on homosexuals. The Netherlands admitted gays into the forces in 1974 but most other western nations began to alter their accession policy in the 1990s. For instance, under the Clinton administration, the US armed forces brought in their contentious 'Don't ask, don't tell policy' in 1993. The policy did not satisfy gay activists who point to the fact that it still implicitly discriminated against gays by normalizing heterosexuality, and that, in practice, it was not workable. In the close environment of the military, in which personnel work together for long periods, it was unrealistic to believe that the sexual orientation of personnel would remain unknown, even if no formal confession were made. As a result, the policy was consistently challenged and was repealed in 2011. Homosexuals will be allowed to serve in the US military openly from 2012. The British forces originally banned homosexuals from serving in the military, affirming this ban in 1996 (as the United States partially liberalized its lesbian and gay policy). However, in 1999, the European Court of Human Rights ruled against this ban and from 2000, the UK Ministry of Defence allowed gay men to serve in all branches of the armed forces. The belief that sexual orientation is irrelevant to military service is perhaps somewhat less universal than the belief in ethnic and racial equality but it is certainly the majoritarian view not only among policy makers, the public, and scholars but, crucially, in the armed forces themselves.

In both the cases of ethnic and racial minorities and gays, former reasons for their exclusion have now been refuted. Individuals and population groups which were once deemed incapable of military service or likely to have a deleterious effect on cohesion among the white heterosexual soldiers have now been incorporated. In each case, the access of ethnic, racial, or gay minorities into the military has been predicated on replacement of social criteria in favour of specific professional qualifications. Black or gay soldiers have been incorporated into the military on the basis that they can carry out their specified role. As long as they are professionally competent, they can serve. The introduction of a task-based cohesion centred on professionalism in favour of social cohesion founded on likeness has been critical in opening up the armed forces for minority groups. With the clear precedence of ethnic, racial, and latterly homosexual accession into the military, the question of whether women could also serve in the infantry has become increasingly prominent.

In his work on gender and the military, Joshua Goldstein begins his analysis with the observation that, historically and cross-culturally, war is an almost exclusively male activity. 'Amazons provide interesting material for the analysis of culture and myth in sexist society, but there is little historical evidence for the participation of women in war.'[2] Thus, '*combat* forces in the world's state armies today include several million soldiers (the exact number depending on the definition of combat), of whom 99.9 per cent are male',[3] while 'among contemporary pre-industrial society, both the very war-prone and the very peaceful ones share a gender division in war with men as the primary (and usually exclusive) fighter'.[4] There have been a number of interesting cases when women disguised as men have fought as combatants, and very occasionally women, such as Jeanne d'Arc or Boudicca, have acted as war leaders, but Goldstein appositely notes that there have been two documented cases of substantial and explicitly sanctioned female involvement in interstate warfighting. Goldstein cites the case of the Dahomey Kingdom's 'Amazon Corps' during the eighteenth and nineteenth centuries. From 1727, this polity recruited, trained, and deployed an all-female combat unit as part of its standing army. The women in this formation were equipped with muskets and swords, were drilled regularly, and, according to western observers, physically resembled men in size, musculature, and demeanour:[5] they were sworn to celibacy on pain of death. The Dahomey example is extremely interesting but, as Goldstein emphasizes, there were some exceptional political circumstances at work. The Kingdom, lasting from 1670 to 1892, was a product of the western slave trade. The Dahomey regime was based on conquering and enslaving neighbouring peoples in order to sell them to European traders. Consequently, the regime depended upon maximizing its military capabilities since its economy was based on warfare itself and the capture of slaves. Accordingly, the incentive of enrichment encouraged a radical use of the female population. Of more immediate relevance, during the Second World War, in the face of catastrophic defeat, the Soviet Union mobilized 800,000 women (8 per cent of its fighting force) not only in combat support roles but a very small number (about 1 per cent of the force) in combat roles; women served as fighter and bomber pilots and crew and also snipers, machine-gunners, and commanders in the infantry.[6] This case will be discussed in more detail below as it is immediately relevant to the question of women's participation in the infantry today. Goldstein, and others, have observed that women also feature prominently in guerrilla warfare as fighters. Again, the special circumstances of guerrilla warfare, where the insurgent forces are typically weak and engage in periodic and irregular activity, have encouraged the use of women in mixed gender units.[7] Yet, Goldstein notes that 'whenever their forces have seized power and become regular armies, women have been excluded from combat'.[8]

For Goldstein, barring some important but exceptional cases, combatant roles have been male, historically and cross-culturally. Other scholars have taken a more normative position. They have maintained not only that women have been excluded from the armed forces but that they should be excluded. Martin van Creveld has been an important but controversial figure here, noting the putative prominence of women in the Israeli Defence Force (IDF): 'both the proponents and the opponents of a greater role for women in western military establishments have frequently made use of that experience in order to press their case.'[9] Van Creveld aims to clarify misconceptions. Accordingly, he has provided a useful historical survey of female participation in the IDF from the first Jewish defence organizations from 1907 to the present.[10] When the Israeli War of Independence ended in 1949, 10,632 women served in the IDF, 'almost all of them in subordinate positions'.[11] Throughout the post-war period from 1949 to 1973, women in the IDF fulfilled roles which were compatible with the gender settlement in western armed forces; they 'became administrators, secretaries, communicators, welfare workers, psychological examiners and medical auxiliaries in non-combat units' or 'instructors whose task was to help students from inside and outside the army to complete their elementary education'.[12] Crucially, van Creveld claims that 'during all the years when it went from one victory to the next the IDF in its treatment of women was retrograde rather than forward looking'.[13] The female pilots, who had flown transport planes and even dropped paratroopers in 1948–9, disappeared.[14] Van Creveld sees 1973 as a watershed for gender relations in the IDF, as it was for the USA.[15] With manpower shortages women were allowed into the IDF in greater numbers, although, as in the west, there were no women in infantry units. Female enfranchisement accelerated yet further after 1985 and, according to van Creveld, 'quantitatively speaking, the feminist experiment was a success'.[16]

However, for van Creveld, the failures of the IDF in Lebanon and against the Intifada were the fault of a still small number of women in the service, none of whom held senior rank. Specifically, van Creveld identifies the physical weakness of women as a cause of the decline of the armed forces. Not only are male soldiers 'often obliged to undertake additional hardship in order to compensate for women's physical weakness',[17] but the performance of the armed forces as a whole is diminished by the inclusion of women: 'There can be no question of the military enabling them [male soldiers] to become all they can be; on the contrary, in the manner of a convoy that can only sail as fast as the slowest ship, they are forced to be *less* than they can be and are likely to be looked down upon by other men.'[18] However, his argument is more interesting than a purely physiological one. Specifically, because women are weaker, 'for them [men] to undergo military training and serve alongside women represents a humiliation'.[19] Not only do male soldiers find service with women demeaning but the presence of women in the armed forces directly undermines the status and reputation of the military in the public domain. Van Creveld draws upon Margaret Mead's observation that 'in any society it is what men do that matters'.[20] Mead claimed, with some justification, that masculine activities and institutions were universally imbued with high status while the feminine domain was routinely denigrated. If the feminine is associated with low status, then the integration of women into a formerly exclusive masculine domain is, according to van Creveld, bound to have deleterious consequences for the armed forces. It will degrade their status, with the result that the talented and strong men, who were once willing to serve in an honourable institution, will leave or refuse to serve. It is an interesting argument, some evidence for which will be discussed below, and it is entirely conceivable that some men in the IDF did indeed feel degraded by having to train and work with women.

Van Creveld opposes the integration of women into the armed forces as a political conservative and Israeli nationalist. His position is shared by many others of his political orientation.[21] However, it is perhaps surprising that his scepticism about the possibility of gender integration in the military is shared by scholars on the opposite side of the political spectrum; namely, radical feminists, like Cynthia Enloe and Jean Bethke Elshtain. Enloe situates the post-1970s integration of women into the armed forces in a larger historical and conceptual framework. She properly regards the armed forces as a central element of state power which she understands, far more contentiously, in purely patriarchal terms. The state represents one of the key institutions by which male domination is secured and the military is, therefore, one of its most potent agents of female suppression. Indeed, according to Enloe, the military invent concepts of national security which are not primarily developed in relation to real external threats but are motivated by the requirement to maintain the internal political—i.e. patriarchal—order: 'the military can use this extraordinary status in relation to the state to define national security. The concept of national security has, in turn, been used to define social order supposedly necessary to ensure that national security. In this circular process, national security can come to mean not only the protection of the state and its citizens from external foes but, perhaps even primarily, the maintenance of social order.'[22] In this enterprise, the state and the military seek to 'militarize' women's lives, thereby subjecting them to intense masculine control through the most diverse and indirect means. The purchase of tinned pasta in the shape of tanks is a form of militarization; it enforces the value of martial patriarchy on women. However, for Enloe, militarization of course primarily affects women who are immediately involved in the armed forces. At that point, 'military official and civilian state authorities... have tried to manoeuvre different groups of women and the ideas about what constitutes "femininity" so that each can serve military objectives'.²³ Women serve the military as 'camp followers' (as wives, mothers, and prostitutes). In every case, the military as a patriarchal institution defines femininity in ways which are convenient to it.

Historically, according to Enloe, in order to preserve patriarchy, women have been excluded from the military, as a pillar of masculine power, and they 'cannot qualify for the entrance to the inner sanctum, combat'.²⁴ Indeed, in a claim very reminiscent of van Creveld, Enloe argues that 'to *allow* women entrance into the essential core of the military world would throw into confusion *all* men's certainty about their male identity'.²⁵ The integration of women into the military presents Enloe with as much difficulty as it does a conservative like van Creveld. Enloe deduces that if the military is an essentially masculinist organization, it should be opposed to female accession. She notes that only severe personnel pressures have forced this patriarchal organization to consent to female accession and she is forced to dismiss women's successful demands for accession as mystification: 'It is a cruel hoax: militarism and the military, those instruments of male ideological and physical domination, are riding on the backs of individual women's genuine desire to find ways to leave oppressive family environments, delay or avoid marriage, acquire "unfeminine" skills.'²⁶ Women are co-opted in an effort to free themselves from the current oppression but they become the victims of a new form of subjection and marginalization in the armed forces. Women remain camp followers, defined by masculine requirements and used to service masculine needs. Enloe takes, not implausibly, the frequent cases of harassment in the armed forces as evidence of this suppression. However, she never proves that women have been subjected to a hoax or that their service in the armed forces must be defined as one of oppression and harassment. She asserts that they remain camp followers, rather in the manner in which van Creveld asserts that women inevitably defile the status of the armed forces.²⁷ With some irony, both argue that women cannot and should not be part of the military.

Jean Bethke Elshtain's work on women and the military is intimately connected to her wider intellectual project on western politics and the public sphere itself. On the basis of her investigation of women in the armed forces, she reaches a sophisticated and sensitive political position in which she advocates a form of 'chastened patriotism'.²⁸ In a 'chastened' political community, men and women will have learnt from the past to generate a liberal but realistic political consensus both about the domestic gender relations and international policy. While pacifist and liberal, 'this civic being does not embrace utopian fantasies of world government or total disarmament'.²⁹ However, while Elsthain eventually reaches a plausible and pragmatic conclusion, for the most part her work adopts a similarly simplistic and polemical position towards the question of female accession. She understands the relation between men and women and the military in terms of Hegel's concept of the 'Beautiful Soul' and the idea of the 'Just Warrior'. On the basis of this, she claims that men are envisioned as the noble defenders of the republic community, protecting political liberty, while women's role is as the domestic carer, fulfilling the role of mother and wife. Elshtain plots the use of these concepts through history from the Greek city state to the present to claim that: 'Beautiful Souls and Just Warriors, though knocked about, continue (in and through rhetoric brought up to date) as ready-made identities which become particularly compelling, or are made possible only for men, in time of war.'[30] Elshtain subsequently complains that certain kinds of 'feminism [which essentialize the difference between men and women] reproduce many assumptions that structure the discourse of realism and just war, respectively, re-creating prototypical characters and figurations'.[31] Elshtain rightly rejects such 'ahistorical abstractions' which take 'a part of some complex situation, event, or idea and mak[e] it "the whole"; moralizing and dogmatizing'.[32] Nonetheless, it is difficult to read Elshtain's narrative about the manifestation of 'the Beautiful Soul' and 'Just Warrior', through history, as anything other than ahistorical abstractions which suggest that gender roles and relations are enduring and ultimately unchanging. Even though her conclusion undercuts their assertion, the rest of the work asserts that the role of women and men in war has been strikingly consistent and, indeed, essentially the same. Women have been excluded from this domain of human activity, monopolized by males, and, as Beautiful Souls, their exclusion will necessarily continue.

Clearly, van Creveld, Enloe, and Elshtain represent important interventions into the topic of gender integration. Moreover, despite the assertiveness of their statements, their arguments against female integration specify two central objections which have been central to the debate. Van Creveld, Enloe, and Elsthain explain female exclusion on physiological and sociological grounds. Women are not physically strong enough to be in the armed forces and they threaten the distinctive masculine cohesion which is essential to combat performance. In line with these central arguments, some of their claims are valid. Some male soldiers do seem to have felt demeaned by the presence of women and numerous female soldiers have been harassed in ways which have marginalized women much more generally from the military. However, their analysis simply is not conceptually or empirically detailed enough to understand the processes of gender integration. Specifically, they are unable to answer the question of how and whether women might be able to be serve as infantry soldiers in a professional military. In order to answer this question, it is necessary to situate the question of female integration into the infantry in the wider context of the history of female accession more generally.

THE HISTORY OF GENDER INTEGRATION

While armed forces, especially standing armies, have in the past typically been male preserves, an important development was observable throughout the twentieth century, especially during the two world wars, which continues and indeed is accelerating today. While certainly still marginalized in terms of roles and numbers, women have increasingly been recruited or conscripted into the armed forces of the state. Thus, in the First World War, women were mobilized as nurses, a function which they continued to fulfil in the Second World War, Korea, and Vietnam. In Britain, women had served as nurses in the British army since the Crimea when the Army Nursing Service was established in 1854 (to become the Queen Alexandra's Imperial Military Nursing Service and, in 1949, the Queen Alexandra's Army Nursing Corps). Following the Haldane Review in 1907, which reformed the army after the difficulties of the Boer War, women were also inducted as nurses into the reserves, the First Aid Nursing Yeomanry and the Voluntary Aid Detachment,[33] and women served in all theatres throughout the First and Second World Wars as nurses. In addition, during the Second World War, 450,000 women were recruited into support functions.[34] Indeed, in Britain, a Women's Royal Auxiliary Corps was created specifically to command women, most of whom served in administrative and secretarial posts, although some women were more actively involved flying planes or driving vehicles outside the combat zone or operating (though not firing) anti-aircraft guns. In the United States, 350,000 women served in the Second World War and women's significant but marginal role in the armed forces was affirmed in 1948 when the Women's Armed Service Integration Act of 1948 established a 2 per cent ceiling on the participation of women in the military; 120,000 US females subsequently served in Korea and 7,000 in Vietnam.[35] France demonstrates a similar trajectory of slow and marginal inclusion from the Second World War to the 1970s in supporting clerical and nursing roles.[36] Up to the 1970s, even during periods of war, women played a minor role in western armed forces, then. Their participation was highly limited in terms of numbers and function. Almost without exception, they served in nursing and administrative roles. Enloe has suggested that, in this way, women in the twentieth century remained (and still remain) as little more than the female 'camp followers' which traditionally accompanied the early modern European army; servicing various masculine needs as wives, prostitutes, servants, cooks, and washers. Enloe is perhaps too disparaging of women's role but the evidence demonstrates that women were in a marginal and very subordinate position.

This marginal status persisted until the 1970s, at which point women's participation in the armed forces began to undergo a manifest change. There were two particularly important factors which precipitated this transformation. First, western militaries began to move away from conscription and towards all-volunteer professional forces. By 1973, America, Australia, Canada, and Britain had abolished conscription. As Mady Segal has noted: 'In general, modern nations that have voluntary systems of recruitment of military personnel (notably Canada, the United Kingdom and the United States) have been increasing women's military roles more rapidly than those with conscription.'[37] With the abolition of conscription, professional armed forces had to attract recruits and since, in the 1970s, 'there were shortages of qualified men',[38] they turned to women to fill that gap. Clearly, this shortage was closely related to economic conditions and Segal observes that: 'In the United States, major growth in the representation of women in the military occurred in the late 1970s, when unemployment declined. When unemployment rose in the early 1980s, the expansion stalled.'[39] Secondly, in the 1970s, the women's liberation movements of the 1960s began to have a direct impact on public understanding of gender and, consequently, on legislation itself. Segal hypothesized that 'the more egalitarian the social values about gender, the greater women's representation in the military'.[40] Following the 'citizenship revolution', 'many nations have enfranchized women in the political system, and cultures increasingly have supported their participation in other social institutions', including notably the military. Equal opportunities legislation in the UK and United States in the 1970s was important in opening the military profession to women. In the United States, the 1948 Armed Services Integration Act which legislated for women's participation in separate and supporting functions was repealed on 20 May 1975 when the House of Representatives passed legislation mandating women's entrance into the armed forces.[41] As a result of this legislation, there was a slow but significant increase in the number of serving female personnel. The proportion of women serving in the United States Army increased from 1.6 to 8.5 per cent between 1973 and 1980.[42] In the United Kingdom, a similar increase in numbers and role is observable in the 1970s and 1980s. In France, there was also a gradual integration of women into the services in the 1970s and the 1980s. The Act of 13 July 1972 eliminated the legal distinction between men and women but this was qualified by the decree of 23 March 1973 which banned women from combat functions followed by further reforms in the 1980s.[43]

From the end of the 1980s and especially in the 1990s, the move towards female integration was notably intensified. Segal notes that Canada's Human Rights Commission Tribunal decision of 1989 was 'directly responsible for breaking down some barriers to women's participation in the armed forces';[44] it demanded complete integration of the military within a decade.[45] In 2000, 11.4 per cent of Canada's armed forces were women in comparison with 9.9 per cent in 1989. In the most recent figures from late 2011, 16.7 per cent of officers and 12.7 per cent of non-commissioned members in the regular Canadian forces are women, though in the combat arms these figures fall to 4.2 per cent and 1.5 per cent for officers and ranks respectively.[46] In the USA, the Gulf War of 1990–1 in which 35,000 women (7 per cent of the force) participated has been widely seen as a watershed in female representation in the military and, by 1996, female participation had reached 13.1 per cent,[47] where it has remained more or less stable; 15 per cent of the armed force is currently female. In the United Kingdom, the MARILYN Review in 1989 explicitly identified women as potential new recruits to fill personnel shortfalls.[48] This was followed in the early 1990s with some important legislative changes. The Women's Royal Army Corps was disbanded following the Conservative government's defence white paper in 1992, *Options for Change*, and an expansion of posts was announced in Parliament on 27 October 1997.[49] There has been a steady increase in the proportion of women in the UK's armed forces since that time from 5.7 per cent in 1997 to 9.6 per cent (7.6 per cent of the army) in 2010. A similar expansion of numbers and roles is observable elsewhere. In France, there were decisive changes in the 1990s when women were allowed to serve on warships, as fighter pilots, and in the infantry. In 2000, all combat arms posts were opened to women and just under 10 per cent of France's armed forces were women in 2006.[50] Women are much less well represented in nations which have retained conscription until recently. Until 1999, Italy was the only NATO country to exclude women entirely from the military when women were finally granted limited accession. Yet, in 2006, there were only 438 women in the armed forces or 0.1 per cent of the force.[51] In Germany, women could only serve as volunteers in the military in the medical and military music services,[52] but, following the European Court of Justice decision on the case of Tanya Kreil, an electronics specialist denied service in the Bundeswehr in 1996, all posts were opened to women.[53] However, although at that point the longer-term target for women was 5.3 per cent in a future standing strength of 289,000 (i.e. 15,000 women), in 2006 only 1.4 per cent of service personnel (or 4,350) were women.[54]

Nevertheless, in most countries, while there has been an increase in absolute numbers since the 1990s, women are restricted in the functions in which they can serve. This is especially the case in the army. In 1997, the UK's Secretary of State for Defence expanded the number of posts which women could fill from 47 per cent to 70 per cent, to include the combat support functions of the Royal Engineers, Royal Artillery, and Royal Electrical and Mechanical Engineers.[55] However, it was ruled that women were not allowed to serve in the infantry and armoured corps. This ban remains in force despite the fact that women have been allowed to serve as combat aviators in all three services. New positions have opened up to women but they are severely under-represented in the combat support branches in the British army. In 2006, 4.8 per cent of personnel in the Royal Artillery were women, 4.8 per cent in the Royal Engineers, and 3 per cent in the Royal Electrical and Mechanical Engineers.[56] Today, women still remain overwhelmingly clustered in intelligence, logistic, and administrative functions. There have been intense debates in the United States about the inclusion of women into the military, accentuated by a series of harassment scandals including the notorious cases of the Tailhook convention and the Aberdeen Training Ground. However, by 2000, 80 per cent of jobs and 95 per cent of career fields were open to women. Yet women remain formally excluded from ground combat roles. Consequently, although there are more women in the US forces than in the UK, they remain compatibly clustered in support roles, especially in the land forces, the US army, and Marine Corps. Until 2011, the Canadian and Danish armies remained the exception in western forces where women are not only formally allowed to serve in all branches but do, in fact, join the infantry, though in both cases very small numbers of women actually serve as infantry soldiers. In 2011, the Australian army opened the infantry to women for the first time. It is unclear what proportion of the infantry will be women but, taking Canada and Denmark as an example, it is likely to be small.

The trajectory of female participation in the armed forces since the Second World War seems to be clear, therefore. From the Second World War to 1970, women played a minor and supporting role in the military in their own independent female corps. From 1970 to the early 1990s, women were increasingly integrated into the military, as the female command and administrative hierarchies were incorporated into male structures, and as functions initially closed to them especially in the domain of combat service support were opened up. Finally, from the early 1990s to the present, women's integration has been forwarded so that the majority of roles in the military have been opened to them including combat support roles and, in some rarer cases, combat roles themselves.[57] Certainly, female participation especially in proportional terms remains minor; in contrast to some ethnic minorities, women constitute just over 50 per cent of the population in all western nations, and yet typically only about 10 per cent of the armed forces are female. Even then, this integration is also highly contentious and fraught with difficulties. However, the process of integration is demonstrable. It raises a critical question, especially in the light of female access to the combat arms in Canada, Denmark, and Australia, whether women might not in the twenty-first century be integrated fully into all western militaries including the infantry. It may be possible that in the coming decades a phenomenon which has occurred only in exceptional circumstances in human history—the use of females as infantry soldiers—might become normalized. If Goldstein is right when he claimed that war is central to cultural definitions of gender—that there is a process of 'reverse causality' whereby 'the war system influences the socialization of children into all gender roles'[58]—the performance of the combat role by women may mark a historic moment in gender definitions, not only in the armed forces but in civilian society itself.

CURRENT OPERATIONS: THE ACCEPTANCE OF WOMEN

The issue of gender integration into the armed forces and specifically women's access to the infantry remains contentious. There is a clear divide between scholars and policy makers in the current era over the issue. Van Creveld and Azar Gat are opposed to women's incorporation into the military and, of course, the infantry while a number of others have advocated the complete integration of the military, which both Canada and Denmark have implemented. As the discussion of the actual numbers of women who are likely to be involved was intended to indicate, any investigation of the question of female accession into the infantry needs to be approached with care and precision and, certainly, blanket statements about what is or is not possible are out of place, not only because the armed forces are changing rapidly and there are evident national differences but also because there is a divide between formal policy and actual reality. The analysis of the gender-integrated military has to delve beyond official policy into the actuality of current practices in the armed forces. It is necessary to gain a realistic picture of the nature of gender relations in the military today, especially as they exist in operations in Afghanistan. It is here, where risks and operational pressures are at their highest, that it might be possible to establish some critical empirical understanding of the new gender balance and its trajectory; if it has one.

As the first major western army to liberalize its accession policies, the Canadian army offers one of the best insights into the possibility of female integration into the infantry. In 1998, nearly ten years after integration had been imposed on the Canadian forces, the army conducted a major study on female integration in the light of some significant difficulties. Significantly, it sought to address the problems which had attended female accession by an affirmation of professionalism. In particular, it dismissed the argument that women and men had different capabilities or dispositions on the basis of which they should be treated differently. The central purpose of the document was to impose a professional ethos on military commanders at all levels. Accordingly, throughout the document, issues, presumed to be gender issues (about women), were reduced to leadership issues about the enforcement of professional standards. Analysing a negative report of gender integration in a Canadian unit in the Balkans, the document asserted, 'These are not gender issues, they are leadership issues!!!'[59] Any apparent problem with female soldiers was, in fact, a product of unprofessional (male) commanders being unable or unwilling to enforce military discipline equally across all their subordinates, both men or women. In order to engender cohesion, it was essential that women were treated and disciplined equally to men. In order to ensure collective performance, female soldiers had to be as competent as their male peers: 'there are no special standards for women.'⁶⁰ The document stressed that women had to be trained and tested properly to an equal standard as male soldiers and that it was a breach of military discipline to 'pass a fault' (i.e. to allow a woman to pass out from training having not successfully completed all the requirements).⁶¹ Once in a unit, women had to be commanded and disciplined in the same way as men. Crucially, commanders who failed to discipline women properly, sometimes out of fear that they would be subject to accusations of sexual harassment, were both censured and reassured: 'No supervisor needs to fear being prosecuted for sexual harassment if they have carried out their duties, and this includes disciplining their subordinates, with integrity and honour.'⁶² Since women were allowed into the infantry, the publication addressed the question of shared bivouacs in which female and males would be living in the closest proximity for long periods, averring in a phrase which summarized the entire document: 'Men and women must learn to live with each other professionally in the field.'⁶³ In order to ensure unit cohesion and effective battlefield performance, the Canadian army has sought to prioritize professionalism, focusing on practical competence and the performance of military drills over social presumptions about gender. In this context, heterosexual fraternization was singled out as one of the most dangerous forms of personal interaction between soldiers since it threatened to reintroduce precisely those cultural values and expectations, perhaps appropriate in civilian life, into the professionalized military.⁶⁴

Over ten years after this report, the organizational culture of the Canadian army has evolved considerably. Certainly, some male soldiers still reject the integration of women but there is ample evidence that the professional competence-based solidarity, which is so critical to female integration, has begun to crystallize. Indeed, many male soldiers have not only fully accepted women into the military but they have also explicitly appealed to professionalism as the criterion of that inclusion. Thus, a senior regimental sergeant major in the Canadian forces, with three decades of experience, saw the entry of women as entirely positive. He had been deployed to Mount Igmar in Bosnia as part of UNPROFOR in 1992. The unit of which he was then part had refused categorically to allow female Canadian soldiers into its base on the grounds that they would jeopardize the cohesion of the unit on a challenging mission. Canadian soldiers and officers were convinced at that point that the entry of women would be corrosive. Yet, he had subsequently watched the increasing integration of women into the army, including deployments to Kandahar, the dangers of which far exceeded Bosnia, without the predicted disaster. On the contrary, as trained professionals, women were completely justified as members of the Canadian Forces and infantry soldiers.⁶⁵ Indeed, in the Canadian army, there is widespread and loudly articulated respect for particular female soldiers and especially for Captain Nichola Goddard and Major Eleanor Taylor. Captain Nichola Goddard was an artillery forward observation officer who deployed to Kandahar in 2006. She was very highly regarded by her colleagues, who were openly devastated when she was killed in a fire-fight, when her armoured vehicle was struck by two rocket-propelled grenades, on 17 May 2006. Similarly Major Eleanor Taylor, a company commander with 1 Royal Canadian Regiment, has been widely identified as having the potential to become the first female infantry battalion commander. Significantly this appointment is not seen as an cynical example of positive discrimination as a result of equal opportunities policy but entirely on her proven merits as an officer; she successfully commanded an infantry company in Kandahar in 2010. Indeed, a male captain not only confirmed her professionalism but actively stated that he would want to serve under her.[66]

There is evidence that professionalism is displacing masculinity as the prime basis of cohesion in the Canadian forces. Significantly, female soldiers and officers have confirmed this development in their own experiences in the armed forces. Although there continued to be resistance from some men, the professional ethos, identified by the army in 1998, has been evident in their career. They had been primarily judged on their own individual competence, not their gender. An infantry captain who had commanded a platoon in Afghanistan was confident of the change:

I don't think there is a huge bias now. Several years ago, there was maybe a bias but there are now no issues. It's not about gender when it's about war, it's about training. It's automatic. It's being instinctual and doing things instinctively. It's all about performance. It's a volunteer army and who dares to put that much restriction on a volunteer force. These people [women] are willing participants.[67]

Indeed, this captain noted that warfare itself might have changed so that it required less traditional masculine qualities. War was no longer 'about a race to the sea. It was about technology. It was no longer a one on one fight but about control and coordination.'[68] Other Canadian female soldiers concurred that professionalism was the central basis of their integration. At the very outset of a discussion, a francophone lance corporal confirmed: 'If you want to do the job, there can be no gender. If I can do the job, I am respected as long as I can pull my weight.' She continued later in the discussion: 'It is about having respect for yourself and your peers. I'm so different at home and at work. In uniform, I am no longer Alice [name changed], I am Trooper Jones [name changed].'[69] A lieutenant who had served in the very demanding and traditionally male role as a forward observation officer in the artillery concluded: 'In the combat arms trades, you are equally accountable, it is task-oriented and task based; it is about meeting the standard.'[70]

Interestingly, some female members noted a marked difference between the regular and reserve forces especially in the 1990s. It might be expected that, as part-time civilian soldiers, reserves might be more open to the presence of women since they worked with women as equals in the civilian sector. Yet, informants had quite the opposite experience. The reserves tended to display the most overt sexism. One lieutenant recorded her experiences: 'In the reserves, I was the brunt of everything. If they could make a joke about women, they would. They would blame women. They would accuse you of having "sand in your ovaries".'[71] She continued: 'In the militia, there was training on the weekends and there was an instructor there whose sole intent was to break females; he was firmly against women being there. He deliberately pushed physical standards above what the girls could achieve, knowing that 98 per cent wouldn't do it and they were verbally abused.'[72] It would be unwise to presume that all the reserves were characterized by this attitude either in the 1990s or now. However, the experience of the female officer is useful because it seems to illustrate not only cultural development in the Canadian army (her subsequent experience in the regulars has been quite different) but also that in a less trained, less professional force, masculinity remains important as a means of generating cohesion. Indeed, it might be possible to suggest that, insecure about his professionalism and the professionalism of his reserve unit, perhaps, this NCO felt compelled to fall back on his masculinity to affirm his own status. For him, the acceptance of women would demonstrate that the reserves were weak and unprofessional. Interestingly, female officers noted that apparently gendered language was still employed as a form of 'hacking' (interpersonal banter) in the regular army. Regular soldiers would ask female soldiers whether they had 'sand in their ovaries' or 'do you want me to carry your purse for you?'[73] Yet, while in the 1990s such remarks were explicitly discriminatory and indeed intended as such, in the current era, they have become the ironic mockery between professional equals. According to these female soldiers, the free and playful use of this gendered language demonstrates precisely that they were considered as soldiers first and not as women; consequently, like hair colour, province of origin, or hockey team, their sex was an irrelevant marker of difference. It could be joked about and used as a source of ribaldry precisely because it did not matter: 'Everyone sees each other as equals and they use it [gender] as ammo.'[74]

The institution of a professional ethos within the Canadian army has, of course, subjected female soldiers to considerable pressures. Since they are mercilessly judged on their performance, with no account accorded to their sex, they must demonstrate their competence. Although every aspect of their performance, especially their leadership, is of major concern for these soldiers, the question of physical strength and fitness is not unnaturally one of the main foci of their attention. It remains a major hurdle for them, especially since a failure of fitness is so palpably obvious; a soldier fails to finish a run or is incapable of picking up heavy equipment. It is because of the physical demands that many women (like men) choose not to join the infantry. One of the problems which face women in the Canadian forces is the differential entry tests for men and women. Initial fitness, before entry to the Canadian forces, is assessed by the 'EXPRES Fitness Test' which is tailored to men and women. In this test, the men and women have differential standards, reflecting their different body shapes, because it has been (apparently) scientifically proven that a woman who passes this test will be able to pass subsequent army fitness tests to the non-gendered standard, especially after they have undergone the appropriate training. The problem with the gendered EXPRES Fitness Test is that not only does it give the impression that women were being favoured but that, once women entered training, they had to overcome a far greater gap between their entry fitness and subsequent infantry tests. They still had to do the basic 13 kilometre load carry, even though they had started from a lower fitness level. The result was that female soldiers had to work very hard to maintain their fitness levels throughout their career.

Yet, it is noticeable that while the strongest men were inevitably stronger than the female soldiers, competent female soldiers could often outperform male soldiers. Many male soldiers are not physically fit and the beer-bellied Canadian senior non-commissioned officer, 'the fat Warrant', whose athletic days are well behind him, is not so much a stereotype as a reality. Thus, in many cases, female Canadian soldiers are not only as professionally competent as their male peers but, determined to sustain their status and perhaps the status of women more generally, they are as fit or fitter than some male soldiers. A lance corporal in the cavalry had worked with a diminutive male lieutenant whom she was stronger than. At one point, when his weakness became embarrassing, she warned him: 'Dude, you have got to work out.'[75] Some have surprised their male counterparts by their ability to march. Nevertheless, some roles, especially in the infantry or certain branches of the artillery, are very challenging on purely physical grounds for women. Realistically assessing their physical capabilities, women often avoid these specialisms and, in training and operations, units will often utilize a division of labour so that the bigger men perform the particularly arduous physical tasks while female soldiers perform other vital duties. One soldier recorded that when she had undergone public order training, she was too small to be useful in the 'baseline' (the shield-wall); she simply could not physically keep aligned with men most of whom were a foot taller than her. Accordingly, she was used to fire CS gas canisters.[76] Clearly, such a division of labour need not break down simply on gender lines; smaller men would also be utilized in different ways.

The fact that professionalism has become the criterion of membership in the Canadian forces, as women have been increasingly integrated into this competence-based order, does not mean that femininity has become irrelevant to interactions between soldiers. On the contrary, it remains an important reference point. The predictably discriminatory aspect of this will be discussed later but the inclusion of women has had a potentially surprising effect on military performance. Van Creveld has asserted that the presence of women must always undermine a military unit. In the case of Canada not only has the presence of competent women not affected combat performance but there is some evidence that their integration may improve the conduct of male soldiers. A well-known female infantry sergeant, who was one of the most respected female NCOs in the Canadian army, with much operational experience including operations in Afghanistan, noted that in the combat zone, gender was not always a determinant of performance. She noted that she had worked with lots of alpha males who 'gave it the big talk'. However, in combat, things were different: 'I saw a dozen guys run away, cringe or cower. They were pulled off the battlefield because they were a danger or a hindrance.'[77] Like everybody else, she had no idea how she would react to enemy fire but, to her surprise, she found it made her angry: 'When I was shot at, I got pissed off and I ran at the fire.'[78] She actively sought to retaliate. The example of a sergeant like this is important. Not only did it constitute professional performance but the fact that she was a woman had a sharp motivating effect on her fellow men. Effectively, they were being shown up as men because she, as a woman, was performing better than them. Other female Canadian soldiers recorded the same dynamic from their own experience. One lieutenant recorded an experience from training when some male soldiers in her unit failed to keep up on a route march and were complaining about the pace. The senior NCO used gender terminology to highlight their failure: 'Stephanie [name changed] is up there humping and you're back here whining.'[79] An infantry captain recorded a similar experience. She described herself as a 'huge advocate of fitness' and ran physical training sessions for her unit, especially because some of the men were overweight; one male soldier lost thirty pounds as a result of these sessions. After hard runs, she recorded how her men would approach and explain their reason for keeping up despite their exhaustion: 'There was no fucking way a woman is going to beat me.'[80] In a genuinely gender-less world, such motivations would be meaningless; it would make no difference whether a man was beaten by another man or by a woman. However, given that gender differences are still very marked in civilian society, it is not surprising that they remain in usage in the military. Yet, they defy van Creveld's claim that women always undermine military performance. In an army based on professional standards, in which the women can be as competent as the men, the presence of female soldiers does not necessarily humiliate the males but may demand higher performances from them because they do not want to be embarrassed by a woman. It is unlikely to be a universal phenomenon, but the presence of strong, skilled women in a unit may actually increase its performance level. In short, it may actually improve its cohesiveness.

There is little doubt that Canada and the Canadian armed forces are unusual. Canada is a bilingual nation with a vocal francophone community which demands political recognition. It is also a very multicultural society with naturalized immigrants from Asia and the Pacific regions. Accordingly,

Canada is a self-consciously liberal and tolerant civil society in which diversity and difference are recognized and respected. It seems plausible to suggest that the advanced gender settlement evident in the armed forces has been substantially facilitated by the wider consensus about diversity and equality. Certainly, some female soldiers suggested that the multiculturalism of Canadian society informed the attitudes of many soldiers.[81] Nevertheless, although certainly privileged, many of the experiences recorded by female Canadian soldiers were replicated in other western militaries. For instance, there was some evidence of successful integration of women into the Bundeswehr on the basis of professional competence which echoes the Canadian experience. As a German naval captain has noted: 'There are some masculine women, there are also some feminine women. In the first instance, they are soldiers however. What counts is that each can do their job.'[82]

In order to demonstrate a potential revision of gender relations in the armed forces in other western countries, it is perhaps useful to examine the most extreme cases, focusing not only on the infantry in general but on elite infantry regiments which have been widely seen as a haven of extreme forms of masculinity. Donna Winslow dissected the often shocking hypermasculinity of Canada's Airborne Regiment in the early 1990s and other scholars have noted an association between military elitism and extreme forms of masculinity. In the early 1980s, an extreme masculine subculture existed in the British Parachute Regiment which had close resonances with Winslow's findings and, indeed, in the Falklands War and in Northern Ireland, the Parachute Regiment was, perhaps not un-coincidentally, involved in a number of abuses. The German Airborne Brigades, themselves heavily masculinized, have similarly been involved in political and training scandals. Sebastian Junger records a comparable culture among the troops of the US 173rd Airborne Brigade, in which bullying and violence was institutionalized as discussed in Chapter 10. Descriptions of the US Marine Corps both in 1991 and 2003 record a similarly aggressive and physical culture,[83] and the British Royal Marines have also developed highly distinctive forms of male-bonding. In a somewhat bizarre scandal exposed on the internet in 2006, initiates to one of the commando units participated in bouts of naked fighting, with their arms wrapped in camping mats. Indeed, this peculiar ritual was only an extension of the Marine's long-standing convention of male nakedness.[84] Elite infantry units seem to demonstrate the most extreme and transgressive masculine cultures which seem to be closely related to their role as assault troops. They have the highest standards of selection, usually involving very intense physical tests, and they are tasked with the most dangerous and difficult missions. Consequently, they have a highly developed esprit de corps, which sometimes encourages arrogance in members of these units and a consequent contempt for others. Transgressive masculine rituals in the informal domain seem to be partly related to this need to assert their distinctiveness through excessive displays of bravado.

In addition, these elite infantry units have been heavily involved in operations to Iraq and Afghanistan. For instance, the UK's Parachute Regiment and Royal Marines have been at the forefront of operations in Afghanistan, their respective brigades deploying three times each to the province. It is true that by the time of the United Kingdom's withdrawal of combat troops by December 2014, some British battalions will have done an equal number of tours to the Marines and paratroopers but this does not invalidate the central claim that these regiments have contributed the most to the Helmand campaign, as their casualties demonstrate. Given the masculinity which has been evident in these elite regiments over a long period, and its enforcement through informal ritual, and the hard physical regimes which characterize them, these organizations seem to represent a good site for investigating whether women can serve as combat soldiers. They represent an extreme and outlying case which may usefully illustrate the possibilities for integration in the infantry more generally. It would, of course, be expected that the members of these elite forces, heavily socialized into a masculine lifeworld, would be most opposed to gender integration. Alternatively, if the male members of these regiments are potentially open to female accession, it would seem to represent a very significant transformation of gender assumptions.

On the basis of Rosabeth Moss Kanter's research on female corporate employment, it would seem plausible to predict that women would find it very difficult to integrate into the overwhelmingly male and very masculinist military especially on combat operations in Iraq and Afghanistan. In her work on corporations, Kanter suggests that the process of Weberian 'rationalization' has in fact always involved and presumed a 'masculine ethic'.[85] In modern western culture, men have been conceived as 'cognitively superior in problem-solving and decision-making' while women have been represented as emotional, sensitive, and caring, in line with their maternal role.[86] Consequently, women have been impeded from participation at the higher levels of management; the 'masculine ethic' has been invoked as an exclusionary principle. For Kanter, male managers engage in 'homosexual reproduction'.[87] In the face of organizational uncertainty and 'the need for smooth communication',[88] male managers prioritize trust and mutual understanding which is primarily presumed on the basis of 'similarity of social background and similarity of organizational experience': 'People who do not "fit in" by *social* characteristics to the homogenous management group tend to be clustered in those parts of management with the least uncertainty.'[89] Consequently, although there are exceptions, women are typically sequestered into human resource departments away from the corporate decision-making process.

In the light of this argument, it might be expected that the pressures which Helmand has exerted on the British armed forces would accentuate the homosociality of the infantry. Given the presumed masculine cultures of Britain's infantry and especially the Parachute Regiment and the Royal Marines, Helmand offers an ideal test-case of the extent to which females might be incorporated into the infantry. British military operations have typically been of company and very occasionally of battalion size, but for those companies and battalions involved, the fighting has been intense. Moreover, the complexity of the physical and political environment in which British troops are operating and the increasing use of improvised explosive devices in place of conventional combat has imposed additional burdens on the British infantry soldier. Helmand is as difficult and dangerous a theatre as any in which British troops have fought or are ever likely to fight. In the face of this uncertainty, Kanter's research from the 1970s would suggest that male soldiers, especially those in regiments which have a highly developed 'masculine ethic', would resort quickly to exclusionary practices in order to maximize not only the organization's success but their individual chance of survival.

In the United Kingdom, as already noted, women are formally excluded from the infantry; they cannot serve, for instance, either as Royal Marines or as paratroopers. However, as a result of the new complexity of operations where new specialist attachments, such as interpreters, dog handlers, bomb disposal experts, are now routinely required on any patrol, women have played an increasingly prominent role on the front line. They have been closely attached to the infantry in the combat zone, working alongside the infantry and in many cases performing traditional infantry roles. The attitudes of a selection of British infantry commanders are perhaps surprising. The Royal Marine major who commanded Lima Company 42 Commando, which acted as the Regional Command South's theatre reserve, and was therefore engaged in a series of intense offensive operations in Afghanistan, noted the new challenges of command in the current era: 'One thing that has changed is the importance of specialist attachments: CIMIC, RMP, CIED [civil-military cooperation, Royal Military Police, Counter-IED], medics and so on. Learning to manage their incorporation into your ORBAT [order of battle] is challenging. How do you command? I know how to command a company but how do you command this organization?'[90] The major added: 'And with that came females.'[91] The potential implication was that females represented an additional command problem and, indeed, the major did admit that having joined the Royal Marines he had probably made certain presumptions about the appropriate place for women in the military. However, his experience with female soldiers had led him to re-evaluate their role. He noted that during his tour, only two soldiers in his company never fired their weapon at the enemy: himself and his radio operator. Every single other member of his company had to fire directly at the enemy, including all his female attachments. The major stressed: 'My one takeaway is that women can play a part on infantry operations.'[92]

His peers offered a similar viewpoint. Another Royal Marine major who had been stationed in Sangin for six months, widely regarded as the most demanding and dangerous posting for British forces between 2006 and 2010, was categorical on the point. He was convinced of the ability of women to participate on the front line from the moment he saw a female medic give one of the Marines, who had lost his lower mandible from a gunshot wound, mouth-to-mouth resuscitation. The female medic kept the Marine alive. He noted that this medic struggled to carry the weight which the Marines typically had to bear but 'the blokes didn't care'.[93] They judged her by reference to the performance of her professional role as a medic, which she accomplished. Indeed, the major reduced the significance of individual strength as a criterion of military aptitude. He noted that some artillery pieces fire very heavy rounds. Formally, in the artillery there are gender-free lifting tests but, because of the weight of the round, many males failed the test; consequently the test was changed and these larger-calibre guns were now crew-loaded. His point was that the physical weakness of the average female soldier in comparison with a male might be irrelevant to military capability because cooperative measures might be introduced to overcome individual shortcomings. He emphasized that the disparity between the physical capability of female attachments and male Marines 'didn't make any difference'.[94] In point of fact, a third Royal Marine major serving in Helmand at the same time also had a 21-year-old female medic, Medical Assistant Class I Kate Nesbitt, attached to his company. Not only was she only the second female to be awarded a Military Cross for her treatment of casualties under fire but he reported that she was able to carry all the weight, some 80 pounds of equipment, in the intense heat of Helmand, even though she is only 5 feet tall. Indeed, perhaps the most striking thing about Nesbitt's extraordinary performance in Helmand was that she has a normal physical appearance. She is a healthy, active young woman but, although she ensured she was physically fit when she deployed, she is not and has never been an elite athlete. Her example suggests that the capabilities of the average woman might be well above those which are typically presumed.

It is not only the Royal Marines who seem to have revised their perception of the place of women in the military. There is some evidence that similar reforms are evident in other infantry regiments in the army, including the Parachute Regiment. Echoing the statements of the Royal Marines, a colour sergeant from the Parachute Regiment said that: 'We had a female medic. She was awesome. She carried the same weight as the blokes. She was doing her job, performing as well or better than the men. Why should sexuality affect cohesion?'[95] Indeed, he observed the potential hypocrisy of excluding women on the grounds of the nefarious potential of sexual relations with male soldiers:

'Gay men don't undermine cohesion. The Parachute Regiment has been full of gay men for years.'[96] It has been well known that one specialist weapons platoon in the Parachute Regiment has become a haven for serving homosexuals, even though this sub-unit is also widely regarded as the best in the British army at its role. He, therefore, regarded it as entirely conceivable that women could serve not only in the infantry but in the Parachute Regiment itself, despite its selection process and reputation. The colour sergeant emphasized only that there had to be equal standards for both men and women: 'as long as she passes the same course'.[97] For this soldier, it would be inappropriate to drop entry standards, but 'if a woman had the same capability, why not?'[98]

Significantly, for this non-commissioned officer, the possibility of women serving in the infantry was intimately related to the increasing professionalization of the British army. It was in effect a potential by-product of the appearance of 'FOB cohesion'. He maintained that as a result of operations and increased training: 'The British army is in a better state than it has ever been.' Indeed, he noted how this professionalism had manifestly improved the quality and maturity of junior soldiers. He had seen a young soldier, returning home after the tour, whom he presumed from his demeanour was a senior corporal with five to eight years' service. In fact, he was a private who had come straight from training out on operations: 'The young blokes are legendary.' Professionalism, where competence, training, and performance are prioritized above all other criteria, has been accentuated by operations in Helmand. In a period of rapid change, even regimental affiliation is becoming less important. For instance, in the past, the Parachute Regiment was notoriously elitist and he noted that even the cooks would be forced to go through airborne selection and training although they would never realistically make a parachute jump on operations or need the physical fitness required of combat paratroopers. Now the question was not whether he was airborne but, 'Does he make good food?' He concluded: 'The younger cadre are much more professional; the senior element are concerned about image.'[99]

Other infantry soldiers who had operated in Helmand had made compatible points on the basis of their recent experiences. For instance, a British army captain working with an infantry unit in Babaji recorded how female medics who had been inserted into the platoons during the course of the deployment had performed excellently and were considered to have been 'worth their weight in gold'. This is a potentially important observation because specialists, including medics, are typically integrated into units on pre-deployment training in order to ensure familiarity and cohesion with colleagues. Yet, in Babaji, unknown females could operate equally well with their male peers precisely because their individual professional skills were at the requisite level.[100] They were integrated on the basis of their professionalism, irrespective of their gender. Other soldiers recorded even higher levels of cohesion during tours: 'There was no difference in attitude between the men and the females and they couldn't wait for the platoon end of tour party.'[101] These statements indicate that cohesion is possible in mixed units, if professionalism becomes the basis of individual and collective assessment.

In the light of the importance of professionalism it is understandable why the Canadian armed forces and Canadian soldiers, including female ones, strongly believe that sexual relations between male and female soldiers are extremely corrosive. They undermine professionalism; 'no matter how competent you are, if you sleep around, you will ruin your reputation, not only your own but of all women.'[102] In fact, given the statement about homosexuality in the Parachute Regiment, it might be argued that sexuality may not be as immediately corrosive of military discipline and cohesion as is often assumed, though heterosexual relations seem to have different implications for the men and women involved from homosexual ones. However, these statements indicate that, minimally, women could be integrated to a far greater degree than ever thought possible in the past. Indeed, one British officer stressed that the objection to the service of women in the infantry was not the product of rational experiment or evidence but, like the Parachute Regiment colour sergeant, was the product of thoughtless traditionalism on the part of senior officers; 'It is anachronistic; it is an old duffer thing.'[103] In fact, while generals may take a more conservative view of the possibility of female integration in the infantry, at the level of senior middle management (half and full colonels), there is a growing consensus that women could serve in the infantry. It is by no means universal but some infantry colonels, who had commanded battle groups in Helmand including significant numbers of women, maintain that women could serve as infantry soldiers, on the basis of their experiences. The combination of advancing professionalization, improved female expectation and performance, and changing gender relations have facilitated a potentially historic change.

The attitudes of male soldiers and Marines will be crucial in challenging the homosociality which has characterized the military up to now. However, in understanding the new position of the women in the armed forces—and assessing the possibility of them serving in the infantry—it is pertinent to consider the experiences of female soldiers themselves; Medical Assistant Class 1 Kate Nesbitt is particularly useful here since she seems to represent the possibility of integration at its most radical. Nesbitt was not originally expected to serve on the front line. Trained as a medic, her original posting was to the medical facility in Camp Bastion, Britain's major base in Helmand, in October 2008. However, having spent two months in theatre in this rear area, a demand for a medic appeared which two male service personnel did not want to accept. Nesbitt volunteered to work in a patrol base, initially for two weeks. Following this period, which was successful, Nesbitt began to go on patrols, eventually being assigned to a Royal Marine unit; the Marines were short of medics and, as Royal Naval personnel, it was appropriate for Nesbitt to be assigned to them in the first instance. While attached to the Marines, she was completely integrated into a troop of Marines. Although she confessed she had 'never dreamt of it', she eventually did patrols and offensive operations with the platoon, clearing and sleeping in compounds on the front line in immediate proximity with her male Marine colleagues, coming repeatedly under fire. Throughout the period, she carried her own equipment, including body armour, weapon, medical kit, and water. While not as heavy as some of the Marine loads, she was carrying some 80 to 85lbs of equipment which, designed for men, was excessively bulky for her frame. Nesbitt emphasized the role which her company commander played in ensuring her acceptance; he communicated her attachment to the company as a matter of fact.[104] She may also have been advantaged since her family had close associations not only with the military but with the Royal Marines: her father and her brothers had all served in the Marines. However, crucially she records that she was completely accepted by the young Marines themselves. She noted that 'before I have arrived, there may have been [concerns about having a female attachment] but not while I was there'. Once she had been attached and demonstrated her ability to cope with the environment physically and to perform her role professionally, her status as a woman became irrelevant, especially when under fire and treating wounded Marines. Then, she was just a medic; not a female medic. In the patrol bases, the Marines provided Nesbitt with a private shower area, created by hanging a poncho, but apart from that small distinction, Nesbitt lived with and like a Marine for the final four months of the tour.

The sharing of these mundane basic conditions seems to have played an important part in Nesbitt's acceptance: 'There were eight of us in the PB [Patrol Base], 24/7 together. I was one of the lads. I slept in the same area, ate at the same time, did everything with the others and no one blinked.'[105] Canadian soldiers have affirmed the point. While civilians (and, indeed, military personnel who have not worked closely with women) presume that the lack of privacy and the different bodily functions must be a problem, female Canadian soldiers suggest that not only is it relatively easy to overcome the minor physical differences by small and discreet measures; they affirm that living together in close proximity transforms the dynamic between men and women. The bond became like that between brothers and sisters since the intimate conditions they were living under, and the demands of the operation, eliminated thoughts of fraternization which were possible in less pressurized and more segregated rear areas. One lieutenant drew a parallel between the experience of living together on exercises and on operations with her coeducational dormitory at university. There, precisely because everyone lived so intimately in dormitories, normal rules of gender privacy were relaxed and no one had relations with each other.[106] She noted the important benefits of living together:

Packs don't take care of people that are segregated. But they will fight for you if you have built a relationship. If you have created that loyalty, if you know the blokes. In Afghanistan, can you imagine if you were segregated. How would you survive? If you were the only female living alone or two females alone. You can't exist with one person—not even if it is your husband. You need a group of people for a healthy mental state.[107]

Indeed, Rosemarie Skaine, in her observations of the US military, has noted the importance of shared living and especially sleeping space. Integrated dormitories certainly pose potential problems in terms of fraternization and especially harassment but Skaine records that despite those potential risks and the cultural opposition to them, there is evidence that they normalize relations between men and women. She cites Brigadier General Mike Hall, former air liaison officer to Central Command during Operation Desert Storm: 'Hall lived in segregated dormitories. Women's dorms were some place else, some place you might like to sneak into.' After he graduated, the dorms became mixed and a whole new culture was formed. What he and his contemporaries had found titillating 'just became everyday life for people in that experience'.[108]

This was certainly Kate Nesbitt's experience. Not only did living with the Marines in patrol bases ensure that she shared the same hardships as them and that the Marines came to recognize her professional contribution to the group but the relationship between Nesbitt and the Marines matured as they coexisted. Indeed, she noted the difference in male–female relations in the patrol base in comparison with at Bastion: 'In the patrol base, you are always on patrol or sleeping or preparing. But there is so much manpower in Bastion. In a patrol base, having a wash meant getting clean and you only had a bucket of water. The only time off you had was eating and you were knackered all the time. In Bastion, the girls wear make up.'[109] Nesbitt observed:

In the patrol bases the relations were so close. We were a nine man team, there wasn't twenty minutes when you weren't thinking about the team. When I returned people said, 'Did you meet anyone nice?' In the patrol bases it wasn't like that. In the Patrol Base, I met eight lads and the ANA. Afterwards, it was so odd not having eight guys that close. Knowing everything about everyone. And you were too close for anything like that to happen and as a medic too busy. We shared everything together such as when we lost Private Kingscott; we all experienced it. And when we weren't working we were thinking of home; writing letters. In Tombstone (the base adjacent to Bastion), there was no guy in my room but in the patrol base, we did everything together. When I first arrived, they felt protective of me and they worried that I was ok but it was more that than girls and boys getting romantic. They were like family, like brothers. In Bastion, the officers were different but in the PBs it was all together. It was very professional. The lads were aware of that. I was under as much pressure as them: I was literally one of the lads.[110]

There are a number of interesting points about Nesbitt's statement. It is important to recognize that while operational pressure minimized the opportunity of fraternization, romantic connections were absent in the patrol bases not simply because there was insufficient time. By working, sleeping, and eating so closely together, the bonds between Nesbitt and the Marines changed. Professional co-dependence changed her status from a potential object of sexual interest into an equal partner. It is noticeable, as Nesbitt alludes, that where the work pressures are less and where women and men are separated from each other in Bastion, fraternization—and some problems stemming from it—has been evident.

In addition to co-location, shared hardship, and the professional dependence of the Marines on her, Nesbitt also records the importance of training. She underwent extensive medical training both in the armed forces and in periods of service in Derriford Hospital, Plymouth, in the United Kingdom to gain more experience of trauma injuries. However, she also underwent standard infantry training. On a patrol, there was no guarantee that she would not have to defend herself or to take the place of an injured colleague. Consequently, she underwent a Royal Naval Military Awareness Course and a two-week army course from which she learnt how to fire a rifle and basic infantry tactics such as section attacks and contact drills. She learnt fire and movement drills, the importance of suppressive fire, and, in theatre, she fired her weapon on a number of occasions. While in Helmand, the Marines also gave her further training on the General Purpose Machine-Gun, 0.5 Browning, and on grenades so that she was capable of a number of roles including riding 'top cover' on vehicles. In each case, the Marines trained her to use these weapons. Consequently, although she claims she became 'one of the lads' and it would be easy to interpret this as evidence of her incorporation into a masculine culture as an honorary male, something more interesting seems to have gone on here. The definition of masculinity which Nesbitt and presumably the Marines themselves used to define membership of the 'lads' in the operational environment did not reference aspects of masculinity which are typical and perhaps primary in the civilian domain—or in the military domain away from the front; namely sexuality. Rather, the masculinity of this group and the possibility of being a member of this masculine group—being one of the lads—relied on professional skills. Masculinity in fact referred to professionalism and as such it was an honourable status which might be conferred on any one, whatever their actual sex.

Although women remain banned from the infantry in the US Army and Marines, some officers have questioned this segregation. For instance, Brigadier General Draude who was a member of the Presidential Commission on the Assignment of Women in the Armed Forces observed: 'We were excluding the majority of the population, not because of lack of desire, or patriotism, or expertise, or character, but because of the way they were born.... Explain to me why it is right to exclude people because of the way they were born—when it's gender, but not race.'[111] Others have affirmed the point on the basis of their operational experience: 'The bottom line is, women came into the fleet, and by the time I got to the USS *Stennis*, this "no women in combat" line was bullcrap. I knew mothers with children flying F-18s into triple-A [anti-aircraft artillery] fire and dropping bombs... Get the best out of these people so they can develop personally and professionally, and get on with it.'[112] In fact, despite the formal legislation, there has been an unplanned and pragmatic accession of women into the infantry. For instance, Erin Solaro, a journalist who had previously served in the armed forces, noted that the official rule on women's exclusion from the infantry was regularly breached by the legalistic method of describing such female soldiers as attached to infantry units. On the basis of this informal integration, there have been a number of successful cases of mixed gender cohesion in combat, facilitated by the professional ethos of the US military; women are judged on the basis of their competence. Solaro describes the women of the First Engineers, 101st Forward Support Battalion, known as the 'lionesses', who were regularly posted to combat units as attachments. The commanding officer of a battalion to which lionesses were attached observed the appearance of women in the combat zone with striking phlegmatism: 'I don't think this is a door-opening experiment, what we've done here. It can't be used as the only case study for women in combat, but it is an interesting chapter.'[113] Solaro cites a number of other examples where women have been operating on the front line with men. One of the most striking examples which she draws upon is of two female military policewomen attached to a Special Operations Forces unit in Parwan in Afghanistan. Their experiences echoed those of Kate Nesbitt's. These women found the SOF teams highly professional in their orientation and willing to accept these female soldiers on a professional basis. Indeed, as a result of cross-training with their SOF partners, they were able to develop some of the requisite skills which they had not learned before and which were not officially open to women.[114] Interestingly, like Nesbitt, these women noted the importance not just of working with male soldiers but living with them: 'When I did a monthlong mission with NAVSOF [Navy Special Operations Forces], I shared a tent with them. It's much better. You lose privacy, but you gain so much more. You're close, you're more cohesive, it's easier.'[115] On the basis of her experiences, one of Solaro's informants wanted to transfer to the Special Operations Forces.

The US Marines formally ban women from infantry operations and some Marine officers affirm this ban rigorously, but there is potentially more openness than might be anticipated. A Marine major, serving in Regional Command South West Afghanistan in 2010, affirmed the exclusion of women from the marine infantry. However, he also noted that the US Marines had developed a female engagement programme with a platoon of specially trained female Marines. Female Marines from this platoon were embedded into combat units and had gone out on patrols and operations with Marine infantry units. In each case, the US Marines had avoided the legal ban on women being attached to the infantry by describing these female Marines as being only 'assigned' to these infantry units. The distinction was partly semantic. He emphasized however that 'the USMC [United States Marines Corps] is pretty tight overall: men and women mix'. Indeed, he noted that 'you stop being a Marine when you say you can't do something because you are a Marine';[116] his point was that if Marines were ordered to form mixed gender teams, they would do it. He had some scepticism whether female integration at this level would work. Indeed, a master sergeant, who had the appearance of a classic masculine Marine, affirmed the point about women, stressing that there was no separation of men and women in basic recruit training: 'Young men and women focus on the task. There is no difference between them; all the qualifications are the same. The kids who join, especially the females, are not there for the bonus. They want to be treated equally.'[117] While maintaining the ban on the infantry, senior United States Marine officers have explicitly emphasized the importance of training and professionalism in integrating women into the Corps. For instance, discussing integrated training including the US Marines' training exercise, the Crucible, Lieutenant General Van Riper observed: 'The key to building effective, cohesive, gender integrated operational units is in creating a training environment that builds progressively to that end.'[118] The result of this has been that 'Marines [male and female] see themselves as members of the same team committed to performing the same tough duties in the same dirty, mentally and physically demanding environment, and from that experience develop an appreciation of each other as professionals.'[119] In training and on operations in the USA, gender barriers seem to be breaking down and women seem to be increasingly accepted by the infantry (if not in the infantry) on the grounds of professionalism. Their competence to perform their role is becoming more important than their sex. Indeed, reflecting this new reality, in early 2012, the Pentagon changed the rules on women's participation in combat roles. While women were still excluded from the combat arms, they could serve on the front line *with* infantry, armour, and Special Operations Forces units.

The de facto integration of women into combat roles is a recent phenomenon for British forces involving very small numbers of women; the situation is very similar in the USA where a legal ban on integration was consistently breached in practice until the new 2012 legislation. Women are incorporated on the basis of their professional competence. Even in Canada and Denmark, which have total accession, the numbers of women in the infantry are very small. Currently, approximately 1 per cent of the officers and 2 to 3 per cent of the NCOs in the combat arms are women. During their tour of Kandahar in 2010, 1 Royal Canadian Regiment had three female soldiers; they were highly regarded. Significantly, one of these, Major Eleanor Taylor, was a company commander (in charge of over 100 soldiers), the others were a captain and a non-commissioned officer; they were performing important functions but in numerical terms they represented some 0.5 per cent of the force.[120] For these militaries, the accession of women is new but, as Joshua Goldstein, noted, there is one obvious historical precedent which may have some useful parallels to changing patterns of women's integration: the Soviet army in the Second World War. In her important study of Soviet soldiers, Krylova records that the authorities and male soldiers themselves were initially opposed to the integration of women; the prime female role was motherhood and women best served the military as 'noncombatant, medical, veterinary and technical help'.[121] However, following the great crisis of October 1941 when the Germans made a 500-kilometre-long breach in Soviet lines leaving the capital defenceless, women were called up. Many of the women were assigned to non-combat functions, freeing men up to fight. For instance, in April 1942, 40,000 women were mobilized into the air force but none went into combat roles: 15,000 took up clerical positions, the other 25,000 were trained as communication specialists, drivers, and armourers.[122] However, by the end of the war 120,000 women had served as combatants in the armed forces.[123] Many of these served in air combat functions but a proportion also eventually served as infantry soldiers. Krylova's analysis of these women is interesting and highly pertinent to the contemporary environment. Soviet women combat soldiers did not, according to Krylova's data, serve as normal riflemen; that remained the preserve of males. However, women were trained as infantry specialists. In particular, they served as signallers, machine-gunners, and snipers. In this way, they could contribute directly to the fighting effort without fundamentally undermining or, indeed, revising gender relations and definitions. In particular, women's participation was facilitated on the basis of 'a non-oppositional though still binary concept of gender'.[124] Women and men could serve on the front line together but women and men were still regarded as fundamentally different and having different functions. Consequently, against initial opposition, Soviet women were incorporated into the combat roles in minority specialist roles where, with training, they were able to perform competently.

The Soviet parallel is illuminating but, on Krylova's interpretation, the current revision of gender may be deeper. Krylova examines the way in which pre-war female parachutists were praised for their skill while their gender became irrelevant. For instance, Nina Kamneva claimed that 'all healthy people could perform such jumps' as long as they had trained themselves properly.[125] However, in the war itself, she claims that gender differentiation remained significant throughout even for those women who were accepted as specialists. They were still women seen as performing a legitimate but different job from men. The fact that they were immediately de-mobilized and their war contribution downplayed by the Soviet regime, which wanted women to return to their legitimate and primary role as mothers after the war, seems to provide evidence for Krylova's argument. Current integration in western armed forces seems to have exceeded a mere accommodation of femininity; the very distinction between men and women, their respective roles and capabilities, seems to have been eroded as professional performance has become the prime reference point. As a result of professionalization, women are not simply allowed to perform male roles as they were in the Soviet army; those roles may themselves no longer be defined primarily in gender terms.

As a result of operations in Iraq and Afghanistan, western forces have incurred significant casualties for the first time in several decades. The commemoration of fallen soldiers has especially since the twentieth century been an important national ritual and the commemoration of the dead of Iraq and Afghanistan has led to the revision of existing practices in western countries.[126] There are many interesting dimensions to this process but one of the most pronounced changes has been the need to remember female soldiers for the first time. One of the most striking aspects of female commemoration is the extraordinarily similar treatment accorded to fallen male and female personnel despite the historically unique experience of having females soldiers (as opposed to civilians) killed. Thus, Canadian female soldiers emphasized that their colleagues who had died had been honoured no differently from male soldiers. Although the public were shocked by the death of the first female soldier, Captain Nichola Goddard, on 17 May 2006, her colleagues grieved for the loss of a much loved and highly proficient comrade: 'I remember the first female casualty, Nichola Goddard. She was highly technically competent. It was not just a woman but a good soldier dying.'[127] Similarly, when Karine Blaise was killed by a massive IED in April 2008, she was commemorated as a soldier, in the first instance, not a female soldier: 'We had a female driver who died in 2008, Karine Blaise. When it happened everyone stopped. This could be you in a year and a half everyone was thinking. She was loved by everyone. We did not see it as a female dying but as a member of the Vingt-deuxs [22nd Royal Regiment].'[128] Supporting the corporal's understanding of her death, her public obituary did not reference her gender but described her as an energetic soldier who gave '100 per cent to every challenge she faced', demonstrating 'qualities of a future leader' and, therefore, 'respected by all members of her squadron'.[129] Even in the highly controversial case of Major Michelle Mendes, where accusations of murder have complicated her commemoration, statements released after her death emphasized her professionalism. Her commanding officer commented that 'She always strove to do her best and was respected for her professional knowledge and work ethic.'[130] A compatible pattern is observable in the United Kingdom. There, too, obituaries for male and female soldiers have been in terms of structure and textual context all but indistinguishable.

For instance, on 19 April 2011, Captain Lisa Head, a bomb disposal expert, was killed defusing an IED in Helmand. Paralleling obituaries for male soldiers, her commanding officer described her personal and professional virtues.

> Captain Lisa Head will be remembered by the officers and soldiers of the Regiment as a passionate, robust and forthright individual who enjoyed life to the full; be it at work, on the sporting field or at the bar. She was totally committed to her profession and rightly proud of being an Ammunition Technical Officer. She took particular pride in achieving the coveted 'High-Threat' status which set her at the pinnacle of her trade. Lisa deployed to Afghanistan with the full knowledge of the threats she would face. These dangers did not faze her as she was a self-assured, highly effective operator and a well liked leader. Methodical and professional in her work, she was always eminently pragmatic and calm under pressure. Having spoken with her prior to deployment she was motivated, enthusiastic and was looking forward to the challenges she would face. Her potential was considerable, and she will be an enormous loss to us all.[131]

It is probably little consolation to their families but these female soldiers have been remembered not as women but primarily as professionals, accorded the full status which goes with that role. Their sex, just like the ethnicity, race, or sexuality of male casualties, has become irrelevant in death.

It may be possible to draw some wider conclusions about this evidence. The experiences of these female soldiers are extremely interesting because, even though most of the women documented in this chapter served with, rather than in, the infantry, they all lived closely with infantry soldiers in the combat zone itself and, in addition to performing their specialist roles, learnt and carried out many tasks designated as infantry roles including firing their weapons in combat. In some cases, there may have been very special circumstances at work. Kate Nesbitt, and other soldiers like Eleanor Taylor, seem to have been able to integrate so well partly because of their own remarkable character. Nevertheless, despite these provisos and the fact that Nesbitt and Taylor are among a minuscule number of women who have served on the front line, they demonstrate categorically that some women do have the capacity to operate as infantry soldiers. They are able to withstand the physical rigours of campaign, can conduct the appropriate military drills, and are able to contribute to the collective performance of the small group, augmenting rather than undermining cohesion. It seems difficult to draw any other conclusion from the experiences of these female warriors and the emergent attitudes of their male colleagues; effective military performance does not necessarily require homosociality. Indeed, perhaps the common presumption that the distinctive solidarity of all male groups is essential to military service may not be true. In alternative cultural and historical circumstances such as those in the professional armies of the twenty-first century, it may be possible for men and women to form competent groups even in the domain of combat itself.

CURRENT OPERATIONS: DISCRIMINATION

It would be easy and perhaps convenient to conclude that a small minority of women can and have been incorporated into the infantry over the past decade or more (in the case of Canada, the Netherlands, Denmark) and that their incorporation represents the displacement of a masculine military culture with a professional ethos. Their experiences represent the beginning of an inevitable process of ever increasing female integration. As the evidence presented above suggests, there seems to have been an important change in gender relations in the armed forces which questions some of the more assertive elements of van Creveld's or Enloe's arguments. Yet, it would be wrong to presume that the accession of women into the combatant role is, has been, or will be, unproblematic. On the contrary the pattern of professional integration is highly uneven across different militaries and within the armed forces of any country.

Rosabeth Moss Kanter's work has already been discussed and other feminist work on organizational sociology is useful in orienting the enquiry into female accession into the military. In particular, while the evidence presented above suggests the possibility of integration, that literature would anticipate the persistence of a strong masculine ethic, which is antithetical to female inclusion, in such an overwhelmingly male institution. Organizational sociologists, such as Joan Acker,[132] Judith Wacjman,[133] and Jeffrey Hearn,[134] have argued that not only do men operate to exclude women deliberately from employment opportunities in a manner consonant with Kanter's analysis but there are underlying cultural presumptions which impede women's accession. For these scholars, the organization itself is invested with cultural significance; it is accorded a sexuality and, of course, because it is male dominated, that sexuality has been masculine. Consequently, the workplace has come to be interpreted as symbolically male, contrasted with a female other which is seen as polluting and dangerous. Organizations have developed a lexicon to identify good and bad practice which until recent changes in legislation was a manifestation of this symbolic gender order. In the light of this symbolic construction of the organization, where the male is potent, the female corrosive, it has become very difficult for individual women to work and be promoted in this environment. It transgresses cultural expectations.

It may be possible to observe a compatible sexuality of the military organization. Because soldiers are predominantly male, the military hierarchy has become infused with gendered symbolic meanings which create a collective reaction against the participation of women. Paul Higate's work on peacekeepers is pertinent here.[135] He has shown how concepts of masculinity inform the self-identity of these soldiers and their collective practices, leading in some instances to cases of sexual exploitation. Higate deliberately explores some of the more extreme aspects of the sexuality of the military organization but his point is much more general. Masculinity and a series of practices associated and legitimized by it are inscribed in military culture. Moreover, the presence of women de-stabilizes these shared masculine understandings and collective definitions on the basis of which male soldiers have organized themselves in the past. Women represent a disruption to the social imaginary and, as such, their entry generates resistance not because they cannot individually perform their role but they threaten to undermine the collective sense of normality and unity in the military organization. Indeed, Judith Stiehm noted the point in her study in the 1980s. She appositely asked: 'How can one distinguish between male culture and military culture?'[136] Indeed, Victoria Basham has convincingly shown that, in Britain, while the Ministry of Defence has introduced formal legislation, the culture of the armed forces remains discriminatory against ethnic minorities, gays, and women.[137]

Despite the fact that women have been active in the combat zone and there have been considerable changes in the attitude towards female service personnel, the overwhelming consensus in the US Army and US Marine Corps is that exclusion of women from the infantry is appropriate. One army major summarized the general position against women's accession into the infantry: 'Why would you voluntarily want to make your units weaker when you are going into combat?'[138] Although progressive views have been recorded, a number of British Royal Marines were also deeply sceptical that women would be able to serve in the infantry. A decorated corporal stated on the basis of what he had seen in Helmand with a female medic attached to his unit that 'there was no place for women on the front-line'; she could not physically cope with the demands of carrying stretchers with large Marines on them.[139] This is not an isolated view. Although he fully recognized that male medics also often failed in combat, another Royal Marine sergeant stressed not the physical weakness of the female medic with whom he had served but her psychological frailty. She blamed herself (wrongly) for the deaths of two Marines struck by an IED while taking cover on a compound roof and emotionally collapsed: 'she went into a ball of chalk.'[140] Clearly, everyone was distressed by the deaths of their colleagues but the sergeant stated that it was simply impossible to allow such traumas to affect performance. In both cases, these observations were made by professional Marines in their twenties who were in no way unintelligent or unthinkingly sexist. Indeed, other American soldiers have explained their opposition to women's incorporation into the infantry. A particularly useful resource is Jason Hartley's statement on the matter. Hartley was a national guardsman called up to serve for a year in Iraq in 2004 but was perhaps unusual in his broadly liberal political

The Female Soldier

views and his own self-consciousness. Nevertheless, his position on women was absolute.

> The topic of women in the infantry is a very tricky one and is the one thing in my personal philosophy that I am sexist about. I see that there are two reasons why women are not allowed in the infantry (or most other combat arms jobs). The average grunt is fairly in touch with his primal self and therefore wants generally only two things: to fuck and to fight, in that order. And the main reason they fight is to be tough and therefore attract more women with whom they can fulfil their desire to fuck. As soon as there are any women within spittin' distance, prime directive number one kicks in, and all things, especially job discipline go straight to hell.[141]

According to Hartley, the male sexual drive eliminates the possibility that men and women can interact with each other professionally; cohesion is therefore impossible within mixed gender groups. James Webb, a retired US Marine officer and former secretary of the navy has made the same point, differentiating ethnic integration from gender integration precisely because of the attraction between the sexes: 'No edict will ever eliminate sexual activity when men and women are thrust together at close quarters.'[142] Hartley also emphasizes the physiological impediments to female accession.

> The second reason women are not (and should not) be in the infantry is as equally based in the primal side of human existence as the first. Women can't do a grunt's job worth a shit. I wish I were wrong about this because it goes against all my notions of gender egalitarianism but it's the truth ... Your average female will not push herself in training the way a man will. I've seen guys in training literally push themselves until they die. I saw a guy finish a PT test two-mile run then collapse and die from a burst aorta ... And I haven't even got to the fact that your average man is much physically stronger than your average female.[143]

To their relative weakness, Hartley appends further physiological disadvantages including menstruation, pregnancy, and breasts.[144] Indeed, even Judith Stiehm noted the physical disparity between men and women. In the early 1980s, the highest women's score on the West Point physical fitness test would have been a man's C- and 87 per cent of women would have failed.[145] Hartley's position represents a common and perhaps dominant view in the United States.

Perhaps most telling are the views of some Canadian infantry officers, precisely because integration has been most advanced and most successful there. Above, it was noted that one Canadian infantry officer openly hoped to serve under Major Eleanor Taylor. He had, nevertheless, regretfully concluded on the basis of his experience and, despite his own political beliefs, that it was best for women not to be in the infantry. This was not because some women could not do the job but that, in his experience, fraternization was almost inevitable. He personally knew of a fellow female officer who had an affair with one of her senior non-commissioned officers and this breach represented such a profound corrosion of professionalism that he now viewed integration as damaging. His desire to serve under Major Taylor seemed to be justified on the grounds that Taylor was a unique female officer.

In the German army, despite the formal integration of women, an exclusionary attitude is predominant in the infantry. An experienced captain in an airborne brigade, who was currently training infantry soldiers at Hammelburg Infantry School, confirmed the excellent performance of women in Kunduz as medics: 'We had four females in the PRT to support the sergeant-major there as medics. They did combat drills too and they were well integrated. The medic who worked for us had excellent skills and she was our first point of contact for us: she was very good. They saved lives.'[146] He even emphasized that in complex stabilization missions, women could perform command functions. The Bundeswehr has employed a female company commander and platoon commander in Kunduz. However, despite their competence, the captain noted: 'But as infantry soldiers—I just can't imagine it.'[147] One of his colleagues, a sergeant major who ran the junior non-commissioned officers' course, which trains twenty-five soldiers, two to three of whom were always female, in infantry tactics, confirmed the point. He had seen numerous cadres go through his training course but in three years (i.e. twelve courses) he had 'only seen one who had the ability to command, who had the charisma to command a section of soldiers'.[148] He explained: 'They have the capability to begin the course but they develop physiological problems and then there is the problem of all living together.'[149] For this sergeant major, women struggled to perform as infantry soldiers because individually they were insufficiently strong and they found it difficult to live in close proximity with males; they could not integrate. In her research into the Bundeswehr, Andrea Jeska has affirmed the point: 'Theoretically women can do everything in the Bundeswehr. In practice, there are military tasks for which women purely anatomically do not have the necessary strength.'[150]

According to Katia Sorin, although the French military integrated fully in 2000, very serious barriers to female accession remain. Even when men and women have the same occupation, they are often assigned different tasks. Female technicians in the navy are often ordered to perform administrative duties: 'Many others deplore assignments to tasks that are seen as strictly feminine, such as bearing a cushion during a ceremony, working in the cloakroom or in reception at a unit open house, being the female delegate at an official dinner, bearing the responsibility for making coffee.'[151] Moreover, female French service personnel believed that they were subject to more stringent assessment than their male peers and that small mistakes would be punished: 'They often have to justify the legitimacy of their presence, their professional competence, or even their family choices.'[152] For Sorin, the atmosphere at the military academy at Saint-Cyr, where officers are trained, was particularly concerning: 'it amounts to a war between the sexes launched by young men and not by the young women's choice, a war that is moreover highly unequal—a handful of young women confronting a hundred or so young men.'[153] Sexual jokes and innuendoes with an explicit aim of degrading and marginalizing females are commonplace. Sorin's examples are perhaps extreme. However, the inclusion of women into the infantry is at best marginal. Very small numbers of women serve in the infantry and never as front-line soldiers; they act as clerks and drivers. In December 2011, a company from 15/2 Infantry Regiment were undergoing validation training at CENZUB. The company, and some additional troops from the regiment, consisted of approximately 200 soldiers; two were women, one the company clerk, the other a driver.

With this low percentage of women, which on Kanter's gender ratio scale confirms that women remain as tokens in the armed forces, harassment of women as a minority group would be expected. Indeed, in the US forces, in particular, harassment has been a recurrent problem. In her important work on integration in the 1980s, Judith Hicks Stiehm recorded extreme forms of bullying, harassment, and sexual abuse (including rape) among the US armed forces.[154] A statement from a female US air force mechanic who served from 1974 and 1980 records not only this abuse but also the presumption that any female who did not make herself available to men was necessarily a lesbian.[155] The acronym WAF (Women's Air Force) was offensively altered to 'We all fuck' to illustrate the point.[156] The Tailhook convention and Aberdeen Proving Ground in the 1990s remain infamous episodes but routine bullying, abuse, and assault was widespread at that time. Carole Burke, for instance, records everyday abuse of women in the US navy in the 1990s who were routinely denigrated for their weight and putative promiscuity with the acronym WUBA (Women Used by All).[157] Similar attitudes are evident in the Bundeswehr. On a Bundesmarine training ship, the *Gorch Foch*, not only are there rumours of extensive fraternization but male trainees play a game called *Eulenschiessen*; they place money in a pot and whoever seduces the most unattractive female colleague on the ship wins the cash.[158] In the promulgation of these vindictive stereotypes, the sexuality of the military organization manifested itself in direct individual aggression.

Harassment remains a serious and continuing issue in the military. The pattern was evident in recent campaigns in Iraq. Solaro describes the actions as those of 'the small percentage of real criminals' or others who 'think their manhood depends upon women's subordination'.[159] Indeed, during her research in Iraq, Solaro felt physically threatened by certain men while staying in transit accommodation.[160] These patterns are evident even in the Canadian military. Although the major threat of rape on operations is deemed to be the Afghan security forces and interpreters of whom Canadian female soldiers were deeply wary (to the point of ensuring that they were obviously armed in their company and were accompanied at night near their positions), sexual abuse is a reality of which they are conscious. Harassment is even more common. One corporal in the artillery recorded a distressing incident of harassment which had occurred early in her career. When she joined her unit as a new soldier with a male colleague, with whom she had trained and whom she considered a friend, she soon discovered that he was spreading malicious lies about her. He claimed that she 'did sexual favours for cigarettes'. It was difficult to know precisely his motivation for this defamation. It is possible he himself was attracted to his victim, who did not reciprocate his interest, or that he felt threatened by her professionally. Whatever his rationale and despite the absurdity of his claims, the rumour was potentially damaging, even disastrous, because it cast aspersions on the character and professionalism of this female soldier. Even if the cigarette rumour itself was not believed, it implied a sexual availability which might be easily accepted. The soldier was herself completely isolated as there were no other females in the unit in whom she could confide. Consequently, she took matters into her own hands and bravely challenged her abuser, physically confronting him: 'I just slammed him.'[161] She physically seized him and warned him that if he did not publicly retract what he had said she would be forced to 'take another direction'; she would have to put in a formal complaint (which she would, of course, win). He recanted his slander.[162] The female soldier's reputation was protected, though not before much personal anguish. The incident demonstrates that even in a force as egalitarian as the Canadian army, discrimination, harassment, and abuse are nevertheless present and they are by no means limited to incompetent female soldiers. Indeed, even when there is not explicit harassment, female Canadian soldiers noted that gender could be used as a form of abuse. Accordingly, while a poor male soldier was just an incompetent or unfit soldier, the sex of female soldiers became a factor in their inabilities; they were inadequate because they were women. Even though there are many fat male soldiers in the army, one officer noted that 'Female soldiers who are overweight are treated significantly different to men. People ask: how did she get into the army?'[163]

At the same time, the consensual sexual pursuit of women is very common. One female British signals officer noticed the problem when she was in Baghdad: 'We had a drama in Baghdad when a girl arrived on her first tour. She was flooded with attention. My regiment was fine; we were used to females and her presence did not give rise to fantasies. But for the Royal Scots, the Irish and the SF, it was "game on".'[164] There was no suggestion of harassment in the account but the episode illustrated the male soldiers found it very difficult to treat an attractive female professionally. This may have been flattering to her but it, in fact, only affirmed male concepts of femininity and the masculine norms of the military.

These is clear evidence of self-consciously instrumental opposition to female integration. However, this masculinization of the armed forces can take an apparently more benign and subtle form. Indeed, organizational sociologists have consistently suggested that it is often the more unacknowledged cultural preconceptions which are most difficult to remove and most powerful in their discriminatory effects precisely because they pervade an organization and the interactions between its members. Consequently, despite professionalization, masculine self-conceptions remain central to the motivation of male soldiers. As a result of intense operations in Iraq and Afghanistan, numerous memoirs have appeared in the last decade. These have proved a useful resource in documenting experiences on operations. However, these memoirs have also illustrated albeit obliquely some of the underlying motivations of professional soldiers. Two popular works stand out here as particularly useful, Craig Mullaney's *The Unforgiving Minute* and Patrick Hennessey's *The Junior Officers' Reading Club*. Both works are written by officers in the US and British armies who were educated at Oxford University and went on to serve in Afghanistan. The authors, as highly educated, self-conscious, and literate individuals, are perceptive and sceptical observers of their wars and the armed forces of which they are members. Nevertheless, the dedications of both books are revealing. Explaining the title of his book, Mullaney cites the final lines of Rudyard Kipling's poem 'If': 'If you can fill the unforgiving minute, With sixty seconds' worth of distance run, Yours is the Earth and everything that's in it, And—which is more—you'll be a Man, my son!' The frontispiece is appropriate because the memoir reveals that for Mullaney military service was not only about proving his manhood but also demonstrating his worth to his estranged father. Following a fire-fight in which one of his men was killed, Mullaney is tortured with remorse that he had failed as a combat leader and as a man; he believed (probably wrongly) that his mistakes caused the death of his soldier.

I sat down, propped my elbows on my knees, and gripped my ears as I sunk my head and sobbed. I closed my eyes and said Our Father. But where was my father now? Who was going to tell me I had done my duty? What would he have done in *my* boots? He never answered my letter. Did he even care anymore what happened to me? Would he be there, like O'Neill's father, to accept my casket?[165]

At the heart of Mullaney's story is a need for recognition from his father. Patrick Hennessey is less revelatory about his motivations for volunteering but his long dedication is perhaps the most illuminating and honest part of the book. Hennessey dedicates his work to his maternal and paternal grandfathers, one a retired cavalry officer and veteran of Normandy, the other an academic who, as a conscientious objector, had driven ambulances during the Second World War. Both had in different ways proved themselves in the crucible of warfare. Hennessey concludes: 'They were towering, inspirational men: a fine soldier and an outstanding scholar. I am neither, but dedicate this book to them to show how I tried to live up to them both.'[166]

It is important not to exaggerate the causal effect of personal and potentially psychological motivations for joining the armed forces with subsequent service or performance in combat. There are a myriad of powerful mediating factors which need to be considered. However, the close but entirely accidental similarity between the motivations of these two officers, one American and one British, is striking and suggests that the need to prove one's masculinity remains an important factor in recruitment. The armed forces and war seem to retain their masculine allure, even for the highly educated. Despite the role which masculinity may play in attracting young men to join up, there is evidence that as a result of professionalization, it has been possible for many of them to accept the accession of women into the military and even into the infantry. Proving themselves to themselves as men seems for many of these individuals to be possible in the company of women. It has become a personalized challenge rather than a communal male enterprise. Nevertheless, there is also unignorable evidence that the armed forces remain masculinized organizations, in symbolic opposition to the female. In their pursuit of masculinity, male soldiers necessarily conceive of women as polluting.

Perhaps surprisingly, in the United Kingdom, these covert techniques of exclusion operate with particular force among the officer corps. In the United Kingdom, British female officers are significantly disadvantaged in terms of promotion because of their exclusion from the combat arms and their under-representation in the combat support arms. Overwhelmingly, the senior ranks in the British army are dominated by officers who are from these arms and it is very difficult to be promoted beyond one-star rank from combat service support branches. This structural difficulty is compounded by the expectations of male officers. For most male officers, especially older men, their only point of reference in terms of women is their wife. Consequently, operating with traditionalist assumptions of the feminine role, they find it difficult to comprehend and command female officers. Female officers, especially those in their thirties, are frequently asked, 'Are you going to get pregnant?' or, more obliquely, 'Are you going the full two years [the normal cycle for an appointment]', or warned, 'Don't get pregnant when you are working for me!'[167] In some cases, senior officers seem to want their female subordinates to act like their wives. 'Senior officers want flattering. But when a junior female officer objects on moral grounds, it is not easy. Indeed, senior officers often expect to flirt with her. One female officer, I knew, used to remove fluff on her commanding officer's shoulder. He loved it and she was reported as an excellent officer.'[168] These unprofessional and often sexualized presumptions about femininity produce anomalously gendered interpretations about professional conduct. Thus, one female informant reported that, as a major, she had developed a very close working relationship with her sergeant major. They operated closely together, including sharing accommodation on exercises, and a (platonic) friendship developed between them. In normal circumstances, where two males were involved, a relationship of this closeness between a company commander and his sergeant major would be regarded as ideal. However, in the context of the professional relationship being between a female and male, members of the battalion began to question whether the two soldiers were not having an affair. Indeed, the commanding officer was reported to have said, 'My wife would not behave like that.'[169] The informant acutely observed that the commanding officer was not necessarily sexist but his only point of reference in terms of females was his wife, a former rural schoolteacher. In this case, the presence of a female in the regiment created confusion among fellow officers about professional and personal relations. For them, there was an unacknowledged but powerful connection between military service and the special bonds it demanded of soldiers and masculinity. Male soldiers could not conceive of these professional relations outside of this gendered vision. They had sexualized the organization and to the detriment of a wholly innocent female major damagingly accused her of undermining professional standards.

There are more obvious examples when femininity is seen as polluting or demeaning for men, in a manner consistent with van Creveld's description. Against expectations, one informant from the British Parachute Regiment recorded that he and by implication his fellow paratroopers were open to the possibility of female infantry soldiers. Other evidence from recent operations suggests a more predictable reaction to women from this regiment. Thus, a female major reported how a colleague of hers, from the RAF, had been attached to one of the parachute battalions during an exercise in Kenya. The female RAF officer was wearing and using only issue kit, including black temperate boots, as might be expected for an individual whose primary role was not in the field; she was also not responsible for what she was issued. However, throughout the exercise the paratroopers mocked her for her equipment and in the end, the adjutant of the battalion (a peer of the female officer) sided with the soldiers and allowed this public mockery to continue. In the infantry and especially the paratroopers, the possession of the correct equipment and the appropriate wearing of it is invested with great importance. Indeed, the term 'ally' has been developed to define appropriate soldierly costume and deportment. The paratroopers typically denigrate any other soldier who fails to fulfil their concept of what is 'ally' in order to highlight their sense of distinction. Certainly, some of this generic status assertion may have been at work with the RAF officer in Kenya but the female informant suggested not implausibly that the mockery was a disguised but nevertheless quite deliberate form of sexism. It involved general abuse of unmilitary deportment to discomfort an attachment whom the paratroopers did not want present not because she was unsoldierly but because she was female. The mockery was a means of cleansing themselves of the pollution of femininity and reaffirming their own masculine solidarities.[170]

Indeed, confirming the point, the informant experienced a very similar reaction among other paratroopers. At university, she had been friends with a male who was now an officer in the Parachute Regiment. They were deploying together to Helmand and had seen each other on pre-deployment training. However, when he found out that the female major had been assigned to his own battle group, 'he looked at me with horror'.[171] The facial expression indicated that the presence of a female in the same battle-space and even, perhaps, the same forward operating base or patrol base would undermine his status. Two fears seem to have been at work here. He seems to have been discomforted by the prospect that he would be forced, by the obligations of friendship, to interact with a non-airborne female in front of his men, undermining his credibility with them. At the same time, he also seems to have been concerned that his experiences in Helmand would be demeaned by the potential presence of a woman, even though she was a friend, because it implied that the environment could not be that harsh. If female soldiers could withstand the pressures of Helmand, paratroopers could hardly claim a distinctive masculine status on the basis of their performance in this theatre. The look of horror was the fear that the very pursuit of masculine status in his own eyes and those of other men which motivated this individual to join the army had been denied. His reaction seems to represent a negative but entirely logical extension of the motivations outlined by Hennessey and Mullaney. Carreiras has recorded a similar response among Portuguese Marines at the prospect of serving with women: 'If women come here, I will step on my beret.'[172] For this Marine, the prime symbol of his masculinity—his green beret—would be defiled by feminine presence.

Perhaps most instructively in terms of the continuing sexuality of the military organization, despite her overwhelmingly unproblematic incorporation, Kate Nesbitt recorded similar incidents and attitudes. Nesbitt recorded one infantry captain with whom she worked who adopted a similar position to her to those articulated above: 'The captain in there was old school. He was not happy about it [having a female medic]. They had a medic on leave and other people had said to him "Kate is good" but he still said no. After I had worked for him, it was reported to me that he had said, "As much as it kills me to say it, she was good. But I wouldn't do it again."'[173] Nesbitt confirmed: 'He did treat me professionally. He was aware of the ANA [Afghan National Army], for instance; he took that into consideration when taking me on patrol. He was old school though. He was not nasty. There are some of these in the army. It is probably because they are so used to working with men for such a long time. They are just not used to working with women.'[174] It is clear from this statement that the male officer was informed by concepts of gender which had nothing to do with performance but to do with arbitrary classifications and finally masculine status. Interestingly, Nesbitt reported that this captain had some photographs of his time in the patrol base, and other soldiers had observed that the presence of a female attachment was potentially dangerous here. If Nesbitt featured in any of the photographs, it would be assumed by both civilian and military viewers that 'it could not have been that hard—there is a girl there: when my wife or girlfriend see it, what will they think if there is a young blond female in it.'[175] In these cases, the presence of women is not rejected because it has any adverse effect on cohesion or performance. On the contrary, the only danger which female deployment poses is to the status of male soldiers in their own eyes and the eyes of their military and civilian colleagues, friends, and families. The presence of women denigrates these men not because women are incapable but because the armed forces remain sexualized organizations in which gendered concepts still inform individual and collective practice. Masculinity is still associated with appropriate military performance on the basis of which men earn respect and status in the armed forces; and in civil society, femininity (and therefore women) is linked to motherhood, domesticity, and weakness. The accession of women threatens their status. Martin van Creveld's argument about the necessity of excluding women from the armed forces may be an extension of his own ideological presumptions which may underestimate the extent of contemporary transformations but his position is manifestly not absurd. On the contrary, despite the significant integration of women into the military especially since 1990, very substantial resistance to female accession remains. Moreover, in many cases, this resistance is not based on a professional assessment of the capabilities of women as soldiers and combatants but on the contrary on masculine cultural presumptions. Even in a professionalized military, concepts of masculinity persist which are used to motivate and unite male soldiers by offering them status in both military and civilian spheres. Women will not be integrated fully into the military until military roles are entirely de-sexualized and the status which accrues to soldiers derives from their performance of a difficult role whatever their gender.

At the end of her work on gender and military, Helena Carreiras declares her pessimism at the prospects of female integration: 'the conclusions of this research entail a somewhat pessimistic view regarding whether greater gender inclusiveness in the military will contribute to change or "shake" the gender regime in the armed forces.'[176] According to Carreiras, women will remain tokens in the armed forces without a major social and cultural transformation of gender definitions and women's place in the workplace. Such a transformation is over the longer term perhaps possible but, as those countries, such as Canada, the Netherlands, and Denmark, which genuinely (and not just formally) allow women into the infantry demonstrate, individual women can be selected and trained as combat soldiers. However, the numbers of women who do actually choose this demanding professional career are likely to be so small in number—probably around 1 per cent of the infantry—that no profound gender re-settlement will be required. Professionalization is likely to be accentuated by this minority accession, where training, competence, and performance are the primary organizational reference points, but the armed forces are likely to remain at least partly sexualized. They are likely to remain inflected with a concept of masculinity, motivating and unifying combat soldiers by the offer of male status. The armed forces have changed profoundly over the last two decades and recent campaigns in Iraq and Afghanistan have only accelerated those changes. Women are performing military functions deemed impossible only a decade ago and the emergence of individuals like Kate Nesbitt represents a significant historical moment for women. Yet, the limitations of that accession have also to be recognized even if the excessive language of Martin van Creveld is to be avoided.[177]

12

The Professional Society

THE SPREAD OF PROFESSIONALISM

Almost 100 years ago, in August 1914, western nations entered a conflict which has been widely regarded as the first industrial war. Four years later, the art of warfare, and, indeed, popular understanding of war and its special horrors, were transformed by the experiences of the trenches—on the Western Front, Gallipoli, and the Isonzo. Military commanders, political leaders, and the population believed the war would be short and could be won by a demonstration of nationalist enthusiasm with young men gallantly dashing at the enemy with rifle and bayonet. The reality proved somewhat different. In the face of mechanized fire, barbed wire, and concrete bunkers, the *élan* of the charge proved hopelessly inadequate. Indeed, the problem posed for commanders was to keep their soldiers moving across the battlefield at all. In the face of a fire-swept zone, most soldiers who were not killed or wounded were typically overcome with inertia; they took cover and failed to participate. The industrial battlefield had set a problem for the mass armies on it which they found very difficult to resolve. As all major combatants quickly realized, complex fire and movement tactics at the platoon level, requiring tactical dispersal, initiative, and great individual and collective skill, were required. The simple lineal tactics of the eighteenth- and nineteenth-century battlefields had become suicidal. Yet, precisely because they were so massive, army commanders recurrently and standardly failed to be able to prepare their troops adequately for battle. The assaults of 1918 were extremely sophisticated in terms of combined arms warfare which used artillery, tanks, and aeroplanes together. Yet, on Operation Michael and the subsequent Allied advance during the Hundred Days, the infantry still assaulted for the most part in mass waves, supported by artillery, tank, and plane.

Combined arms warfare became ever more sophisticated in the Second World War and, of course, in Korea and Vietnam, but the infantry problem remained, as S. L. A. Marshall famously noted. Certainly, there were elite units especially in airborne and Marine regiments which underwent extensive training and were capable of executing complex tactics from 1915 onwards.

However, for the most part from 1914 to 1973, the mass infantry relied on the morale of its soldiers to fight. Armies appealed to the masculinity and patriotism of their men to bind them together and obligate each other to fight. It is an extraordinary fact but, united in these nationalist fraternities, infantry soldiers were repeatedly willing to engage in wave assaults, knowing that they had a very good chance of being killed. Yet, they would charge into the steel storm rather than face dishonour. At other times, when the assault stalled and inertia seized the majority of the soldiers, they would rely on exemplary individuals, who had predominantly already been recognized by the army and by the soldiers as bearing heightened responsibility, to continue the assault. A mass–individual dynamic characterized infantry combat in this period. Battles were typically decided not by the mass, which remained inactive, but by very small numbers of combatants on each side. On Omaha Beach on D-Day, Easy Red sector, one of the most bloody and hard-fought landing zones, was defended almost solely by a single machine-gunner, Heins Severloh, in Widerstandsnest 62. Between approximately 7 a.m. in the morning, when the first wave of the 1st Division—and eventually elements of the 29th Division—waded ashore in front of this position on the bluff until 3.00 p.m., he fired some 12,000 rounds from his MG 42 as well as a further 400 rounds from two rifles he had used, with very little assistance from his colleagues; at the outset of the action, Severloh was aware of no other machine-guns firing except his own.[1] It is difficult to estimate how many Severloh personally killed and wounded, although estimates of between 1,000 and 3,000 killed have been propounded. Significantly, Severloh was himself eventually forced to abandon his position by US soldiers creeping around the draw to outflank him to the west: 'From 1500, I saw only 250 meters to my west, ever greater numbers of Americans climbing the steep heights in long lines together.'[2] However, from Balkosi's account, it is clear that in the first instance very small numbers of US soldiers were involved in these infiltrations. Meanwhile, hundreds of uninjured (but traumatized) troops from the 29th and 1st Infantry Divisions remained inactive on the sand seeking cover from Severloh's fire behind beach obstacles and the tank wall. This pattern was entirely typical of twentieth-century warfare at the platoon level.

A mythology, indistinguishable from national memories, has sacralized the citizen soldier of the twentieth century and above all of the Great Generation which fought the Second World War, fundamentally distorting the problem of cohesion and the realities of combat in that era. This mythology not only obstructs the attempt to assess combat performance from 1914 to 1973 objectively, but it also obscures the very great difference between the citizen army of the twentieth century and the professional army of the twenty-first. It is often assumed that the kind of solidarity displayed by professional soldiers today matches that of the citizen armies in these conflicts and that their conduct under fire is more or less the same. Today, professional soldiers themselves often doubt their worth in comparison with their grandfathers who fought in the Second World War. They surely confuse the clarity and assumed nobility (at least for the Allies) of the political objectives of this conflict with the quality of the combat performance of the soldiers who actually fought it. In fact, as a result of intense and realistic training, professional soldiers are typically able to conduct the tactics which often eluded their predecessors. The tactics which were recognized from the First World War onwards are now routinely executed in combat because they have become automatic in training. Indeed, it is not only that professional soldiers now train more than their citizen ancestors. Training itself has been transformed with the investment in facilities and the introduction of more sophisticated instructional methods, often derived from sports science. Especially with the introduction of close-quarters battle training, the physical and psychological aspects of combat performance have been subjected to a level of scrutiny which is quite new. The result is that the infantry platoon, although organized and equipped in ways which would have been recognizable to the soldiers of 1917, has attained a far higher level of cohesion than the platoons which fought at Ypres or Vimy Ridge. The professional platoon is now capable of complex, coordinated fire and movement. Moreover, the cohesion of the professional army has been strengthened by the appearance of a perhaps ironically impersonal competence-based ethos in which the criterion of membership is no longer race, ethnicity, or nationality or, in some cases, even gender but mere ability. The basis of comradeship and solidarity has become professionalism itself. In this context, it is becoming possible for a small number of women to fight as infantry soldiers.

In the industrial era, western nations have typically been at the forefront of military developments. Their priority on the battlefield has been at least a partial reflection of their economic and political supremacy. That may now be changing: 'Western states' defence budgets are under pressure and their military procurement is constrained. But in other regions—notably Asia and the Middle East—defence spending and arms acquisitions are booming. There is persuasive evidence that a global redistribution of military power is under way.'[3] Nevertheless, while the military balance may be currently shifting away from western Europe and America, to Brazil, China, and Asia, its forces, primarily due to America's continuing military supremacy, remain the model which others are following. Consequently, although there are inevitably major cultural and organizational differences between western forces and the emergent powers, the armed forces in Asia and South America have substantially imitated western military posture and methods. At the highest level, they are seeking to procure the kind of precision-guided munitions, surveillance assets, and network-centric capabilities which have been at the basis of US dominance since the end of the Cold War. Perhaps, even more decisively, non-western nations have also moved away from conscription towards a professional all-volunteer military. Consequently, although the effects have been highly differentiated, the infantries of these nations have begun to engage in the kind of training and to develop the skills and solidarities evident in western professional forces.

The failure of the Soviet army in Afghanistan has been widely regarded as one of the central reasons for the collapse of the Soviet regime in 1989. Yet, despite the end of the communist system, many of the problems which attended the old Soviet army persisted in the Russian army in the 1990s as operations in Chechnya and the widespread institutionalization of systematic bullying (*dedovshchina*) attest.[4] Even up to the early part of the last decade, military reforms were still, for the most part, ineffective. Steven Miller, for instance, has demonstrated the way in which political, institutional, and financial obstacles impeded any attempt at reform.[5] Today, 190,000 soldiers in the 395,000-strong Russian army are still conscripted and many of the problems of the Soviet era remain evident. Nevertheless, as threats have changed, Russia has sought to restructure its forces in a manner which parallels western transformations.[6] In particular, the Russian army replaced its old divisions—twenty-three were disbanded—with more deployable and flexible brigades at the end of 2009, including twenty-eight new multi-role air assault and aviation formations.[7] In order to increase the responsiveness and effectiveness of the armed forces, the number of airborne regiments has been increased from two to three, although there are problems with the availability of transport planes.[8] The Russian army is trying to create smaller but more proficient forces, concentrating resources on more elitist units, as western forces have done over the last two decades. In order to increase the professionalism of the army, a new process for training non-commissioned officers has been introduced with an aspiration of creating a professional cadre, though the programme has been difficult to implement.[9] The Russian army is very distinct from genuinely professional western forces but some attempt has been made to implement some of the techniques and skills evident in NATO armies.

The Chinese army of 1.6 million soldiers is a half-conscript, half-professional force and still has some of the characteristics of a mass twentieth-century force. However, China has implemented very significant changes to the army, aided by a growth in the defence budget of 7.5 per cent in 2010. In particular, in the last decade, there has been an effort to improve the reaction capacities of the People's Liberation Army (PLA). Like Russia, China disbanded its division in favour of brigades in the late 1990s with the mechanized brigade identified as the main operational unit, five out of seven of which are regarded as elite, though this process was halted in 2003 due to a lack of officers.[10] The PLA is currently experimenting with 'special mission battalions' drawn from infantry or mechanized brigades and capable of rapid reaction, primarily against internal insurgencies. The development of these battle groups has been intensified by the publication of a new military training manual in 2006 and they have been seen to highlight the importance of elite troops. It is noticeable that the more highly trained units, such as the elite 58th Mechanized Brigade, have been more active than less trained units. Although the connection with the people remains politically important, the PLA has, in fact, become increasingly professional. Indeed, since the late 1970s, the PLA has downsized from 4.5 million to its current size; many experts noting a trend towards professionalization since that time.[11] Interestingly, Lyle Goldstein has demonstrated the salience of the Falklands War to China strategists and defence planners.[12] This war has been seen as a potential model for the Chinese forces, especially but not exclusively in terms of a possible amphibious assault on Taiwan. Various lessons about air and maritime power have been derived from the conflict but, especially in terms of ground troops, the Chinese have recognized the importance of training. The most significant conclusion on ground combat in the Falklands is that a 'smaller, higher quality force can defeat a larger force'.[13] One commentator observed: 'in numbers the [Argentine forces] far exceeded those of the English ground forces sent to the Falklands, but in the quality [of the troops] there was obviously no way [the two forces] could be mentioned in the same breath';[14] in night fighting and mountain warfare, British troops were far superior. Significantly, 'Chinese analysts argue that training was a critical advantage for British ground forces in all aspects of the war.'[15] The central lesson of the Falklands seems to be the importance of professionalism to current military operations especially in terms of training.

As the fastest growing economy in Latin America and an increasingly important global actor, it is noticeable that some of these patterns (of reactivity and professionalism) are also evident in Brazil. As with Russia and China, the Brazilian army still employs national service, including some 70,000 conscripts out of a force of 190,000.[16] The defence priority for the army is the protection of the Amazon and the maintenance of internal stability, for which selected brigades have been developed with enhanced equipment, mobility, and training. Three light infantry brigades of this kind are stationed in the Amazon.[17] Clearly, while professionalization is evident in the armies of these three major powers, it would be unwise to presume that professionalism at the platoon level has or will necessarily take the form which it has among the western forces described in this book. Yet, it is conceivable—perhaps even likely—that the intensification of training regimes will generate the kind of practices and solidarities which are emerging in the west. Yet, local civil and military cultures will, of course, contour and inform the specific character of this professionalism in China, Russia, and Brazil as it does in each western nation. A deeper investigation of the character of professionalism in the infantry of these nations is likely to be highly illuminating.

THE NATIONAL MISSION

The emergence of a professional military represents a historic change for the west—and, indeed, perhaps globally—fundamentally altering the dynamics of combat at the platoon level. However, it is also important to recognize certain continuities. In his work on the American soldier, Charles Moskos maintained that while Americans—and perhaps the soldiers of western democracies more widely—disparaged overt and aggressive displays of patriotism or attempts at indoctrination, the belief that they were fighting for something which they communally valued was important to their combat performance. For Moskos, the role of 'latent ideology' was not to be underestimated in explaining combat motivation, although it was easy for the social scientists to overlook it, taking the disdain of front-line soldiers for idealism too literally. One of the critical reasons for US defeat in Vietnam was precisely because there was no sense of national mission or purpose to unite the troops and to explain their suffering. Vietnam demonstrated what would happen to an army without a latent ideology. An army without a sense of national purpose collapses.

This book has suggested that a very significant amendment is required to Moskos's observations. While the major conflicts of the twentieth century seemed to demonstrate that the citizen army required a latent ideology and this ideology was drawn upon both directly and obliquely in combat to encourage performance, this dependence is not so obvious in a professional army. In a professional army, combat soldiers are motivated not in the first instance by their masculinity, nationality, or ethnicity but by an ethos of professionalism. They are animated by a sense of professional pride. As professionalism becomes a central unifying and motivating force in itself, it would seem plausible to suggest that political motivations, justifications, and indoctrination may also become less important to the combat soldier of the twenty-first century. Indeed, although his concept of this contrast between the postmodern and professional soldier may be overstated, Fabrizio Battistelli usefully recorded the decline of political motivation as a reason for service after the end of the Cold War.[18] Indeed, despite the increased levels of danger on operations since 2001, the concept of national mission or latent ideology seems to have played a lesser role than in the twentieth century. It is very noticeable that the opposition to the Iraq War after 2004 was numerically, if not qualitatively, compatible with that of the Vietnam War. By 2006, 60 per cent of the US population were recorded as opposing the Iraq War;[19] during the Vietnam War, that percentage of opposition was reached only in April 1968, increasing thereafter, especially once the USA declared its withdrawal in 1970. Yet, there was no collapse of morale in the US Army or Marines in Iraq. They continued to fight despite widespread popular disillusionment and opposition. A sense of professional not patriotic duty seems to have substantially displaced the national mission as the central means of sustaining cohesion. It might be suggested that professionalism may have become the latent ideology of the professional soldier.

This displacement is very obvious in the European, Canadian, and Australian militaries who have not fought since 2001 to avenge the Twin Towers. Although NATO invoked Article 5 of the Washington Treaty in response to the 9/11 attacks, there was little popular sense of a genuine existential threat. This is quite different in the United States. There, the majority of its political leaders and population and, of course, the armed forces have understood themselves to be at war since 9/11. George Bush's invocation of the term 'the global war on terror' may have been regarded as inaccurate and inappropriate by some commentators and may have little resonance in Europe, Australia, and Canada but for the United States, the 9/11 attacks have been defined as acts of war; and continue to be so. Since that day in September 2001, the United States has understood itself to have been at war and this declaration of war has been central to the performance of its armed forces. Indeed, many citizens have felt impelled by a sense of duty to serve their country. Thus, following 9/11, a significant number of service personnel who had retired, resigned, or were part of the reserves volunteered to return to full-time duty. Perhaps even more strikingly, citizens who had never been in the armed forces decided to volunteer for military service following those attacks. The most famous example here was the case of Pat Tillman, the professional American football player, who gave up his multi-million dollar sporting career explicitly to defend the United States. He was eventually selected for the Rangers and was killed in a friendly fire incident (which the Pentagon clumsily tried to cover up) in Afghanistan on 22 April 2004.[20] There is no equivalent example in Europe or Canada. Indeed, in the UK, a well-known Manchester United and ex-England footballer, Paul Scholes, claimed on his retirement in July 2011 that England players were not even motivated to play for their country in international tournaments. It is unlikely that they would be willing to fight for it. Illustrating this sense of duty, many US service-persons bear elaborate tattoos which feature the stars and stripes. In one case, a US Marine captain had an impressive design of the national flag on his upper arm below which the words 'For Honor' were inscribed. These tattoos physically embody—and communicate—an individual and collective sense of mission, which has become a self-conscious reference point for serving personnel.

This duty is repeatedly articulated in conversation. In 2009, General Stanley McChrystal, having taken over command of ISAF, created a series of Counter-Insurgency Advisory, Assistance, and Training (CAAT) teams which were attached to all the Regional Commands. These teams provided tactical advice on counter-insurgency and were a means of disseminating the US approach as articulated in Field Manual 3–24 to deployed forces. They were also used so that McChrystal could gain a better understanding of operations and difficulties at ground level. Typically, these teams consisted of retired and very well-pensioned ex-Special Forces officers and senior non-commissioned officers with significant operational experience. In November 2009, a CAAT team was deployed into Regional Command South in the headquarters at Kandahar Airfield. One of the members of this team was a retired army Special Forces warrant officer, two of whose sons were in the US Special Forces and had served or were currently serving in Afghanistan. The individual spent significant amounts of time in the field with operating troops and, indeed, in Helmand was in a vehicle which was struck by an IED. Eight months later in June 2011, the retired warrant officer was still in theatre having taken only minimal leave, despite having a wife at home. The members of the CAAT teams were certainly earning significant salaries but it was clear that his motivation went well beyond the financial. For instance, when asked why he had stayed in Kandahar so long, he replied with complete sincerity: 'Last time I looked we were at war.'

A sense of national mission is very prominent in the professional forces of the United States, then. Indeed, this concept of duty distinguishes US forces from their professional NATO partners still further through its heavy religious inflection. As the sociologist Grace Davie has shown,[21] the role of religion in American civil society demonstrates a quite different pattern from all other western democracies. For those nations, the secularization thesis, one of the most recent articulations of which has been made by Charles Taylor,[22] is broadly sustainable. According to this thesis, formal religion, specifically Christianity, has become less important in the west as a result of the industrial revolution and especially since the early decades of the twentieth century as the state began to provide social services and security for its citizens. Yet, the United States stands out against this general process of secularization. The United States remains a deeply religious and Christian country, with church attendance remaining high. There are a number of reasons which have been proposed to explain continuing American religiosity, one of the most convincing of which is the relatively inadequate provision of state social services. In the absence of state provision, churches of a bewildering variety of denominations have tended to perform a communal welfare role. Clearly, the intense religiosity displayed by many Americans cannot be reduced to economic dependence, but the lack of public welfare provides an institutional context in which the church has remained a social focus for dispersed communities. As a focus, it remains a forum (which most European and Canadian churches now lack) to communicate its teachings and guidance to its congregation and to unite them into a community.

The religiosity of the United States is apparent within their armed forces. US soldiers were traditionally viewed as sceptical towards religion in the Second World War, as they were towards political ideology. The US Army and Marines in Vietnam seem to have been actively godless. Along with heightened political and ethnic consciousness, atheism was evident in the military. Indeed, as works like Michael Herr's *Dispatches* suggest, a nihilism was evident among American troops where biblical motifs, among others, were employed to indicate the senselessness of the war—or to revel in it. The post-Vietnam reforms of the US armed forces primarily involved professionalization. For instance, the US army's Training and Doctrine Command (TRADOC) sought to identify the most effective way of fighting a modern war and began to organize and train the force to execute this 'manoeuvrist' approach. However, alongside the purely technical matter of improving the combat performance of the army and Marines, professionalization also involved the institution of a body of ethics which would prevent any repetition of mutinies and atrocities which had been a feature of Vietnam. From the 1970s onwards, with increasing force in the 1990s, the US Army and the US Marines sought to impose an honour code on their troops. In the US Marines, the Commandant General Charles Krulak was at the forefront of this promotion of honour as a way of eliminating indiscipline, bullying, and criminal activities among the troops with his 'Marine Corps Values Program'.[23] Religion became a central part of this new ethical training, albeit accidentally. Especially, among the officer corps, Christianity has become extremely important. In sharp contrast to their NATO peers, US officers are openly religious and are often committed Christians, attending not only church on Sundays but bible classes on a daily or weekly basis. General James Mattis, for instance, is reported to wake at 3 a.m. every morning to spend one hour in devotion, reading the Bible, before a period of academic study followed by a daily run. Clearly, not all officers are religious but the professional US military is generally infused with a religiosity which is both unusual in comparison with its allies and provides officers with a shared culture. This religiosity is closely related to the US concept of duty, honour, and national mission to create a warrior ethos by which the American military as an institution understands itself. Many serving personnel actively see themselves not only as fulfilling their civic obligations but that those commitments are divinely inspired. American soldiers do not generally conceive themselves as crusaders; most are not interested in occupying the territories they invade. However, they have a tendency to see themselves as religiously motivated warriors who are defending not just the USA but the west as a democratic and Christian civilization against radical Muslims. Samuel Huntington's thesis of the clash of civilizations is crude and polemical but its religious conception of the current global order has some resonances with the US military's own self-perception.

This fusion of nationalism and religiosity is very obvious in the so-called Warrior's Creed which was developed in the 1990s and which now features prominently in US infantry doctrine:

I am an American Soldier.

I am a Warrior and a member of a team. I serve the people of the United States and live the Army Values.

I will always place the mission first.

I will never quit.

I will never leave a fallen comrade.

I am disciplined, physically and mentally tough, trained and proficient in my warrior tasks and drills. I always maintain my arms, my equipment and myself.

I am an expert and a professional.

I stand ready to deploy, engage and destroy the enemies of the United States of American in close combat. I am a guardian of freedom and the American way of life.

I am an American Soldier.[24]

The concept of a creed deliberately suggests that the soldiers are joining a religious order whose duty is to defend the United States. It would be very difficult to ignore the monastic and patriotic resonances of this oath. Yet, even here, the concept of professionalism is central to the fulfilment of this vocation. Indeed, while the oath is bracketed by statements of national religious faith, the main content consists entirely of references to performance. Particularly noticeable is the presence of the words 'warrior tasks' and 'drills' which the soldier is *trained* to perform at the heart of the text. Textually, then the concept of professional proficiency has priority; it is given a meaning and purpose by the duty of national defence. Perhaps as a result of their sense of national mission, strengthened by religious conviction, the US forces have proved themselves most willing to endure the difficulties and dangers of current operations in Afghanistan and Iraq. However, they have only been able to endure in Iraq and Afghanistan because of their professionalism. Even American soldiers, fighting for the honour of their country and strengthened by their religious faith, finally rely on professional competences to perform in battle.[25]

CIVIL–MILITARY RELATIONS: OTHER FORMS OF PROFESSIONALISM

Since the publication of Huntington and Janowitz's studies in 1957 and 1960, the issue of civil–military relations, and, above all, civilian control of the military, has been of deep concern to social scientists studying the armed forces. Many of the debates have focused on the concepts of objective and subjective control originally propounded by Huntington and Janowitz. In *The Soldier and the State*, Huntington described and, indeed, advocated the objective control of the military as the most conducive way of generating armed forces capable of both defending the nation (under the direction of the state) while remaining relatively de-politicized especially in party terms; 'The greatest service they can render is to remain true to themselves, to serve with silence and courage in the military way.'[26] Janowitz, by contrast, suggested that by the 1950s, the armed forces had become such a large technocratic and managerial organization involved in constabulary operations that it was increasingly becoming more like a modern civilian corporation than a traditional warfighting body. Consequently, he claimed that it was coming under increasing subjective control as it integrated ever more closely with the standards and goals of civilian society. Civilian control necessarily followed the civilianization of the military. Both concepts are intriguing and they have proved extremely fruitful in thinking about civil–military relations in the twentieth century and especially in the 1950s and 1960s when America and most other western powers possessed conscripted military forces. In the twenty-first century, when these citizen forces, on which the arguments of Huntington and Janowitz were predicated, have disappeared to be replaced by all-volunteer professional armies, it may be necessary to reconsider the paradigm of civil–military relations.

Clearly, the central argument of this book is that professionalization fundamentally changes the armed forces as an institution; this transformation is likely to have considerable impact on the military's relation to its host society. However, civilian society has itself also undergone profound changes, the significance of which need to be considered in understanding the revision of civil–military relations in the current era. Indeed, one of the criticisms which might be made of both Huntington's and Janowitz's work is that their vision of civilian society is under-developed and, in places (such as Huntington's polemical descriptions of civilian society as a new Babylon), crude. Above all, they presume a unity to civil society which does not seem warranted. While it is true that civil society is distinct from the military in that civilians are not equipped with advanced weaponry with which to kill their enemies, civilians form themselves into highly distinctive groups which are perhaps as different from each other as military personnel are different from civil society as a whole. The fact that civilian groups share a common lack on the monopoly of violence has encouraged social scientists to impose a false similarity upon them. Clearly, no attempt can be made here to resolve the issue of civil–military relations in western societies in the twentieth century but, perhaps by considering more closely the nature of civilian society as it globalizes, it might be possible to move towards a more adequate understanding of this relationship between a professional military and its host society.

As the argument up to this point has sought to establish, professionalism has become a defining principle for western armies in the twentieth century, encompassing both the skills which are required to operate in battle and the distinctive status honour which ensures that individual soldiers are obliged to fulfil their duties. It seems that a similar professionalization may be evident in the armies of Asia, China, and Latin America too. A potentially global military transformation seems to be taking place as the citizen army is replaced by the professional force. The armed forces are clearly distinctive in comparison with civil society and with other professions. No other organization enjoys a legitimate monopoly on violence in the way that the armed forces do. However, for all the distinctiveness of the military, it is worth recognizing the wider historical significance of the professionalization of the armed forces for it seems to parallel important changes in civilian society too. Professionalization may not be confined to the military but may be a widespread social phenomenon indicating the appearance of quite novel forms of social solidarity, which are likely to have serious repercussions in terms of civil–military relations. Specifically, the move from a poorly trained, under-performing, but homogeneous mass to highly specialized, diverse, but increasingly cohesive teams which has been evident in the armed forces seems to be apparent in other sectors.

Clearly, there are any number of areas of social life in which this move from general participation to specialist engagement might be documented. However, since the west's global dominance in the past two centuries has been primarily based on its industrial power, it seems to be appropriate to consider manufacturing in order to illustrate this shift from unskilled mass to specialized teams. Moreover, since industrial sociologists have often taken the car industry as a bell-wether of industrial production, this sector would seem to provide an apposite example, especially since the appearance of the skilled professionalized team has been particularly apparent in this industry. Early to mid-twentieth-century industry was dominated by mass production techniques. These techniques had been elaborated by F. W. Taylor in his famous work *Scientific Management*, published in 1911, and were widely observable in industry in the twentieth century, especially in car manufacture, which has often been seen by social scientists and historians as illustrative of wider trends in manufacturing. Indeed, Henry Ford had already begun to implement 'scientific management' in his factories before the publication of Taylor's work, giving his name to a new production paradigm. Henry Ford and Fordism, more generally, advanced the principle of line production to new heights. By the 1930s, huge factories appeared in America and Europe, organizing production on a lineal basis. Raw materials were assembled at one end of the factory and the production process was broken down into sequential stages on a relentless assembly line. A mass workforce was employed to work particular stations, each individual ascribed an exclusive function on the assembly line. Fordist manufacture represented a huge leap in productivity. As Richard Overy has discussed,[27] Fordist mass production techniques were critical to the Allied victory in the Second World War. For instance, Ford's 900-acre Willow Run Plant, near Detroit, was, at the height of its productive capacities in 1944, able to manufacture 650 B-24 Liberator bombers (with 1,550,000 parts) a month:[28] over twenty a day.

However, despite its achievements, there were also some obvious shortcomings to mass industrial production. The Fordist factory was very hierarchical. Individual workers were assigned their position under management, mediated by various grades of middle management from the shop-floor upwards. These managers checked the quality of work and disciplined workers if they were failing to reach the required standards. The individual was stripped of all autonomy or the requirement to apply judgement and the work on the production line was by necessity extremely monotonous. Individual workers performed the same simple task over and over again. As a result, the workforce typically invested little in their work. In his study of the Ford plant in Halewood in the late 1960s, Huw Beynon records this automated disengagement of the workers very clearly: 'When I am here my mind's a blank—you *make* it go blank.'[29] Indeed, Beynon's workers repeatedly told a joke about 'the man who left Ford to work in a sweet-factory where he had to divide up the reds from the blues, but he left because he couldn't take the decision-making'.[30] Some of the more perceptive trade union stewards, who were a central focus on Beynon's work, noted that the centralization of the production process actively encouraged withdrawal on the part of the workers: 'As a shop steward I see the main problem to be the centralization of power on both the management and trade union side. This is bad for the operator as well because both the steward and the operator get frustrated when things get taken over their heads. You need to have much more autonomy for shop-floor unionism.'[31] As a result, workers noted the way in which the production line and the system of management actively encouraged passivity on their part: 'I place the car off the hoist, I've been doing that for three years now. With the line you've got to adapt yourself to the speed. Some rush and get a break. I used to try and do that but the job used to get out of hand. I just amble along now.'[32]

The Fordist factory not only demanded minimal intellectual and personal engagement by the worker, then, but it could be a place of low levels of productivity as individual workers were able to shirk. The workforce often displayed mass apathy. British car manufacture in the 1960s and 1970s demonstrated a culture of inactivity and carelessness very clearly and perhaps at its most extreme.[33] However, within this atomized and alienating process of production, the myth of the Stakhanovite also emerged in Soviet Russia to inspire the workers. The Stakhanovite was a virtuosic worker who, fired by communist zeal, exceeded normal scales of output. Whether such individuals existed in reality in the Soviet Union is open to question but the myth usefully illustrates that the high-performing individual was identified by the regime as the ideal solution for mass passivity. He was the example. The Stakhonovite myth seems to have been less evident in the west. Beynon, however, does record individuals in Ford's plants in Britain who were committed to their work, sought promotion, and who, dismayed by their co-workers' laxness, acted as informants for foremen, often becoming targets for practical jokes and abuse by their colleagues.[34] Interestingly, while Beynon's work included glimpses of the more productive individual workers surrounded by a recalcitrant mass, bored, alienated, and exhausted, the oppressive conditions of mass manufacture also engendered individual acts of resistance and deviance, of often extraordinary sophistication and creativity. These individuals, perhaps, represented the British car industry's true Stakhanovites; geniuses not of production but disruption. As one worker noted there was a direct correlation between the stifling work conditions and these acts of insubordination: 'The atmosphere you get in here is so completely false. Everyone is downcast and fed up. You can't even talk about football. You end up doing stupid things. Childish things—playing tricks on each other.'[35] In the British factory of the late 1960s and early 1970s, perhaps partly explaining the nation's industrial decline, the best examples of Stakonovite virtuosity were reserved, then, not for the assembly line but for practical jokes away from it.

Beynon brilliantly captures the more elaborate and amusing of these acts of defiance, even if his interpretations are somewhat more earnest that the spirit in which they were carried out.[36] However, some individuals could influence formal industrial relations. According to Beynon, one militant worker, called Kenny, 'hated Ford's and loved a fight'; 'he hated being told what to do'.[37] Accordingly, when he was working on the line, putting 'bits and pieces' into hooks at the front of the line, he decided to fill every other hook and to leave big gaps.[38] Kenny recorded: 'The foreman went mad. Berserk he went. He started jumping on and off the line, running down the line filling up the hooks . . . Well the lads caught on and they started leaving empty hooks. He was going crazy. Then we got hold of Eddie [Roberts, a senior steward] to complain that the foreman was working. We did that every day. The situation is a lot better now. In fact we've got one of the easiest sections in the plant. It can be done see.'[39] The deviance demonstrated by workers in the Halewood plant in Britain in the late 1960s might be unusual and extreme but Beynon's work usefully illustrates with rich ethnographic accounts both the passivity and disengagement of the mass industrial workforce, punctuated by individuals a small minority of whom demonstrated greater commitment but whose most colourful characters reserved their virtuosity for acts of defiance and subversion. Stripped of control over their workplace, the reactions of both the disengaged mass and the rare extraordinary individual, like Mick

Donnelly, seem to be compatible with each other. These individuals might be seen as the Audie Murphys of the assembly line; virtuosic individuals in a mass of unmotivated and unskilled non-participants.

In the 1970s, as a result of global economic turbulence and increased competitiveness, mass production went through a crisis which demanded a fundamental reform of the Fordist system. Specifically, as shorter but higher-quality production runs became the norm and computer technology took over many of the assembling jobs on the line, the old mass labour began to be replaced by a small, more skilled—more professional—workforce. Like the army, industry downsized. John Atkinson was one of the first to notice that flexible specialization replaced mass, uniform labour with a dual workforce in which there is a small, highly skilled core.[40] Crucially, the teams in this core took greater responsibility for their contribution to the production process, cooperating with each other horizontally and seeking to innovate collectively.[41] Jürgens has particularly emphasized the importance of highly trained and skilled groups and group relations in Japanese manufacturing.[42] Writing in the late 1980s, Christine Lane has been sceptical about claims by Piore and Sabel that industry was undergoing a second industrial divide. She maintained that not only has the pursuit of flexible specialization differed in Britain and Germany but that 'Taylorist practise is still widespread'.[43] Nevertheless, while her qualifications are well taken, especially in the 1980s, she emphasizes the emergence of the *Anlagenführer* (work-station leader) in the car industry. At Volkswagen's bodyshop at Wolfburg the number of *Anlagenführer* increased from 150 to 290 in the 1980s, and in the trim and final assembly area of Hall 54, where previously none had existed, there were eighty *Anlagenführer*.[44] The *Anlagenführer* had far greater responsibility than the old foreman and was tasked with coordinating a team of specialist workers at a particular station. The role indicated the emergence of specialist work-teams. The creation of these teams eliminated gaps between production, inspection, and maintenance with the team performing all the roles.[45] In engaging the worker more, the team seems to fulfil one of the key principles of flexible specialization,[46] because the emergence of production teams allows for more responsive and de-centralized management of the manufacturing process with suggestions, recommendations, and innovations coming from the shop-floor itself.[47] Whether scholars prefer the term flexible specialization, post-Fordism, or neo-Taylorism, and disputes continue about the extent of the changes, all are agreed that industrial production has undergone a profound transformation in the last four decades. The old industrial hierarchy with management directing a mass of low-skilled, low-performing workers, fulfilling discrete sequential functions on the assembly line, has been substantially replaced by a flatter organization in which production is organized into skilled and cohesive teams. The mass workforce, with its high-performance or highly disruptive individuals, has been replaced by skilled and cohesive teams. The pressures which have impelled the appearance of production teams in industry are distinct and the kinds of team which have emerged are of course quite different from elite military squads. Yet, there are similarities between the two spheres. Industry and the armed forces have replaced low-skilled mass with more highly skilled specialists operating as well-drilled teams.

The appearance of specialized teams seems to be evident in many other spheres. Sport has always been a significant activity in modern western society, as indeed, it has in all societies. However, as a result of the de-regulation of the media in the 1980s and the appearance of global media companies, sport has undergone a profound transition. It has been radically commercialized and become an increasingly important focus of public attention. Indeed, some sociologists talk about the 'sportification' of society. Not only is sport an increasingly important social practice, then, but sport has some parallels to combat. Sport, like combat, involves teams engaged in a competitive struggle with each other. Consequently, it may be pertinent to explore the way in which changes in sport over the last three decades, as a result of globalization, may have primarily involved a process of professionalization which parallels developments in the military. For most of the twentieth century, as professional sport developed, sporting performance assumed an identifiable form. Individual stars dominated the action. Even in team games, such as football, the individual dominated play. In football, there are some obvious examples of such players: Alfredo di Stefano, Puskas, Pele, George Best, and Bobby Charlton could all be regarded as individual virtuosos. Their team-mates depended upon their individual skill. For instance, before the 1968 European Cup Final, as the Manchester United team waited in the tunnel at Wembley before going onto the pitch, many members of the team showed visible fear. It was the most important game of their lives and was hugely significant for the club which had lost a team in an air crash at Munich ten years before travelling to this competition. At this point, Bobby Charlton declared that if anyone found themselves under pressure on the pitch, they should pass the ball to him. In this way, he demonstrated precisely the practical and moral dominance of the individual in this era of football. On the pitch, his play was characterized by extraordinary individual goals, in which Charlton struck the ball from distance with great force.

In the 1970s, new training techniques and new tactics began to appear. As a result, the best football teams began to develop a greater level of coordination than previously, allowing the ball to be passed much more quickly and accurately between players in slick interplay. The Ajax team was, of course, the most famous example of the rise of collective expertise in football. In the early 1970s, Ajax dominated the European Cup with a technique known as 'total football'; players rotated into different positions as circumstances dictated. Their star player, Johann Cruyff, was technically equal or even superior to his immediate predecessors, such as Puskas or Charlton, but his skill was an element of refined team play. Crucially, Ajax and the national team sought to create space collectively through passing and movement.[48]

Collective expertise has been substantially developed since the 1970s and, by the early twenty-first century, highly drilled teams have come to dominate sport. In the last decade, Zinedine Zidane has been one of the most obvious examples of a supreme virtuoso; his individual mastery of tackling, close-ball control, and shooting is undeniable. Yet, even discounting the critical role which a system of trainers, nutritionists, coaches, physiologists, and psychologists played in developing these skills, he was always integrated into and augmented by a wider system of play both for France and for Real Madrid. Zidane's goal in the Champions League final against Bayer Leverkusen in 2002 demonstrates precisely this integration of individuality within a highly developed team very well. In the injury time of the first half, Zidane received a precise, floating pass from Roberto Carlos just outside Leverkusen's area, as Real's forwards opened the defence up with some simultaneous diagonal runs. Their movement gave Zidane the opportunity to strike; he hit a clean volley straight into the goal. Zidane's individual brilliance is not to be questioned but it was facilitated by the precisely coordinated actions of his team-mates who made the space for him; 'He cast his presence over a Real Madrid machine in which all moving parts appear to be interchangeable.'[49]

This move from individual star to cohesive team is particularly obvious in cricket, which is the most individual of all team games. Before cricket began to be commercialized in the 1970s, the game was dominated by individual players, like Don Bradman, Gary Sobers, or Harold Larwood. At that time, with the introduction of one-day cricket, sponsors, greater television exposure, and increased rewards for winning, more systematic approaches began to be introduced to training and playing which began to turn the cricket team with its loose confederation of individual bowlers and batsmen into a cohesive squad. This is particularly obvious in the field where the bowling side's players now operate as a highly practised unit seeking to suppress the run-rate through collective effort.

Significantly, the importance of teamwork is evident in even apparently individualistic sports today. Indeed, the move to teamwork in even the most individualistic sports perhaps exemplifies the new salience of collective virtuosity most clearly. Even in the unlikely example of surfing, where individuals necessarily ride their boards alone and the entire sport is inflected with a counter-cultural ethos, a collective dimension has been introduced in order to improve performance in the last two decades, predominantly as a result of the commercialization of the sport. In the 1960s, Greg Noll dominated big wave riding in Hawaii. Like Best, Pele, or Puskas, he was part of an elite group which together mutually encouraged its members to ever greater performances. Surfers usually paddled out to the line-up and surfed together. Film footage shows that very frequently, especially on bigger waves, surfers took off together, riding down the face (or falling off) together. It seems that the physical co-presence of fellow surfers in the water and, indeed, on the same wave motivated surfers, both inspiring them with confidence and adding a competitive edge to their surfing. The sport at this point was disorganized, anarchic, and amateurish. However, on 4 December 1966, when Noll famously rode Makaha Point in a storm swell on probably the biggest wave to have ever been ridden up to that point, he was the only surfer in the water. Initially, Noll and his fellow surfers had paddled out together but, seeing the seriousness of the situation, all Noll's colleagues retreated to the safety of the beach. Noll calculated that if a surfer was caught in the impact zone where the waves broke, he would have an 80 per cent chance of drowning.[50] As he sat on his board, determining what to do, Noll recorded that he would never have forgiven himself if he had not at least tried to ride one of these huge waves: 'A lot of conflicts raced around my head. My chance of a lifetime—am I going to blow it or do something about it? I've got family, kids, and people I care about a great deal—is this goddamn wave really worth risking my life? I felt kind of crazy even considering it.'[51] Noll's existential doubts were countered by a strong competitive ethos: 'I analyzed the situation a little longer and gave myself better than a 50 per cent chance of surviving one of these monsters. I figured I had an edge, since all my adult surfing had been devoted to big waves. My motivation was competitive. Deep down I always wanted to catch a bigger wave than anyone else had ever ridden. Finally, after a lifetime of working up to it, the time had come. The chances of this type of surf coming again might be another dozen years away and out of my grasp. It was now or, quite possibly, never.'[52] Noll was alone in the water when he finally made his decision. However, his surfing rivals and friends were central to his understanding of the predicament he found himself in as to whether to paddle for a big wave or not. It seems highly likely that the passive presence of his surf community on the beach, many of whom had admitted defeat in these waves, watching Noll provided an immediate if potentially unacknowledged motivation for Noll. The shame of returning to the beach and his colleagues without having at least attempted a wave seems to have been a significant factor in his decision to go. He eventually selected a wave, paddled for it and did, in fact, get up on his board only to be swallowed as it broke on him; he eventually scrambled to the shore to be greeted by his fellow surfers as a hero. In this way, his ride seemed to echo York's individual actions at the Argonne or Murphy's actions at Holtzwihr, while the passive support of Noll's friends on the beach seems to accord with York and Murphy's comrades.

In the late 1990s, big wave riding adapted into a team sport. In order to ride the biggest breaks, surfers used jet-skis to tow surfers into the waves and to rescue them from them. The use of jet-skis seemed to simplify surfing, making it relatively easy for surfers to catch waves for which they no longer need to paddle or to execute the difficult and critical manoeuvre of changing from prone to standing position at the very peak of the wave. Indeed, there is currently some reaction to tow-in surfing precisely because those supported by jet-skis are able to slide onto waves with little effort, displacing surfers paddling themselves. Despite its advantages tow-in surfing is not, in fact, easy especially in the very large waves for which it was developed. To negotiate routes in and out of the swell, to release a surfer into a wave accurately, and above all to rescue people from the impact zone, the jet-ski pilot needs significant skill. Tow-surfing also requires practice to be able to ride a surf board while being towed. Moreover, tow-in surfing in big waves requires proper coordination between the jet-ski pilots and the surfer.

In the 1990s, Laird Hamilton was one of the surfers at the forefront of these developments and he began to ride waves of 80 feet, approximately twice the height of even the biggest waves ridden by Greg Noll. Crucially, Hamilton's ability to ride these breaks was not reliant merely on his own evident individual skill but on the collective competence of his support team mounted on jet-skis (the 'Strap Crew', consisting of Dave Kalama, Darrick Doerner, Buzzy Kerbox, and a number of other elite surfers): 'The reason I am able to ride the waves I do and do what I do is because I have partners like Dave and Darrick. I am only arriving at this level because I am being driven by these guys to this level.'[53] Specifically: 'You've got to have eyes in the back of your head and I've got eyes—I got Dave and Darrick. They see what I need to see.'[54] His teammates described how they performed this role: 'I'll just kind of sit right on the crest of the shoulder as I can see what Laird is doing and what is behind us,'[55] while Rush Randall noted that 'It's a three man operation so that Laird and Kalama are paired up and I'll be in the channel for safety.' Interestingly, while individual and collective skill is essential to surviving and performing in large waves, teamwork generates a moral solidarity in the jet-ski crew which assists the individual surfer even when they are physically alone: 'When you are under the water and you know, "Ok, I am here by myself underwater" but I know that someone is up there, doing something to help me right now even if they can't help me, the confidence that is instilled by believing in that person buys you time. It gives you confidence to make it to the surface.'[56] Gerry Lopez, a former professional world champion, emphasized the point: 'It makes survival a whole different story than if you were swimming around in the water with no one but yourself.' There may be a purely psychological dimension here of not feeling alone but the confidence inspired by others is also practical; in trained teams like the Strap Crew, fallen surfers know that once they surface, they will be rescued quickly by a colleague on a jet-ski. Crucially, the relaxation that this shared competence instils in the individual reduces the amount of oxygen that the submerged surfer uses, preventing panic, and therefore demonstrably improving their individual chances of survival. The team physically empowers the individual. It should be noted that effecting a rescue out of the impact zone can be very hazardous for jet-ski pilots, who may be caught up in subsequent waves. However, good jet-ski pilots will risk themselves for colleagues and their motivation for mounting a rescue and the way they understand it are intriguing. Laird Hamilton is very illuminating on this point: 'If one of those guys goes down I will put myself on the line every time and each one of those guys will put themselves on the line for guys they don't even know—for guys they might not even like. It's part of their personality, part of their nature so that when they go home at night, they sleep well.'[57] Hamilton attributes this requirement to rescue the drowning surfer to individual personality but, in fact, he describes more specifically a collective ethos which has emerged around tow-in surfing. In the dangerous domain of big wave surfing, individual survival requires the performance of particular acts, especially rescuing, which must be executed without thought and hesitation independently of personal relationships. Despite the very strong bonds which clearly unite big wave surf teams, like the Strap Crew, Hamilton describes the emergence of a potentially impersonal collective ethos, which is quite distinct from Noll's individualistic attempts to prove himself better than the rest of his peers. Big waves demand a different collective mentality; it relies not on an active individual encouraged by peers in the line-up or on the beach but requires a genuine team of which the individual surfer, even under deluge of a breaking wave, is still a member. With their teamwork, reliance on instinctively performed drills, Hamilton's Strap Crew seem to echo Junger's 1st Platoon on the Gatigal. When he rode Makaha Point, Noll's fellow surfers watching from the beach seem to have motivated him but they could not have possibly helped him if he had been trapped in the impact zone, as well he knew. He was alone. The displacement of the exemplary individual and under-performing mass by the professionalized team is in no way limited to the infantry platoon, then. It is a widespread social phenomenon. It is evident in industry and sport. Even in academia, one of the most apparently solitary professions, the development of research teams and clusters all competing for reputation and funding has been notable in the last thirty years. Across a wide range of social activities, small groups have begun to specialize to demonstrate a quite different form of solidarity—and changed capability—from their predecessors.

THE DECLINE OF COMMUNITY

In his widely read book *Bowling Alone*, Robert Putnam seeks to analyse the decline of American civic life in the last three decades of the twentieth century. For Putnam, one of the most important collective goods which any society possesses is what he terms 'social capital'. Social capital refers to the dense bonds of reciprocity between citizens which encourage them to give and receive help from each other independently of the market. A strong society, according to Putnam, is one whose members are willing to band together in communities and associations providing not only essential services, like social and health care, for each other, but also a vibrant public life in which individuals interact and communicate with each other.[58] Putnam rejects the notion that modern mass society involves a necessary weakening of communal bonds; 'it is emphatically not my view that community bonds in America have weakened steadily throughout history.'[59] Rather, Putnam claims there are 'up and downs in civic engagement'; it is 'a story of collapse *and* renewal'.[60] However, for Putnam, the United States has been going through a collapse of civic engagement or of social capital since the 1970s.

Putnam's rejection of a declinist account of civic participation is potentially optimistic for he manifestly believes that the current decline might be reversed; there is no historical necessity to it. However, the fact that the current decline is now a long-term process connected with inevitable features of modern society is taken by Putnam to indicate that the decline of civic participation may be pathological. It is the product of long-term developments which have made communal associations gradually irrelevant. Putnam suggests that a number of different indices demonstrate this decline in civic participation; voting rates have declined from 62 per cent in 1960 to 48.9 per cent in 1996, voluntary associations have tripled in number since 1975 but are now a tenth of their size, while the number of men and women who took any leadership role in any local organizations was reduced by 50 per cent between 1974 and 1994.[61] Similarly, while sports watching has increased, participation rates have declined, even with the fitness boom of the 1990s. Putnam notes the increase in voluntary work but claims that it is 'not really an exception to the broader generational decline in social capital'.[62]

For Putnam, the decline of community or social capital is fundamentally a generational phenomenon. The high point of twentieth-century social capital coincided with the early and middle adulthood of the generation which was born in the 1920s and 1940s. This 'Great' Generation experienced the Depression, Roosevelt's New Deal, the Second World War, and post-war affluence. It was, therefore, imbued with the spirit of civic duty and the manifest benefits which the performance of such duty brought; 'virtually no cohort in America is more engaged or more tolerant than those born around 1940–45.'[63] By contrast, for both the baby-boom generation, born between 1946 and 1964, and the X Generation, born between 1965 and 1985, the benefits of civic participation were much less clear. However, Putnam, despite the regular criticism of the X Generation, reserves his scorn for the baby boomers; the X Generation only 'accelerated the tendencies to individualism found among boomers'.[64] One of the decisive factors in the decline of civic participation according to Putnam has been the decline of the family and the baby boomers' denial of their parental and spousal responsibilities.

There are many problems with Putnam's account. While there are good reasons to explain why the 'Great' Generation has been more involved in civic organizations than their children and grandchildren, Putnam's celebration of this generation does not seem entirely justified. The achievements of the post-war settlement are clear but they also involved high levels of authoritarianism (especially in southern, central, and eastern Europe), and even democratic regimes in those decades are now widely regarded as fundamentally racist and sexist. The emergence of civil rights, feminist, and gay rights movements from the late 1950s casts doubt on Putnam's assertion that this generation was tolerant. More importantly, while Putnam's evidence that a certain kind of communal association is declining is plausible, he seems to overstate the case. Specifically, Putnam favours formal voluntary institutions in which citizens gather to address a suite of needs. He admits that not all social networks have atrophied. On the contrary, 'thin, single stranded, surf-by interactions are gradually replacing dense, multi-stranded, well-exercised bonds. More of our social connectedness is of one-shot, special purpose and self-oriented.'[65] By contrast, 'Large groups with local chapters, long histories, multiple objectives and diverse constituencies are being replaced by more evanescent, single-purpose organizations, smaller groups.'[66] He concludes, 'place-based social capital is being supplanted by function-based social capital'.[67] This is an important passage for, here, Putnam revises his argument fundamentally. He is no longer arguing that there is no social capital or that all community has gone but only that the basis of public association has changed; social capital is being generated in different ways. Association has become more temporary, issue-based, and functional. Perhaps Putnam is right to prefer traditional forms of organization and perhaps they are more effective but his judgement seems based on personal preference rather than evidence; the adjectives 'thin' and 'dense' which he freely uses are manifestly not neutral terms but deliberately applaud one form of association while denigrating the other—without justifying why. Yet, in an increasingly mobile, globalized, and diverse society in which gender and social hierarchies have been substantially revised perhaps the formal associations of the Great Generation have become obsolete and archaic. In a globalizing society, it is conceivable that the more transient social movement may be a more effective way of generating social capital. By contrast, local chapters and formal organizations may indeed have created stable, multiple bonds between their members, but perhaps only because they were already socially similar to each other. The formal organization perhaps was better adapted to the more homogeneous and, in fact, discriminatory society of the immediate post-war period.

Professionalization, not only in the military, but across civil society more widely, may be important in the light of these changing patterns of solidarity of which Putnam speaks. Certainly, professionalism is by no means the only or the most important form of solidarity in western society in the twenty-first century. As Putnam and new social movement theorists, like Melucci, Touraine, or Maffesoli, have noted, a bewildering variety of new associations, often voluntary, informal, and contingent, have appeared in the last few decades.

Moreover, the nuclear family and especially the relationship between parents and their children has been radically reformed so that older and younger generations share intimacies which were unknown between the parents and children of previous generations. In a diverse and complex globalizing society, there are naturally multiple forms of solidarity suspending the individual in overlapping, interconnected, and increasingly transnational social networks. The professional status group constitutes only one stratum of this highly complex geography of interlinked groups and associations. However, as the mass labour force has been replaced by specialist technicians, the professional group, no longer limited to the traditional professions, seems to be an increasingly prominent feature in this globalizing landscape. In particular, the principle of professionalism seems to be an increasingly important means of collective definition.

In his recent work *The Civil Sphere*,[68] Jeffrey Alexander has explored the shared normative basis of complex, globalizing democracies. Although his intellectual influences are diverse, his work might be read as an attempt to modernize Durkheim and Talcott Parsons for the diverse and diffuse social orders of the twenty-first century. Certainly, Alexander is a long way from the simple, even mechanical, concept of Parsonian norms. Nevertheless, not implausibly, he suggests that even globalizing societies require some shared sphere in which social and cultural presumptions are debated and established. This is the 'civil sphere' which includes but extends well beyond formal political institutions and the traditionally conceived public sphere: 'civil society is not merely an institutional realm. It is also a realm of structured, socially established consciousness, a network of understandings creating structures of feeling that permeate social life and run just below the surface of strategic institutions and self-conscious elites.'[69] Crucially, this dense and often unacknowledged subterranean world of 'socially established consciousness' not only articulates the complex networks of which society is comprised but it underpins and informs the exercise of power. Alexander does not in any way deny the influence of social, political, and economic elites but their authority is necessarily constrained and channelled by wider normative presumptions of legitimacy, which are not defined by them but by the civil sphere generally: 'these dynamics [of social power] are everywhere subject, not only in principle but also in practice, to the communicative and regulatory powers of the civil sphere.'[70] Alexander explores the historical development of the civil sphere in western and, especially, American society, seeking to show how its norms have been structured by a series of binary oppositions. Alexander does not discuss professionalism as part of this civil sphere. He is more concerned with considering the ethical dimensions of this space. Nevertheless, it seems plausible to suggest that professionalism may be an increasingly important part of the civil sphere in globalizing societies. It seems possible that professionalism has become a shared cultural reference point, potentially uniting individuals even though they are employed in quite different spheres. According to Durkheim, a division of labour was possible and could avoid its fragmentation only if a collective conscience had been developed which sensitized individuals to their interdependence. Individuals could be different only if their communal allegiance had already been established. In the absence of this sense of community, the division of labour would necessarily drive individuals and groups apart. The concept of professionalism may serve the purpose which Alexander identifies for the civil sphere, preceding, underpinning, and informing social interaction, political debate, and the exercise of power and, crucially, unifying the multiple networks which characterize contemporary society. Professionalism may be a central integrative element in the civil sphere.

There seems to be considerable evidence that the concept of professionalism has become increasingly salient in public culture, as a collective reference point. In the past two decades, the de-regulated global media have increasingly focused on human interest stories in order to generate and sustain audiences. As Richard Sennett decried in the 1970s, the personal has become public with a resulting corruption of the public sphere. Citizens have become personalities. The proliferation of reality television shows in which ordinary people or celebrities are exposed to constant surveillance is a prime example of this personalization of public life and the publicizing of the private. Indeed, the current phone-hacking scandal which is engulfing Rupert Murdoch's News Corporation in the United Kingdom might be seen only as the extreme and criminal extension of a much wider trend. However, although personalization is very evident, the new media have also increasingly focused on individual skill to attract the public. Thus, every major television network in western societies substantially relies on programmes in which ordinary citizens or celebrities compete with each other to demonstrate their competences in a specific area of activity: cooking, dancing, singing, conducting, adventure challenges, or business. The most successful of these programmes typically fuse a fascination with the personal and displays of professional expertise, training, and development. The UK-conceived programme 'The Apprentice', where candidates perform a series of entrepreneurial tasks to be hired by a well-known English businessman, Alan Sugar, most explicitly tests for professional business aptitude but the UK and US versions of 'The X Factor' are, in fact, similarly selecting for professional performance skills. These programmes celebrate professional skills and, in the case of talent shows for ordinary citizens, they are means of selecting unknown individuals for professional employment in the music industry.

Putnam is unlikely to approve of the new salience of professionalism as a means of generating social solidarity because, as we have seen, professionalism is almost the exact opposite of the dense multiply stranded relations between citizens which Putnam advocates. Professionalism is not based on 'likeness' and it does not necessarily involve deep personal commitment to fellow professionals, though such commitments can develop. Rather professionalism involves the generation of cooperative relations around discrete areas of expertise; it is highly specific and limited, although in established professions, these relations involve complex performances and therefore demand dense cooperative ties. Moreover, as the military profession demonstrates, professional expertise increasingly transcends national borders. It is becoming transnational, allowing individuals from different countries, increasingly working together in the same multinational corporations or internationalizing public sector organizations, to cooperate with each other. Professionalism may be an increasingly important part of our future. It is unlikely to be the sole basis for community but it seems plausible to suggest that professionalism has become a central means of generating flexible, contingent, and transnational solidarities which reflect the increasingly globalized world in which we now exist. This world does not seem to be as comfortable as the stable hierarchies of the 1950s which Putnam wishes to idealize. It is potentially more anonymous but, in its flexibility and its inclusiveness, it is perhaps more adaptable than the rigid hierarchies and orders favoured by the Great Generation.

It has become a standard and sometimes platitudinous claim in history and the social sciences that warfare in any era reflects the nature of the society which wages it. This is, of course, inevitably the case. The claim becomes interesting however when the precise ways in which warfighting reflects wider patterns of social organization are illustrated. Hans Delbrück's seminal work on the history of warfare is a prime example here.[71] He shows how systems of political organization from antiquity to the Napoleonic Wars generated specific types of military force and, therefore, determined the character of warfare in any epoch. For Delbrück, the emergence of the infantry at the beginning of the modern era was the unique achievement of Swiss democracy and 'was the contribution of the Swiss to world history'.[72] Similarly, western armed forces of the twenty-first century seem to reflect their host societies. Yet, without detailed exposition, this statement is vacuous. However, once professionalism is identified as the defining characteristic of today's armies and the distinct qualities and requirements of professionalism are recognized, the connection between the way western democracies fight wars and the way they organize themselves more generally begins to emerge, in a perhaps suggestive way. The professional army is not merely defined by the fact that the soldiers in it are paid for their voluntary service. Professionalism refers above all to a complex of competences and a distinctive corporate identity which binds the members of the military together, committing them mutually to their duties. Cohesion—or combat performance—in the military is no longer based on 'likeness'. The potentially warm solidarity celebrated nostalgically as the 'Great Generation' which experienced that cohesion in the Second World War has gone, while its failings and exclusions are conveniently forgotten. In its place, the colder and more impersonal bonds of professionalism have appeared. This may be personally regrettable but the evidence suggests that it has produced a more competent military. Moreover, in the professional armed forces, it is possible in the extreme environment of combat to see a new kind of community which is not limited only to the military. On the contrary, the professional community increasingly displayed by western military forces seems to have much wider resonance in contemporary society. A community based on competence and skill, not on social origin or similarity, seems to reflect and affirm today's reordered, diverse, globalizing societies.

For the past ten years, western industrial powers have been fighting a war in Afghanistan. In stark contrast to the wars they fought with and against each other in the twentieth century, these wars are not conflicts fought by citizens in defence of their nations. These wars have arisen rhizomatically along the new fissures of the global order and they are fought by professionals; small groups of specialists. These specialists often appear unconnected to the rest of civil society precisely because so few citizens have any connection with the military today. Indeed, Huntington might perhaps argue there is an even greater divide between the Spartan armed forces of the west and the Babylonian societies which they defend than when he wrote *The Soldier and the State*. Yet, against Huntington's concept of objective control, it might in fact be possible to claim that through their professionalism, soldiers today are as closely connected with their host society as the often badly trained citizens of the great wars of the twentieth century. A civilian may have as much in common with a soldier fighting in Kandahar as they do with a trained barrister, a radiologist equipped with the latest technology, or a physicist in a university laboratory and engaged in arcane professional practices. Citizens in twenty-first-century society may be divided by their professional expertise but perhaps, ironically, also united by it; their vocation may be the one thing which they now also share. The professional military is, indeed, increasingly specialist (and Spartan) as Huntington argued but it is not brought under objective political and social control because its soldiers have a duty to defend a society from which they are different, as Huntington thought. Its arcane expertise may be precisely what connects it to its host society and to other civilians; in their execution of their highly distinctive specialist duties, soldiers perform like civilian professionals. They, like their civilian peers, are imbued with skills and galvanized by the same pursuit of excellence. There is a paradoxical convergence on a shared ethos of professionalism, even as professional expertise becomes ever more specialist. Yet, this convergence on professionalism does not seem to mean that the military is coming under subjective military control. The armed forces do not seem to be civilianizing, as Janowitz claimed. Ironically, and against Janowitz's concept of subjective control, the armed forces seem to be converging with the standards and expectations in professional civilian society insofar as they distance themselves from it and dedicate themselves solely to their vocation. The military may be under increasing subjective and objective control and simultaneously freed from either. It might be more accurate to say that the military is coming under professional control of an increasingly diverse—yet vocationally unified—civilian society.

Western citizens are generally ignorant of the war in Afghanistan and have even less idea about how the military conducts operations there. Nevertheless, in their increasing prioritization of skills, performance, and competence and their concomitant indifference to social homogeneity and 'likeness', civilians may be strangely similar to infantry soldiers fighting in the towns and villages of southern Afghanistan, even though they apparently have nothing in common with them. Western nations seem to be fighting wars professionally because, as societies, they are becoming increasingly professionalized. Professionalism, for all its failings and inadequacies, may be becoming both the means of competing and the basis of solidarity in the increasingly diverse and fluidly globalizing world in which all are now living. Indeed, the prospects for those outside professional society, especially following the Credit Crunch and recession, are potentially bleak.

Notes

PRELIMS

1. Footnote 18, Anthony King, 'The Existence of Group Cohesion in the Armed Forces', *Armed Forces & Society* 33(4) 2007: 645.

CHAPTER 1

1. E. K. Taylor 2009. *America's Army and the Language of Grunts*, Bloomington, Ind.: Author House, 347. The quotation has a number of variants.
2. H. Ferguson 2004. 'The Sublime and Subliminal: modern identities and the aesthetics of combat', *Theory Culture and Society* 21(3): 1–33; E. Leed 1979. *No Man's Land*. Cambridge: Cambridge University Press, 19–21.
3. E. Jünger 2003. *Storm of Steel*. Harmondsworth: Penguin, 101.
4. Jünger, *Storm of Steel*, 254.
5. Jünger, *Storm of Steel*, 31.
6. Jünger, *Storm of Steel*, 256.
7. J. Romains 2000. *Verdun*. Translated by Gerard Hopkins. London: Prion, 93.
8. Romains, *Verdun*, 93.
9. H. Barbusse 2003. *Under Fire*. Translated by Robin Buss. Harmondsworth: Penguin, 136.
10. E. Blunden 1982. *Undertones of War*. Harmondsworth: Penguin, 67.
11. L. Wyn Griffiths 1981. *Up to Mametz*. Norwich: Gliddon Books, 232.
12. J. C. Dunn 1987. *The War the Infantry Knew 1914–1919: a chronicle of service in France and Belgium with the Second Battalion, His Majesty's Twenty-Third Foot, the Royal Welch Fusiliers founded on personal records, recollections and reflections*. London: Cardinal, 315–16.
13. Dunn, *The War the Infantry Knew*, 318.
14. Dunn, *The War the Infantry Knew*, 318.
15. J. R. R. Tolkien 1977. *The Lord of the Rings: the return of the king*. London: George Allen and Unwin, 186.
16. J. R. R. Tolkien 1977. *The Lord of the Rings: the two towers*. London: George Allen and Unwin, 206.
17. J. Edmonds and C. Falls 1940. *Military Operations: France and Belgium 1917*. London: Macmillan and Co, 113.
18. Jünger, *Storm of Steel*, 128.
19. Jünger, *Storm of Steel*, 232.
20. Jünger, *Storm of Steel*, 33.
21. Jünger, *Storm of Steel*, 236.
22. Jünger, *Storm of Steel*, 232.
23. G. Sajer 1993. *The Forgotten Soldier*. London: Orion, 239.
24. Sajer, *The Forgotten Soldier*, 257.

25. N. Mailer 1949. *The Naked and the Dead*. London: Allan Wingate, 212.
26. Mailer, *The Naked and the Dead*, 213.
27. J. Heller 1994. *Catch-22*. London: Vintage, 501.
28. Heller, *Catch-22*, 503.
29. Heller, *Catch-22*, 504.
30. Heller, *Catch-22*, 504.
31. Heller, *Catch-22*, 52.
32. Heller, *Catch-22*, 510–11.
33. Heller, *Catch-22*, 513–14.
34. Heller, *Catch-22*, 519.
35. The corrosion of human community is central to the war paintings of Otto Dix and Paul Nash. Dix's 'War' triptych or Nash's 'Menin Road' and 'We are making a new World', explore the elimination of humanity by industrial warfare. The same theme of war as a hell which annihilates human community is central to Benjamin Britten's *War Requiem* which specifically equates the First World War with Judgement Day and uses Wilfred Owen's poem 'Strange Meeting', which includes a descent to the Underworld.
36. P. Fussell 1975. *The Great War and Modern Memory*. Oxford: Oxford University Press, 335.
37. Fussell, *The Great War and Modern Memory*, 115; Leed, *No Man's Land*, 115.
38. Indeed, although writers and artists were drawn to the image of war as a 'God of Pain', the ideal of the community remains central to most of these representations which decry its loss. Thus, Barbusse's critique of war is predicated on his faith in human society. His anger at the First World War is precisely directed at the loss of the community. He recognizes that the cause of the war is not the disintegration of all society but rather the corruption of noble communal ideals into aggressive nationalism: 'Out of patriotism—which is respectable as long as it remains in the realm of feelings and the arts, just like the love of family or one's province, which are equally sacred—they make a utopian, non-viable concept... Morality is delightful—and they pervert it. How many crimes have they made into virtues, with a single word, by calling them "national"?' (Barbusse, *Under Fire*, 216). In the extraordinary, if not bizarre, ending of the book, the patriotic moral community is resurrected. The soldiers arise together out of the flooded trenches to reject the war and to announce 'an understanding between democracies, between immense powers, a rising of the peoples of the world' (Barbusse, *Under Fire*, 318). 'Between two masses of mirky clouds a tranquil ray shines out' (Barbusse, *Under Fire*, 319) to signify this new world community. Otto Dix's War Triptych depicts a similarly idealized community represented by the spectral dead soldiers helping each other from the battlefield. These are idealized communities formed after war in opposition to the presumed nihilism of combat.
39. T. Ashworth 1980. *Trench Warfare 1914–1918: the live and let live system*. London: Macmillan, 13.
40. Ashworth, *Trench Warfare 1914–1918*, 13.
41. Ashworth, *Trench Warfare 1914–1918*, 14–15.
42. Ashworth, *Trench Warfare 1914–1918*, 32.
43. Ashworth, *Trench Warfare 1914–1918*, 38.

44. Ashworth, *Trench Warfare 1914–1918*, 38.
45. Ashworth, *Trench Warfare 1914–1918*, 39.
46. H. Strachan 2006. 'Training, Morale and Modern War', *Journal of Contemporary History* 41(2) April: 213.
47. Ashworth, *Trench Warfare 1914–1918*, 48–9.
48. Ashworth, *Trench Warfare 1914–1918*, 71.
49. R. Collins 2008. *Violence: a microsociological theory*. Princeton: Princeton University Press, 39–82, 105–6.
50. Collins, *Violence*, 85–7.
51. Collins, *Violence*, 85.
52. Collins, *Violence*, 85.
53. Collins, *Violence*, 87; P. Caputo 1999. *A Rumour of War*. London: Pimlico, 304.
54. Collins, *Violence*, 95.
55. Collins, *Violence*, 97.
56. John Hockey's now classic study of 'squaddies' provides a very perceptive sociology of garrison life, rather than combat. It is particularly useful on the dynamics of deviance and work avoidance among soldiers (J. Hockey 1986. *Squaddies: Portrait of a subculture*. Exeter: University of Exeter). More recently, Hockey has explored the experiential phenomenology of soldiering e.g. J. Hockey 2009. 'Switching on: sensory work in the infantry', *Work, Employment and Society* 23(3): 477–93. See also Charles Kirke's formal model of cohesion aimed at describing the daily lives of soldiers, not specifically combat performance. (C. Kirke 2011. *Red Coat, Green Machine: continuity and change in the British army 1700–2000*. London: Continuum; C. Kirke 2009. 'Group Cohesion, Culture and Practice', *Armed Forces & Society* 35(4): 745–53.
57. N. Machiavelli 1965. *The Art of War*. Indianapolis: Bobbs Merrill, 7–43.
58. N. Machiavelli 1999. *The Prince*. Translated by G. Bull. London: Penguin, 39 (italics added).
59. Machiavelli, *The Prince*, 40.
60. Machiavelli, *The Prince*, 41.
61. K. von Clausewitz 1982. *On War*. Translated by M. Howard and P. Paret. Princeton: Princeton University Press, 187.
62. T. Parsons 1949. *The Structure of Social Action*. New York: Free Press.
63. Parsons, *The Structure of Social Action*, 88.
64. Parsons, *The Structure of Social Action*, 91.
65. Parsons, *The Structure of Social Action*, 91.
66. Durkheim's own biography connects him to the central theme of this book; his son, André, was killed in the First World War.
67. É. Durkheim 1976. *The Elementary Forms of the Religious Life*. Translated by J. Ward Swain. Old Woking: Unwin Brothers, 5.
68. Durkheim, *The Elementary Forms of the Religious Life*, 5.
69. Durkheim, *The Elementary Forms of the Religious Life*, 5.
70. Durkheim, *The Elementary Forms of the Religious Life*, 5.
71. Durkheim, *The Elementary Forms of the Religious Life*, 6.
72. Durkheim, *The Elementary Forms of the Religious Life*, 7.
73. Jünger, *Storm of Steel*, 1.

74. K. Marlantes 2011. *What it is Like to Go to War*. New York: Atlantic Monthly Press, 7.
75. Grand Quartier General 1916. *Manuel du chef de section d'infanterie*. Paris: Imprimerie Nationale, 16.
76. War Office. 1921. *Infantry Training*, ii: *War*. London: His Majesty's Stationery Office, 78.
77. P. Savage and R. Gabriel 1976. 'Cohesion and Disintegration in the American Army: an alternative perspective', *Armed Forces & Society* 2(3): 364.
78. G. Siebold and D. Kelly 1988. *Development of the Platoon Cohesion Questionnaire*. Alexandria, Va.: US Army Research Institute for the Behavioral and Social Sciences, Manpower and Personnel Laboratory.
79. D. Marlowe 1979. 'Cohesion, Anticipated Breakdown, and Endurance in Battle'. Unpublished draft, Washington: Walter Reed Army Institute of Research.
80. R. MacCoun 1993. 'What is Known about Unit Cohesion and Military Performance', in National Defence Research Institute, *Sexual Orientation and US Military Personnel Policy: options and assessment*. Washington: RAND, 303.
81. I. Bloch 1992. *Is War now Impossible?* Aldershot: Gregg Revivals.
82. P. Kindsvatter 1995. *American Soldiers: ground combat in the World Wars, Korea and Vietnam*. Lawrence, Kan.: University of Kansas Press.
83. J. Black 2010. *War: a short history*. London: Continuum.
84. Black, *War*, 163–7.
85. It is also acknowledged that territorial, militia, or paramilitary forces have played an important part in warfare and have been used frequently by western and non-western states. Some of the arguments forwarded here may be potentially applicable to forces of this kind but citizen soldiers of this type are not explicitly discussed.
86. The contribution of Russian generals to developments at the operational level before and during the Second World War and then again with Marshall Ogarkov's reforms in the 1970s are fully recognized. However, the Russian army is excluded from dedicated treatment here because, despite their expertise in sniping and infiltration in the Second World War, and the fact that they were key combatants in the First and Second World Wars and the Cold War, their infantry were typically used throughout the twentieth century as a mass, unskilled, and conscript force. Despite recent reforms, they predominantly remain such a force today (see Chapter 12).

CHAPTER 2

1. M. Janowitz and E. Shils 1948. 'Cohesion and Disintegration in the Wehrmacht in World War II', *Public Opinion Quarterly* Summer: 281.
2. Janowitz and Shils, 'Cohesion and Disintegration', 286.
3. Janowitz and Shils, 'Cohesion and Disintegration', 291.
4. Janowitz and Shils, 'Cohesion and Disintegration', 304–6.
5. Janowitz and Shils, 'Cohesion and Disintegration', 281.
6. Janowitz and Shils, 'Cohesion and Disintegration', 315.
7. Janowitz and Shils, 'Cohesion and Disintegration', 284.

8. Janowitz and Shils, 'Cohesion and Disintegration', 284.
9. Janowitz and Shils, 'Cohesion and Disintegration', 284.
10. In fact, although it is now normally overlooked, Janowitz and Shils fully recognized the importance of political indoctrination in their discussions of the Wehrmacht. However, in their conclusion, the cohesion of the primary group is prioritized, somewhat unjustifiably given the manifest importance of political motivation which their own evidence provides.
11. N. Kinzer Stewart 1991. *Mates and Muchachos: unit cohesion in the Falklands/Malvinas War.* New York: Brasseys Inc, 16–17.
12. Kinzer Stewart, *Mates and Muchachos*, 17–18.
13. D. Henderson 1985. *Cohesion: the human element.* Washington: National Defence University Press, 155.
14. Henderson, *Cohesion*, 152.
15. Henderson, *Cohesion*, 18.
16. Henderson, *Cohesion*, 55.
17. Henderson, *Cohesion*, 18.
18. Henderson, *Cohesion*, 18.
19. Henderson, *Cohesion*, 19.
20. Henderson, *Cohesion*, 19.
21. L. Wong et al. 2003. *Why They Fight: combat motivation in the Iraq War.* Carlisle Barracks, Pa.: Strategic Studies Institute, US Army War College.
22. Wong et al., *Why They Fight*, 17.
23. Wong et al., *Why They Fight*, 9.
24. Wong et al., *Why They Fight*, 10.
25. Wong et al., *Why They Fight*, 10.
26. Wong et al., *Why They Fight*, 10–11.
27. Wong et al., *Why They Fight*, 11.
28. Wong et al., *Why They Fight*, 13.
29. Wong et al., *Why They Fight*, 12–13.
30. A. Tziner and Y. Vardi 1982, 'Effects of Command Style and Group Cohesiveness on the Performance Effectiveness of Self-Selected Tank Crews', *Journal of Applied Psychology* 67(6): 773.
31. J. Griffith 1989. 'The Army's New Unit Personnel Replacement and its Relationship to Unit Cohesion and Social Support', *Military Psychology* 1(1): 20–1.
32. Griffith, 'The Army's New Unit Personnel Replacement', 30–2.
33. G. Siebold 2007. 'The Essence of Military Cohesion', *Armed Forces and Society* 33 (2): 288.
34. D. Segal and M. Kestnbaum 2002. 'Professional Closure in the Military Market: a critique of pure cohesion', in D. Snider (ed.), *The Future of the Army Profession.* New York: Primis.
35. L. Festinger, S. Schachter, and K. Back 1950. *Social Pressures in Informal Groups.* New York: Harpers and Brothers.
36. M. Hogg 1992. *The Social Psychology of Group Cohesiveness: from attraction to social identity.* New York: Harvester Wheatsheaf, 19.
37. Festinger et al., *Social Pressures in Informal Groups*, 164.
38. L. Festinger 1950. 'Informal Social Communication', *Psychological Review* 57: 274.

39. Festinger et al., *Social Pressures in Informal Groups*, 10.
40. A. Lott and B. Lott 1965. 'Group Cohesion and Interpersonal Attraction: a review of relationships with antecedent and consequent variables', *Psychological Bulletin* 64(4): 259.
41. Lott and Lott, 'Group Cohesion and Interpersonal Attraction', 259.
42. M. Shaw 1976. *Group Dynamics: the psychology of small group relations*. New York: McGraw Hill, 197.
43. Shaw, *Group Dynamics*, 197.
44. A. Zander 1982. *Making Groups Effective*. London: Jossey Bass, 4; C. Evans and K. Dion 1991. 'Group and Performance: a meta-analysis', *Small Group Research* 22(7) May: 175; L. Berkowitz 1954. 'Group Standards, Cohesiveness and Productivity', *Human Relations* 7: 509.
45. N. Gross and W. Martin 1952. 'On Group Cohesiveness', *American Journal of Sociology* 57(6) May: 553.
46. Hogg, *The Social Psychology of Group Cohesiveness*, 1.
47. I. Janis 1982. *Group Think: psychological studies of policy decisions and fiascoes*. Boston: Houghton Miffin Company, 8.
48. MacCoun, 'What is Known about Unit Cohesion and Military Performance', 295.
49. L. Rosen, K. Knudson, and P. Fancher 2002. 'Cohesion and the Culture of Hypermasculinity in the US Army Units', *Armed Forces and Society* 29(3) Spring: 325–52.
50. D. Winslow 1997. *The Canadian Airborne Regiment: a socio-cultural inquiry*. Ottawa: Minister of Public Works and Government Services, Canada, 123–4.
51. Winslow, *The Canadian Airborne Regiment*, 97.
52. Winslow, *The Canadian Airborne Regiment*, 101.
53. C. Hamner 2011. *Enduring Battle: American soldiers in three wars, 1776–1945*. Lawrence, Kan.: University of Kansas Press, 175.
54. J. G. Fuller 1990. *Troop Morale and Popular Culture in the British and Dominion Armies 1914–18*. Oxford: Clarendon Press, 27–9.
55. Fuller, *Troop Morale and Popular Culture*, 22–6.
56. Fuller, *Troop Morale and Popular Culture*, 169.
57. Fuller, *Troop Morale and Popular Culture*, 174.
58. Hogg, *The Social Psychology of Group Cohesiveness*, 58.
59. R. Martens and J. Peterson 1971. 'Group Cohesiveness as a Determinant of Success and Membership Satisfaction in Team Performance', *International Review of Sport Psychology* 6: 50.
60. Martens and Peterson, 'Group Cohesiveness as a Determinant of Success', 56–7.
61. D. M. Landers and G. Lüschen 1974. 'Team Performance Outcome and Cohesiveness of Competitive Coacting Groups', *International Review of Sport Sociology* 9: 57–69.
62. Hogg, *The Social Psychology of Group Cohesiveness*, 143.
63. C. Moskos 1975. 'The American Combat Soldier in Vietnam', *Journal of Social Issues* 31(4): 37.
64. C. Moskos 1970. *The American Enlisted Man: the rank and file in today's military*. New York: Russell Sage Foundation, 145.
65. Moskos, 'The American Combat Soldier in Vietnam', 37.

66. Moskos, *The American Enlisted Man*, 145.
67. Moskos, *The American Enlisted Man*, 145.
68. Cited in MacCoun, 'What is Known about Unit Cohesion and Military Performance', 306.
69. Lott and Lott, 'Group Cohesion and Interpersonal Attraction', 261.
70. MacCoun, 'What is Known about Unit Cohesion and Military Performance'.
71. MacCoun, 'What is Known about Unit Cohesion and Military Performance', 291.
72. MacCoun, 'What is Known about Unit Cohesion and Military Performance', 291.
73. E. Kier 1998. 'Homosexuality in the US Military: open integration and combat effectiveness', *International Security* 23(2): 17.
74. Kier, 'Homosexuality in the US Military', 19.
75. Kier, 'Homosexuality in the US Military', 21.
76. MacCoun, 'What is Known about Unit Cohesion and Military Performance', 293.
77. MacCoun, 'What is Known about Unit Cohesion and Military Performance', 293.
78. MacCoun, 'What is Known about Unit Cohesion and Military Performance', 313.
79. Parsons, *The Structure of Social Action*, 24.
80. In this way, the study is closely related to historical research on the related question of morale. Baynes, Fuller, Watson, and Fennell have all sought to identify the generation or collapse of morale by reference to a wide range of institutional factors.
81. Janowitz and Shils, 'Cohesion and Disintegration in the Wehrmacht in World War II', 281.
82. Parsons, *The Structure of Social Action*, 44.
83. Hogg, *The Social Psychology of Group Cohesiveness*, 54.
84. Hogg, *The Social Psychology of Group Cohesiveness*, 54.
85. Hogg, *The Social Psychology of Group Cohesiveness*, 54.
86. S. Biddle 2004. *Military Power*. Princeton: Princeton University Press.
87. Headquarters Department of the Army 2007. *Field Manual 3–21.8. The Infantry Rifle Platoon and Squad*. Washington, 1–2.

CHAPTER 3

1. S. L. A. Marshall 1983. *Ambush: the battle of Dau Tieung*. New York: Jove, 3.
2. R. Spiller 1988. 'SLA Marshall and the Ratio of Fire', RUSI *Journal* 133/December: 69.
3. Spiller, 'SLA Marshall and the Ratio of Fire', 69.
4. S. L. A. Marshall 2000. *Men against Fire*. Norman, Okla.: University of Oklahoma Press, 56.
5. S. L. A. Marshall 1968. *Bird*. New York: Warner Books, 57.
6. Marshall, *Bird*, 23.
7. D. Grossman 1996. *On Killing: the psychological costs of learning to kill in war and society*. Boston: Back Bay Books.
8. Marshall, *Bird*, 44–7.
9. P. Kindsvatter 1995. *American Soldiers: ground combat in the World Wars, Korea and Vietnam*. Lawrence, Kan.: University of Kansas Press, 222–3.

10. F. Smoler 1989. 'The Secret of the Soldiers who wouldn't Shoot', *American Heritage* 40: 7.
11. Spiller, 'SLA Marshall and the Ratio of Fire', 67.
12. S. L. A. Marshall 1986a. *Night Drop*. New York: Jove, 89–90, 103.
13. Smoler, 'The Secret of the Soldiers who wouldn't Shoot', 3.
14. Spiller, 'SLA Marshall and the Ratio of Fire', 68.
15. Spiller, 'SLA Marshall and the Ratio of Fire', 68.
16. Spiller, 'SLA Marshall and the Ratio of Fire', 68.
17. Spiller, 'SLA Marshall and the Ratio of Fire', 68–9.
18. Spiller, 'SLA Marshall and the Ratio of Fire', 69.
19. Spiller, 'SLA Marshall and the Ratio of Fire', 69.
20. Spiller, 'SLA Marshall and the Ratio of Fire', 69.
21. J. Whiteclay Chambers II 2003. 'SLA Marshall's *Men against Fire*: new evidence regarding fire ratios', *Parameters*: 117.
22. Whiteclay Chambers II, 'SLA Marshall's *Men against Fire*', 119.
23. Whiteclay Chambers II, 'SLA Marshall's *Men against Fire*', 119–20.
24. Spiller, 'SLA Marshall and the Ratio of Fire', 63–5; Whiteclay Chambers, 'SLA Marshall's *Men against Fire*', 121.
25. See also P. Mansoor 1999. *The GI Offensive in Europe*. Lawrence, Kan.: University of Kansas, 262–7.
26. Marshall, *Bird*, 56.
27. Marshall, *Bird*, 56.
28. S. L. A. Marshall 2001. *Island Victory: the battle of Kwajalein Atoll*. Lincoln, Nebr.: University of Nebraska Press, 1.
29. Marshall, *Island Victory*, 1.
30. The Island episode has itself been questioned by Smoler who claimed to have examined Marshall's work *The Capture of Makin*. He asserts that there is no specific reference to this Japanese assault and its repulse by small numbers of firers from the 3rd Battalion, 165th Infantry Regiment. Rather than too little shooting, the text records 'much aimless shooting by "trigger-happy" men' (S. L. A. Marshall 1990. *The Capture of Makin*. Washington: Centre of Military History United States Army, 94; Smoler, 'The Secret of the Soldiers who wouldn't Shoot', 6). Yet, this description refers not to the defensive action which occurred on the third night of the invasion, 'the Fight on Saki Night' of 22–3 November, but to an action on the morning of 21 November. Following a difficult first night, an engineer falsely identified the presence of a significant force of Japanese in the jungle which precipitated a 'wave of shooting hysteria' (Marshall, *The Capture of Makin*, 94; Smoler, 'The Secret of the Soldiers who wouldn't Shoot', 6). 'When the engineer admitted that he had seen no enemy but merely "had heard firing", shouted orders to cease fire proved ineffectual. Direct orders to each individual soldier were necessary.' This incident is also recorded verbatim in the US Army's official history of the operation (K. Roberts Greenfield 1955. *The United States Army in World War II*. Washington: Office of the Chief of Military History, Department of the Army, 108). On this evidence, Smoler concludes that the performance of US troops was creditable. The important incident on the third night of the Island fight is, in fact, described at length in the *Capture of Makin*, and only by ignoring this

event can Smoler sustain his claim about performance levels (Marshall, *The Capture of Makin*, 114, 118–19).
31. Roberts Greenfield, *The United States Army in World War II*, 122–4.
32. Roberts Greenfield, *The United States Army in World War II*, 118–22.
33. Roberts Greenfield, *The United States Army in World War II*, 122.
34. Roberts Greenfield, *The United States Army in World War II*, 123.
35. Roberts Greenfield, *The United States Army in World War II*, 123.
36. Roberts Greenfield, *The United States Army in World War II*, 123.
37. Roberts Greenfield, *The United States Army in World War II*, 123.
38. Roberts Greenfield, *The United States Army in World War II*, 123.
39. Roberts Greenfield, *The United States Army in World War II*, 123.
40. Roberts Greenfield, *The United States Army in World War II*, 125.
41. Roberts Greenfield, *The United States Army in World War II*, 124.
42. Marshall, *Island Victory*, 82.
43. Marshall, *Island Victory*, 66.
44. The actions of the 2nd Battalion, 506nd Parachute Infantry Regiment have been recorded by Stephen Ambrose (*Band of Brothers*) and immortalized in the HBO film based on this book.
45. Marshall, *Men against Fire*, 72.
46. Marshall, *Men against Fire*, 72, emphasis added.
47. Marshall, *Night Drop*, 332.
48. Marshall, *Night Drop*, 332.
49. Marshall, *Night Drop*, 334.
50. Marshall, *Night Drop*, 334.
51. R. Glenn, 2000. *Reading Athena's Dance Card: men against fire in Vietnam.* Annapolis, Md.: Naval Institute Press, 135.
52. Collins, *Violence*, 95; Marshall, *Men against Fire*, 183.
53. Marshall, *Night Drop*, 131.
54. Marshall, *The Capture of Makin*, 94.
55. S. Ambrose 1997. *Citizen Soldiers*. New York: Touchstone, 125.
56. Ambrose, *Citizen Soldiers*, 128.
57. Marshall, *Men against Fire*, 56.
58. Indeed, while mistakes and generally poor performances are evident in Marshall's work on the Second World, Korean, and Vietnam Wars, it is certainly difficult to extract even a rough figure of one in four firers from his narratives. For instance, in his account of the Battle of Kwajalein Atoll there are approximately forty specific references to individual firers (or weapon users), who in most cases are named, and four references to general collective firing by squads or platoons. In his account, therefore, if the collective firing is assumed generously to involve soldiers who were not mentioned in any of the other firings, then Marshall records about 200 US soldiers using their weapons in the fight of the Kwajalein Atoll. This battle primarily involved six companies from two infantry regiments, the 184th and the 32nd from the 7th Infantry Division; approximately 1,200 men in all. On the basis of this evidence, a figure of one in six firers is generated. However, once again, although Marshall's descriptions are themselves compelling with all actions of the significant 'firers' discussed, there is simply not enough detail about

soldiers who played a subordinate role. Many of these soldiers may indeed have fired a weapon; but they were just not recorded doing so. The account of the Kwajalein Atoll broadly seems to affirm Marshall's observation about the poor performance of the US infantry but the one in four figure cannot be definitely derived from the evidence.

59. D. Lee 2006. *Up Close and Personal: the reality of close-quarter fighting in World War II*. London: Greenhill, 265.
60. Collins, *Violence*; Grossman, *On Killing*; R. Henriksen 2007. 'Warriors in Combat: what makes people actively fight in combat', *Journal of Strategic Studies* 30(2): 187–223; K. Jordan 2002. 'Right for the Wrong Reasons', *Journal of Military History* 66: 135–62.
61. Grossman, *On Killing*, 13.
62. Grossman, *On Killing*, 21.
63. Collins, *Violence*, 53.
64. L. Truscott 1990. *Mission Command: a personal story*. Novato, Calif.: Presidio, 534–5.
65. Truscott, *Mission Command*, 534.
66. Truscott, *Mission Command*, 534.
67. G. Patton 1995. *War as I Knew it*. Boston: Houghton Mifflin, 338.
68. Patton, *War as I Knew it*, 338.
69. See Chapter 5.
70. R. Palmer, B. I. Wiley, and W. Keast 1948. *The Army Ground Forces: the procurement and training of ground combat troops*. Washington: History Division, Department of the Army, 451.
71. J. Balkosi 1999. *Beyond the Beachhead: the 29th Infantry Division in Normandy*. Mechanicsburg, Pa.: Stackpole, 87–9.
72. Balkosi, *Beyond the Beachhead*, 90.
73. Patton, *War as I Knew it*, 341.
74. H. Leinbaugh and J. Campbell, 1985. *The Men of Company K*. New York: William Morrow and Company, 284.
75. Leinbaugh and Campbell, *The Men of Company K*, 43.
76. Leinbaugh and Campbell, *The Men of Company K*, 232.
77. Leinbaugh and Campbell, *The Men of Company K*, 88.
78. Leinbaugh and Campbell, *The Men of Company K*, 159.
79. Palmer et al., *The Army Ground Forces*.
80. Palmer et al., *The Army Ground Forces*, 3.
81. Palmer et al., *The Army Ground Forces*, 9–10.
82. Palmer et al., *The Army Ground Forces*, 10.
83. Palmer et al., *The Army Ground Forces*, 19–20.
84. Palmer et al., *The Army Ground Forces*, 19, letter to General McNair 31 July 1942.
85. Palmer et al., *The Army Ground Forces*, 50.
86. Palmer et al., *The Army Ground Forces*, 50.
87. Palmer et al., *The Army Ground Forces*, 4.
88. Palmer et al., *The Army Ground Forces*, 50.
89. Palmer et al., *The Army Ground Forces*, 50.
90. Palmer et al., *The Army Ground Forces*, 51.

91. Palmer et al., *The Army Ground Forces*, 51.
92. See Chapter 7 for a full discussion of the problem in the citizen army of the twentieth century. At this point, the generally poor performance of citizen armies is claimed rather than proven in order to set up the book's arguments.
93. T. H. Place 2000. *Military Training in the British Army 1940-44*. London: Routledge, 51.
94. Place, *Military Training in the British Army*, 79.
95. Place, *Military Training in the British Army*, 79.
96. D. French 2000. *Raising Churchill's Army: the British army and the war against Germany 1919-1945*. Oxford: Oxford University Press, 71-2.
97. Place, *Military Training in the British Army*, 79.
98. S. Jary 1988. *18 Platoon*. Taunton: Light Infantry Office, 7.
99. S. Jary 2002. 'Funeral Oration for Corporal Douglas Proctor (1917-2002)', in *Serve to Lead*. Sandhurst: Royal Military Academy, 94.
100. Lee, *Up Close and Personal*, 186.
101. R. Graves 1966. *Goodbye to All That*. London: Cassell, 161.
102. Graves, *Goodbye to All That*, 161-2.
103. Glenn, *Reading Athena's Dance Card*, 136.
104. Jordan, 'Right for the Wrong Reasons', 147-56.
105. Glenn, *Reading Athena's Dance Card*.
106. Glenn, *Reading Athena's Dance Card*, 5.
107. Glenn, *Reading Athena's Dance Card*, 39.
108. Glenn, *Reading Athena's Dance Card*, 52-8.
109. Glenn, *Reading Athena's Dance Card*, 66.
110. Glenn, *Reading Athena's Dance Card*, 79.
111. Glenn, *Reading Athena's Dance Card*, 42.
112. Glenn, *Reading Athena's Dance Card*, 39, 47.
113. Marshall, *Bird*, 103-4.
114. E. Murphy 2007. *Dak To*. New York: Ballantine, 95.
115. Murphy, *Dak To*, 95.
116. Murphy, *Dak To*, 96.
117. Glenn, *Reading Athena's Dance Card*, 3.

CHAPTER 4

1. Marshall, *Men against Fire*, 42.
2. Marshall, *Men against Fire*, 43.
3. Marshall, *Men against Fire*, 42.
4. G. Mosse 1985. *Nationalism and Sexuality*. New York: Howard Fertig.
5. T. Dupuy 1977. *A Genius for War*. London: Macdonald and Janes; M. van Creveld 2010. *Fighting Power*. Westport, Conn.: Praeger, 5-9.
6. Janowitz and Shils. 'Cohesion and Disintegration in the Wehrmacht in World War II'.
7. K. Theweleit 1987. *Male Fantasies: women, floods, bodies, history*. Cambridge: Polity, 402.
8. Theweleit, *Male Fantasies*, 33.

9. SLAM Papers AHEC MS#B649: Training, Regular Army NCOs, 3.
10. W. Reibert 1940. *Der Dienstunterricht im Heere*. Berlin: E. S. Mittler und Sohn, 1.
11. Reibert, *Der Dienstunterricht im Heere*, 42.
12. Heeres Division 130/2a. 1922. *Ausbildungsvorschrift für die Infanterie, heft 2a: Die Schützencompanie*. Berlin: Offene Worte, para 84, 39.
13. Heeres Division 130/2a. 1942. *Ausbildungsvorschrift für die Infanterie*, para 412, 165–6.
14. H. V. Dicks, 'The Psychological Foundations of the Wehrmacht', February 1944 (WO 241/1), 5.
15. Dicks, 'The Psychological Foundations of the Wehrmacht', 5.
16. Dicks, 'The Psychological Foundations of the Wehrmacht', 6.
17. Sajer, *The Forgotten Soldier*, 301.
18. E. Dinter 1985. *Hero or Coward*. London: Frank Cass, 169–70.
19. P. Simkins 1988. *Kitchener's Army: the raising of the new armies, 1914–16*. Manchester: Manchester University Press, 82.
20. Simkins, *Kitchener's Army*, 174.
21. Dinter, *Hero or Coward*, 41–9.
22. A. Kellett 1982. *Combat Motivation: the behaviour of soldiers in battle*. The Hague: Kluwer/Nijhoff Publishing, 23.
23. War Office. 1944. *Infantry Training. Part VIII: fieldcraft, battle drill, section and platoon tactics*, 57–8.
24. War Office, *Infantry Training. Part VIII*, 58.
25. S. Stouffer, A. Lumsdaine, M. Harper Lumsdaine, R. Williams, M. Brewster Smith, I. L, Janis, S. Star, and L. Cottrell 1949. *The American Soldier*, ii: *Combat and its Aftermath*. Princeton: Princeton University Press, 131, 134.
26. Stouffer et al., *The American Soldier*, ii. 134.
27. R. Little 1964. 'Buddy Relations and Combat Performance', in M. Janowitz (ed.), *The New Military*. New York: Russell Sage Foundation, 198.
28. Little, 'Buddy Relations and Combat Performance', 198.
29. Little, 'Buddy Relations and Combat Performance', 218.
30. Little, 'Buddy Relations and Combat Performance', 217.
31. Little, 'Buddy Relations and Combat Performance', 199.
32. Little, 'Buddy Relations and Combat Performance', 201.
33. Little, 'Buddy Relations and Combat Performance', 205.
34. G. Pratten 2010. *Australian Battalion Commanders in the Second World War*. Cambridge: Cambridge University Press.
35. Pratten, *Australian Battalion Commanders in the Second World War*, 41–6.
36. J. G. Fuller, *Troop Morale and Popular Culture in the British and Dominion Armies 1914–18*.
37. B. Gammage 1975. *The Broken Years: the Australian soldier in the Great War*. Harmondsworth: Penguin; J. Ross 1985. *The Myth of the Digger: the Australian soldier in two world wars*. Alexandria, NSW: Hale and Iremonger; M. Johnston 2002. *At the Frontline: experience of Australian soldiers in World War II*. Cambridge: Cambridge University Press.
38. Dinter, *Hero or Coward*, 44–5.
39. Simkins, *Kitchener's Army*, 123.

40. Little, 'Buddy Relations and Combat Performance', 218.
41. Little, 'Buddy Relations and Combat Performance', 219.
42. Little, 'Buddy Relations and Combat Performance', 219.
43. Moskos, *The American Enlisted Man*, 156.
44. Moskos, *The American Enlisted Man*, 154.
45. R. Eisenhart 1975. 'You Can't Hack it Little Girl: A Discussion of the Covert Psychological Agenda of Modern Combat Training', *Journal of Social Issues* 31(4): 15.
46. Eisenhart, 'You Can't Hack it Little Girl', 15.
47. Eisenhart, 'You Can't Hack it Little Girl', 16.
48. Eisenhart, 'You Can't Hack it Little Girl', 16–17.
49. Eisenhart, 'You Can't Hack it Little Girl', 17.
50. Eisenhart, 'You Can't Hack it Little Girl', 17.
51. W. Arkin and L. Dobrofksy, 1978. 'Military Socialization and Masculinity', *Journal of Social Issues* 34(1): 154.
52. Arkin and Dobrofksy, 'Military Socialization and Masculinity', 158.
53. Arkin and Dobrofksy, 'Military Socialization and Masculinity', 160.
54. Arkin and Dobrofksy, 'Military Socialization and Masculinity', 162.
55. W. Cockerham 1978. 'Attitudes towards Combat among US Army Paratroopers', *Journal of Political and Military Sociology* 6 (Spring): 4.
56. Cockerham, 'Attitudes towards Combat among US Army Paratroopers', 4. Cockerham's important intervention will be discussed at greater length in Chapter 6. However, while initially emphasizing the centrality of masculinity to airborne training and the generation of primary groups, the central theme of Cockerham's article is that, in fact, genuine combat cohesion and effective tactical performance has little to do with masculinity; it is a product of training and professionalism.
57. G. Aran 1974. 'Parachuting', *American Journal of Sociology* 80(1): 123–52.
58. Aran, 'Parachuting', 150.
59. Aran, 'Parachuting', 148.
60. S. Wesbrook 1980. 'The Potential for Military Disintegration', in S. Sarkesian (ed.), *Combat Effectiveness and Cohesion*. London: Sage, 257.
61. Wesbrook, 'The Potential for Military Disintegration', 257.
62. Wesbrook, 'The Potential for Military Disintegration', 259.
63. Wesbrook, 'The Potential for Military Disintegration', 274.
64. Simkins, *Kitchener's Army*, 187.
65. Simkins, *Kitchener's Army*, 32.
66. Simkins, *Kitchener's Army*, 167–8.
67. P. Hart 2006. *The Somme*. London: Cassell, 104.
68. Hart, *The Somme*, 106.
69. General Staff 1918. SS 143. *The Training and Employment of Platoons 1918*, 4–5.
70. Hart, *The Somme*, 104.
71. War Office 1932. *Infantry Training*, i: *Training*. London: His Majesty's Stationery Office, 10, italics added.
72. War Office 1929. *Field Service Regulations*, ii: *Operations*. London: His Majesty's Stationery Office, 2–3.

73. M. Jolly 1997. 'Love Letters vs Letters Carved in Stone: Gender, Memory and the Forces Sweetheart Exhibition', in M. Evans and K. Lunn (eds.), *War and Memory in the Twentieth Century*. Oxford: Berg, 112.
74. Jolly, 'Love Letters vs Letters Carved in Stone', 116.
75. Jolly, 'Love Letters vs Letters Carved in Stone', 116.
76. J. Granatstein 2004. *Canada's Army: waging war and keeping the peace*. Toronto: University of Toronto Press, 74.
77. Granatstein, *Canada's Army*, 75.
78. Granatstein, *Canada's Army*, 66.
79. Granatstein, *Canada's Army*, 74, 181.
80. H. Strachan 2001. *The First World War*. Oxford: Oxford University Press, 129, 151.
81. Strachan, *The First World War*, 156.
82. Strachan, *The First World War*, 156.
83. Ministère de la Guerre, État Major de l'Armée. 1944. *Règlement de l'infanterie, deuxieme partie: combat*. Paris: Charle Lavauzelle et Cie, 23.
84. Ministère de la Guerre, *Règlement de l'infanterie*, 23.
85. Ministère de la Guerre, *Règlement de l'infanterie*, 23–4.
86. Ministère de la Guerre, *Règlement de l'infanterie*, 24.
87. Ministère de la Guerre, *Règlement de l'infanterie*, 24.
88. Ministère de la Guerre, *Règlement de l'infanterie*, 25.
89. Ministère de la Guerre, *Règlement de l'infanterie*, 27.
90. Ministère de la Defence Nationale et de la Guerre 1938. *Règlement de l'infanterie, premier partie: instruction*. Paris: Imprimerie Nationale, 26.
91. *Règlement de l'infanterie, deuxieme partie (1944)*, 15.
92. M. Thomas 1998. *The French Empire at War: 1940–45*. Manchester: Manchester University Press, 75.
93. A. Beevor 2010. *D-Day*. London: Penguin, 387.
94. Beevor, *D-Day*, 388.
95. Moskos, *The American Enlisted Man*, 152, 155.
96. Moskos, *The American Enlisted Man*, 155.
97. Moskos, 'The American Combat Soldier in Vietnam', 27.
98. Moskos, 'The American Combat Soldier in Vietnam', 36.
99. Moskos, 'The American Combat Soldier in Vietnam', 30–1.
100. Kindsvatter, *American Soldiers*, 153.
101. J. Sadkovich 1991. 'Of Myths and Men: Rommel and the Italians in North Africa 1940–42', *International History Review* 13(2) May: 282–340.
102. See also L. Ceva 1990. 'The North African Campaign 1940–43: a reconsideration', *Journal of Strategic Studies* 13(1): 84–103.
103. B. Sullivan 2010. 'The Italian Armed Forces, 1918–40', in A. Millett and W. Murray (eds.), *Military Effectiveness*, ii: *The Interwar Period*. Cambridge: Cambridge University Press, 169.
104. Sullivan, 'The Italian Armed Forces, 1918–40'; B. Sullivan 1997. 'The Italian Soldier in Combat, June 1940–September 1943: myths, realities and explanations', in P. Addison and A. Calder, *Time to Kill: the soldier's experience of the war in the west, 1939–45*. London: Pimlico; M. Knox 1982. *Mussolini Unleashed*

1939–41: politics and strategy in Fascist Italy's last war. Cambridge: Cambridge University Press; M. Knox 2010. 'The Italian Armed Forces, 1940–43', in A. Millett and W. Murray (eds.), *Military Effectiveness,* iii: *The Second World War.* Cambridge: Cambridge University Press; H. Boog, W. Rahn, R. Stumpf, and B. Wegner 2001. *Germany and the Second World War,* vi: *The Global War: widening of the conflict into a world war and the shift of initiative 1941–43.* Oxford: Clarendon; L. Ceva and G. Rochat 2001. 'Italy', in M. R. D. Foot (ed.), *The Oxford Companion to World War II.* Oxford: Oxford University Press; J. Gooch 1982. 'Italian Military Competence', *Journal of Strategic Studies* 5(2): 257–65; B. Stegemann 1995. 'The Italian Conduct of the War in the Mediterranean and North Africa', in G. Schreiber, B. Stegemann, and D. Vogel, *Germany and the Second World War,* iii: *The Mediterranean, South East Europe and North Africa.* Oxford: Clarendon.
105. A. Stanislav 1982. 'Causes of the Low Morale in the Italian Armed Forces in the Two World Wars', *Journal of Strategic Studies* 5(2): 254.
106. Stanislav, 'Causes of the Low Morale in the Italian Armed Forces in the Two World Wars', 254.
107. Gooch, 'Italian Military Competence', 263.
108. Gooch, 'Italian Military Competence', 264.
109. C. Davies 1982. 'Itali sunt imbelles', *Journal of Strategic Studies* 5(2): 267.
110. Davies, 'Itali sunt imbelles', 268.
111. Davies, 'Itali sunt imbelles', 268.
112. Davies, 'Itali sunt imbelles', 268.
113. M. Thompson 2004. *The White War: life and death on the Italian Front 1915–1919.* London: Faber and Faber, 47–9.
114. Thompson, *The White War,* 36, 69, 93.
115. Thompson, *The White War,* 180–4.
116. Ceva, 'The North African Campaign 1940–43', 101.
117. C. D'Este, 1983. *Decision in Normandy.* London: Pan, 273.
118. M. Hastings 1984. *Overlord.* London: Macmillan, 57.
119. Hastings, *Overlord,* 57.
120. D'Este, *Decision in Normandy,* 272.
121. D'Este, *Decision in Normandy,* 277; Hastings, *Overlord,* 57.
122. D'Este, *Decision in Normandy,* 273.
123. D'Este, *Decision in Normandy,* 274; Hastings, *Overlord,* 173.
124. A. Watson 2011. 'Fighting for Another Fatherland: the Polish minority in the German army, 1914–18', *English Historical Review* 126(522) October: 1148.
125. Watson, 'Fighting for Another Fatherland', 1150.
126. The Wehrmacht seems to have been aware of the problem and similar strategies seem to have been used by some German commanders in the Second World War. Lieutenant General Badinski's 276 Infantry Division in Normandy consisted of numerous 'Volksdeutsch' who were only half-heartedly committed to the war. Badinski's solution was to provide them with increased personal care and support, treat them the same as ethnic Germans, and to integrate them into units. K. Badinski 1947. 'Bericht über den Kampfeinsatz der 276 I.D. in der Normandie vom 20.6–20.8.44', Militär-Archiv Freiburg im Bresau, ZA1/877, 8.

127. Badinski, 'Bericht über den Kampfeinsatz der 276 I.D. in der Normandie vom 20.6–20.8.44', 1156–64.
128. Badinski, 'Bericht über den Kampfeinsatz der 276 I.D. in der Normandie vom 20.6–20.8.44', 1165.
129. Badinski, 'Bericht über den Kampfeinsatz der 276 I.D. in der Normandie vom 20.6–20.8.44', 1165; D. Segal and M. Kestnbaum 2002. 'Professional Closure in the Military Market: a critique of pure cohesion', in D. Snider (ed.), *The Future of the Army Profession*. New York: Primis, 445.
130. S. Wessely 2006. 'Twentieth Century Theories on Combat Motivation and Breakdown', *Journal of Contemporary History* 41(2): 277.
131. The Bartov-Janowitz and Shils debate has certain similarities with the famous debate between Daniel Goldhagen and Christopher Browning with regard to the Holocaust. For Browning the participation of German soldiers in mass killings in eastern Europe after 1941 should be primarily understood in terms of localized group dynamics; ordinary men were driven to outrageous acts by immediate social pressures to which all are potentially susceptible. For Goldhagen, the Germans were special, indoctrinated with a long-standing 'eliminationist' philosophy accentuated and rationalized by Adolf Hitler and the Nazi party and brought to fruition after 1942.
132. O. Bartov 1992. *Hitler's Army*. Oxford: Oxford University Press, 5.
133. Bartov, *Hitler's Army*, 49.
134. O. Bartov 1992. 'The Conduct of War: Soldiers and the Barbarization of Warfare', *Journal of Modern History* 64, Supplement (December): S36–7.
135. Bartov, *Hitler's Army*, 115.
136. S. Fritz 1995. *Frontsoldaten: the German soldier in World War II*. Lexington, Ky.: University of Kentucky Press.
137. Fritz, *Frontsoldaten*, 82.
138. Fritz, *Frontsoldaten*, 199.
139. Fritz, *Frontsoldaten*, 199.
140. Fritz, *Frontsoldaten*, 200.
141. Fritz, *Frontsoldaten*, 206.
142. Fritz, *Frontsoldaten*, 206.
143. Fritz, *Frontsoldaten*, 266; M. Geyer 1986. 'German Strategy in the Age of Machine Warfare, 1914–45', in P. Paret, G. Craig, and F. Gilbert, *Makers of Modern Strategy*. Princeton: Princeton University Press, 596–7.
144. Fritz, *Frontsoldaten*, 217.
145. W. Balck 1922. *Development of Tactics, World War*. Translated by H. Bell. Fort Leavenworth, Va.: General Service Schools Press, 294–5.
146. W. Wette 2005. *Die Wehrmacht*. Verlag: Fisher Taschenbuch, 68, 53–5.
147. Wette, *Die Wehrmacht*, 52–3.
148. Wette, *Die Wehrmacht*, 74. Interestingly, Wette notes that while full Jews could not serve in the Wehrmacht, individuals designated as half Jewish (with two Jewish grandparents) were allowed to serve as soldiers (but not officers) if they volunteered. However, Jews could 'aryanise' themselves by declaring themselves German and renouncing Judaism. In fact, many Jewish soldiers falsified their papers: 2,000–3,000 Jews and perhaps 150,000–200,000 part Jews served in the

Wehrmacht; hundreds of officers and twenty generals were part Jewish (Wette, *Die Wehrmacht*, 84–6).
149. Reibert, *Der Dienstunterricht*, 7.
150. Reibert, *Der Dienstunterricht*, 7.
151. Reibert, *Der Dienstunterricht*, 7–27.
152. Reibert, *Der Dienstunterricht*, 8.
153. Reibert, *Der Dienstunterricht*, 22.
154. Reibert, *Der Dienstunterricht*, 23.
155. Reibert, *Der Dienstunterricht*, 25.
156. Reibert, *Der Dienstunterricht*, 27.
157. Reibert, *Der Dienstunterricht*, 29.
158. Nationalist ideology was affirmed by new insignia and above all the swastika which appeared on Wehrmacht flags and uniforms (Reibert, *Der Dienstunterricht*, 31).
159. A. Holl 2005. *An Infantryman in Stalingrad*. Translated by J. Mark and N. Page. Sydney: Leaping Horseman Books, 164.
160. Holl, *An Infantryman in Stalingrad*, 203.
161. J. Förster 1997. 'Motivation and Indoctrination in the Wehrmacht, 1933–45', in Addison and Calder, *Time to Kill*, 266.
162. Förster, 'Motivation and Indoctrination in the Wehrmacht, 1933–45', 267–9.
163. Some memoirs written long after the war record the scepticism of German soldiers towards this political indoctrination. Bidermann, for instance, records his opposition to Hitler who was seen in contrast to the Fatherland and who is satirized as 'the Greatest War Lord of all Time' (H. Bidermann 2002. *In Deadly Combat: a German soldier's memoir of the Eastern Front*. Lawrence, Kan.: BCA, 117, 152, 227). Political scepticism is typical among front-line soldiers. However, most evidence suggests that Bidermann was in the minority. Moreover, it is possible that Bidermann has underlined his opposition to Hitler during the war post factum, in the light of changed political circumstances. Indeed, the very emphasis which he places on Hitler shows the prominence of the Führer in the minds of the Landser, and Bidermann himself emphasized his duty to his fatherland: 'The code of honour, however, long inherent in the German soldier who stood to protect the fatherland with weapon in hand, remained within his consciousness' (Bidermann, *In Deadly Combat*, 152–3).
164. See Chapter 5.
165. T. Schulte 1997. 'The German Soldier in Occupied Russia', in Addison and Calder, *Time to Kill*, 277.
166. In fact, Janowitz and Shils always recognized the importance of political ideology to cohesion in the Wehrmacht and discuss the issue at length. In their conclusion, they unjustifiably diminished its role, and readers seem to have subsequently (and not unreasonably) read their article as prioritizing the 'pure' cohesion of the primary group.
167. Ambrose, *Citizen Soldiers*, 346–50.
168. S. Stouffer, E. Suchman, L. DeVinney, S. Star, and R. Williams 1949. *The American Soldier*, i: *Adjustment during Army Life*. Princeton: Princeton University Press, 583.

169. Stouffer et al., *The American Soldier*, i. 595.
170. Stouffer et al., *The American Soldier*, i. 596.
171. B. Nalty 1989. *Strength for the Fight: a history of black Americans in the military*. London: Collier Macmillan, 120.
172. Nalty, *Strength for the Fight*, 153; S. Mershon and S. Schlossman 1998. *Foxholes and Colour Lines*. Baltimore: Johns Hopkins University Press, 61.
173. Nalty, *Strength for the Fight*; Mershon and Schlossman, *Foxholes and Colour Lines*, 44–54.
174. U. Lee. 2004. *The Employment of Negro Troops*. Honolulu: University Press of the Pacific, 145.
175. Lee, *The Employment of Negro Troops*, 155.
176. Lee, *The Employment of Negro Troops*, 244.
177. Lee, *The Employment of Negro Troops*, 341.
178. Lee, *The Employment of Negro Troops*, 589.
179. Lee, *The Employment of Negro Troops*, 589.
180. Lee, *The Employment of Negro Troops*, 350.
181. Moskos, *The American Enlisted Man*, 128.
182. Moskos, 'The American Combat Soldier in Vietnam', 34.
183. Moskos, 'The American Combat Soldier in Vietnam', 34.
184. See K. Marlantes 2010. *Matterhorn*. London: Corvus, for an interesting fictional treatment of fragging.
185. D. Killingray with M. Plant 2010. *Fighting for Britain: African soldiers in the Second World War*. Woodbridge: James Currey, 84.
186. Killingray with Plant, *Fighting for Britain*, 13, 84.
187. Killingray with Plant, *Fighting for Britain*, 85.
188. Killingray with Plant, *Fighting for Britain*, 157.
189. Killingray with Plant, *Fighting for Britain*, 157.
190. Killingray with Plant, *Fighting for Britain*, 157.
191. Killingray with Plant, *Fighting for Britain*, 129–33.
192. Killingray with Plant, *Fighting for Britain*, 131, 129.
193. Killingray with Plant, *Fighting for Britain*, 133.
194. D. Marston 2003. *Phoenix from the Ashes: the Indian army in the Burma campaign*. Westport, Conn.: Praeger, 15.
195. Marston, *Phoenix from the Ashes*, 47.
196. T. Barkawi 2005. *Globalization and War*. Oxford: Rowman and Littlefield, 85–7; T. Barkawi 2006. 'Culture and Combat in the Colonies: the Indian army in the Second World War', *Journal of Contemporary History* 41(2): 332.
197. Barkawi, 'Culture and Combat in the Colonies', 332.
198. Marston, *Phoenix from the Ashes*, 224.
199. Marston, *Phoenix from the Ashes*, 225.
200. Marston, *Phoenix from the Ashes*, 225.
201. Marston, *Phoenix from the Ashes*, 226.
202. Barkawi, 'Culture and Combat in the Colonies', 355.
203. Barkawi, 'Culture and Combat in the Colonies', 355.
204. M. Alexander 2011. 'Colonial Minds Confounded: French colonial troops in the Battle of France, 1940', in M. Thomas (ed.), *The French Colonial Mind*, ii:

   Violence, Military Encounters and Colonialism. Lincoln, Nebr.: University of Nebraska Press, 250.
205. Alexander, 'Colonial Minds Confounded', 251–2.
206. Alexander, 'Colonial Minds Confounded', 252.
207. Alexander, 'Colonial Minds Confounded', 253.
208. Alexander, 'Colonial Minds Confounded', 264.
209. Alexander, 'Colonial Minds Confounded', 276.
210. A. Horne 2006. *A Savage War of Peace*. New York: New York Review Book, 139.

CHAPTER 5

1. J. English and B. Gudmundsson 1994. *On Infantry: the military profession*. Westport, Conn.: Praeger, 4.
2. English and Gudmundsson, *On Infantry*, 4; B. Gudmundsson 1989. *Stormtroop Tactics: innovation in the German army, 1914–18*. Westport, Conn.: Praeger, 8.
3. J. Bourke 2000. *An Intimate History of Killing*. London: Granta, 92–3.
4. A. Du Picq 2006. *Battle Studies*. Translated by J. Greely and R. Cotton. Charleston, SC: Bibliobazaar, 136.
5. See also M. Howard 1986. 'Men against Fire: the doctrine of the offensive in 1914', in P. Paret, G. Craig, and F. Gilbert, *Makers of Modern Strategy*. Princeton: Princeton University Press.
6. Bourke, *An Intimate History of Killing*, 89–90.
7. S. Bidwell and D. Graham 2004. *Fire-Power*. Barnsley: Pen and Sword, 52. T. Travers 1987. *The Killing Ground: the British army, the Western Front and the emergence of modern war 1900–1918*. Barnsley: Pen and Sword, 67.
8. P. Porter 2009. *Military Orientalism*. London: C. Hurst and Co, 97.
9. Porter, *Military Orientalism*, 103–4.
10. War Office 1914. *Infantry Training: 4-company organisation*. London: His Majesty's Stationery Office, 134.
11. War Office, *Infantry Training: 4-company organisation*, 146.
12. Travers, *The Killing Ground*, 67.
13. War Office 1909. *Field Service Regulations*. London: His Majesty's Stationery Office, 119.
14. P. Hart 2006. *The Somme*. London: Cassell, 117.
15. R. Prior and T. Wilson 2005. *The Somme*. New Haven: Yale University Press, 112–13.
16. Prior and Wilson, *The Somme*, 114.
17. Prior and Wilson, *The Somme*, 114–15.
18. Travers, *The Killing Ground*, 95.
19. S. Sassoon 2000. *Memoirs of an Infantry Officer*. London: Faber and Faber, 5–6; R. Graves 1966. *Goodbye to All That*. London: Cassell, 210.
20. B. Liddell Hart 1965. *The Memoirs of Captain Basil Liddell Hart*. London: Cassell, 19.
21. General Staff 1917. SS 143. *Instructions for the Training of Platoons for Offensive Action, 1917*, 11.
22. General Staff 1917. SS 185. *Assault Training*. London: Harrison and Sons, 3.

23. M. Humphries 2007. 'Old Wine in New Bottles: a comparison of British and Canadian preparations for the Battle of Arras', in G. Hayes, A. Iarocci, and M. Bechthold (eds.), *Vimy Ridge: a Canadian reassessment*. Waterloo, Ontario: Wilfred Laurier University Press, 72–3.
24. General Staff, *The Training and Employment of Platoons 1918*, 6.
25. J. Granatstein 2004. *Canada's Army: waging war and keeping the peace*. Toronto: University of Toronto Press, 82.
26. Granatstein, *Canada's Army*, 115.
27. Granatstein, *Canada's Army*, 116.
28. M. Goya 2004. *La Chair et l'acier: l'armée française et l'invention de la guerre moderne (1914–18)*. Paris: Tallandier, 20.
29. Goya, *La Chair et l'acier*, 45.
30. Goya, *La Chair et l'acier*, 48.
31. Goya, *La Chair et l'acier*, 52.
32. Goya, *La Chair et l'acier*, 50–1.
33. Goya, *La Chair et l'acier*, 50–1.
34. Goya, *La Chair et l'acier*, 19.
35. Goya, *La Chair et l'acier*, 19. I am grateful to Paul Michaelson for this translation.
36. Goya, *La Chair et l'acier*, 58.
37. Goya, *La Chair et l'acier*, 58–9.
38. Goya, *La Chair et l'acier*, 63.
39. Goya, *La Chair et l'acier*, 63.
40. R. Doughty 2005. *Pyrrhic Victory: French strategy and operations in the Great War*. Cambridge, Mass.: Belknap, 27.
41. Doughty, *Pyrrhic Victory*, 28.
42. Goya, *La Chair et l'acier*, 35–8.
43. Doughty, *Pyrrhic Victory*, 75.
44. Raths, *Vom Massensturm zur Stosstrupptaktic*, 30.
45. Raths, *Vom Massensturm zur Stosstrupptaktic*, 30.
46. Raths, *Vom Massensturm zur Stosstrupptaktic*, 31; Gudmundson, *Stormtroop Tactics*, 22.
47. Raths, *Vom Massensturm zur Stosstrupptaktic*, 31.
48. General Staff, War Office 1909. *Education and Training of the German Infantry*. London: His Majesty's Stationery Office, 42.
49. Raths, *Vom Massensturm zur Stosstrupptaktic*, 37–8; D. Showalter 2004. *Tannenberg*. Washington: Brassey's, 123.
50. Raths, *Vom Massensturm zur Stosstrupptaktic*, 42.
51. Raths, *Vom Massensturm zur Stosstrupptaktic*, 33–4.
52. Showalter, *Tannenberg*, 239.
53. English and Gudmundsson, *On Infantry*, 11.
54. W. Balck 1922. *Development of Tactics, World War*. Translated by H. Bell. Fort Leavenworth, Va.: General Service Schools Press, 28.
55. Thompson, *The White War*, 2.
56. Thompson, *The White War*, 228.
57. Thompson, *The White War*, 229.
58. Thompson, *The White War*, 230–8.

59. F. Kiesling 1996. *Arming against Hitler*. Lawrence, Kan.: University Press of Kansas.
60. R. Doughty 1985. *The Seeds of Disaster: the development of French army doctrine, 1919–1939*. Hamden, Conn.: Archon Books, 86.
61. War Office 1931. *Infantry Training*, ii: *War*. London: His Majesty's Stationery Office, 10.
62. War Office, *Infantry Training*, ii. 13.
63. War Office, *Infantry Training*, ii. 13.
64. *Army Training Memorandum. No. 35: War* August 1940. GS Publications, 12.
65. *Army Training Memorandum. No. 35: War*, 12.
66. Place, *Military Training in the British Army 1940–44*, 35.
67. Place, *Military Training in the British Army 1940–44*, 35.
68. J. Ellis 1990. *World War II: the sharp end*. London: Windrow and Greene, 94.
69. Lee, *Up Close and Personal*, 104.
70. Lee, *Up Close and Personal*, 105.
71. S. Bull and G. Rottman 2008. *Infantry Tactics of the Second World War*. Wellingborough: Osprey, 25.
72. Patton, *War as I Knew it*, 411; Bull and Rottman, *Infantry Tactics of the Second World War*, 25.
73. G. Harrison 1993. *United States Army in World War II. European Theatre of Operations: cross-channel attack*. Washington: Center of Military History, 359.
74. Harrison, *United States Army in World War II*, 359.
75. Ambrose, *Citizen Soldiers*, 44–5. Cole was awarded a posthumous medal of honour for this action; he was killed on 18 September 1944 in Holland during Operation Market Garden.
76. Ambrose, *Citizen Soldiers*, 379.
77. C. Cameron 1994. *American Samurai: myth, imagination and the conduct of battle in the First Marine Division, 1941–51*. Cambridge: Cambridge University Press, 120–2.
78. G. Linderman 1997. *The World within War*. New York: Free Press, 165.
79. E. Sledge 2010. *With the Old Breed*. London: Ebury, 78.
80. Sledge, *With the Old Breed*, 79.
81. Patton, *War as I Knew it*, 338.
82. Patton, *War as I Knew it*, 338.
83. Patton, *War as I Knew it*, 338.
84. Patton, *War as I Knew it*, 340.
85. Patton, *War as I Knew it*, 339.
86. Patton, *War as I Knew it*, 339.
87. e.g. E. Rommel 2009. *Infantry Attacks*. Newbury: Greenhill Books, 36.
88. See Chapter 6.
89. Patton, *War as I Knew it*, 411.
90. R. Citino 1999. *The Path to Blitzkrieg*. London: Lynne Reiner, 126.
91. R. Appleman 1990. *Ridgway Duels for Korea*. College Station, Tex.: Texas A & M University, 210.
92. Appleman, *Ridgway Duels for Korea*, 210.

93. S. Sandler 1999. *The Korean War*. London: University College London Press, 161.
94. Sandler, *The Korean War*, 161; Fehrenbach, *This Kind of War*, 391–2; A. Bevin 1998. *Korea*. New York: Hippocrene Books, 446.
95. Sandler, *The Korean War*, 161.
96. H. Moore and J. Galloway 2002. *We Were Soldiers Once...and Young*. London: Corgi, 143.
97. Moore and Galloway, *We Were Soldiers Once...and Young*, 144.
98. A. Garland 1985. *Infantry in Vietnam*. New York: Jove, 106.
99. Garland, *Infantry in Vietnam*, 133.
100. Garland, *Infantry in Vietnam*, 134.
101. V. Bramley 1995. *Two Sides of Hell*. London: Bloomsbury, 149.
102. Bramley, *Two Sides of Hell*, 137, 139.
103. M. Adkin 2003. *Goose Green: a battle is fought to be won*. London: Cassell, 242–7; S. Fitz-Gibbon 2001. *Not Mentioned in Despatches: the history and mythology of the Battle of Goose Green*. Cambridge: Lutterworth Press, 74–133.
104. Adkin, *Goose Green*, 246.
105. Adkin, *Goose Green*, 247. Junior Brecon is the British army training course for section commanders; like the Royal Marines' equivalent course (see Chapters 8 and 9), it focuses primarily on the section attack.
106. Adkin, *Goose Green*, 247; Fitz-Gibbon, *Not Mentioned in Dispatches*, 74–133; N. van der Bijl 1999. *Nine Battles to Stanley*. Barnsley: Pen and Sword, 134.
107. Place, *Military Training in the British Army 1940–44*.
108. Place, *Military Training in the British Army 1940–44*, 53.
109. E. Kier 1999. *Imagining War: French and British military doctrine between the wars*. Princeton: Princeton University Press, 130.
110. Commander-in-Chief, Home Forces 1942. *The Instructor's Handbook on Fieldcraft and Battle Drill*. Military Library Research Ltd, 83.
111. Commander-in-Chief, Home Forces, *The Instructor's Handbook on Fieldcraft and Battle Drill*, 46.
112. War Office 1921. *Infantry Training*, ii: *War*. London: His Majesty's Stationery Office, 80.
113. T. Skeyhill and R Wheeler (eds.) 2004. *Sergeant York and the Great War*. San Antonio, Tex.: Vision Forum, 19–23.
114. Skeyhill and Wheeler (eds.), *Sergeant York and the Great War*, 28–9.
115. Skeyhill and Wheeler (eds.), *Sergeant York and the Great War*, 155–6.
116. Skeyhill and Wheeler (eds.), *Sergeant York and the Great War*, 156.
117. Skeyhill and Wheeler (eds.), *Sergeant York and the Great War*, 157.
118. Skeyhill and Wheeler (eds.), *Sergeant York and the Great War*, 160.
119. Skeyhill and Wheeler (eds.), *Sergeant York and the Great War*, 161.
120. Skeyhill and Wheeler (eds.), *Sergeant York and the Great War*, 162.
121. Skeyhill and Wheeler (eds.), *Sergeant York and the Great War*, 169.
122. C. Whiting 2000. *American Hero: the life and death of Audie Murphy*. York: Eskdale, 16.
123. A. Murphy 2002. *To Hell and Back*. New York: Picador, 237.
124. Murphy, *To Hell and Back*, 238.

125. Murphy, *To Hell and Back*, 239.
126. Murphy, *To Hell and Back*, 240.
127. Whiting, *American Hero*, 16.
128. Whiting, *American Hero*, 17.
129. Murphy, *To Hell and Back*, 241–2.
130. Marshall, *Night Drop*, 198–203.
131. Harrison, *United States Army in World War II*, 281.
132. Marshall, *Night Drop*, 198.
133. Marshall, *Night Drop*, 200.
134. Ambrose, *Citizen Soldiers*, 20.
135. Ambrose, *Citizen Soldiers*, 125.
136. Hastings, *Overlord*, 130.
137. Hastings, *Overlord*, 130: Beevor, *D-Day*, 129–30.
138. Hastings, *Overlord*, 130: Beevor, *D-Day*, 130.
139. Hastings, *Overlord*, 130.
140. J. Sheldon and N. Cave 2007. *The Battle for Vimy Ridge—1917*. Barnsley: Pen and Sword, 81.
141. Sheldon and Cave, *The Battle for Vimy Ridge*, 81.
142. Granatstein, *Canada's Army*.
143. Granatstein, *Canada's Army*, 207–8.
144. G. Pratten 2010. *Australian Battalion Commanders in the Second World War*. Cambridge: Cambridge University Press, 156.
145. Pratten, *Australian Battalion Commanders in the Second World War*, 159.
146. G. Williamson 1991. *Infantry Aces of the Reich*. London: BCA, 73.
147. Williamson, *Infantry Aces of the Reich*, 73.
148. Williamson, *Infantry Aces of the Reich*, 105.
149. Williamson, *Infantry Aces of the Reich*, 106.
150. Williamson, *Infantry Aces of the Reich*, 62.
151. Fehrenbach, *This Kind of War*, 124.
152. His citation reads: 'The enemy sent wave after wave of fanatical troops against his platoon which held a key terrain feature on "Heartbreak Ridge". Valiantly defending its position, the unit repulsed each attack until ammunition became practically exhausted and it was ordered to withdraw to a new position. Voluntarily remaining behind to cover the withdrawal, Pfc. Pililaau fired his automatic weapon into the ranks of the assailants, threw all his grenades and, with ammunition exhausted, closed with the foe in hand-to-hand combat, courageously fighting with his trench knife and bare fists until finally overcome and mortally wounded. When the position was subsequently retaken, more than 40 enemy dead were counted in the area he had so valiantly defended.' <http://www.history.army.mil/html/moh/koreanwar/html>.
153. See Chapter 6.
154. N. Warr 1997. *Phase Line Green*. New York: Ivy Books, 137–8.
155. Warr, *Phase Line Green*, 138.
156. Warr, *Phase Line Green*, 138.
157. Warr, *Phase Line Green*, 138–9.
158. Warr, *Phase Line Green*, 139.

159. Warr, *Phase Line Green*, 140.
160. <http://militarytimes.com/citations-medals-awards/recipient.php?recipientid=23848>.
161. e.g. Lieutenant Joe Marm's actions at LZ X-Ray. See Moore and Galloway, *We Were Soldiers Once... and Young*, 146.
162. Marlantes, *Matterhorn*, 475.
163. Marlantes, *Matterhorn*, 476.
164. Marlantes, *Matterhorn*, 477.
165. Personal communication, 28 July 2011.
166. <http://militarytimes.com/citations-medals-awards/recipient.php?recipientid=4191>.
167. This is confirmed in his recent memoir which describes the actual events on which the novel is based. See Marlantes, *What it is Like to Go to War*.

CHAPTER 6

1. Gudmundsson, *Stormtroop Tactics*, 8; Balck, *Development of Tactics, World War*.
2. Gudmundsson, *Stormtroop Tactics*, 9.
3. War Office 1914. *Infantry Training: 4-company organisation*. London: His Majesty's Stationery Office, 134.
4. War Office 1909. *Field Service Regulations*. London: His Majesty's Stationery Office, 119.
5. Citino, *The Path to Blitzkrieg*, 15.
6. H. Delbrück 1975. *The History of the Art of War within the Framework of Political History*, i: *Antiquity*. London: Greenwood, 272–82.
7. Delbrück, *The History of the Art of War within the Framework of Political History*, i. 273–4.
8. Delbrück, *The History of the Art of War within the Framework of Political History*, i. 272.
9. Delbrück, *The History of the Art of War within the Framework of Political History*, i. 273.
10. Delbrück, *The History of the Art of War within the Framework of Political History*, i. 273.
11. Delbrück, *The History of the Art of War within the Framework of Political History*, i. 273.
12. Delbrück, *The History of the Art of War within the Framework of Political History*, i. 274.
13. E. Rommel 2009. *Infantry Attacks*. Newbury: Greenhill Books, 9.
14. Rommel, *Infantry Attacks*, 9.
15. Rommel, *Infantry Attacks*, 9.
16. Rommel, *Infantry Attacks*, 10–11.
17. Rommel, *Infantry Attacks*, 156.
18. Rommel, *Infantry Attacks*, 13.
19. G. Wynne 1940. *If Germany Attacks*. London: Faber and Faber, 147–8; T. Lupfer 1981. *The Dynamics of Doctrine*. Fort Leavenworth, Kan.: Combat Studies

Institute, US Army Command and General Staff College; Gudmundson, *Stormtroop Tactics*, 46–50.
20. Gudmundson, *Stormtroop Tactics*, 50–1.
21. Gudmundson, *Stormtroop Tactics*, 149–50.
22. *Ausbildungsvorschrift für die Fusstruppen im Kriege*. January 1917. Reichsdruckerei, 78.
23. *Ausbildungsvorschrift für die Fusstruppen im Kriege*, 68–9.
24. *Ausbildungsvorschrift für die Fusstruppen im Kriege*, 68.
25. *Ausbildungsvorschrift für die Fusstruppen im Kriege*, 83.
26. *Ausbildungsvorschrift für die Fusstruppen im Kriege*, 85–6.
27. *Ausbildungsvorschrift für die Fusstruppen im Kriege*, 86.
28. *Ausbildungsvorschrift für die Fusstruppen im Kriege*, 123.
29. *Ausbildungsvorschrift für die Fusstruppen im Kriege*, 79.
30. *Ausbildungsvorschrift für die Fusstruppen im Kriege*, 139.
31. *Ausbildungsvorschrift für die Fusstruppen im Kriege*, 140.
32. *Ausbildungsvorschrift für die Fusstruppen im Kriege*, 140.
33. *Ausbildungsvorschrift für die Fusstruppen im Kriege*, 140.
34. *Ausbildungsvorschrift für die Fusstruppen im Kriege*, 140.
35. *Ausbildungsvorschrift für die Fusstruppen im Kriege*, 143.
36. J. Corum 1992. *The Roots of Blitzkrieg*. Lawrence, Kan.: University of Kansas Press, 31: Citino, *The Path to Blitzkrieg*, 9.
37. Citino, *The Path to Blitzkrieg*, 9.
38. Corum, *The Roots of Blitzkrieg*, 10–11.
39. Citino, *The Path to Blitzkrieg*, 11.
40. Citino, *The Path to Blitzkrieg*, 18.
41. Heeres Division 130/2a 1922. *Ausbildungsvorschrift für die Infanterie, heft 2a: Die Schützencompanie*. Berlin: Offene Worte, paragraph *39*, 17.
42. *Ausbildungsvorschrift für die Infanterie*, paragraph *42*, 18.
43. *Ausbildungsvorschrift für die Infanterie*, paragraph *270*, 113.
44. Heeres Division 130/2a 1942. *Ausbildungsvorschrift für die Infanterie, heft 2a: Die Schützencompanie*. Berlin: Offene Worte; Heeres Division 130/2a 1941. *Ausbildungsvorschrift für die Infanterie, heft 2a: Die Schützencompanie*. Berlin: Offene Worte; Heeres Division 130/2a 1938. *Ausbildungsvorschrift für die Infanterie, heft 2a: Die Schützencompanie*. Berlin: Reichsdruckerie; Heeres Division 130/2b 1936. *Ausbildungsvorschrift für die Infanterie, heft 2: Die Schützencompanie*. Berlin: Offene Worte; Heeres Division 130/2a 1922. *Ausbildungsvorschrift für die Infanterie, heft 2a: Die Schützencompanie*. Berlin: Offene Worte.
45. Heeres Division 130/2b, *Ausbildungsvorschrift*, paragraph *330*, 5–6.
46. Heeres Division 130/2a, *Ausbildungsvorschrift*, paragraphs *230*, *422*, 96–101, 165–7.
47. Heeres Division 130/2a, *Ausbildungsvorschrift*, paragraphs *246–310*, 106–27.
48. Heeres Division 130/2a, *Ausbildungsvorschrift*, paragraph *251*, 108.
49. Heeres Division 130/2a, *Ausbildungsvorschrift*, Bild *19*, 109; Reibert, *Der Dienstunterricht*, 273.
50. Heeres Division 130/2a, *Ausbildungsvorschrift*, paragraph *268*, 113.
51. Reibert, *Der Dienstunterricht*, 275.

Notes

52. Heeres Division 130/2a, *Ausbildungsvorschrift*, paragraph *271*, 114.
53. Heeres Division 130/2a, *Ausbildungsvorschrift*, paragraph *294*, 121.
54. Heeres Division 130/2a, *Ausbildungsvorschrift*, paragraphs *292–303*, 121–5; Reibert, *Der Dienstunterricht*, 279.
55. Reibert, *Der Dienstunterricht*, 280.
56. Reibert, *Der Dienstunterricht*, 273.
57. Reibert, *Der Dienstunterricht*, 281.
58. Heeres Division 130/2a, *Ausbildungsvorschrift*, paragraphs *307–8*, 126; Reibert, *Der Dienstunterricht*, 281. Reibert's version includes a few very minor differences; for instance, the assault is not launched with a 'hurra' but with a bugle signalling 'Attack'.
59. Reibert, *Der Dienstunterricht*, 237.
60. Goya, *La Chair et l'acier*, 199.
61. Goya, *La Chair et l'acier*, 261.
62. Goya, *La Chair et l'acier*, 261.
63. Goya, *La Chair et l'acier*, 261.
64. Goya, *La Chair et l'acier*, 265.
65. Goya, *La Chair et l'acier*, 265. In his compelling analysis of the development of German stormtroop tactics in the First World War, Hew Strachan has suggested that these innovations were impelled by a requirement to reduce casualties by an army fighting on two fronts. The evidence cited here suggests other combatants were also driven to adopt equivalent tactics for similar reasons. The tactical convergence also suggests the mutual influence of belligerents on each other was very important to the dissemination of fire and movement tactics.
66. A. Laffargue 1917. *The Attack in Trench Warfare*. New York: D. Van Nostrand Company, 2–3.
67. Laffargue, *The Attack in Trench Warfare*, 31.
68. Laffargue, *The Attack in Trench Warfare*, 36.
69. Laffargue, *The Attack in Trench Warfare*, 50.
70. Laffargue, *The Attack in Trench Warfare*, 52.
71. Grand Quartier General 1916. *Manuel du chef de section d'infanterie*. Paris: Imprimerie Nationale, 51–2; Grand Quartier General. 1918. *Instruction sur le combat offensif des petits unités*. Paris: Imprimerie Nationale, 37.
72. Grand Quartier General, *Instruction sur le combat offensif des petits unités*, 41–2, croquois 1.
73. Grand Quartier General, *Instruction sur le combat offensif des petits unités*, 41, croquois 1.
74. Grand Quartier General, *Manuel du chef de section d'infanterie*, 374; Grand Quartier General, *Instruction sur le combat offensif des petits unités*, 26.
75. Grand Quartier General, *Manuel du chef de section d'infanterie*, 374; Grand Quartier General, *Instruction sur le combat offensif des petits unités*, 42–4, 46–7, 96–7.
76. Grand Quartier General, *Manuel du chef de section d'infanterie*, 66, 409.
77. Grand Quartier General, *Manuel du chef de section d'infanterie*, 67.
78. Grand Quartier General, *Instruction sur le combat offensif des petits unités*, 43.
79. W. Philpott 2010. *Bloody Victory*. London: Abacus, 205, 241.

80. Philpott, *Bloody Victory*, 444.
81. F. Culmann 1921. *Cours de tactique générale d'après l'expérience de la Grande Guerre*. Paris: Charles Lavauzelle, 479.
82. P. Griffith 1994. *Battle Tactics of the Western Front*. New Haven: Yale University Press, 7.
83. Griffith, *Battle Tactics of the Western Front*, 8.
84. Granatstein, *Canada's Army*, 82, 111.
85. Griffith, *Battle Tactics of the Western Front*, 54–6.
86. Griffith, *Battle Tactics of the Western Front*, 57.
87. Griffith, *Battle Tactics of the Western Front*, 59.
88. General Staff, *Instructions for the Training of Platoons for Offensive Action, 1917*, 11.
89. General Staff, *Instructions for the Training of Platoons for Offensive Action, 1917*, 13.
90. 'An intelligent use of the ground frequently enables forward movement to be made without loss' (General Staff, *Instructions for the Training of Platoons for Offensive Action, 1917*, 12).
91. General Staff, *Instructions for the Training of Platoons for Offensive Action, 1917*, Appendix: VIII–IX.
92. General Staff, *Instructions for the Training of Platoons for Offensive Action, 1917*, Appendix: XI.
93. General Staff, *Instructions for the Training of Platoons for Offensive Action, 1917*, Appendix: X.
94. General Staff 1917. SS 185. *Assault Training*. London: Harrison and Sons, 3.
95. General Staff, *Assault Training*, 4.
96. General Staff, *The Training and Employment of Platoons 1918*, 13.
97. General Staff, *The Training and Employment of Platoons 1918*, 13.
98. Liddell Hart, *The Memoirs of Captain Basil Liddell Hart*, 43.
99. Liddell Hart, *The Memoirs of Captain Basil Liddell Hart*, 45.
100. War Office 1921. *Infantry Training*, ii: *War*. London: His Majesty's Stationery Office, 18–19.
101. War Office, *Infantry Training*, ii. 78, 80.
102. War Office 1944. *Infantry Training. Part VIII: Fieldcraft, battle drill, section and platoon tactics*, 56; W. Owen 2004. 'Platoon and Section Manoeuvre 1917–2003', *Army Doctrine and Training News* 20: 41.
103. War Office, *Infantry Training*, 53; Commander-in-Chief, Home Forces, *The Instructor's Handbook on Fieldcraft and Battle Drill*, 60–1.
104. War Department 1914. *US Army Field Service Regulations*. Washington: Government Printing Office, 95.
105. War Department, War Plans Division 1918. *Instructions for the Offensive Combat of Small Units*. Washington: Government Printing Office, 13.
106. War Department, War Plans Division, *Instructions for the Offensive Combat of Small Units*, 13.
107. War Department, War Plans Division, *Instructions for the Offensive Combat of Small Units*, 22.

108. War Department, War Plans Division, *Instructions for the Offensive Combat of Small Units*, 22.
109. War Department, War Plans Division, *Instructions for the Offensive Combat of Small Units*, 23.
110. Infantry Journal 1939. *Infantry in Battle*. Washington: Infantry Journal Inc, 230.
111. English and Gudmundsson, *On Infantry*, 126–9.
112. War Department 1944. *Field Manual 7-10: Rifle Company, Infantry Regiment*. Washington: Government Printing Office, 47–8.
113. D. Webster 2008. *Parachute Infantry*. New York: Bantam Dell, 26.
114. Balkosi, *Beyond the Beachhead*, 51.
115. Balkosi, *Beyond the Beachhead*, 53.
116. Grand Quartier General, *Manuel du chef de section d'infanterie*, 412.
117. Grand Quartier General, *Manuel du chef de section d'infanterie*, 411.
118. Grand Quartier General, *Instruction sur le combat offensif des petits unités*, 29–30.
119. Jünger, *Storm of Steel*, 214.
120. Jünger, *Storm of Steel*, 214.
121. Reibert, *Der Dienstunterricht*, 153–60.
122. Reibert, *Der Dienstunterricht*, 237.
123. Reibert, *Der Dienstunterricht*, 237.
124. Lee, *Up Close and Personal*, 85.
125. Baynes, *Far from a Donkey*, 150.
126. Baynes, *Far from a Donkey*, 150.
127. Place, *Military Training in the British Army 1940-44*, 53.
128. Commander-in-Chief, Home Forces, *The Instructor's Handbook on Fieldcraft and Battle Drill*, 39.
129. Commander-in-Chief, Home Forces, *The Instructor's Handbook on Fieldcraft and Battle Drill*, 1.
130. Commander-in-Chief, Home Forces, *The Instructor's Handbook on Fieldcraft and Battle Drill*, 62.
131. Commander-in-Chief, Home Forces, *The Instructor's Handbook on Fieldcraft and Battle Drill*, 1.
132. Commander-in-Chief, Home Forces, *The Instructor's Handbook on Fieldcraft and Battle Drill*, 63–70.
133. War Office, *Infantry Training*, 59–61.
134. War Office, *Infantry Training*, 60.
135. Place, *Military Training in the British Army 1940-44*, 49.
136. Place, *Military Training in the British Army 1940-44*, 49–51.
137. T. Copp 2007. *The Brigade*. Mechanicsburg, Pa.: Stackpole, 23: J. English 2009. *The Canadian Army in the Normandy Campaign*. Mechanicsburg, Pa.: Stackpole. 71–3.
138. Place, *Military Training in the British Army 1940-44*, 51.
139. Place, *Military Training in the British Army 1940-44*, 52.
140. Place, *Military Training in the British Army 1940-44*, 56.
141. Place, *Military Training in the British Army 1940-44*, 53.

142. Copp, *The Brigade*, 23: English, *The Canadian Army in the Normandy Campaign*, 71–3.
143. English, *The Canadian Army in the Normandy Campaign*, 75.
144. Place, *Military Training in the British Army 1940–44*, 54.
145. Place, *Military Training in the British Army 1940–44*, 56.
146. Lee, *Up Close and Personal*, 121.
147. D. Gilchrist 1960. *Castle Commando*. London: Oliver and Boyd; J. Dunning 2000. *It had to be Tough*. Romsey: James Dunning Publications; J. Ladd 1978. *Commandos and Rangers of World War II*. London: Macdonald and Jane's; H. St George Saunders 1949. *The Green Beret*. London: Michael Joseph.
148. Indeed, David Stirling, the founder of the SAS, went through the Commando School at Achnacarry before deploying to North Africa.
149. Lee, *Up Close and Personal*, 78.
150. Lee, *Up Close and Personal*, 84.
151. Lee, *Up Close and Personal*, 86.
152. War Office 1942. *Notes from Theatres of War No. 11: destruction of a German battery by No. 4 Commando during the Dieppe Raid*. GS Publications, 3.
153. War Office, *Notes from Theatres of War No. 11*, 3.
154. War Office, *Notes from Theatres of War No. 11*, 3.
155. War Office, *Notes from Theatres of War No. 11*, 18.
156. Lee, *Up Close and Personal*, 107.
157. Lee, *Up Close and Personal*, 86.
158. Headquarters Department of the Army, 1959. Field Manual 7-10, *Rifle Company, Infantry and Airborne Division Battle Groups*, 114.
159. S. L. A. Marshall 1983. *Ambush: the battle of Dau Tieung*. New York: Jove, 83.
160. L. McAulay 1987. *The Battle of Long Tan*. London: Arrow, 25.
161. B. Grandin, H. Smith, G. Kendall, D. Buick, D. Sabben, M. Stanley, and A. Roberts 2004. *The Battle of Long Tan*. Crows Nest, NSW: Allen and Unwin, 176–7.
162. Laffargue, *The Attack in Trench Warfare*, 17.
163. Laffargue, *The Attack in Trench Warfare*, 70.
164. Laffargue, *The Attack in Trench Warfare*, 70–1.
165. Laffargue, *The Attack in Trench Warfare*, 71–2.
166. Grand Quartier General, *Manuel du chef de section d'infanterie*, 13; Grand Quartier General, *Instruction sur le combat offensif des petits unités*, 8, 26–8.
167. Culmann, *Cours de tactique générale d'après l'expérience de la Grande Guerre*, 487.
168. Culmann, *Cours de tactique générale d'après l'expérience de la Grande Guerre*, 488.
169. Culmann, *Cours de tactique générale d'après l'expérience de la Grande Guerre*, 489.
170. Culmann, *Cours de tactique générale d'après l'expérience de la Grande Guerre*, 489.

171. Culmann, *Cours de tactique générale d'après l'expérience de la Grande Guerre*, 489.
172. Culmann, *Cours de tactique générale d'après l'expérience de la Grande Guerre*, 489.
173. Gudmundsson, *Stormtroop Tactics*, 49–50.
174. Jünger, *Storm of Steel*, 221.
175. Jünger, *Storm of Steel*, 222. A better translation for 'ribbons' (*Bänder*, in the original German; E. Jünger 1941. *In Stahlgewittern*. Berlin: E. S. Mittler und Sohn, 246) is 'tape'. For divisional rehearsals for both German and Allied armies, the lines of enemy trenches were typically marked out by strips of white tape secured to posts since it was too onerous a task to dig rehearsal trenches of the requisite scale.
176. Gudmundsson, *Stormtroop Tactics*, 92.
177. Military Attaché Report, 'Information on German Manoeuvres 13 May 1926', US Military Intelligence Reports: Germany 1919–41, Reel 14, 0411: 25.
178. Baynes, *Far from a Donkey*, 136.
179. War Office, *Infantry Training*, 79.
180. War Office, *Infantry Training*, 86–7.
181. Commander-in-Chief, Home Forces, *The Instructor's Handbook on Fieldcraft and Battle Drill*, 97.
182. War Department, War Plans Division, *Instructions for the Offensive Combat of Small Units*, 12.
183. War Department, War Plans Division, *Instructions for the Offensive Combat of Small Units*, 16.
184. B. Rawling 1992. *Surviving Trench Warfare*. Toronto: University of Toronto Press, 47.
185. Rawling, *Surviving Trench Warfare*, 48.
186. Rawling, *Surviving Trench Warfare*, 50.
187. Lee, *Up Close and Personal*, 90.
188. Truscott, *Mission Command*, 33. Perhaps significantly, for the disastrous Operation Market Garden in September 1944, there were no models available for the Arnhem area, the crucial bridge in which the British 1 Airborne Division failed to hold despite their best efforts (WO 205/77). Clearly, the operation did not fail because of the lack of models but their absence was evidence of the haste and even carelessness with which that operation was planned and executed and which precipitateness was certainly responsible for the catastrophe.
189. S. Ambrose 2003. *Pegasus Bridge*. London: Pocket Books, 68–5.
190. Ambrose, *Pegasus Bridge*, 77; 5 Parachute Brigade Operations in Normandy WO/223/17.
191. Balkosi, *Beyond the Beachhead*, 61.
192. Balkosi, *Beyond the Beachhead*, 128.
193. Balkosi, *Beyond the Beachhead*.

CHAPTER 7

1. Infantry Journal 1939. *Infantry in Battle*, 226.
2. Infantry Journal, *Infantry in Battle*, 225–6.
3. R. Doughty 1985. *The Seeds of Disaster: the development of French army doctrine, 1919–1939*. Hamden, Conn.: Archon Books, 60–71.
4. Doughty, *The Seeds of Disaster*, 28–30.
5. Kiesling, *Arming against Hitler*, 64–6.
6. Doughty, *The Breaking Point*, 105.
7. Doughty, *The Breaking Point*, 199.
8. Kiesling, *Arming against Hitler*, 67.
9. Doughty, *The Breaking Point*, 197–9.
10. Doughty, *The Breaking Point*, 198.
11. Doughty, *The Breaking Point*, 200.
12. Dunn, *The War the Infantry Knew*, 354.
13. Dunn, *The War the Infantry Knew*, 422.
14. Griffith, *Battle Tactics of the Western Front*, 59.
15. Baynes, *Far from a Donkey*, 145.
16. Dunn, *The War the Infantry Knew*, 337.
17. J. Edmonds and W. Miles 1938. *Military Operations: France and Belgium 1916*. London: Macmillan and Co, 570.
18. Edmonds and Miles, *Military Operations*, 571.
19. Edmonds and Miles, *Military Operations*, 571.
20. Edmonds and Miles, *Military Operations*, 572.
21. J. Edmonds and C. Falls 1940. *Military Operations: France and Belgium 1917*. London: Macmillan and Co, 13.
22. Edmonds and Falls, *Military Operations*, 555.
23. P. Kennedy 2010. 'Britain in the First World War', in A. Millett and W. Murray (eds.), *Military Effectiveness*, i: *The First World War*. Cambridge: Cambridge University Press, 31, 70.
24. Place, *Military Training in the British Army 1940–44*, 63–79.
25. Place, *Military Training in the British Army 1940–44*, 63–79.
26. Place, *Military Training in the British Army 1940–44*, 67.
27. Place, *Military Training in the British Army 1940–44*, 68.
28. D'Este, *Decision in Normandy*.
29. Reports on Operations in Normandy 6 June–27 August (27 June 1945, WO 106/4466).
30. Hastings, *Overlord*, 177–9; D'Este, *Decision in Normandy*, 282–3.
31. Rawling, *Surviving Trench Warfare*, 22–3.
32. Rawling, *Surviving Trench Warfare*, 35.
33. Rawling, *Surviving Trench Warfare*, 35.
34. Rawling, *Surviving Trench Warfare*, 123.
35. A. Godefroy 2007. 'The 4th Canadian Division: "trenches should never be saved"', in G. Hayes, A. Iarocci, and M. Bechthold (eds.), *Vimy Ridge: a Canadian reassessment*. Waterloo, Ontario: Wilfred Laurier University Press, 217–18.
36. Godefroy, 'The 4th Canadian Division: "trenches should never be saved"', 221.

37. English, *Lament for an Army*, 18.
38. English, *Lament for an Army*, 17.
39. J. English 2009. *The Canadian Army in the Normandy Campaign*. Mechanicsburg, Pa.: Stackpole, 5.
40. C. P. Stacey 1960. *The Victory Campaign: the operations in north-west Europe 1944-5*. Ottawa: the Queen's Printer and Controller for Stationery, 276; Copp, *The Brigade*, 207. The Germans seem to have concurred. Stacey recorded one German report on the Allied infantry: 'The morale of the enemy infantry is not high. It depends largely on artillery and air support. In case of a well placed concentration of fire from our own artillery the infantry will often leave its positions and retreat hastily. Whenever enemy is engaged with force he usually retreats or surrenders' (Stacey, *The Victory Campaign*, 274).
41. Stacey, *The Victory Campaign*, 274-5.
42. Stacey, *The Victory Campaign*, 275.
43. Stacey, *The Victory Campaign*, 277.
44. Stacey, *The Victory Campaign*, 275.
45. Copp, *The Brigade*, 207-8; T. Copp 2004. *Fields of Fire*. Toronto: University of Toronto Press, 7.
46. Copp, *Fields of Fire*, 267.
47. Copp, *Fields of Fire*, 112-14.
48. Copp, *Fields of Fire*, 140, 260-1.
49. Copp, *Fields of Fire*, 266.
50. English, *The Canadian Army in the Normandy Campaign*, 106.
51. Copp, *Fields of Fire*, 106.
52. Copp, *Fields of Fire*, 112.
53. Copp, *Fields of Fire*, 112.
54. Copp, *Fields of Fire*, 115.
55. Copp, *Fields of Fire*, 113.
56. English, *The Canadian Army in the Normandy Campaign*, 90-6, 99-107, 243.
57. English, *The Canadian Army in the Normandy Campaign*, 89.
58. English, *The Canadian Army in the Normandy Campaign*, 106.
59. Granatstein, *Canada's Army*, 255, 277-8, 256.
60. English, *The Canadian Army in the Normandy Campaign*, 181-2.
61. Copp, *Fields of Fire*, 152.
62. Copp, *The Brigade*, 72.
63. Copp, *The Brigade*, 64-9.
64. Copp, *The Brigade*, 74-6.
65. Copp, *The Brigade*, 79.
66. English, *The Canadian Army in the Normandy Campaign*, 193.
67. English, *The Canadian Army in the Normandy Campaign*, 193; Copp, *The Brigade*, 67-86; Copp, *Fields of Fire*, 173-6; Granatstein, *Canada's Army*, 268.
68. English, *The Canadian Army in the Normandy Campaign*, 83.
69. English, *The Canadian Army in the Normandy Campaign*, 183.
70. English, *The Canadian Army in the Normandy Campaign*, 186.

71. J. Gooch 2010. 'Italy during the First World War', in A. Millett and W. Murray (eds.), *Military Effectiveness*, i: *The First World War*. Cambridge: Cambridge University Press, 180.
72. Gooch, 'Italy during the First World War', 180.
73. Gooch, 'Italy during the First World War', 180.
74. Gooch, 'Italy during the First World War', 180.
75. Sadkovich, 'Of Myths and Men: Rommel and the Italians in North Africa 1940–42', 298.
76. M. Knox 2010. 'The Italian Armed Forces, 1940–43', in A. Millett and W. Murray (eds.), *Military Effectiveness*, iii: *The Second World War*. Cambridge: Cambridge University Press, 166.
77. R. Stumpf 2001. 'The War in the Mediterranean 1942–1943', in H. Boog, W. Rahn, R. Stumpf, and B. Wegner, *Germany and the Second World War*, vi: *The Global War: widening of the conflict into a world war and the shift of initiative 1941–43*. Oxford: Clarendon, 653.
78. B. Stegemann 1995. 'The Italian Conduct of the War in the Mediterranean and North Africa', in G. Schreiber, B. Stegemann, and D. Vogel, *Germany and the Second World War*, iii: *The Mediterranean, South East Europe and North Africa*. Oxford: Clarendon, 761.
79. Thompson, *The White War*, 5.
80. Gooch, 'Italy during the First World War', 181–2.
81. Gooch, 'Italy during the First World War', 182.
82. B. Sullivan 2010. 'The Italian Armed Forces, 1918–40', in A. Millett and W. Murray (eds.), *Military Effectiveness*, ii: *The Interwar Period*. Cambridge: Cambridge University Press, 201.
83. Sullivan, 'The Italian Armed Forces, 1918–40', 201.
84. Sullivan, 'The Italian Armed Forces, 1918–40', 202.
85. M. Knox 1982. *Mussolini Unleashed 1939–41: politics and strategy in Fascist Italy's last war*. Cambridge: Cambridge University Press, 251.
86. Sadkovich's claim that the Italians lost Cyrenaica in only a month less than Germany would lose both Cyrenaica and Tripolitania may be misleading. The Afrika Corps, with its important Italian formations, faced a much larger British force in 1942–3 than Graziani in 1940–1.
87. Knox, 'The Italian Armed Forces, 1940–43', 163–4.
88. Stumpf, 'The War in the Mediterranean 1942–1943', 767.
89. Stumpf, 'The War in the Mediterranean 1942–1943', 778.
90. Stumpf, 'The War in the Mediterranean 1942–1943', 778.
91. D. Trask 1993. *The AEF and Coalition Warmaking, 1917–18*. Lawrence, Kan.: University of Kansas Press, 3, 6, 167–8.
92. Trask, *The AEF and Coalition Warmaking*, 61.
93. Trask, *The AEF and Coalition Warmaking*, 19.
94. Trask, *The AEF and Coalition Warmaking*, 19; P. Braim 1987. *The Test of Battle*. Newark, Del.: University of Delaware Press, 48.
95. Trask, *The AEF and Coalition Warmaking*, 65, 70.
96. Trask, *The AEF and Coalition Warmaking*, 71.
97. Trask, *The AEF and Coalition Warmaking*, 72.

98. Braim, *The Test of Battle*, 58.
99. Braim, *The Test of Battle*, 58.
100. Braim, *The Test of Battle*, 119.
101. Braim, *The Test of Battle*, 98.
102. Trask, *The AEF and Coalition Warmaking*, 151.
103. Trask, *The AEF and Coalition Warmaking*, 151.
104. Balkosi, *Beyond the Beachhead*, xiv.
105. Balkosi, *Beyond the Beachhead*, xiv.
106. Balkosi, *Beyond the Beachhead*, 138.
107. Balkosi, *Beyond the Beachhead*, 155–6.
108. Balkosi, *Beyond the Beachhead*, 186.
109. Hastings, *Overlord*, 291.
110. C. MacDonald 1986. *Korea: the war before Vietnam*. London: Macmillan Press, 203.
111. A. Bevin 1998. *Korea*. New York: Hippocrene Books, 82, 107.
112. Fehrenbach, *This Kind of War*, 115: Hastings, *The Korean War*, 82–4.
113. S. Sandler 1999. *The Korean War*. London: University College London Press, 68.
114. Sandler, *The Korean War*, 68.
115. W. Donnelly 2000. 'Thunderbirds in Korea: the US 45th Infantry Division, 1950–52', *Journal of Military History* 64(4): 1128.
116. Donnelly, 'Thunderbirds in Korea', 1138–9.
117. Macdonald, *Korea*, 215.
118. R. Appleman 1989. *Disaster in Korea*. College Station, Tex.: Texas A & M University, 102.
119. Fehrenbach, *This Kind of War*, 466; Macdonald, *Korea*, 254; Hastings, *The Korean War*, 348.
120. W. Westmoreland 1976. *A Soldier Reports*. Garden City, NY: Doubleday and Company, 228.
121. S. Stanton 1985. *The Rise and Fall of an American Army*. Novato, Calif.: Presidio, 343.
122. B. Rogers 1974. *Cedar Falls-Junction City*. Washington: Department of the Army, 148.
123. Moore and Galloway, *We Were Soldiers Once... and Young*, 243.
124. Moore and Galloway, *We Were Soldiers Once... and Young*, 244.
125. S. Stanton 1999. *The 1st Cavalry Division in Vietnam*. Novato, Calif.: Presidio, 62.
126. Moore and Galloway, *We Were Soldiers Once... and Young*, 251.
127. Stanton, *The Rise and Fall of an American Army*, 87.
128. Stanton, *The Rise and Fall of an American Army*, 86.
129. Stanton, *The Rise and Fall of an American Army*, 87.
130. Stanton, *The Rise and Fall of an American Army*, 88.
131. Stanton, *The Rise and Fall of an American Army*, 114; S. L. A. Marshall 1967. *Battles in the Monsoon: campaigning in the Central Highlands Vietnam, Summer 1966*. Nashville: Battery Press, 19–40.
132. Marshall, *Battles in the Monsoon*, 74–86.
133. Stanton, *The Rise and Fall of an American Army*, 116.
134. Westmoreland, *A Soldier Reports*, 296.

135. A. Bradford 1994. *Some even Volunteered*. Westport, Conn.: Praeger, 14.
136. Bradford, *Some even Volunteered*, 14.
137. Bradford, *Some even Volunteered*, 14.
138. Stanton, *The Rise and Fall of an American Army*, 271.
139. Stanton, *The Rise and Fall of an American Army*, 272.
140. Westmoreland, *A Soldier Reports*, 376.
141. Westmoreland, *A Soldier Reports*, 378.
142. Stanton, *The Rise and Fall of an American Army*, 294.
143. E. King 1973. *The Death of an Army*. New York: Saturday Review Press, 70.
144. A. Millett 1980. *Semper Fidelis: the history of the United States Marines Corps*. New York: Macmillan, 598.
145. Millett, *Semper Fidelis*, 602.
146. Gudmundsson, *Stormtroop Tactics*, 84.
147. T. Lupfer 1981. *The Dynamics of Doctrine*. Fort Leavenworth, Kan.: Combat Studies Institute, US Army Command and General Staff College, 44.
148. Gudmundsson, *Stormtroop Tactics*, 85.
149. Balck, *Development of Tactics, World War*, 173.
150. Gudmundsson, *Stormtroop Tactics*, 142.
151. Balck, *Development of Tactics, World War*, 173.
152. Rommel, *Infantry Attacks*, 18, 54.
153. Griffith, *Battle Tactics of the Western Front*, 8, 60, 195.
154. Griffith, *Battle Tactics of the Western Front*, 195.
155. Infantry Journal, *Infantry in Battle*, 116.
156. Jünger, *Storm of Steel*, 232.
157. Jünger, *Storm of Steel*, 275.
158. Copp, *Fields of Fire*, 7–8.
159. H. Cole 1993. *The Ardennes: Battle of the Bulge*. Washington: Center of Military History, United States Army, 80.
160. Cole, *The Ardennes*, 80.
161. A. Kershaw 2005. *The Longest Winter*. Harmondsworth: Penguin, 84–114; Ambrose, *Citizen Soldiers*, 192–4.
162. Ambrose, *Citizen Soldiers*, 194.
163. Ambrose, *Band of Brothers*, 147–52.
164. D. Glantz with J. House 2009. *The Stalingrad Trilogy*, ii: *Armageddon in Stalingrad: September–November 1942*. Lawrence, Kan.: University of Kansas Press, 23.
165. A. Holl 2005. *An Infantryman in Stalingrad*. Translated by J. Mark and N. Page. Sydney: Leaping Horseman Books, 50, 92–4.
166. Holl, *An Infantryman in Stalingrad*, 94.
167. Holl, *An Infantryman in Stalingrad*, 93.
168. H. Kissel 1956. 'Panic in Battle', *Military Review* July: 97–8.
169. Kissel, 'Panic in Battle', 99.
170. Kissel, 'Panic in Battle', 98, 99–100.
171. Kissel, 'Panic in Battle', 102.
172. K.-H. Frieser 2005. *The Blitzkrieg Legend*. Annapolis, Md.: Naval Institute Press, 31–3.
173. Frieser, *The Blitzkrieg Legend*, 29.

174. Military Attaché Report, 'The Experiences of an Infantry Battalion in the Attack on Belgium', US Military Intelligence Reports: German 1919–1941, 27 December 1940: 2–4.
175. Military Attaché Report, 'The Experiences of an Infantry Battalion in the Attack on Belgium', 6.
176. Clearly, pressures from other theatres impeded these preparations very severely.
177. E. Maurer 1946. 'Schilderung der Operationen gegen die Truppen der US-Armee: die Kämpfe der 253 Inf. Div.', Bundesarchiv-Militär-archiv Freiburg-im-Breisgau, ZA1/1711, 20. Maurer wrongly designates the 709 as the 706.
178. Maurer, 'Schilderung der Operationen gegen die Truppen der US-Armee', 21.
179. Maurer, 'Schilderung der Operationen gegen die Truppen der US-Armee', 21.
180. Maurer, 'Schilderung der Operationen gegen die Truppen der US-Armee', 23.
181. K. Badinski 1947. 'Bericht über den Kampfeinsatz der 276 I.D. in der Normandie vom 20.6–20.8.44', Bundesarchiv-Militär-archiv Freiburg-im-Breisgau, ZA1/877, 7–8.
182. Badinski, 'Bericht über den Kampfeinsatz der 276 I.D. in der Normandie vom 20.6–20.8.44', 25.
183. F. von der Heydte 1948. 'Das Fallschirmjägerregiment 6 in der Normandie I: Die Kämpfe in Raum von Carentan', Bundesarchiv-Militär-archiv Freiburg-im-Breisgau, ZA1/1188, 12.
184. von der Heydte, 'Das Fallschirmjägerregiment 6 in der Normandie I', 13.
185. von der Heydte, 'Das Fallschirmjägerregiment 6 in der Normandie I', 14.
186. von der Heydte, 'Das Fallschirmjägerregiment 6 in der Normandie I', 28.
187. von der Heydte, 'Das Fallschirmjägerregiment 6 in der Normandie I', 32.
188. von der Heydte, 'Das Fallschirmjägerregiment 6 in der Normandie I', 33.
189. Beevor, *D-Day*, 40.
190. Jary, *18 Platoon*, 17.
191. Jary, *18 Platoon*, 17.
192. Jary, *18 Platoon*, 17.
193. R. Weighley 1981. *Eisenhower's Lieutenants*. London: John Wiley and Son, 128.
194. Weighley, *Eisenhower's Lieutenants*, 128.
195. US Military Intelligence Reports: Germany 1919–41. Military Attaché Report 'Demonstration at the Infantry School', 24 April 1939, Reel 21 0135: 3.
196. US Military Intelligence Reports: Germany 1919–41 Military Attaché Report 7 February 1940, 'Demonstration at the Infantry School: Infantry Battalion in Attack, Practical Demonstration at Döberitz', 27 January 1940, Reel 21, 0138: 5.
197. US Military Intelligence Reports: Germany 1919–1941. Military Attaché Report 17,210, 'Wartime Training of a German Infantry Division', 16 April 1940 Reel 21 0149, 26.
198. Military Attaché Report 17,210, 'Wartime Training of a German Infantry Division', 16 April 1940, 9.
199. Military Attaché Report 17,210, 'Wartime Training of a German Infantry Division', 13. Military Intelligence reports in the early 1930s record the formation of 'Einheitsgruppe' which replaced the rifle squad/section (US Military Intelligence Reports: Germany 1919–41. Military Attaché report 'Supporting Fire and Fire action of Unit Groups', 24 January 1933, Reel 20, 0021, 1). Instead of four

rifle sections and a machine-gun section, the platoon was to be organized as three sections each built around its own machine-gun.

200. Military Attaché Report 17,210, 'Wartime Training of a German Infantry Division', 16 April 1940, 13.
201. 'This fire was continued for about five minutes when elements of two and four men each in the left (6th) company advanced by rushes in the same manner as taught in our own service' (Military Attach Report 7 February 1940, 'Demonstration at the Infantry School: Infantry Battalion in Attack, Practical Demonstration at Döberitz' 27 January 1940, 3).
202. J. Mark 2006. *Island of Fire*. Sydney: Leaping Horseman Books, 134.
203. Holl, *An Infantryman in Stalingrad*, 112.
204. Glantz with House, *The Stalingrad Trilogy*, ii. 380.
205. Glantz with House, *The Stalingrad Trilogy*, ii. 453.
206. Glantz with House, *The Stalingrad Trilogy*, ii. 453.
207. Glantz with House, *The Stalingrad Trilogy*, ii. 229.
208. Glantz with House, *The Stalingrad Trilogy*, ii. 461.
209. Glantz with House, *The Stalingrad Trilogy*, ii. 650.
210. Holl, *An Infantryman in Stalingrad*, 13.
211. Mark, *Island of Fire*, 231.
212. Mark, *Island of Fire*, 39, 72–4.
213. Mark, *Island of Fire*, 101.
214. Mark, *Island of Fire*, 173.
215. Mark, *Island of Fire*, 173–4.
216. Mark, *Island of Fire*, 288.
217. Commander-in-Chief, Home Forces, *The Instructor's Handbook on Fieldcraft and Battle Drill*, 180.
218. Commander-in-Chief, Home Forces, *The Instructor's Handbook on Fieldcraft and Battle Drill*, 183.
219. Commander-in-Chief, Home Forces, *The Instructor's Handbook on Fieldcraft and Battle Drill*, 183.
220. B. Condell and D. Zabecki 2009. *On the German Art of War: Truppenführung. Germany Army Manual for Unit Command in World War II*. Mechanicsburg, Pa.: Stackpole, 100–1.
221. Military Attaché Report, Infantry Battalion in the Attack: practical demonstration at Doberitz 27 January 1940, 5.
222. Military Attaché Report 'The Manoeuvres of the German 4th Division', October 1925, Reel 14, 0234, 44.
223. Military Attaché Report 17,210, 'Wartime Training of a German Infantry Division', 16 April 1940, 26.
224. McAulay, *The Battle of Long Tan*; Grandin et al., *The Battle of Long Tan*.
225. McAulay, *The Battle of Long Tan*, 180–1.
226. Grandin et al., *The Battle of Long Tan*, 132.
227. Grandin et al., *The Battle of Long Tan*, 44.
228. G. McKay 1996. *Delta Four*. St Leonards, NSW: Allen and Unwin, 4.

229. McKay, *Delta Four*, 189.
230. B. Liddell Hart 1999. *Thoughts on War*. Padstow: Spellmount, 254.
231. Liddell Hart, *Thoughts on War*, 254–5.
232. Liddell Hart, *Thoughts on War*, 255.
233. M. Alexander 1992. *The Republic in Danger*. Cambridge: Cambridge University Press, 37–8.
234. Alexander, *The Republic in Danger*, 41.
235. B. Liddell Hart (ed.) 1987. *The Rommel Papers*. New York: Da Capo Press, 468.
236. Liddell Hart (ed.), *The Rommel Papers*, 487.
237. T. Pakenham 1982. *The Boer War*. London: Macdonald.
238. E. Luttwak 1995. 'Toward Post-Heroic Warfare', *Foreign Affairs* 74(3): 109–22.

CHAPTER 8

1. Kindsvatter, *American Soldiers*.
2. Granatstein, *Canada's Army*, 406: English, *Lament for an Army*.
3. M. Martin 1977. 'Conscription and the Decline of the Mass Army in France, 1960–75', *Armed Forces & Society* 3(3): 355–406; C. Kelleher 1978. 'Mass Armies in the 1970s: the debate in Western Europe', *Armed Forces & Society* 5(1): 3–30; J. Van Doorn (ed.) 1968. *Armed Forces and Society*. The Hague: Mouton; K. Haltiner 2003. 'The Decline of the Mass Army', in G. Caforio (ed.), *Handbook of the Sociology of the Military*. London: Kluwer/Plenum; K. Haltiner 1998. 'The Definite End of the Mass Army in Western Europe?', *Armed Forces & Society* 25 (1): 7–36.
4. J. McKenna 1997. 'Towards the Army of the Future: Domestic Politics and the End of Conscription in France', *West European Politics* 20(4) October: 133; P. Bratton 2002. 'France and the Revolution in Military Affairs', *Contemporary Security Policy* 23(2) August: 92.
5. Bratton, 'France and the Revolution in Military Affairs', 92.
6. McKenna, 'Towards the Army of the Future', 136.
7. C. Moskos 1988. *Soldiers and Sociology*. Fort Belvoir, Va.: United States Army Research Institute for Behaviour and Social Sciences, 58–9.
8. Moskos, *Soldiers and Sociology*, 60.
9. Moskos, *Soldiers and Sociology*, 61–2.
10. Moskos, *Soldiers and Sociology*, 69.
11. C. Dandeker 1994. 'New Times for the Military', *British Journal of Sociology* 45(4): 645–7; C. Dandeker 1998.'A Farewell to Arms? The military and the nation-state in a changing world', in James Burk (ed.), *The Adaptive Military*. London: Transaction; P. Manigart 2004. 'Restructuring of the Armed Forces', in Giuseppe Caforio (ed.), *Handbook of the Sociology of the Military*. London: Kluwer/Plenum, 331.
12. R. Henriksen 2007. 'Warriors in Combat: what makes people actively fight in combat', *Journal of Strategic Studies* 30(2): 214.
13. M. Weber 1978. *Economy and Society: an outline of interpretive sociology*, vol. ii. Translated by G. Rother and C. Wittich. Berkeley: University of California Press, 935.

14. M. Weber 1978. *Economy and Society: an outline of interpretive sociology*, vol. i. Translated by G. Rother and C. Wittich. Berkeley: University of California Press, 341–2.
15. Weber, *Economy and Society: an outline of interpretive sociology*, ii. 935.
16. Weber, *Economy and Society: an outline of interpretive sociology*, i. 342.
17. R. Bendix 1960. *Max Weber*. London: Heinemann, 270.
18. Weber, *Economy and Society: an outline of interpretive sociology*, i. 342.
19. Weber, *Economy and Society: an outline of interpretive sociology*, i. 342.
20. Weber, *Economy and Society: an outline of interpretive sociology*, i. 344.
21. Weber, *Economy and Society: an outline of interpretive sociology*, i. 344.
22. R. Collins 1979. *The Credential Society*. London: Academic Press, 49.
23. A. Abbott 1988. *The System of Profession*. Chicago: University of Chicago Press.
24. Collins, *The Credential Society*, 49.
25. Abbott, *The System of Profession*.
26. R. Collins 2000. *The Sociology of Philosophies*. Cambridge, Mass.: Belknap Press.
27. S. Huntington 1957. *The Soldier and the State*. Cambridge, Mass.: Belknap Press, 7.
28. Huntington, *The Soldier and the State*, 8.
29. Huntington, *The Soldier and the State*, 9, 10.
30. Huntington, *The Soldier and the State*, 11.
31. Huntington, *The Soldier and the State*, 11.
32. Huntington, *The Soldier and the State*, 11.
33. Huntington, *The Soldier and the State*, 13.
34. Huntington, *The Soldier and the State*, 12–13.
35. Huntington, *The Soldier and the State*, 13.
36. Huntington, *The Soldier and the State*, 12.
37. Huntington, *The Soldier and the State*, 14.
38. Huntington, *The Soldier and the State*, 15.
39. Huntington, *The Soldier and the State*, 15.
40. Huntington, *The Soldier and the State*, 15.
41. B. McCoy 2007. *The Passion of Command*. Quantico, Va.: Marine Corps Association.
42. M. Janowitz 1960. *The Professional Soldier*. New York: Free Press, 5.
43. Janowitz, *The Professional Soldier*, 8–9.
44. Janowitz, *The Professional Soldier*, 9–10.
45. Janowitz, *The Professional Soldier*, 10.
46. Janowitz, *The Professional Soldier*, 11–15.
47. Janowitz, *The Professional Soldier*, 15.
48. Janowitz, *The Professional Soldier*, 15.
49. Huntington, *The Soldier and the State*, 465.
50. Headquarters Department of the Army 2007. *Field Manual 3-21.8. The Infantry Rifle Platoon and Squad*. Washington, 1.8–9.
51. Strachan, 'Training, Morale and Modern War', 217.
52. Strachan, 'Training, Morale and Modern War', 222–3.
53. Strachan, 'Training, Morale and Modern War', 225.
54. Strachan, 'Training, Morale and Modern War', 227.

55. Headquarters Department of the Army, *Field Manual 3-21.8. The Infantry Rifle Platoon and Squad*, 1.7.
56. Headquarters Department of the Army, *Field Manual 3-21.8. The Infantry Rifle Platoon and Squad*, 1-7.
57. Headquarters Department of the Army, *Field Manual 3-21.8. The Infantry Rifle Platoon and Squad*, 1-8.
58. Headquarters Department of the Army 2006. *ARTEP 7-1-Drill. Warrior Battle Drills.* Washington, 2.1.
59. Headquarters Department of the Army, *ARTEP 7-1-Drill. Warrior Battle Drills*, 2.1.
60. Headquarters Department of the Army, *ARTEP 7-1-Drill. Warrior Battle Drills*, 2.2.
61. Headquarters Department of the Army, *Field Manual 3-21.8. The Infantry Rifle Platoon and Squad*, 7-51.
62. Heeres Division 130/2a, *Ausbildungsvorschrift für die Infanterie, heft 2a: Die Schützencompanie*, paragraphs 307-8, 126.
63. Fieldnotes, 20 July 2009.
64. Fieldnotes, 2 June 2011.
65. USMC captain, personal interview, 3 June 2011.
66. USMC captain, personal communication, 3 June 2011.
67. S. Junger 2010. *War*. London: Fourth Estate, 117.
68. Junger, *War*, 119.
69. Junger, *War*, 120.
70. Junger, *War*, 121.
71. McCoy, *The Passion of Command*, 6.
72. McCoy, *The Passion of Command*, 7.
73. McCoy, *The Passion of Command*, 31.
74. McCoy, *The Passion of Command*, 33.
75. McCoy, *The Passion of Command*, 31.
76. R. Lowry 2010. *New Dawn: the battles for Fallujah*. New York: Savas Beattie, 146-3, 306; D. Camp 2009. *Operation Phantom Fury*. Minneapolis: Zenith, 229-31.
77. D. Bellavia 2008. *House to House*. London: Simon and Schuster, 227-67.
78. Lowry, *New Dawn*, 306-7; <http://militarytimes.com/citations-medals-awards/recipient.php?recipientid=3836>.
79. Lowry, *New Dawn*, 150.
80. Lowry, *New Dawn*, 150.
81. Lowry, *New Dawn*, 150.
82. Lowry, *New Dawn*, 150.
83. Bellavia, *House to House*, 271.
84. P. Harclerode 2000. *Secret Soldiers*. London: Cassell, 377.
85. Globe and Laurel 2010. 'Commando Training Centre RM'. July/August, 288-9.
86. Globe and Laurel, 'Commando Training Centre RM', 288.
87. Headquarters Department of the Army, *Field Manual 3-21.8. The Infantry Rifle Platoon and Squad*, 7.40-1.
88. Globe and Laurel, 'Commando Training Centre RM', 288.

89. Globe and Laurel, 'Commando Training Centre RM', 288.
90. Fieldnotes, 18 October 2011.
91. Canadian army senior non-commissioned officer, personal communication, 18 October 2011.
92. USMC captain, personal interview, 2 June 2011.
93. Canadian Army senior non-commissioned officer, personal communication, 18 October 2011.
94. Commando Training Centre Royal Marines 2011. *Close Quarters Battle Instructor: course manual*, 5-1.
95. Commando Training Centre Royal Marines 2011. *Close Quarters Battle Instructor: course manual*, 5-2.
96. Canadian senior non-commissioned officer, CQB instructor, personal communication, 18 October 2011.
97. Commando Training Centre Royal Marines, *Close Quarters Battle Instructor*, 5.1, 5.3.
98. Commando Training Centre Royal Marines, *Close Quarters Battle Instructor*, 5-4.
99. Commando Training Centre Royal Marines, *Close Quarters Battle Instructor*, 6-2.
100. Fieldnotes, 20 May 2011.
101. Fieldnotes, 9–10 May 2011.
102. Fieldnotes, 18 October 2011.
103. Simunitions are advanced paint-ball rounds used by the Royal Marines and British army in CQB training. The rounds imitate the velocity and accuracy of a proper round at close range, impacting on contact to spray the target with paint. The rounds are an excellent way of determining the accuracy of the shooting.
104. Fieldnotes, 27 July 2011.
105. Commando Training Centre Royal Marines, *Close Quarters Battle Instructor*, 13.7–8.
106. Commando Training Centre Royal Marines, *Close Quarters Battle Instructor*, 12-1.
107. Commando Training Centre Royal Marines, *Close Quarters Battle Instructor*, 12-1, 19-2.
108. Royal Marines corporals, personal interview, 30 September 2010.
109. Headquarters Department of the Army, *Field Manual 3–21.8. The Infantry Rifle Platoon and Squad*, 7–41.
110. French Army, warrant officer, personal interview, 8 December 2011; Fieldnotes, 7 December 2011.
111. Fieldnotes, 20 July 2011
112. Fieldnotes, 20–1 July 2011.
113. Fieldnotes, 20 July 2011.
114. The body armour to body armour drill is not practised by the Canadian army.
115. Commando Training Centre Royal Marines, *Close Quarters Battle Instructor*, 12-3.
116. Commando Training Centre Royal Marines, *Close Quarters Battle Instructor*, 12-3.

117. The Canadian army do not employ IBT but prefer to maintain section and platoon integrity with designated commanders retaining control throughout. Accordingly, while the No. 2 always commands the stack in the Royal Marines no matter what his rank, the Canadian army position the section commander and his second in command at No. 3 in their respective fire teams. The section commander takes a position in the second stack, commanding both his fire team or stack and his second in command's fire team. The section commander's fire team follows and supports the other assaulting team into rooms (Canadian army senior non-commissioned officer, urban operations instructor, personal communication, 18 October 2011).
118. Commando Training Centre Royal Marines, *Close Quarters Battle Instructor*, 12.2–3.
119. Fieldnotes, 9 September 2010.

CHAPTER 9

1. W. Cockerham 1978. 'Attitudes towards Combat among US Army Paratroopers', *Journal of Political and Military Sociology* 6 (Spring): 1–15.
2. Cockerham, 'Attitudes towards Combat among US Army Paratroopers', 12.
3. Cockerham, 'Attitudes towards Combat among US Army Paratroopers', 13.
4. P. Bourne 1970. *Men, Stress and Vietnam*. Boston: Little Brown.
5. Fehrenbach, *This Kind of War*, 246.
6. Cockerham, 'Attitudes towards Combat among US Army Paratroopers', 13.
7. S. Biddle 2004. *Military Power*. Princeton: Princeton University Press; P. Smith 2008. 'Meaning and Military Power: moving on from Foucault', *Journal of Power* 1(3): 275–93; Henriksen, 'Warriors in Combat: what makes people actively fight in combat'.
8. E. Coss 2010. *All for the King's Shilling: the British soldier under Wellington, 1808–1814*. Norman, Okla.: University of Oklahoma Press, 3.
9. Coss, *All for the King's Shilling*, 86–122.
10. Coss, *All for the King's Shilling*, 97.
11. Coss, *All for the King's Shilling*, 46.
12. Coss, *All for the King's Shilling*, 175.
13. Coss, *All for the King's Shilling*, 175.
14. Coss, *All for the King's Shilling*, 175–6.
15. Coss, *All for the King's Shilling*, 197.
16. Coss, *All for the King's Shilling*, 198.
17. J. Lynn 1984. *The Bayonets of the Republic*. Urbana, Ill.: University of Illinois Press.
18. Coss, *All for the King's Shilling*, 173.
19. Coss, *All for the King's Shilling*, 168–72.
20. Coss, *All for the King's Shilling*, 172.
21. Coss, *All for the King's Shilling*, 171.
22. Coss, *All for the King's Shilling*, 163.
23. Coss, *All for the King's Shilling*, 159.
24. Coss, *All for the King's Shilling*, 165.
25. Coss, *All for the King's Shilling*, 166.

26. Coss, *All for the King's Shilling*, 166.
27. Coss, *All for the King's Shilling*, 166.
28. Coss, *All for the King's Shilling*, 166.
29. Coss, *All for the King's Shilling*, 165; Lynn, *The Bayonets of the Republic*, 216.
30. Lynn, *The Bayonets of the Republic*, 31.
31. Lynn, *The Bayonets of the Republic*, 210.
32. Henriksen 'Warriors in Combat: what makes people actively fight in combat', 210.
33. H. Delbrück 1990. *The History of the Art of War within the Framework of Political History*, iii: *Medieval Warfare*. London: Greenwood.
34. Smith, 'Meaning and Military Power: moving on from Foucault', 282.
35. Smith, 'Meaning and Military Power: moving on from Foucault', 283.
36. Joint Warfare Publication 0-01. 2001. *British Defence Doctrine*. Shrivenham: Joint Doctrine and Concepts Centre, 4.6.
37. McCoy, *The Passion of Command*, 36.
38. McCoy, *The Passion of Command*, 37.
39. 'The Mind at War: Understanding, Preparing and Treating Combat Stress', 12–13 October, Annual Gregg Centre for the Study of War and Society-Combat Training Centre Fall Conference, University of New Brunswick, Canada. Fieldnotes, 12 October 2011.
40. Sergeant, Canadian army, Fieldnotes, 15 October 2011.
41. e.g. warrant officer, Canadian army, personal communication, 17 October 2011.
42. Personal interviews, major, Royal Marines, 13 May 2010, major, Canadian army, 17 June 2010.
43. Major, Royal Marines, personal interview, 5 May 2010.
44. Major, Royal Marines, personal interview, 5 May 2010.
45. Major, Royal Marines, personal interview, 5 May 2010.
46. E. Southby-Tailyour 2010. *Helmand Assault*. London: Ebury, 219–32.
47. Southby-Tailyour, *Helmand Assault*, 228.
48. Southby-Tailyour, *Helmand Assault*, 228.
49. Captain, Bundeswehr, personal interview, 9 December 2010.
50. Captain, Bundeswehr, personal interview, 9 December 2010.
51. Captain, Bundeswehr, personal interview, 12 December 2010.
52. Captain, Bundeswehr, personal interview, 9 December 2010.
53. Bryan McCoy has made a similar point: '*Everything is training. Never miss an opportunity to train.* Training does *not* stop in theatre. Make a list of your unit tasks and battle drills that you are most dissatisfied with and use that as your start point for "opportunity training"' (McCoy, *The Passion of Command*, 36).
54. Fieldnotes, 4 June 2011.
55. McCoy, *The Passion of Command*, 39.
56. Major, British army, personal communication, 7 December 2011.
57. C. Spence 1998. *Sabre Squadron*. London: Penguin, 251; P. Ratcliffe 2001. *Eye of the Storm*. London: Michael O'Mara, 354.
58. Captain, Canadian army, personal communication, 15 October 2011.
59. Captain, Canadian army, personal communication, 15 October 2011.
60. Captain, Canadian army, personal communication, 15 October 2011.
61. Fieldnotes, 4 June 2010.

62. Captain, Royal Marines, personal interview, 16 July 2010.
63. Captain, Royal Marines, personal interview, 16 July 2010.
64. Lieutenant colonel, French army, personal communication, 12 December 2012.
65. French major, French army, personal communication, 12 December 2011.
66. Fieldnotes, 12 December 2011.
67. É. Durkheim 1984. *The Division of Labour in Society*. Translated by W. D. Halls. Basingstoke: Macmillan, 84.
68. I. Kant *Critique of Pure Reason*. 1929. Translated by N. Kemp Smith. London: Macmillan.
69. É. Durkheim 1965. *Sociology and Philosophy*. Translated by D. Pocock. London: Cohen and West, 62.
70. Durkheim, *Sociology and Philosophy*, 36, 45.
71. Durkheim, *The Elementary Forms of the Religious Life*, 9.
72. Durkheim, *The Elementary Forms of the Religious Life*, 9.
73. Durkheim, *The Elementary Forms of the Religious Life*, 10.
74. Durkheim, *The Elementary Forms of the Religious Life*, 11.
75. Durkheim, *The Elementary Forms of the Religious Life*, 11.
76. Durkheim, *The Elementary Forms of the Religious Life*, 11.
77. Headquarters Department of the Army, *Field Manual 3-21.8. The Infantry Rifle Platoon and Squad*, 5-18.
78. S. Naylor 2005. *Not a Good Day to Die*. Harmondsworth: Penguin.
79. Naylor, *Not a Good Day to Die*, 94-5.
80. Naylor, *Not a Good Day to Die*, 94-5, 159-2.
81. Naylor, *Not a Good Day to Die*, 161, 330-1, 335.
82. The British Task Force in Helmand was involved in intense fighting from numerous platoon houses and had mounted several significant operations by this stage. However, these operations were British rather than NATO ones and were actively eschewed by the NATO commander, General David Richards, who regarded them as inappropriate.
83. B. Horn 2010. *No Lack of Courage*. Toronto: Dundum Press, 79.
84. Horn, *No Lack of Courage*, 79.
85. Horn, *No Lack of Courage*, 81.
86. Sond Chara means Red Dagger in Pashtun, a reference to the 3 Commando Brigade's insignia.
87. Sergeant, Royal Marines, personal interview, 6 September 2011.
88. J. Searle 1995. *The Construction of Social Reality*. London: Allen Lane.
89. Searle, *The Construction of Social Reality*, 24.
90. Searle, *The Construction of Social Reality*, 34.
91. Searle, *The Construction of Social Reality*, 82-3.
92. Searle, *The Construction of Social Reality*, 43.
93. Searle, *The Construction of Social Reality*, 37-43.
94. H. Garfinkel 2002. *Ethnomethodology's Program: working out Durkheim's aphorism*. Oxford: Rowman and Littlefield, 65.
95. H. Garfinkel 1967. *Studies in Ethnomethodology*. Cambridge: Polity, 1.
96. Junger, *War*, 121.
97. <http://news.bbc.co.uk/1/hi/scotland/8249485.stm>.

98. P. Langer 2010. 'Studying Violence: methodological considerations of doing research on combat experiences in Afghanistan', paper presented to 11th Biennial Conference of ERGOMAS, 14 June 2010, 3.
99. Langer, 'Studying Violence: methodological considerations of doing research on combat experiences in Afghanistan', 3. Translation amended by author from original Langer translation.
100. Langer, 'Studying Violence: methodological considerations of doing research on combat experiences in Afghanistan', 4.
101. Langer, 'Studying Violence: methodological considerations of doing research on combat experiences in Afghanistan', 4.
102. Brigadier, German army, cited in E. Sanger 2012. 'Using Historical Experience: the British and the Bundeswehr in Afghanistan', Ph.D. thesis, European University Institute.
103. <http://www.irwin.army.mil/Pages/default.aspx>.
104. The facility burnt down in October 2011 after very brief usage.
105. C. Bellamy 2001. 'Combining Combat Readiness and Compassion', *NATO Review* 49(2): 9–11.
106. Sergeant, Royal Marines, personal interview, 14 December 2010.
107. Sergeant, Royal Marines, personal interview, 14 December 2010.
108. Fieldnotes, 9–10 May 2010.
109. G. Girton 1986. 'Kung Fu: toward a praxiological hermeneutic of the martial arts', in H. Garfinkel (ed.), *Ethnomethodological Studies of Work*. London: Routledge and Kegan Paul, 65.
110. Girton, 'Kung Fu: toward a praxiological hermeneutic of the martial arts', 66.
111. Sergeant, Royal Marines, personal interview, 14 December 2010.
112. Sergeant, Royal Marines, personal interview, 14 December 2010.
113. Sergeant, Royal Marines, personal interview, 14 December 2010.
114. Commando Training Centre Royal Marines, *Close Quarters Battle Instructor*, 12–13.
115. Commando Training Centre Royal Marines, *Close Quarters Battle Instructor*, 12–14.
116. Fieldnotes, 21 July 2011.
117. Fieldnotes, 22 July 2010.
118. Description taken from observation of close-quarters battle demonstration, Commando Training Centre Lympstone, 26 October 2010. Discussion with two Royal Marine corporals, qualified Modern Urban Combat instructors, 30 September 2010.
119. Senior non-commissioned officer, urban operations instructor, Canadian army, personal communication, 18 October 2011. The colour-coding seems to have a wide purchase in the Canadian army. During an instructional session at the Infantry School, staff aimed to play the new Canadian forces video 'Resilience: the Warrior's edge' which has very similar content to the Royal Marines' own Mindset lecture. However, in this class, the video would not work and a sergeant joked with her colleague who was struggling to operate the audio-visual equipment: 'Are you in the yellow? Are you going orange? I can help you' (Fieldnotes, 15 October 2011).

120. M. Csikszentmihalyi 1992. *Flow*. London: Rider; M. Csikszentmihalyi 1996. *Optimal Experience*. Cambridge: Cambridge University Press.
121. Fieldnotes, 20–2 July 2011.
122. McCoy, *The Passion of Command*, 32.
123. McCoy, *The Passion of Command*, 32.
124. McCoy, *The Passion of Command*, 32.
125. Fieldnotes, 15 October 2011.
126. Fieldnotes, 15 October 2011.
127. Fieldnotes, 12 October 2011.
128. Fieldnotes, 12 October 2011.
129. Fieldnotes, 22 September 2009.
130. Fieldnotes, 22 September 2009.
131. Fieldnotes, 11 December 2010.
132. Fieldnotes, 11 December 2011.
133. T. Boudreau 2008. *Packing Inferno*. Port Townsend, Wash.: Feral House, 160.
134. Boudreau, *Packing Inferno*, 161.
135. Boudreau, *Packing Inferno*, 17.
136. Boudreau, *Packing Inferno*, 17.

CHAPTER 10

1. É. Durkheim 1952. *Suicide*. Translated by J. Spalding and G. Simpson. London: Routledge and Kegan Paul, 389.
2. É. Durkheim 1957. *Professional Ethics and Civic Morals*. London: Routledge and Kegan Paul, 23.
3. Durkheim, *Professional Ethics and Civic Morals*, 23–4.
4. e.g. A. Abbott 1988. *The System of Profession*. Chicago: University of Chicago Press.
5. Huntington, *The Soldier and the State*, 10.
6. Huntington, *The Soldier and the State*, 16.
7. Huntington, *The Soldier and the State*, 17–18.
8. Huntington, *The Soldier and the State*, 18.
9. W. Heefner 2010. *Dogface Soldier: the life of General Lucian K. Truscott*. Columbia, Miss.: University of Missouri Press, 291.
10. Janowitz, *The Professional Soldier*, x.
11. R. Collins 2004. *Interaction Ritual Chains*. Princteon: Princeton University Press, 91.
12. W. McNeill 1995. *Keeping Together in Time: dance and drill in human history*. Cambridge, Mass.: Harvard University Press, 1.
13. McNeill, *Keeping Together in Time*, 2.
14. McNeill, *Keeping Together in Time*, 2.
15. V. Bramley 1991. *Excursion to Hell*. London: Pan, 170.
16. R. Horsfall 2002. *Fighting Scared*. London: Cassell, 88.
17. S. Preece 2004. *Among the Marines*. Edinburgh: Mainstream, 160.
18. Junger, *War*, 240.
19. Junger, *War*, 237.

20. Junger, *War*, 246.
21. Junger, *War*, 23.
22. Junger, *War*, 23.
23. Junger, *War*, 79.
24. Junger, *War*, 79.
25. Junger, *War*, 234.
26. Junger, *War*, 160.
27. Junger, *War*, 160.
28. Junger, *War*, 160–1.
29. Junger, *War*, 161.
30. Junger, *War*, 74.
31. Captain, German army, personal interview, 9 December 2010.
32. Captain, German army, personal interview, 9 December 2010.
33. Captain, German army, personal interview, 12 December 2012.
34. U. Ben-Shalom, Z. Lehrer, and E. Ben-Ari 2005. 'Cohesion during Military Operations: a field study on combat units in the Al-Aqsa Intifada', *Armed Forces and Society* 32(1): 63–79.
35. Ben-Shalom et al., 'Cohesion during military operations', 73.
36. The OPTAG training team which I interview consisted of a colour sergeant, a sergeant, and a captain. Interview at Camp Bastion, 27 June 2010.
37. OPTAG (Operational Tactics Advisory Group) team, personal interview, 27 June 2010.
38. A multiple is a small unit of about twelve soldiers, often divided into three small teams (bricks), which have often been favoured by the British army in counter-insurgency operations especially in cities.
39. OPTAG team, personal interview, 27 June 2010.
40. OPTAG team, personal interview, 27 June 2010.
41. OPTAG team, personal interview, 27 June 2010.
42. Durkheim, *The Elementary Forms of the Religious Life*.
43. Durkheim, *The Division of Labour in Society*, 84.
44. Durkheim, *The Elementary Forms of the Religious Life*, 84.
45. Durkheim, *The Division of Labour in Society*, 85.
46. Durkheim, *The Division of Labour in Society*, 85.
47. Durkheim, *The Division of Labour in Society*, 85.
48. The Royal Marines use the term troop instead of platoon.
49. Major, Royal Marines, personal interview, 13 May 2010.
50. Ambrose, *Citizen Soldiers*, 277.
51. D. Frankel and M. Salomon 2002. *Band of Brothers*, Warner Home Video, Episode 4 Replacements.
52. Beevor, *D-Day*, 258.
53. Johnston, *At the Frontline*, 168.
54. Johnston, *At the Frontline*, 168.
55. Johnston, *At the Frontline*, 168.
56. Johnston, *At the Frontline*, 169.
57. Johnston, *At the Frontline*, 169.
58. Johnston, *At the Frontline*, 169.

59. McKay, *Delta Four*, 178.
60. McKay, *Delta Four*, 178.
61. Thompson, *The White War*, 266.
62. Ambrose, *Citizen Soldiers*, 343–4.
63. Thompson, *The White War*, 269–70.
64. Thompson, *The White War*, 261.
65. Thompson, *The White War*, 273–4.
66. Thompson, *The White War*, 273–4.
67. Thompson, *The White War*, 264–5.
68. Thompson, *The White War*, 227.
69. Thompson, *The White War*, 275.
70. Thompson, *The White War*, 276.
71. Thompson, *The White War*, 275.
72. Thompson, *The White War*, 275.
73. Sullivan, 'The Italian Soldier in Combat', 179.
74. Truscott, *Mission Command*, 544.
75. Truscott, *Mission Command*, 544.
76. Truscott, *Mission Command*, 544.
77. Sullivan, 'The Italian Soldier in Combat', 178.
78. Strachan, 'Training, Morale and Modern War', 215.
79. H. Severloh 2010. *WN 62: Erinnerungen an Omaha Beach Normandie, 6. June 1944*. Garbsen: Hek Creativ, 26.
80. Back-waters (Haff) refers to the rivers and fens which predominate in that region of Poland.
81. H. Kuss 2008. *Schiksal einer Generation*. Selent: Pour la Mérite, 194.
82. Mark, *Island of Fire*, 125.
83. Beevor, *D-Day*, 392.
84. C. Jennings and A. Weale 1996. *Green-Eyed Boys*. London: HarperCollins, 165.
85. Jennings and Weale, *Green-Eyed Boys*, 167. The sergeant seems to have been from the machine-gun platoon from Support Company (Bramley, *Excursion to Hell*, 176).
86. Sergeant, Royal Marines, personal interview, 6 September 2011.
87. Major, Royal Marines, email communication, 30 September 2011.
88. Jennings and Weale, *Green-Eyed Boys*, 166.
89. Bramley, *Excursion to Hell*, 1991.
90. D. Gaunt 2007. 'Vikings VM'ingS in the "Gan"', *Globe and Laurel* March/April: 107.
91. Sergeant, Royal Marines, personal interview, 22 July 2011.
92. <http://www.cbc.ca/news/canada/montreal/story/2011/07/21/montreal-menard-hearing-721.html>.
93. McCoy, *The Passion of Command*, 54.
94. McCoy, *The Passion of Command*, 54.
95. McCoy, *The Passion of Command*, 54.
96. McCoy, *The Passion of Command*, 55.
97. McCoy, *The Passion of Command*, 55.
98. i.e. 3rd Battalion, 26 Brigade.
99. R. Eckhold 2010. *Fallschirmjäger in Kunduz*. DirektDrukerei Günther, 120.

100. Eckhold, *Fallschirmjäger in Kunduz*, 120.
101. Eckhold, *Fallschirmjäger in Kunduz*, 267–9.
102. See A. King 2011. *The Transformation of Europe's Armed Forces*. Cambridge: Cambridge University Press.
103. Eckhold, *Fallschirmjäger in Kunduz*, 181–8, 149.
104. Eckhold, *Fallschirmjäger in Kunduz*, 212–16.
105. Eckhold, *Fallschirmjäger in Kunduz*, 291.
106. J. Frederick 2010. *Black Hearts*. London: Macmillan

CHAPTER 11

1. Army Lessons Learned Centre 1998. 'Lessons Learned: leadership in a mixed gender environment', *Dispatches* 5(2). Canada: National Defence, 3.
2. J. Goldstein 2004. *War and Gender*. Cambridge: Cambridge University Press, 19.
3. Goldstein, *War and Gender*, 10.
4. Goldstein, *War and Gender*, 21.
5. Goldstein, *War and Gender*, 61–4.
6. Goldstein, *War and Gender*, 22, 64–72.
7. Goldstein, *War and Gender*, 83.
8. Goldstein, *War and Gender*, 83.
9. M. van Creveld 2000. 'Armed but not Dangerous: women in the Israeli military', *War in History* 7: 82.
10. Van Creveld, 'Armed but not Dangerous: women in the Israeli military', 82.
11. Van Creveld, 'Armed but not Dangerous: women in the Israeli military', 87.
12. Van Creveld, 'Armed but not Dangerous: women in the Israeli military', 89.
13. Van Creveld, 'Armed but not Dangerous: women in the Israeli military', 90.
14. Van Creveld, 'Armed but not Dangerous: women in the Israeli military', 89.
15. Van Creveld, 'Armed but not Dangerous: women in the Israeli military', 91.
16. Van Creveld, 'Armed but not Dangerous: women in the Israeli military', 95.
17. Van Creveld, 'Armed but not Dangerous: women in the Israeli military', 13.
18. M. van Creveld 2000. 'Less than We Can Be: men, women and the modern military', *Journal of Strategic Studies* 23(4): 12.
19. Van Creveld, 'Less than We Can Be', 12.
20. Van Creveld, 'Less than We Can Be', 12.
21. A. Gat 2000. 'Female Participation in War: bio-cultural interactions', *Journal of Strategic Studies* 23(4): 21–31.
22. C. Enloe 2000. *Maneuvers*. Berkeley: University of California Press, 46.
23. Enloe, *Maneuvers*, 36.
24. C. Enloe 1983. *Does Khaki become you?* London: Pluto, 15.
25. Enloe, *Does Khaki become you?*, 15, italics in original.
26. Enloe, *Does Khaki become you?*, 16; Enloe, *Maneuvers*, 48.
27. Enloe confesses her ignorance of the armed forces (and, therefore, the possibilities for women's accession) when she declares: 'We know all too little about the internal workings of NATO—how decisions are made, where compromises are hammered out, what items never get on the agenda, how public and private arms manufacturers make their wishes felt, how civilian politicians and military

professionals get on together. What we do know is that, with the small and interesting exception of the Senior Women Officers Commission, the NATO elite is an all-male club' (Enloe, *Does Khaki become you?*, 130–1). In point of fact, much was known about both the political and military dynamics of NATO and there was a large scholarly and practitioner literature on the organization.
28. J. Bethke Elsthain 1987. *Women and War*. New York: Basic Books, 252.
29. Elsthain, *Women and War*, 259.
30. Elsthain, *Women and War*, 148.
31. Elsthain, *Women and War*, 237.
32. Elsthain, *Women and War*, 240.
33. R. Woodward and T. Winter 2007. *Sexing the Soldier*. London: Routledge, 24.
34. H. Carreiras 2006. *Gender and the Military*. London: Routledge, 8.
35. E. Solaro 2006. *Women in the Line of Fire*. Emeryville, Calif.: Seal Press, 11–12; Enloe, *Maneuvers*, 280.
36. K. Sorin 2006. 'Women in the French Forces: integration versus conflict', in F. Pinch, A. MacIntyre, P. Browne, and A. Okros, *Challenge and Change in the Military: gender and diversity issues*. Kingston, Ontario: National Defence Academy Press, 79–80.
37. M. Wechsler Segal 1995. 'Women's Military Roles Cross-Nationally: past, present and future', *Gender and Society* 9(6): 765.
38. Segal, 'Women's Military Roles Cross-Nationally: past, present and future', 765.
39. Segal, 'Women's Military Roles Cross-Nationally: past, present and future', 767.
40. Segal, 'Women's Military Roles Cross-Nationally: past, present and future', 769.
41. J. Stiehm 1981. *Bring me Men and Women*. Berkeley: University of California Press, 10.
42. R. Skaine 1999. *Women at War*. Jefferson, NC: McFarland and Company, 61. Judith Hicks Stiehm noted a 'backlash' against women's representation in the US armed forces in the early 1980s as a result of improving (male) recruitment and a more conservative political environment (1989. *Arms and the Enlisted Women*. Philadephia: Temple University Press, 47–9, 159).
43. Sorin, 'Women in the French Forces', 80–1.
44. Segal, 'Women's Military Roles Cross-Nationally', 769.
45. Carreiras, *Gender and the Military*, 109; G. Harries-Jenkins 2006. 'Institution to Occupation to Diversity: gender in the military today', in Pinch et al., *Challenge and Change in the Military: gender and diversity issues*, 31.
46. Department of National Defence 2011. *Canadian Forces Employment Equity Report: 2010–2011*. September.
47. Skaine, *Women at War*, 64, 61.
48. Woodward and Winter, *Sexing the Soldier*, 33.
49. Woodward and Winter, *Sexing the Soldier*, 41–2.
50. Sorin, 'Women in the French Forces', 81–2.
51. Harries-Jenkins, 'Institution to Occupation to Diversity', 32.
52. Harries-Jenkins, 'Institution to Occupation to Diversity', 33.
53. Harries-Jenkins, 'Institution to Occupation to Diversity', 33.
54. Carreiras, *Gender and the Military*, 99.
55. Woodward and Winter, *Sexing the Soldier*, 42.

56. Woodward and Winter, *Sexing the Soldier*, 43.
57. Sorin divides the process of accession in the French military into four periods: Second World War to 1951, 1951–72, 1972–81, 1981 to the present (Sorin, 'Women in the French Forces', 80–1). Jeanne Holm records the process of gender integration in the period 1970 to 1990 which at that point she saw as an 'unfinished revolution' (J. Holm 1982. *Women in the Military*. Novato, Calif.: Presidio).
58. Goldstein, *War and Gender*, 410.
59. Army Lessons Learned Centre, 'Lessons Learned: leadership in a mixed gender environment', 9.
60. Army Lessons Learned Centre, 'Lessons Learned: leadership in a mixed gender environment', 5.
61. Army Lessons Learned Centre, 'Lessons Learned: leadership in a mixed gender environment', 9.
62. Army Lessons Learned Centre, 'Lessons Learned: leadership in a mixed gender environment', 7.
63. Army Lessons Learned Centre, 'Lessons Learned: leadership in a mixed gender environment', 23.
64. Army Lessons Learned Centre, 'Lessons Learned: leadership in a mixed gender environment', 10, 12, 19.
65. Regimental sergeant major, Canadian army, personal communication, 12 October 2011.
66. Infantry captain, Canadian army, personal communication, 12 October 2011.
67. Infantry captain, Canadian infantry, personal interview, 19 October 2011.
68. Infantry captain, Canadian infantry, personal interview, 19 October 2011.
69. Lance corporal, Canadian army, personal interview, 17 October 2011.
70. Captain A, Canadian army, personal interview, 17 October 2011.
71. Captain B, Canadian army, personal interview, 17 October 2011.
72. Captain B, Canadian army, personal interview, 17 October 2011.
73. Captain A, Canadian army, personal interview, 17 October 2011.
74. Captain A, Canadian army, personal interview, 17 October 2011.
75. Lance corporal, Canadian army, personal interview, 17 October 2011.
76. Lance corporal, Canadian army, personal interview, 17 October 2011.
77. Infantry sergeant, Canadian army, communication, 15 October 2011.
78. Infantry sergeant, Canadian army, communication, 15 October 2011.
79. Captain B, Canadian army, personal interview, 17 October 2011.
80. Captain B, Canadian army, personal interview, 19 October 2011.
81. Captain A, Canadian army, personal interview, 19 October 2011.
82. A. Jeska 2010. *Wir sind Kind Mädchenverein: Frauen in der Bundeswehr*. Munich: Diana, 87.
83. A. Swofford 2003. *Jarhead*. New York: Scribner; E. Wright 2009. *Generation Kill*. London: Corgi.
84. V. Basham 2009. 'Harnessing Social Diversity in the British Armed Forces: the limitations of "management" approaches', *Commonwealth and Comparative Politics* 47(4): 423; V. Basham 2009. 'Effecting Discrimination: operational effectiveness and harassment in the British Armed Forces', *Armed Forces & Society* 35(4): 736.

85. R. Kanter 1977. *Men and Women of the Corporation*. New York: Basic Books, 22.
86. Kanter, *Men and Women of the Corporation*, 22–5.
87. Kanter, *Men and Women of the Corporation*, 48.
88. Kanter, *Men and Women of the Corporation*, 55.
89. Kanter, *Men and Women of the Corporation*, 55.
90. Major, Royal Marines, personal interview, 5 May 2010.
91. Major, Royal Marines, personal interview, 5 May 2010.
92. Major, Royal Marines, personal interview, 5 May 2010.
93. Major, Royal Marines, personal interview, 13 May 2010.
94. Major, Royal Marines, personal interview, 13 May 2010.
95. Colour sergeant, Parachute Regiment, personal interview, 27 June 2010.
96. Colour sergeant, Parachute Regiment, personal interview, 27 June 2010.
97. Colour sergeant, Parachute Regiment, personal interview, 27 June 2010.
98. Colour sergeant, Parachute Regiment, personal interview, 27 June 2010.
99. Colour sergeant, Parachute Regiment, personal interview, 27 June 2010.
100. Captain, British army, personal interview, 27 June 2010.
101. Sergeant, British army, personal interview, 27 June 2010.
102. Captain A, Canadian army, personal interview, 17 October 2011.
103. Captain, British army, personal interview, 27 June 2010.
104. Medical Assistant Class 1 Kate Nesbitt, personal interview, 14 December 2010.
105. Medical Assistant Class 1 Kate Nesbitt, personal interview, 14 December 2010.
106. Captain B, Canadian army, personal interview, 17 October 2011.
107. Captain B, Canadian army, personal interview, 17 October 2011.
108. Skaine, *Women at War*, 169–70.
109. Medical Assistant Class 1 Kate Nesbitt, personal interview, 14 December 2010.
110. Medical Assistant Class 1 Kate Nesbitt, personal interview, 14 December 2010.
111. Solaro, *Women in the Line of Fire*, 185.
112. Solaro, *Women in the Line of Fire*, 297.
113. Solaro, *Women in the Line of Fire*, 100.
114. Solaro, *Women in the Line of Fire*, 115–21.
115. Solaro, *Women in the Line of Fire*, 122.
116. Major, USMC, personal interview, 27 June 2010.
117. Master sergeant, USMC, personal interview, 27 June 2010.
118. P. Van Riper 1997. 'Gender Integrated/Segregated Training', *Marine Corps Gazette* November: 65.
119. Van Riper, 'Gender Integrated/Segregated Training', 65.
120. It is interesting that this low percentage figure indicating that 1% of the infantry might be female correlates with the physical performance of women. A UK MOD Report based on extensive physiological testing concluded: 'approximately 1 per cent of women can equal the performance of the average man...The study concluded that about 0.1 per cent of the female applicants and 1 per cent of trained female soldiers would reach the required standards to meet the demands of these roles' (Ministry of Defence. 2002. *Women in the Armed Forces*. London: Directorate of Service Personnel Policy Service Conditions, 4).
121. A. Krylova 2010. *Soviet Women in Combat*. Cambridge: Cambridge University Press, 73–83.

122. Krylova, *Soviet Women in Combat*, 151–2.
123. Krylova, *Soviet Women in Combat*, 169.
124. Krylova, *Soviet Women in Combat*, 13.
125. Krylova, *Soviet Women in Combat*, 64.
126. A. King 2010. 'The Afghan War and "Postmodern" Memory: commemoration and the dead of Helmand', *British Journal of Sociology* 61(1): 1–25.
127. Captain A, Canadian army, personal interview, 17 October 2011.
128. Lance corporal, Canadian army, personal interview, 17 October 2011.
129. <http://www.ctv.ca/war/#109>.
130. <http://www.ctv.ca/CTVNews/TopStories/20090424/afghanistan_death_090424/>.
131. <http://www.mod.uk/DefenceInternet/DefenceNews/MilitaryOperations/CaptainLisaJadeHeadDiesOfWoundsSustainedInAfghanistan.htm>.
132. J. Aker 1990. 'Hierarchies, jobs and bodies', *Gender and Society* 4(2): 139–58.
133. J. Wacjman 1998. *Managing like a Man*. Cambridge: Polity; J. Wacjman 1996. 'Women and men managers', in R. Crompton, D. Gallie, and K. Purcell (eds.), *Changing Forms of Employment*. London: Routledge.
134. J. Hearn and W. Parking 1987. *Sex at Work*. Brighton: Wheatsheaf Books; J. Hearn, D. L. Sheppard, P. Tancred-Sherriff, and G. Burrell (eds.) 1989. *The Sexuality of the Organisation*. London: Routledge.
135. P. Higate 2007. 'Peacekeepers, Masculinities and Sexual Exploitation', *Men and Masculinities* 10(1): 99–119.
136. Stiehm, *Bring me Men and Women*, 65.
137. V. Basham 2008. 'Everyday Gendered Experiences and the Discursive Construction of Civilian and Military Identities in Britain', *Nordic Journal of Masculinity Studies* 3(2): 150–66; 'Harnessing Social Diversity in the British Armed Forces: the limitations of "management" approaches'; 'Effecting Discrimination: operational effectiveness and harassment in the British Armed Forces'.
138. Major, US army, personal interview, 15 March 2010.
139. Fieldnotes, 20 May 2011.
140. Sergeant, Royal Marines, personal interview, 6 September 2011.
141. J. Hartley 2005. *Just Another Soldier*. London: New York, 93.
142. J. Buckley 1997. 'The Unit Cohesion Factor', *Marine Corps Gazette* November: 69.
143. Hartley, *Just Another Soldier*, 94.
144. Hartley, *Just Another Soldier*, 95.
145. Stiehm, *Bring me Men and Women*, 166.
146. Captain, Airborne Brigade, Bundeswehr, personal interview, 9 December 2010.
147. Captain, Airborne Brigade, Bundeswehr, personal interview, 9 December 2010.
148. Sergeant major, Hammelburg Infantry School, personal interview, 9 December 2010.
149. Sergeant major, Hammelburg Infantry School, personal interview, 9 December 2010. See also G. Kümmel 2002. 'Complete Access: Women in the Bundeswehr and Male Ambivalence', *Armed Forces & Society*, 28 (4), 555–73.
150. Jeska, *Wir sind Kind Mädchenverein: Frauen in der Bundeswehr*, 13.
151. Sorin, 'Women in the French Forces', 88.
152. Sorin, 'Women in the French Forces', 88.

153. Sorin, 'Women in the French Forces', 88.
154. Stiehm, *Arms and the Enlisted Women*, 250.
155. Interestingly, this informant also noted that genuinely gay women in the Women's Air Force at that time in one of the barracks at which she was stationed formed themselves into an aggressive lesbian gang, known as the Rat Pack, which was itself guilty of harassing other women (Stiehm, *Arms and the Enlisted Women*, 248–9). The existence of lesbian gangs in the armed forces is still an issue today but remains an under-investigated area.
156. Stiehm, *Arms and the Enlisted Women*, 250.
157. C. Burke 1996. 'Pernicious Cohesion', in J. Stiehm (ed.), *It's our Military too: women and the U.S. military*. Philadelphia: Temple University Press, 208.
158. J. Zajcek 2010. *Unter Soldatinnen*. Munich: Piper, 68, 91.
159. Solaro, *Women in the Line of Fire*, 38.
160. Solaro, *Women in the Line of Fire*, 39–40.
161. Corporal, Canadian army, personal interview, 17 October 2011.
162. Victoria Basham describes how malicious rumours of this type about a female's sexual availability are very common in the British army ('Everyday Gendered Experiences and the Discursive Construction of Civilian and Military Identities in Britain', 156–7).
163. Captain B, Canadian army, personal interview 17 October 2011.
164. Major A, British army, personal interview, 13 May 2010.
165. C. Mullaney 2009. *The Unforgiving Minute*. New York: Penguin, 296.
166. P. Hennessey 2009. *The Junior Officers' Reading Club*. London: Allen Lane, xiii.
167. Major A, British army, personal interview, 13 May 2010; Basham, 'Everyday Gendered Experiences and the Discursive Construction of Civilian and Military Identities in Britain', 157.
168. Major A, British army, personal interview, 13 May 2010.
169. Major A, British army, personal interview, 13 May 2010.
170. Major B, British army, personal interview, 13 May 2010.
171. Major B, British army, personal interview, 13 May 2010.
172. Carreiras, *Gender and the Military*, 183.
173. Medical Assistant Class 1 Kate Nesbitt, personal interview, 14 December 2010.
174. Medical Assistant Class 1 Kate Nesbitt, personal interview, 14 December 2010.
175. Medical Assistant Class 1 Kate Nesbitt, personal interview, 14 December 2010.
176. Carreiras, *Gender and the Military*, 206.
177. In their recent report for the United States Marines Corps, Annemarie Randazzo-Matsel et al. come to a similar conclusion about the scale of female accession into the infantry. (See A. Randazzo-Matsel, J. Schulte, and J. Yopp 2012. *Assessing the Implications of Possible Changes to Women in Service Restrictions: practices of foreign militaries and other organizations* CNA analysis and solutions.)

CHAPTER 12

1. Severloh, *WN 62*, 59–71.
2. Severloh, *WN 62*, 70.

3. International Institute for Strategic Studies 2011. *The Military Balance 2011*. London: Routledge, executive summary: 1.
4. M. Kramer 2006. 'Book Review: *The Russian Military: power and policy* edited by Steven E. Miller and Dmitri V. Trenin', *Comparative Politics* 4(4): 794.
5. S. Miller 2004. 'Introduction. Moscow's military power: Russia's search for security in an age of transition', in S. Miller and D. Trenin (eds.), *The Russian Military*. Cambridge, Mass.: MIT Press.
6. See King, *The Transformation of Europe's Armed Forces*.
7. International Institute for Strategic Studies, *The Military Balance 2011*, 175.
8. International Institute for Strategic Studies, *The Military Balance 2011*, 177.
9. International Institute for Strategic Studies, *The Military Balance 2011*, 178.
10. International Institute for Strategic Studies, *The Military Balance 2011*, 199–200.
11. A. Scobell 2005. 'China's Evolving Civil–Military Relations', *Armed Forces & Society* 31(2): 227.
12. L. Goldstein 2008. 'China's Falklands Lesson', *Survival* 50(3): 65–82.
13. Goldstein, 'China's Falklands Lesson', 76.
14. Goldstein, 'China's Falklands Lesson', 76.
15. Goldstein, 'China's Falklands Lesson', 77.
16. International Institute for Strategic Studies, *The Military Balance 2011*, 361.
17. International Institute for Strategic Studies, *The Military Balance 2011*, 347.
18. F. Battistelli 1997. 'Peacekeeping and the Postmodern Soldier', *Armed Forces & Society* 23(3): 467–84.
19. <http://articles.cnn.com/2006-08-09/us/iraq.poll_1_opinion-research-corporation-poll-iraq-war-poll-respondents?_s=PM:US>.
20. J. Krakauer 2010. *Where Men Win Glory*. New York: Anchor.
21. G. Davie 2002. *Europe: the exceptional case*. London: Darton, Longman and Todd.
22. C. Taylor 2007. *The Secular Age*. Cambridge, Mass.: Belknap Press.
23. D. Lance 2001. 'Marine Corps Value Program in 2001: where are we now?', *Marine Corps Gazette* September; M. Karcher 2001. 'Passing the Torch: retired Marines instil Corps' values in youth', *Leatherneck* September.
24. Headquarters Department of the Army, *Field Manual 3-21.8. The Infantry Rifle Platoon and Squad*, 1.10.
25. Imitating the American army and in the face of intense operations in Kandahar, the Canadian forces have recently institutionalized their own concept of a 'warrior ethos' which is built on three pillars of intellect, physical ability, and resilience. The warrior ethos includes an ethical dimension but primarily focuses upon and encourages the 'professional bearing'; soldiers are instructed to develop the three pillars of their competence continually in order to prepare for modern day operations.
26. Huntington, *The Soldier and the State*, 466.
27. Overy, *Why the Allies Won*, 232–42.
28. Overy, *Why the Allies Won*, 240.
29. H. Beynon 1973. *Working for Ford*. London: Allen Lane, 109.
30. Beynon, *Working for Ford*, 109.
31. Beynon, *Working for Ford*, 98.
32. Beynon, *Working for Ford*, 118.

33. R. Church 1995. *The Rise and Decline of the British Motor Industry*. Cambridge: Cambridge University Press.
34. Beynon, *Working for Ford*, 85–6.
35. Beynon, *Working for Ford*, 118.
36. Beynon, *Working for Ford*, 85–6.
37. Beynon, *Working for Ford*, 131.
38. Beynon, *Working for Ford*, 131.
39. Beynon, *Working for Ford*, 131.
40. J. Atkinson 1988. 'Recent Changes in the Internal Labour Market Structure in the UK', in W. Buitelaar (ed.), *Technology and Work*. Aldershot: Avebury.
41. Atkinson, 'Recent Changes in the Internal Labour Market Structure in the UK'; U. Jürgens 1989. 'The Transfer of Japanese Management Concepts in the International Automobile Industry', in S. Wood (ed.), *The Transformation of Work?* London: Unwin Hyman; U. Jürgens, K. Dohse, and T. Malsch 1986. 'New Production Concepts in West German Plants', in S. Tolliday and J. Zeitlin (eds.), *The Automobile Industry and its Workers*. Cambridge: Polity, 265; S. Vallas 1999. 'Rethinking Post-Fordism: The Meaning of Workplace Flexibility', *Sociological Theory* 17(1): 68–101; V. Smith 1997. 'New Forms of Work Organization', *Annual Review of Sociology* 23: 321; B. Harrison 1994. *Lean and Mean: the changing landscape of corporate power in the age of flexibility*. New York: Basic Books, 194ff.; M. Piore and S. Sabel 1984. *The Second Industrial Divide*. New York: Basic Books; R. Murray 1990. 'Fordism and Post-Fordism', in S. Hall and M. Jacques (eds.), *New Times*. London: Lawrence and Wishart; C. Lane 1988. 'Industrial Change in Europe: the pursuit of flexible specialization in Britain and West Germany', *Work, Employment and Society* 2(2): 141–68; S. Wood 1989. 'The Transformation of Work', in S. Wood (ed.), *The Transformation of Work?* London: Unwin Hyman; A. Amin (ed.) 1995. *Post-Fordism: a reader*. Oxford: Blackwell; W. Hutton 1996. *The State We're In*. London: Vintage.
42. Jürgens, 'The Transfer of Japanese Management Concepts in the International Automobile Industry', 206.
43. Lane, 'Industrial Change in Europe: the pursuit of flexible specialization in Britain and West Germany', 162.
44. Jürgens et al., 'New Production Concepts in West German Plants', 265.
45. Jürgens et al., 'New Production Concepts in West German Plants', 269–70.
46. Harrison, *Lean and Mean*, 10.
47. Harrison, *Lean and Mean*, 10.
48. D. Winner 2001. *Brilliant Orange*. London: Bloomsbury.
49. <http://news.bbc.co.uk/sport1/hi/football/champions_league/1990408.stm>.
50. G. Noll and A. Gabbard 1999. 'The Last Wave', in J. Long and H.-V. Sponholz (eds.), *The Big Drop*. Guilford, Conn.: Globe Pequot Press, 51.
51. Noll and Gabbard, 'The Last Wave', 52.
52. Noll and Gabbard, 'The Last Wave', 51–2.
53. Laird Hamilton in S. Peralta 2005. *Riding Giants*. Sony Pictures Home Entertainment.
54. Hamilton in Peralta, *Riding Giants*.
55. Dave Kalama in Peralta, *Riding Giants*.

56. Kalama in Peralta, *Riding Giants*.
57. Hamilton in Peralta, *Riding Giants*.
58. R. Putnam 2000. *Bowling Alone*. London: Simon and Schuster, 23–8.
59. Putnam, *Bowling Alone*, 25.
60. Putnam, *Bowling Alone*, 25.
61. Putnam, *Bowling Alone*, 31–2, 49, 60.
62. Putnam, *Bowling Alone*, 132.
63. Putnam, *Bowling Alone*, 357.
64. Putnam, *Bowling Alone*, 259.
65. Putnam, *Bowling Alone*, 183–4.
66. Putnam, *Bowling Alone*, 184.
67. Putnam, *Bowling Alone*, 184.
68. J. Alexander 2006. *The Civil Sphere*. Oxford: Oxford University Press.
69. Alexander, *The Civil Sphere*, 54.
70. Alexander, *The Civil Sphere*, 110.
71. Delbrück, *The History of the Art of War within the Framework of Political History*, iv: *The Dawn of Modern Warfare*, iii: *Medieval Warfare*, ii: *The Barbarian Invasions*.
72. Delbrück, *The History of the Art of War within the Framework of Political History*, iii. 590–2, 648–56; iv. 86.

Bibliography

Abbott, A. 1988. *The System of Profession*. Chicago: University of Chicago Press.
Adkin, M. 2007. *Goose Green: a battle is fought to be won*. London: Phoenix.
Aker, J. 1990. 'Hierarchies, Jobs and Bodies', *Gender and Society* 4(2): 139–58.
Alexander, J. 2006. *The Civil Sphere*. Oxford: Oxford University Press.
Alexander, M. 1992. *The Republic in Danger*. Cambridge: Cambridge University Press.
—— 2011. 'Colonial Minds Confounded: French colonial troops in the Battle of France, 1940', in M. Thomas (ed.), *The French Colonial Mind*, ii: *Violence, Military Encounters and Colonialism*. Lincoln, Nebr.: University of Nebraska Press.
Ambrose, S. 1997. *Citizen Soldiers*. New York: Touchstone.
—— 2001. *Band of Brothers*. London: Pocket Books.
—— 2003. *Pegasus Bridge*. London: Pocket Books.
Amin, A. (ed.). 1995. *Post-Fordism: a reader*. Oxford: Blackwell.
Appleman, R. 1989. *Disaster in Korea*. College Station, Tex.: Texas A & M University.
—— 1990. *Ridgway Duels for Korea*. College Station, Tex.: Texas A & M University.
Aran, G. 1974. 'Parachuting', *American Journal of Sociology* 80(1): 123–52.
Arkin, W. and Dobrofksy, L. 1978. 'Military Socialization and Masculinity', *Journal of Social Issues* 34(1): 151–66.
Army Lessons Learned Centre 1998. 'Lessons Learned: leadership in a mixed gender environment', *Dispatches* 5(2). Canada: National Defence.
Army Training Memorandum. No. 35: War. August 1940. GS Publications.
Ashworth, T. 1980. *Trench Warfare 1914–1918: the live and let live system*. London: Macmillan.
Atkinson, J. 1988. 'Recent Changes in the Internal Labour Market Structure in the UK', in W. Buitelaar (ed.), *Technology and Work*. Aldershot: Avebury.
Ausbildungsvorschrift für die Fusstruppen im Kriege. January 1917. Reichsdruckerei.
Badinski, K. 1947. 'Bericht über den Kampfeinsatz der 276 I.D. in der Normandie vom 20.6-20.8.44', Bundesarchiv-Militär-archiv Freiburg-im-Breisgau, ZA1/877, 1–75.
Balck, W. 1922. *Development of Tactics, World War*. Translated by H. Bell. Fort Leavenworth, Va.: General Service Schools Press.
Balkosi, J. 1999. *Beyond the Beachhead: the 29th Infantry Division in Normandy*. Mechanicsburg, Pa.: Stackpole.
Barbusse, H. 2003. *Under Fire*. Translated by Robin Buss. Harmondsworth: Penguin.
Barkawi, T. 2005. *Globalization and War*. Oxford: Rowman and Littlefield.
—— 2006. 'Culture and Combat in the Colonies: the Indian army in the Second World War', *Journal of Contemporary History* 41(2): 325–55.
Bartov, O. 1992. *Hitler's Army*. Oxford: Oxford University Press.
—— 1992. 'The Conduct of War: soldiers and the barbarization of warfare', *Journal of Modern History* 64, Supplement (December): S32–45.
Basham, V. 2008. 'Everyday Gendered Experiences and the Discursive Construction of Civilian and Military Identities in Britain', *Nordic Journal of Masculinity Studies* 3(2): 150–66.

Basham, V. 2009. 'Harnessing Social Diversity in the British Armed Forces: the limitations of "management" approaches', *Commonwealth and Comparative Politics* 47(4): 411–29.
—— 2009. 'Effecting Discrimination: operational effectiveness and harassment in the British Armed Forces', *Armed Forces & Society* 35(4): 728–44.
Battistelli, F. 1997. 'Peacekeeping and the Postmodern Soldier', *Armed Forces & Society* 23(3): 467–84.
Baynes, J. 1988. *Morale: a study of men and courage*. Garden Park City, NY: Avery Publishing Group.
—— 1995. *Far from a Donkey: the life of General Sir Ivor Maxse*. London: Brassey's.
Beevor, A. 2010. *D-Day*. London: Penguin.
Bellamy, C. 2001. 'Combining Combat Readiness and Compassion', *NATO Review* 49 (2): 9–11.
Bellavia, D. 2008. *House to House*. London: Pocket Book.
Ben-Ari, E. 2001. 'Tests of Soldiering, Trials of Manhood', in E. Ben-Ari and Z. Rosenheck (eds.), *War, Politics and Society in Israel*. New Brunswick, NJ: Transaction Publishers.
Bendix, R. 1960. *Max Weber*. London: Heinemann.
Ben-Shalom, U., Lehrer, Z., and Ben-Ari, E. 2005. 'Cohesion during Military Operations: a field study on combat units in the Al-Aqsa Intifada', *Armed Forces and Society* 32(1): 63–79.
Berkowitz, L. 1954. 'Group Standards, Cohesiveness and Productivity', *Human Relations* 7: 509–19.
Bevin, A. 1998. *Korea*. New York: Hippocrene Books.
Beynon, H. 1973. *Working for Ford*. London: Allen Lane.
Biddle, S. 2004. *Military Power*. Princeton: Princeton University Press.
Bidermann, H. 2002. *In Deadly Combat: a German soldier's memoir of the Eastern Front*. Lawrence, Kan.: BCA.
Bidwell, S. and Graham, D. 2004. *Fire-Power*. Barnsley: Pen and Sword.
Black, J. 2010. *War: a short history*. London: Continuum.
Bloch, I. 1992. *Is War now Impossible?* Aldershot: Gregg Revivals.
Blunden, E. 1982. *Undertones of War*. Harmondsworth: Penguin.
Boog, H., Rahn, W., Stumpf, R. I., and Wegner, B. 2001. *Germany and the Second World War VI. The Global War: widening of the conflict into a world war and the shift of initiative 1941–43*. Oxford: Clarendon.
Boudreau, T. 2008. *Packing Inferno*. Port Townsend, Wash.: Feral House.
Bourke, J. 2000. *An Intimate History of Killing*. London: Granta.
Bourne, P. 1970. *Men, Stress and Vietnam*. Boston: Little Brown.
Bradford, A. 1994. *Some even Volunteered*. Westport, Conn.: Praeger.
Braim, P. 1987. *The Test of Battle*. Newark, Del.: University of Delaware Press.
Bramley, V. 1991. *Excursion to Hell*. London: Pan.
—— 1995. *Two Sides of Hell*. London: Bloomsbury.
Bratton, P. 2002. 'France and the Revolution in Military Affairs', *Contemporary Security Policy* 23(2) August: 87–112.
Browning, C. 1992. *Ordinary Men: Reserve Police Battalion 101 and the final solution in Poland*. New York: HarperCollins.

Buckley, J. 1997. 'The Unit Cohesion Factor', *Marine Corps Gazette* November: 66–70.
Bull, S. and Rottman, G. 2008. *Infantry Tactics of the Second World War*. Wellingborough: Osprey.
Burke, C. 1996. 'Pernicious Cohesion', in J. Stiehm (ed.), *It's our Military too: women and the U.S. military*. Philadelphia: Temple University Press.
Cameron, C. 1994. *American Samurai: myth, imagination and the conduct of battle in the First Marine Division, 1941–51*. Cambridge: Cambridge University Press.
Camp, D. 2009. *Operation Phantom Fury*. Minneapolis: Zenith.
Caputo, P. 1999. *A Rumour of War*. London: Pimlico.
Carreiras, H. 2006. *Gender and the Military*. London: Routledge.
Ceva, L. 1990. 'The North African Campaign 1940–43: a reconsideration', *Journal of Strategic Studies* 13(1): 84–103.
—— and Rochat, G. 2001. 'Italy', in M. R. D. Foot (ed.), *The Oxford Companion to World War II*. Oxford: Oxford University Press.
Church, R. 1995. *The Rise and Decline of the British Motor Industry*. Cambridge: Cambridge University Press.
Citino, R. 1999. *The Path to Blitzkrieg*. London: Lynne Reiner.
Clausewitz, K. von. 1982. *On War*. Translated by M. Howard and P. Paret. Princeton: Princeton University Press.
Cockerham, W. 1978. 'Attitudes towards Combat among US Army Paratroopers', *Journal of Political and Military Sociology* 6 (Spring): 1–15.
Coker, C. 2007. *Warrior Ethos: military culture and the war on terror*. London: Routledge.
Cole, H. 1993. *The Ardennes: Battle of the Bulge*. Washington: Center of Military History, United States Army.
Collins, R. 1979. *The Credential Society*. London: Academic Press.
—— 2000. *The Sociology of Philosophies*. Cambridge, Mass.: Belknap Press.
—— 2004. *Interaction Ritual Chains*. Princeton: Princeton University Press.
—— 2008. *Violence: a microsociological theory*. Princeton: Princeton University Press.
Commander-in-Chief, Home Forces. 1942. *The Instructor's Handbook on Fieldcraft and Battle Drill*. Military Library Research Ltd.
Commando Training Centre Royal Marines 2011. *Close Quarters Battle Instructor: course manual*.
Condell, B. and Zabecki, D. 2009. *On the German Art of War: Truppenfuhrung. German army manual for unit command in World War II*. Mechanicsburg, Pa.: Stackpole.
Copp, T. 2004. *Fields of Fire*. Toronto: University of Toronto Press.
—— 2007. *The Brigade*. Mechanicsburg, Pa.: Stackpole.
Corum, J. 1992. *The Roots of Blitzkrieg*. Lawrence, Kan.: University of Kansas Press.
Coss, E. 2010. *All for the King's Shilling: the British soldier under Wellington, 1808–1814*. Norman, Okla.: University of Oklahoma Press.
Csikszentmihalyi, M. 1992. *Flow*. London: Rider.
—— 1996. *Optimal Experience*. Cambridge: Cambridge University Press.
Culmann, F. 1921. *Cours de tactique générale d'après l'expérience de la Grande Guerre*. Paris: Charles Lavauzelle.

Dandeker, C. 1994. 'New Times for the Military', *British Journal of Sociology* 45(4): 637–54.

—— 1998. 'A Farewell to Arms? The military and the nation-state in a changing world', in James Burk (ed.), *The Adaptive Military*. London: Transaction.

Davie, G. 2002. *Europe: the exceptional case*. London: Darton, Longman and Todd.

Davies, C. 1982. 'Itali sunt imbelles', *Journal of Strategic Studies* 5(2): 266–9.

Delbrück, H. 1975. *The History of the Art of War within the Framework of Political History*, i: *Antiquity*. London: Greenwood.

—— 1990. *The History of the Art of War within the Framework of Political History*, iv: *The Dawn of Modern Warfare*. Translated by W. Renfroe. Lincoln, Nebr.: University of Nebraska Press.

—— 1990. *The History of the Art of War within the Framework of Political History*, iii: *Medieval Warfare*. London: Greenwood.

—— 1990. *The History of the Art of War within the Framework of Political History*, ii: *The Barbarian Invasions*. Translated by W. Renfroe. Lincoln, Nebr.: University of Nebraska Press.

Department of National Defence 2011. *Canadian Forces Employment Equity Report: 2010-2011*. September.

D'Este, C. 1983. *Decision in Normandy*. London: Pan.

Dicks, H. V. 1944. 'The Psychological Foundations of the Wehrmacht', February (WO 241/1).

Dinter, E. 1985. *Hero or Coward*. London: Frank Cass.

Donnelly, W. 2000. 'Thunderbirds in Korea: the US 45th Infantry Division, 1950–52', *Journal of Military History* 64(4): 1113–39.

Doughty, R. 1985. *The Seeds of Disaster: the development of French army doctrine, 1919-1939*. Hamden, Conn.: Archon Books.

—— 1990. *The Breaking Point: Sedan and the fall of France*. Hamden, Conn.

—— 2005. *Pyrrhic Victory: French strategy and operations in the Great War*. Cambridge, Mass.: Belknap.

Dunn, J. C. 1987. *The War the Infantry Knew 1914-1919: a chronicle of service in France and Belgium with the Second Battalion, His Majesty's Twenty-Third Foot, the Royal Welch Fusiliers founded on personal records, recollections and reflections*. London: Cardinal.

Dunning, J. 2000. *It Had to be Tough*. Romsey: James Dunning Publications.

Du Picq. A. 2006. *Battle Studies*. Translated by J. Greely and R. Cotton. Charleston, SC: Bibliobazaar.

Dupuy, T. 1977. *A Genius for War*. London: Macdonald and Janes.

Durkheim, É. 1952. *Suicide*. Translated by J. Spalding and G. Simpson. London: Routledge and Kegan Paul.

—— 1957. *Professional Ethics and Civic Morals*. London: Routledge and Kegan Paul.

—— 1965. *Sociology and Philosophy*. Translated by D. Pocock. London: Cohen and West.

—— 1976. *The Elementary Forms of the Religious Life*. Translated by J. Ward Swain. Old Woking: Unwin Brothers.

—— 1984. *The Division of Labour in Society*. Translated by W. D. Halls. Basingstoke: Macmillan.

Eckhold, R. 2010. *Fallschirmjäger in Kunduz*. Chemnitz: DirektDrukerei Günther.
Edmonds, J. and Falls, C. 1940. *Military Operations: France and Belgium 1917*. London: Macmillan and Co.
—— and Miles, W. 1938. *Military Operations: France and Belgium 1916*. London: Macmillan and Co.
Eisenhart, R. 1975. 'You Can't Hack it Little Girl: a discussion of the covert psychological agenda of modern combat training', *Journal of Social Issues* 31(4): 13–24.
Ellis, J. 1990. *World War II: the sharp end*. London: Windrow and Greene.
Elsthain, J. Bethke 1987. *Women and War*. New York: Basic Books.
English, J. 1998. *Lament for an Army*. Toronto: Canadian Institute of International Affairs.
—— 2009. *The Canadian Army in the Normandy Campaign*. Mechanicsburg, Pa.: Stackpole.
—— and Gudmundsson, B. 1994. *On Infantry: the military profession*. Westport, Calif.: Praeger.
Enloe, C. 1983. *Does Khaki become you?* London: Pluto.
—— 2000. *Maneuvers*. Berkeley: University of California Press.
Evans, C. and Dion, K. 1991. 'Group and Performance: a meta-analysis', *Small Group Research* 22(7) May: 175–86.
Fehrenbach, T. 1963. *This Kind of War*. New York: Macmillan.
Fennell, J. 2011. *Combat and Morale in the North African Campaign: the Eighth Army and the path to El Alamein*. Cambridge: Cambridge University Press.
Ferguson, H. 2004. 'The Sublime and Subliminal: modern identities and the aesthetics of combat', *Theory Culture and Society* 21(3): 1–33.
Festinger, L. 1950. 'Informal Social Communication', *Psychological Review* 57: 271–82.
—— Schachter, S., and Back, K. 1950. *Social Pressures in Informal Groups*. New York: Harpers and Brothers.
Fitz-Gibbon, S. 2001. *Not Mentioned in Despatches: the history and mythology of the Battle of Goose Green*. Cambridge: Lutterworth Press.
Forster, J. 1997. 'Motivation and Indoctrination in the Wehrmacht, 1933–45', in P. Addison and A. Calder, *Time to Kill: the soldier's experience of the war in the west, 1939–45*. London: Pimlico.
Frankel, D. and Salomon, M. 2002. *Band of Brothers*. Warner Home Video.
Frederick, J. 2010. *Black Hearts*. London: Macmillan.
French, D. 2000. *Raising Churchill's Army: the British army and the war against Germany 1919–1945*. Oxford: Oxford University Press.
—— 2005. *Military Identities: the regimental system, the British army and the British people, 1870–2000*. Oxford: Oxford University Press.
Frieser, K.-H. 2005. *The Blitzkrieg Legend*. Annapolis, Md.: Naval Institute Press.
Fritz, S. 1995. *Frontsoldaten: the German soldier in World War II*. Lexington, Ky.: University of Kentucky Press.
Fuller, J. G. 1990. *Troop Morale and Popular Culture in the British and Dominion Armies 1914–18*. Oxford: Clarendon Press.
Fussell, P. 1975. *The Great War and Modern Memory*. Oxford: Oxford University Press.

Gammage, B. 1975. *The Broken Years: The Australian Soldier in the Great War*. Harmondsworth: Penguin.
Garfinkel, H. 1967. *Studies in Ethnomethodology*. Cambridge: Polity.
—— 2002. *Ethnomethodology's Program: working out Durkheim's aphorism*. Oxford: Rowman and Littlefield.
Garland, A. 1985. *Infantry in Vietnam*. New York: Jove.
Gat, A. 2000. 'Female Participation in War: bio-cultural interactions', *Journal of Strategic Studies* 23(4): 21–31.
Gaunt, D. 2007. 'Vikings VM'ingS in the "Gan"', *Globe and Laurel* March/April: 107–8.
General Staff 1917. SS 143. *Instructions for the Training of Platoons for Offensive Action, 1917*.
—— 1917. SS b185. *Assault Training*. London: Harrison and Sons.
—— 1918. SS 143. *The Training and Employment of Platoons 1918*.
General Staff, War Office 1909. *Education and Training of the German Infantry*. London: His Majesty's Stationery Office.
—— 1941. *German Infantry in Action (minor tactics)*.
Geyer, M. 1986. 'German Strategy in the Age of Machine Warfare, 1914–45', in P. Paret, G. Craig, and F. Gilbert, *Makers of Modern Strategy*. Princeton: Princeton University Press.
Gilchrist, D. 1960. *Castle Commando*. London: Oliver and Boyd.
Girton, G. 1986. 'Kung Fu: toward a praxiological hermeneutic of the martial arts', in H. Garfinkel (ed.), *Ethnomethodological Studies of Work*. London: Routledge and Kegan Paul.
Glantz, D. with House, J. 2009. *The Stalingrad Trilogy*, ii: *Armageddon in Stalingrad: September–November 1942*. Lawrence, Kan.: University of Kansas Press.
Glenn, R. 2000. *Reading Athena's Dance Card: men against fire in Vietnam*. Annapolis, Md.: Naval Institute Press.
—— 2000. 'Introduction' to S. L. A. Marshall, *Men against Fire*. Norman, Okla.: University of Oklahoma Press.
Globe and Laurel 2010. 'Commando Training Centre RM', July/August: 288–90.
Godefroy, A. 2007. 'The 4th Canadian Division: "trenches should never be saved"', in G. Hayes, A. Iarocci, and M. Bechthold (eds.), *Vimy Ridge: a Canadian reassessment*. Waterloo, Ontario: Wilfred Laurier University Press.
Goldhagen, D. 1997. *Hitler's Willing Executioners*. London: Abacus.
Goldstein, J. 2004. *War and Gender*. Cambridge: Cambridge University Press.
Goldstein, L. 2008. 'China's Falklands Lesson', *Survival* 50(3): 65–82.
Gooch, J. 1982. 'Italian Military Competence', *Journal of Strategic Studies* 5(2): 257–65.
—— 2010. 'Italy during the First World War', in A. Millett and W. Murray (eds.), *Military Effectiveness*, i: *The First World War*. Cambridge: Cambridge University Press.
Goya, M. 2004. *La Chair et l'acier: l'armée française et l'invention de la guerre moderne (1914–18)*. Paris: Tallandier.
Granatstein, J. 2004. *Canada's Army: waging war and keeping the peace*. Toronto: University of Toronto Press.

Grandin, B., Smith, H., Kendall, G., Buick, B., Sabben, D., Stanley, M., and Roberts, A. 2004. *The Battle of Long Tan*. Crows Nest, NSW: Allen and Unwin.
Grand Quartier General 1916. *Manuel du chef de section d'infanterie*. Paris: Imprimerie Nationale.
—— 1918. *Instruction sur le combat offensif des petits unités*. Paris: Imprimerie Nationale.
Graves, R. 1966. *Goodbye to All That*. London: Cassell.
Griffith, J. 1989. 'The Army's New Unit Personnel Replacement and its Relationship to Unit Cohesion and Social Support', *Military Psychology* 1(1): 17–34.
Griffith, P. 1994. *Battle Tactics of the Western Front*. New Haven: Yale University Press.
Gross, N. and Martin, W. 1952. 'On Group Cohesiveness', *American Journal of Sociology* 57(6) May: 546–64.
Grossman, D. 1996. *On Killing: the psychological costs of learning to kill in war and society*. Boston: Back Bay Books.
Gudmundsson, B. 1989. *Stormtroop Tactics: innovation in the German Army, 1914–18*. Westport, Calif.: Praeger.
Haltiner, K. 1998. 'The Definite End of the Mass Army in Western Europe?', *Armed Forces & Society* 25(1): 7–36.
—— 2003. 'The Decline of the Mass Army', in G. Caforio (ed.), *Handbook of the Sociology of the Military*. London: Kluwer/Plenum.
Hamner, C. 2011. *Enduring Battle: American soldiers in three wars, 1776–1945*. Lawrence, Kan.: University of Kansas Press.
Harclerode, P. 2000. *Secret Soldiers*. London: Cassell.
Harries-Jenkins, G. 2006. 'Institution to Occupation to Diversity: gender in the military today', in F. Pinch, A. MacIntyre, P. Browne, and A. Okros, *Challenge and Change in the Military: gender and diversity issues*. Kingston, Ontario: National Defence Academy Press.
Harrison, B. 1994. *Lean and Mean: the changing landscape of corporate power in the age of flexibility*. New York: Basic Books.
Harrison, G. 1993. *United States Army in World War II. European Theatre of Operations: cross-channel attack*. Washington: Center of Military History, United States Army.
Hart, P. 2006. *The Somme*. London: Cassell.
Hartley, J. 2005. *Just Another Soldier*. London: Harper.
Hastings, M. 1984. *Overlord*. London: Macmillan.
—— 1987. *The Korean War*. London: Michael Joseph.
Hauser, W. 1973. *America's Army in Crisis*. Baltimore: Johns Hopkins University Press.
Headquarters Department of the Army 2006. *ARTEP 7-1-Drill. Warrior Battle Drills*. Washington.
—— 2007. *Field Manual 3-21.8. The Infantry Rifle Platoon and Squad*. Washington.
Hearn, J. and Parking, W. 1987. *Sex at Work*. Brighton: Wheatsheaf Books.
—— Sheppard, D., Tancred-Sherriff, P., and Burrell, G. (eds.) 1989. *The Sexuality of the Organisation*. London: Routledge.

Heefner, W. 2010. *Dogface Soldier: the life of General Lucian K. Truscott*. Columbia, Miss.: University of Missouri Press.
Heeres Division 130/2a 1922. *Ausbildungsvorschrift für die Infanterie, heft 2a: Die Schützenkompanie*. Berlin: Offene Worte.
—— 1938. *Ausbildungsvorschrift für die Infanterie, heft 2a: Die Schützenkompanie*. Berlin: Reichsdruckerie.
—— 1941. *Ausbildungsvorschrift für die Infanterie, heft 2a: Die Schützenkompanie*. Berlin: Offene Worte.
—— 1942. *Ausbildungsvorschrift für die Infanterie, heft 2a: Die Schützenkompanie*. Berlin: Offene Worte.
Heeres Division 130/2b 1936. *Ausbildungsvorschrift für die Infanterie, heft 2: Die Schützenkompanie*. Berlin: Offene Worte.
Heller, J. 1994. *Catch-22*. London: Vintage.
Henderson, D. 1985, *Cohesion: the human element*. Washington: National Defence University Press.
Hennessey, P. 2009. *The Junior Officers' Reading Club*. London: Allen Lane.
Henriksen, R. 2007. 'Warriors in Combat: what makes people actively fight in combat', *Journal of Strategic Studies* 30(2): 187–223.
Hetherington, T. and Junger, S. 2010. *Restrepo*. DVD. Studio: Dogwoof.
Heydte, F. von der. 1948. 'Das Fallschirmjägerregiment 6 in der Normandie I: die Kämpfe in Raum von Carentan', Bundesarchiv-Militär-archiv Freiburg-im-Breisgau, ZA1/1188, 1–110.
Higate, P. 2007. 'Peacekeepers, Masculinities and Sexual Exploitation', *Men and Masculinities* 10(1): 99–119.
Hockey, J. 1986. *Squaddies: portrait of a subculture*. Exeter: University of Exeter.
—— 2009. 'Switching on: sensory work in the infantry', *Work, Employment and Society* 23(3): 477–93.
Hogg, M. 1992. *The Social Psychology of Group Cohesiveness: from attraction to social identity*. New York: Harvester Wheatsheaf.
Holl, A. 2005. *An Infantryman in Stalingrad*. Translated by J. Mark and N. Page. Sydney: Leaping Horseman Books.
Holm, J. 1982. *Women in the Military*. Novato, Calif.: Presidio.
Holmes, R. 1994. *Firing Line*. London: Pimlico.
—— 2001. *Redcoat*. London: Harper Collins.
—— 2001. *The Western Front*. London: BBC.
Horn, B. 2010. *No Lack of Courage*. Toronto: Dundum Press.
Horne, A. 2006. *A Savage War of Peace*. New York: New York Review Book.
Horsfall, R. 2002. *Fighting Scared*. London: Cassell.
Howard, M. 1986. 'Men against Fire: the doctrine of the offensive in 1914', in P. Paret, G. Craig, and F. Gilbert, *Makers of Modern Strategy*. Princeton: Princeton University Press.
Humphries, M. 2007. 'Old Wine in New Bottles: a comparison of British and Canadian preparations for the Battle of Arras', in G. Hayes, A. Iarocci, and M. Bechthold (eds.), *Vimy Ridge: a Canadian reassessment*. Waterloo, Ontario: Wilfred Laurier University Press.
Huntington, S. 1957. *The Soldier and the State*. Cambridge, Mass.: Belknap Press.

Hutton, W. 1996. *The State We're In*. London: Vintage.
Infantry Journal 1939. *Infantry in Battle*. Washington: Infantry Journal Inc.
International Institute for Strategic Studies 2011. *The Military Balance 2011*. London: Routledge.
Janis, I. 1982. *Group Think: psychological studies of policy decisions and fiascoes*. Boston: Houghton Miffin Company.
Janowitz, M. 1960. *The Professional Soldier*. New York: Free Press.
—— 1975. *Military Conflict*. London: Sage.
—— and Shils, E. 1948. 'Cohesion and Disintegration in the Wehrmacht in World War II', *Public Opinion Quarterly* Summer: 280–315.
Jary, S. 1988. *18 Platoon*. Taunton: Light Infantry Office.
—— 2002. 'Funeral Oration for Corporal Douglas Proctor (1917–2002)', in *Serve to Lead*. Sandhurst: Royal Military Academy.
Jennings, C. and Weale, A. 1996. *Green-Eyed Boys*. London: HarperCollins.
Jeska, A. 2010. *Wir sind Kind Mädchenverein: Frauen in der Bundeswehr*. Munich: Diana.
Johnston, M. 2002. *At the Frontline: experience of Australian soldiers in World War II*. Cambridge: Cambridge University Press.
Joint Warfare Publication 0-01 2001. *British Defence Doctrine*. Shrivenham: Joint Doctrine and Concepts Centre.
Jolly, M. 1997. 'Love Letters vs Letters Carved in Stone: gender, memory and the forces sweetheart exhibition', in M. Evans and K. Lunn (eds.), *War and Memory in the Twentieth Century*. Oxford: Berg.
Jones, J. 1987. *The Thin Red Line*. Glasgow: William Collins and Sons and Co.
Jordan, K. 2002. 'Right for the Wrong Reasons', *Journal of Military History* 66: 135–62.
Jünger, E. 1941. *In Stahlgewittern*. Berlin: E. S. Mittler und Sohn.
—— 2003. *Storm of Steel*. Harmondsworth: Penguin.
Junger, S. 2010. *War*. London: Fourth Estate.
Jürgens, U. 1989. 'The Transfer of Japanese Management Concepts in the International Automobile Industry', in S. Wood (ed.), *The Transformation of Work?* London: Unwin Hyman.
—— Dohse, K., and Malsch, T. 1986. 'New Production Concepts in West German Plants', in S. Tolliday and J. Zeitlin (eds.), *The Automobile Industry and its Workers*. Cambridge: Polity.
Kant, I. *Critique of Pure Reason*. 1929. Translated by N. Kemp Smith. London: Macmillan.
Kanter, R. 1977. *Men and Women of the Corporation*. New York: Basic Books.
Karcher, M. 2001. 'Passing the Torch: retired marines instil Corps' values in youth', *Leatherneck* September.
Keegan, J. 1976. *The Face of Battle*. Harmondsworth: Penguin.
Kelleher, C. 1978. 'Mass Armies in the 1970s: the debate in Western Europe', *Armed Forces & Society* 5(1): 3–30.
Kellett, A. 1982. *Combat Motivation: the behaviour of soldiers in battle*. The Hague: Kluwer/Nijhoff Publishing.
Kennedy, P. 2010. 'Britain in the First World War', in A. Millett and W. Murray (eds.), *Military Effectiveness*, i: *The First World War*. Cambridge: Cambridge University Press.

Kershaw, A. 2005. *The Longest Winter*. Harmondsworth: Penguin.
Kier, E. 1998. 'Homosexuality in the US Military: open integration and combat effectiveness', *International Security* 23(2): 5–39.
—— 1999. *Imagining War: French and British military doctrine between the wars*. Princeton: Princeton University Press.
Kiesling, E. 1996. *Arming against Hitler*. Lawrence, Kan.: University Press of Kansas.
Killingray, D. with Plant, M. 2010. *Fighting for Britain: African soldiers in the Second World War*. Woodbridge: James Currey.
Kindsvatter, P. 1995. *American Soldiers: ground combat in the world wars, Korea and Vietnam*. Lawrence, Kan.: University of Kansas Press.
King, A. 2006. 'The Word of Command: communication and cohesion in the military', *Armed Forces and Society* 32(1): 493–512.
—— 2010. 'The Afghan War and "Postmodern" Memory: commemoration and the dead of Helmand', *British Journal of Sociology* 61(1): 1–25.
—— 2011. *The Transformation of Europe's Armed Forces*. Cambridge: Cambridge University Press.
King, E. 1973. *The Death of an Army*. New York: Saturday Review Press.
Kinzer Stewart, N. 1991. *Mates and Muchachos: unit cohesion in the Falklands/Malvinas War*. New York: Brasseys Inc.
Kirke, C. 2009. 'Group Cohesion, Culture and Practice', *Armed Forces & Society* 35(4): 745–53.
—— 2011. *Red Coat, Green Machine: continuity and change in the British army 1700–2000*. London: Continuum.
Kissel, H. 1956. 'Panic in Battle', *Military Review* July: 96–107.
Knox, M. 1982. *Mussolini Unleashed 1939–41: politics and strategy in Fascist Italy's last war*. Cambridge: Cambridge University Press.
—— 2010. 'The Italian Armed Forces, 1940–43', in A. Millett and W. Murray (eds.), *Military Effectiveness*, iii: *The Second World War*. Cambridge: Cambridge University Press.
Krakauer, J. 2010. *Where Men Win Glory*. New York: Anchor.
Kramer, M. 2006. 'Book Review: *The Russian Military: power and policy* edited by Steven E. Miller and Dmitri V. Trenin', *Comparative Politics* 4(4): 794–5.
Krylova, A. 2010. *Soviet Women in Combat*. Cambridge: Cambridge University Press.
Kuss, H. 2008. *Schiksal einer Generation*. Selent: Pour la Mérite.
Ladd, J. 1978. *Commandos and Rangers of World War II*. London: Macdonald and Jane's.
Laffargue, A. 1917. *The Attack in Trench Warfare*. New York: D. Van Nostrand Company.
Lance, D. 2001. 'Marine Corps Value Program in 2001: where are we now?', *Marine Corps Gazette* September.
Landers, D. M. and Luschen, G. 1974. 'Team Performance Outcome and Cohesiveness of Competitive Coacting Groups', *International Review of Sport Sociology* 9: 57–69.
Lane, C. 1988. 'Industrial Change in Europe: the pursuit of flexible specialization in Britain and West Germany', *Work, Employment and Society* 2(2): 141–68.
Langer, P. 2010. 'Studying Violence: methodological considerations of doing research on combat experiences in Afghanistan', paper presented to 11th Biennial Conference of ERGOMAS, 14 June.

Lee, D. 2006. *Up Close and Personal: the reality of close-quarter fighting in World War II*. London: Greenhill.
Lee, U. 2004. *The Employment of Negro Troops*. Honolulu: University Press of the Pacific.
Leed, E. 1979. *No Man's Land*. Cambridge: Cambridge University Press.
Leinbaugh, H. and Campbell, J. 1985. *The Men of Company K*. New York: William Morrow and Company.
Liddell Hart, B. 1933. *The Future of Infantry*. London: Faber and Faber.
—— 1965. *The Memoirs of Captain Basil Liddell Hart*. London: Cassell.
—— (ed.) 1987. *The Rommel Papers*. New York: Da Capo Press.
—— 1999. *Thoughts on War*. Padstow: Spellmount.
Linderman, G. 1997. *The World within War*. New York: Free Press.
Little, R. 1964. 'Buddy Relations and Combat Performance', in M. Janowitz (ed.), *The New Military*. New York: Russell Sage Foundation.
Lott, A. and Lott, B. 1965. 'Group Cohesion and Interpersonal Attraction: A Review of Relationships with Antecedent and Consequent Variables', *Psychological Bulletin* 64 (4): 259–309.
Lowry, R. 2010. *New Dawn*. New York: Savas Beattie.
Lupfer, T. 1981. *The Dynamics of Doctrine*. Fort Leavenworth, Kan.: Combat Studies Institute, US Army Command and General Staff College.
Luttwak, E. 1995. 'Toward Post-Heroic Warfare', *Foreign Affairs* 74(3): 109–22.
Lynn, J. 1984. *The Bayonets of the Republic*. Urbana, Ill.: University of Illinois Press.
McAulay, L. 1987. *The Battle of Long Tan*. London: Arrow.
MacCoun, R. 1993. 'What is Known about Unit Cohesion and Military Performance', in National Defence Research Institute, *Sexual Orientation and US Military Personnel Policy: options and assessment*. Washington: RAND.
—— Kier, E., and Belkin, A. 2006. 'Does Social Cohesion Determine Motivation in Combat? An old question with an old answer', *Armed Forces and Society* 32(4) July: 646–54.
McCoy, B. 2007. *The Passion of Command*. Quantico, Va.: Marine Corps Association.
MacDonald, C. 1986. *Korea: the war before Vietnam*. London: Macmillan Press.
Machiavelli, N. 1965. *The Art of War*. Indianapolis: Bobbs Merrill.
—— 1999. *The Prince*. Translated by G. Bull. London: Penguin.
McKay, G. 1996. *Delta Four*. St Leonards, NSW: Allen and Unwin.
McKenna, J. 1997. 'Towards the Army of the Future: domestic politics and the end of conscription in France', *West European Politics* 20(4) October: 125–45.
McNeill, W. 1995. *Keeping Together in Time: dance and drill in human history*. Cambridge, Mass.: Harvard University Press.
Mailer, N. 1949. *The Naked and the Dead*. London: Allan Wingate.
Manigart, P. 2004. 'Restructuring of the Armed Forces', in Giuseppe Caforio (ed.), *Handbook of the Sociology of the Military*. London: Kluwer/Plenum.
Mansoor, P. 1999. *The GI Offensive in Europe: the triumph of American Infantry Division, 1941–1945*. Lawrence, Kan.: University of Kansas.
Mark, J. 2006. *Island of Fire*. Sydney: Leaping Horseman Books.
Marlantes, K. 2010. *Matterhorn*. London: Corvus.
—— 2011. *What it is Like to Go to War*. New York: Atlantic Monthly Press.

Marlowe, D. 1979. 'Cohesion, Anticipated Breakdown, and Endurance in Battle'. Unpublished draft, Washington: Walter Reed Army Institute of Research.
Marshall, S. L. A. 1952. *The River and the Gauntlet*. New York: Warner Books.
—— 1967. *Battles in the Monsoon: campaigning in the Central Highlands Vietnam, Summer 1966*. Nashville: Battery Press.
—— 1968. *Bird*. New York: Warner Books.
—— 1983. *Ambush: the battle of Dau Tieung*. New York: Jove.
—— 1990. *The Capture of Makin*. Washington: Centre of Military History United States Army.
—— 1986. *Night Drop*. New York: Jove.
—— 1986. *Pork Chop Hill*. New York: Jove.
—— 2000. *Men against Fire*. Norman, Okla.: University of Oklahoma.
—— 2001. *Island Victory: the battle of Kwajalein Atoll*. Lincoln, Nebr.: University of Nebraska Press.
Marston, D. 2003. *Phoenix from the Ashes: the Indian army in the Burma campaign*. Westport, Conn.: Praeger.
Martens, R. and Peterson, J. 1971. 'Group Cohesiveness as a Determinant of Success and Membership Satisfaction in Team Performance', *International Review of Sport Psychology* 6: 50–9.
Martin, M. 1977. 'Conscription and the Decline of the Mass Army in France, 1960–75', *Armed Forces & Society* 3(3): 355–406.
Maurer, E. 1946. 'Schilderung der Operationen gegen die Truppen der US-Armee: de Kämpfe der 253 Inf. Div.', Bundesarchiv-Militär-archiv Freiburg-im-Breisgau, ZA1/1711, 2–44.
Mershon, S. and Schlossman, S. 1998. *Foxholes and Colour Lines*. Baltimore: Johns Hopkins University Press.
Millett, A. 1980. *Semper Fidelis: the history of the United States Marines Corps*. New York: Macmillan.
Ministère de la Défense Nationale et de la Guerre 1938. *Règlement de l'infanterie, première partie: instruction*. Paris: Imprimerie Nationale.
Ministère de la Guerre, État Major de l'Armée 1944. *Règlement de l'infanterie, deuxième partie: combat*. Paris: Charle Lavauzelle et Cie.
Ministry of Defence 2002. *Women in the Armed Forces*. London: Directorate of Service Personnel Policy Service Conditions.
Miller, F. D. 1991. *Reflections of a Warrior*. New York: Pocket Books.
Miller, S. 2004. 'Introduction. Moscow's Military Power: Russia's search for security in an age of transition', in S. Miller and D. Trenin (eds.), *The Russian Military*. Cambridge, Mass.: MIT Press.
Moore, H. and Galloway, J. 2002. *We Were Soldiers Once...and Young*. London: Corgi.
Moskos, C. 1970. *The American Enlisted Man: the rank and file in today's military*. New York: Russell Sage Foundation.
—— 1975. 'The American Combat Soldier in Vietnam', *Journal of Social Issues* 31(4): 25–37.
—— 1988. *Soldiers and Sociology*. Fort Belvoir, Va.: United States Army Research Institute for Behaviour and Social Sciences.

—— Allen Williams, J. and Segal, D. 2000. *The Postmodern Military*. Oxford: Oxford University Press.
—— and Burk, J. 2003. 'The Postmodern Military', in J. Burk (ed.), *The Adaptive Military*. London: Transaction.
Mosse, G. 1985. *Nationalism and Sexuality*. New York: Howard Fertig.
Mullaney, C. 2009. *The Unforgiving Minute*. New York: Penguin.
Murphy, A. 2002. *To Hell and Back*. New York: Picador.
Murphy, E. 2007. *Dak To*. New York: Ballantine.
Murray, R. 1990. 'Fordism and Post-Fordism', in S. Hall and M. Jacques (ed.), *New Times*. London: Lawrence and Wishart.
Nalty, B. 1989. *Strength for the Fight: a history of black Americans in the military*. London: Collier Macmillan.
Naylor, S. 2005. *Not a Good Day to Die*. Harmondsworth: Penguin.
Newsome, B. 2003. 'The Myth of Intrinsic Combat Motivation', *Journal of Strategic Studies* 26(4): 24–46.
Noll, G. and Gabbard, A. 1999, 'The Last Wave', in J. Long and H.-V. Sponholz (eds.), *The Big Drop*. Guilford, Conn.: Globe Pequot Press.
Overy, R. 2006. *Why the Allies Won*. London: Pimlico.
Owen, W. 2004. 'Platoon and Section Manoeuvre 1917–2003', *Army Doctrine and Training News* 20: 39–44.
Pakenham, T. 1982. *The Boer War*. London: Macdonald.
Palmer, R., Wiley, B. I., and Keast, W. 1948. *The Army Ground Forces: the procurement and training of ground combat troops*. Washington: History Division, Department of the Army.
Parsons, T. 1949. *The Structure of Social Action*. New York: Free Press.
Patton, G. 1995. *War as I Knew it*. Boston: Houghton Mifflin.
Peralta, S. 2005. *Riding Giants*. Sony Pictures Home Entertainment.
Perkins, H. 2002. *The Rise of Professional Society*. London: Routledge.
Philpott, W. 2010. *Bloody Victory*. London: Abacus.
Piore, M. and Sabel, S. 1984. *The Second Industrial Divide*. New York: Basic Books.
Place, T. H. 2000. *Military Training in the British Army 1940–44*. London: Routledge.
Porter, P. 2009. *Military Orientalism*. London: C. Hurst and Co.
Pratten, G. 2010. *Australian Battalion Commanders in the Second World War*. Cambridge: Cambridge University Press.
Preece, S. 2004. *Among the Marines*. Edinburgh: Mainstream.
Prior, R. and Wilson, T. 2005. *The Somme*. New Haven: Yale University Press.
Putnam, R. 2000. *Bowling Alone*. London: Simon and Schuster.
Randazzo-Matsel, A., Schulte, J. and Yopp, J. 2012. *Assessing the Implications of Possible Changes to Women in Service Restrictions: practices of foreign militaries and other organizations* CNA analysis and solutions. Report, United States Marines Corps.
Ratcliffe, P. 2001. *Eye of the Storm*. London: Michael O'Mara.
Raths, R. 2009. *Vom Massensturm zur Stosstrupptaktic*. Potsdam: Rombach.
Rawling, B. 1992. *Surviving Trench Warfare*. Toronto: University of Toronto Press.
Reibert, W. 1940. *Der Dienstunterricht im Heere*. Berlin: E. S. Mittler und Sohn.

Remarque, E. M. 1982. *All Quiet on the Western Front*. Translated by A. W. Wheen. London: Random House.
Roberts Greenfield, K. 1955. *The United States Army in World War II*. Washington: Office of the Chief of Military History, Department of the Army.
Rogers, B. 1974. *Cedar Falls-Junction City*. Washington: Department of the Army.
Romains, J. 2000. *Verdun*. Translated by Gerard Hopkins. London: Prion.
Rommel, E. 2009. *Infantry Attacks*. Newbury: Greenhill Books.
Rosen, L., Knudson, K., and Fancher, P. 2002. 'Cohesion and the Culture of Hypermasculinity in the US Army Units', *Armed Forces and Society* 29(3) Spring: 325–52.
Ross, J. 1985. *The Myth of the Digger: the Australian soldier in two world wars*. Alexandria, NSW: Hale and Iremonger.
Sadkovich, J. 1991. 'Of Myths and Men: Rommel and the Italians in North Africa 1940–42', *International History Review* 13(2) May: 282–340.
St George Saunders, H. 1949. *The Green Beret*. London: Michael Joseph.
Sajer, G. 1990. *The Forgotten Soldier*. London: Orion.
Sandler, S. 1999. *The Korean War*. London: University College London Press.
Sanger, E. 2012. 'Using Historical Experience: the British and the Bundeswehr in Afghanistan', Ph.D. thesis, European University Institute.
Sassoon, S. 2000. *Memoirs of an Infantry Officer*. London: Faber and Faber.
Savage, P. and Gabriel, R. 1976. 'Cohesion and Disintegration in the American Army: an alternative perspective', *Armed Forces & Society* 2(3): 340–70.
—— and Gabriel, R. 1978. *Crisis in Command*. New York: Hill and Wang.
Schulte, T. 1997. 'The German Soldier in Occupied Russia', in P. Addison and A. Calder, *Time to Kill: the soldier's experience of the war in the west, 1939–45*. London: Pimlico.
Scobell, A. 2005. 'China's Evolving Civil–Military Relations', *Armed Forces & Society* 31(2): 227–44.
Searle, J. 1995. *The Construction of Social Reality*. London: Allen Lane.
Segal, D. and Kestnbaum, M. 2002. 'Professional Closure in the Military Market: a critique of pure cohesion', in D. Snider (ed.), *The Future of the Army Profession*. New York: Primis.
Segal, M. Wechsler. 1995. 'Women's Military Roles Cross-Nationally: past, present and future', *Gender and Society* 9(6): 757–75.
Severloh, H. 2010. *WN 62: Erinnerungen an Omaha Beach Normandie, 6. June 1944*. Garbsen: Hek Creativ.
Shaw, M. 1976. *Group Dynamics: the psychology of small group relations*. New York: McGraw Hill.
Sheldon, J. and Cave, N. 2007. *The Battle for Vimy Ridge—1917*. Barnsley: Pen and Sword.
Showalter, D. 2004. *Tannenberg*. Washington: Brassey's.
Siebold, G. 2007. 'The Essence of Military Cohesion', *Armed Forces and Society* 33(2): 286–95.
—— and Kelly, D. 1988. *Development of the Platoon Cohesion Questionnaire*. Alexandria, Va.: US Army Research Institute for the Behavioural and Social Sciences, Manpower and Personnel Laboratory.

Simkins, P. 1988. *Kitchener's Army: the raising of the new armies, 1914–16*. Manchester: Manchester University Press.
Skaine, R. 1999. *Women at War*. Jefferson, NC: McFarland and Company.
Skeyhill, T. and Wheeler, R. (eds.). 2004. *Sergeant York and the Great War*. San Antonio, Tex.: Vision Forum.
Sledge, E. 2010. *With the Old Breed*. London: Ebury.
Smith, L. 1994. *Between Mutiny and Obedience*. Princeton, NJ: Princeton University Press.
Smith, P. 2008. 'Meaning and Military Power: moving on from Foucault', *Journal of Power* 1(3): 275–93.
Smith, V. 1997. 'New Forms of Work Organisation', *Annual Review of Sociology* 23: 315–39.
Smoler, F. 1989. 'The Secret of the Soldiers who wouldn't Shoot', *American Heritage* 40: 36–45. <http//www.americanheritage.com/content/secret-soldiers-who-didn%E2%80%99tshoot (page number refer to the web-site)>.
Solaro, E. 2006. *Women in the Line of Fire*. Emeryville, Calif.: Seal Press.
Sorin, K. 2006. 'Women in the French Forces: integration versus conflict', in F. Pinch, A. MacIntyre, P. Browne, and A. Okros, *Challenge and Change in the Military: gender and diversity issues*. Kingston, Ontario: National Defence Academy Press.
Southby-Tailyour, E. 2010. *Helmand Assault*. London: Ebury.
Spence, C. 1998. *Sabre Squadron*. London: Penguin.
Spiller, R. 1988. 'SLA Marshall and the Ratio of Fire', RUSI *Journal* 133 (December): 63–71.
Stacey, C. P. 1960. *The Victory Campaign: the operations in north-west Europe 1944–5*. Ottawa: the Queen's Printer and Controller for Stationery.
Stanislav, A. 1982. 'Causes of the Low Morale in the Italian Armed Forces in the Two World Wars', *Journal of Strategic Studies* 5(2): 248–56.
Stanton, S. 1985. *The Rise and Fall of an American Army*. Novato, Calif.: Presidio
—— 1999. *The 1st Cavalry Division in Vietnam*. Novato, Calif.: Presidio.
Stegemann, B. 1995. 'The Italian Conduct of the War in the Mediterranean and North Africa', in G. Schreiber, B. Stegemann, and D. Vogel, *Germany and the Second World War*, iii: *The Mediterranean, South East Europe and North Africa*. Oxford: Clarendon.
Stiehm, J. 1989. *Arms and the Enlisted Women*. Philadelphia: Temple University Press.
—— 1981. *Bring me Men and Women*. Berkeley: University of California Press.
Stouffer, S., Lumsdaine, A., Harper Lumsdaine, M., Williams, R., Brewster Smith, M., Janis, I. L., Star, S., and Cottrell, L. 1949. *The American Soldier*, ii: *Combat and its Aftermath*. Princeton: Princeton University Press.
—— Suchman, E., DeVinney, L., Star, S., and Williams, R. 1949. *The American Soldier*, i: *Adjustment during Army Life*. Princeton: Princeton University Press.
Strachan, H. 2001. *The First World War*. Oxford: Oxford University Press.
—— 2006. 'Training, Morale and Modern War', *Journal of Contemporary History* 41 (2) April: 211–27.
Stumpf, R. 2001. 'The War in the Mediterranean 1942–1943', in H. Boog, W. Rahn, R. Stumpf, and B. Wegner, *Germany and the Second World War*, vi: *The Global War: widening of the conflict into a world war and the shift of initiative 1941–43*. Oxford: Clarendon.

Sullivan, B. 1994. 'Caporetto: causes, recovery and consequences', in G. Andreopoulous and H. Selesky (eds.), *The Aftermath of Defeat*. New Haven: Yale University Press.

—— 1997. 'The Italian Soldier in Combat, June 1940–September 1943: myths, realities and explanations', in P. Addison and A. Calder, *Time to Kill: the soldier's experience of the war in the west, 1939–45*. London: Pimlico.

—— 2010. 'The Italian Armed Forces, 1918–40', in A. Millett and W. Murray (eds.), *Military Effectiveness*, ii: *The Interwar Period*. Cambridge: Cambridge University Press.

Swofford, A. 2003. *Jarhead*. New York: Scribner.

Taylor, C. 2007. *The Secular Age*. Cambridge, Mass.: Belknap Press.

Taylor, E. K. 2009. *America's Army and the Language of Grunts*, Bloomington, Ind.: Author House.

Theweleit, K. 1987. *Male Fantasies: women, floods, bodies, history*. Cambridge: Polity.

Thomas, M. 1998. *The French Empire at War: 1940–45*. Manchester: Manchester University Press.

Thompson, M. 2004. *The White War: life and death on the Italian Front 1915–1919*. London: Faber and Faber.

Tolkien, J. R. R. 1977. *The Lord of the Rings: the two towers*. London: George Allen and Unwin.

—— 1977. *The Lord of the Rings: the return of the king*. London: George Allen and Unwin.

Trask, D. 1993. *The AEF and Coalition Warmaking, 1917–18*. Lawrence, Kan.: University of Kansas Press.

Travers, T. 1987. *The Killing Ground: the British army, the Western Front and the emergence of modern war 1900–1918*. Barnsley: Pen and Sword.

—— 2005. *How the War was Won*. Barnsley: Pen and Sword.

Truscott, L. 1990. *Mission Command: a personal story*. Novato, Calif.: Presidio.

Tziner, A. and Vardi, Y. 1982, 'Effects of Command Style and Group Cohesiveness on the Performance Effectiveness of Self-Selected Tank Crews', *Journal of Applied Psychology* 67(6): 769–75.

US Military Intelligence Reports: Germany 1919–1941. Military Attaché Report, 'The Experiences of an Infantry Battalion in the Attack on Belgium', 27 December 1940, Reel 24, 0356: 1–6.

US Military Intelligence Reports: Germany 1919–1941. Military Attaché Report 17,210, 'Wartime Training of a German Infantry Division', 16 April 1940 Reel 21 0149: 1–27.

US Military Intelligence Reports: Germany 1919–41 Military Attaché Report 7 February 1940, 'Demonstration at the Infantry School: Infantry Battalion in Attack, Practical Demonstration at Döberitz', 27 January 1940, Reel 21, 0138: 1–6.

US Military Intelligence Reports: Germany 1919–41. Military Attaché Report 'Demonstration at the Infantry School', 24 April 1939, Reel 21 0135: 1–3.

US Military Intelligence Reports: Germany 1919–41. Military Attaché report 'Supporting Fire and Fire Action of Unit Groups', 24 January 1933, Reel 20, 0021: 1.

US Military Intelligence Reports: Germany 1919–41. Military Attaché Report 'Information on German Manoeuvres', 13 May 1926, Reel 14, 0411: 25.

US Military Intelligence Reports: Germany 1919–41. Military Attaché Report 'The Manoeuvres of the German 4th Division', October 1925, Reel 14, 0234: 1–44.
Vallas, S. 1999. 'Re-thinking Post-Fordism: The Meaning of Workplace Flexibility', *Sociological Theory* 17(1): 68–101.
Van Creveld, M. 2000. 'Armed but not Dangerous: women in the Israeli military', *War in History* 7: 82–98.
—— 2000. 'Less than We Can Be: men, women and the modern military', *Journal of Strategic Studies* 23(4): 1–20.
—— 2010. *Fighting Power*. Westport, Conn.: Praeger.
Van der Bijl, N. 1999. *Nine Battles to Stanley*. Barnsley: Pen and Sword.
Van Doorn, J. (ed.) 1968. *Armed Forces and Society*. The Hague: Mouton.
Van Riper, P. 1997. 'Gender Integrated/Segregated Training', *Marine Corps Gazette* November: 64–5.
Vaughan, D. and Schum, W. 2001. 'Motivation and US Narrative Accounts of the Ground War in Vietnam', *Armed Forces and Society* 28(1) Fall: 7–31.
Von Zugbach, R. 1988. *Power and Prestige in the British Army*. Aldershot: Gower.
Wacjman, J. 1996. 'Women and Men Managers', in R. Crompton, D. Gallie, and K. Purcell (eds.), *Changing Forms of Employment*. London: Routledge.
—— 1998. *Managing like a Man*. Cambridge: Polity.
War Department 1914. *US Army Field Service Regulations*. Washington: Government Printing Office.
—— 1944. *Field Manual 7–10: rifle company, infantry regiment*. Washington: Government Printing Office.
—— 1959. *Field Manual 7–10: rifle company, infantry and airborne division battle groups*. Washington: Government Printing Office.
War Department, War Plans Division 1918. *Instructions for the Offensive Combat of Small Units*. Washington: Government Printing Office.
War Office 1909. *Field Service Regulations*. London: His Majesty's Stationery Office.
—— 1914. *Infantry Training: 4-company organisation*. London: His Majesty's Stationery Office.
—— 1921. *Infantry Training*, ii: *War*. London: His Majesty's Stationery Office.
—— 1929. *Field Service Regulations*, ii: *Operations*. London: His Majesty's Stationery Office.
—— 1931. *Infantry Training*, ii: *War*. London: His Majesty's Stationery Office.
—— 1932. *Infantry Training*, i: *Training*. London: His Majesty's Stationery Office.
—— 1938. *Infantry Section Leading*. London: His Majesty's Stationery Office.
—— 1942. *Notes from Theatres of War No. 11: destruction of a German battery by No. 4 Commando during the Dieppe Raid*. GS Publications
—— 1944. *Infantry Training. Part VIII: fieldcraft, battle drill, section and platoon tactics*.
Warr, N. 1997. *Phase Line Green*. New York: Ivy Books.
Watson, A. 2008. *Enduring the Great War: combat, morale and collapse in the German and British armies, 1914–18*. Cambridge: Cambridge University Press.
—— 2011. 'Fighting for Another Fatherland: the Polish minority in the German army, 1914–18', *English Historical Review* 126(522) October: 1137–66.

Weber, M. 1978. *Economy and Society: an outline of interpretive sociology*, volume i. Translated by G. Rother and C. Wittich. Berkeley: University of California Press.
—— 1978. *Economy and Society: an outline of interpretive sociology*, volume ii. Translated by G. Rother and C. Wittich. Berkeley: University of California Press.
Webster, D. 2008. *Parachute Infantry*. New York: Bantam Dell.
Weighley, R. 1981. *Eisenhower's Lieutenants*. London: John Wiley and Son.
Wesbrook, S. 1980. 'The Potential for Military Disintegration', in S. Sarkesian (ed.), *Combat Effectiveness and Cohesion*. London: Sage.
Wessely, S. 2006. 'Twentieth Century Theories on Combat Motivation and Breakdown', *Journal of Contemporary History* 41(2): 269–86.
Westmoreland, W. 1976. *A Soldier Reports*. Garden City, NY: Doubleday and Company.
Wette, W. 2005. *Die Wehrmacht*. Verlag: Fisher Taschenbuch.
Whiteclay Chambers II, J. 2003. 'SLA Marshall's *Men against Fire*: new evidence regarding fire ratios', *Parameters* 33: 113–21.
Whiting, C. 2000. *American Hero: the life and death of Audie Murphy*. York: Eskdale.
Williamson, G. 1991. *Infantry Aces of the Reich*. London: BCA.
Winner, D. 2001. *Brilliant Orange*. London: Bloomsbury.
Winslow, D. 1997. *The Canadian Airborne Regiment: a socio-cultural inquiry*. Ottawa: Minister of Public Works and Government Services, Canada.
Wong, L., Kolditz, T., Millen, R., and Potten, T. 2003. *Why They Fight: combat motivation in the Iraq War*. Carlisle Barracks, Pa.: Strategic Studies Institute, US Army War College.
Wood, S. 1989. 'The Transformation of Work', in S. Wood (ed.), *The Transformation of Work?* London: Unwin Hyman.
Woodward, R. and Winter, T. 2007. *Sexing the Soldier*. London: Routledge.
Wright, E. 2009. *Generation Kill*. London: Corgi.
Wyn Griffiths, L. 1981. *Up to Mametz*. Norwich: Gliddon Books.
Wynne, G. 1940. *If Germany Attacks*. London: Faber and Faber.
Zajcek, J. 2010. *Unter Soldatinnen*. Munich: Piper.
Zander, A. 1982. *Making Groups Effective*. London: Jossey Bass.

Index

Note: page numbers in *italics* refer to figures.

9/11 attacks 425
Abbott, Andrew 216
accountability 301–2
Achnacarry battle school 154–6
Afghanistan
   42 Commando Royal Marines 274–7
   Canadian forces 209
   close-quarters battle 240
   Gatigal ambush 230–2, 302–3
   Operation Aabi Toorah 2B: 276–7
   platoons 17–18
   tactics in 19
   Taliban 275, 297, 298–9
AGCT (Army General Classification Test), USA 54
AGF (Army Ground Forces), USA 54–5
Al Kut, Iraq 232–3, 328, 372
Alexander, Harold 152–3
Alexander, Jeffrey 441–2
All Quiet on the Western Front (Remarque) 2
Alpini, Italy 176, 177
Amazon Corps, Kingdom of Dahomey 378
Ambrose, Stephen 188
Ambush (Marshall) 41
American Civil War 129–30
American Expeditionary Force 178–9
Anderson, Charles 121
Andrevski, Stanislav 81
anomie 72, 339–40
anti-Semitism 88–9
ANZAC forces 374
Aran, G. 72
Arditi, Italy 176, 177
Arkin, William 72
Armed Services Integration Act, USA 384
Army General Classification Test (AGCT), USA 54
Army Ground Forces (AGF), USA 54–5

Art of War, The (Machiavelli) 13
Ashworth, Tony 9–10
Assault Training (UK manual) 140, 143–4
Assheton, Charles 121
Atkinson, John 432
Attack in Trench Warfare, The (Laffargue) 138
attitudinal conditioning 325–6, 328
Ausbildung der Schützengruppe (A.d.S., Training Regulation for the Rifle Squad) 134–5
Ausbildungsvorschrift für die Fusstruppen im Kriege (German manual) 17, 133–4
Ausbildungsvorschrift für die Infanterie (AVI, *Training Rules for Infantry*) 65, 134–7
Australian forces
   bayonet charges 103, 121
   close-quarters battle (CQB) techniques 241
   conscription 209–10
   deviant cohesion in First World War 32
   drills 156–7
   gender integration 386
   individual action 121
   mass formations 198–200
   mateship 68–9, 198, 359–61
   platoons 17
   professional solidarity 353
   wives and families, status of 69
Australian Task Force: in Vietnam 198–200, 209–10

baby-boom generation 439
Back, Kurt 28–9
Badinski, Kurt 190–1, 460 n.127
Balck, Wilhelm 88, 167, 186
Balkosi, Joseph 52, 148, 162, 180
Bannerman, Alexander 99–100

banzai charges 110
Barbusse, Henri 2, 447 n.38
Barkawi, Tarak 95
Barnard Castle battle school 153
BARs (Browning Automatic Rifles) 148
Bartov, Omar 86–7
Basham, Victoria 408
Battistelli, Fabrizio 424
Battle Drill (Utterson-Kelso) 152–3
battle drills 152–3, 208–65
   Battle Drill 07-3-D3991—React to Contact: task steps and performance measures 225–6
   close-quarters battle (CQB) 237–65
   current US doctrine 225–6
   Gatigal ambush 230–2
   importance of 266
   Al Kut, Iraq 232–3
   mnemonics 226–9
   professional armies 208–11
   professionalism, concept of 211–22
battle formations: Greek phalanxes/Roman legions 131
battle preparation 157–63
   British forces 159–60
   Canadian forces 160–1
   First World War 157–61
   French forces 157–8
   German forces 158–9
   model building 282–91, 295, 296
      Rehearsal of Concept (ROC) drills 284–91, *285*, *286*, *287*, *292*, 295, 296, 299
   and moral obligation 295–6, 338
   Second World War 160, 161–2
   standardization of orders 281
   training 280–99
   US forces 160, 162
Bayonet (US field manual) 109
bayonet charges 98–114, 205, 206–7
   assessment of 114–15
   Australian forces 103, 121
   British forces 98–103, 108–9, 115, 130
   Canadian forces 103
   as citizen army manoeuvre 101–2
   and cohesion 98–100
   Falklands War 115
   First World War 100–2, 103, 112, 114

French forces 103–5, 112–13
German forces 105–6, 112
Iraq war 115
Italian forces 106–7
Korean War 112–13
moral significance of 205
and morale 99–100, 101–3, 110–11
Prussia 130
Second World War 108–12, 114, 121
US forces 109–12, 113
Vietnam War 113
Beevor, Antony 191, 359
Bellamy, Christopher 319
Bellavia, David 234–6, 239
Belzar, Joseph 56
Ben Shalom, Uzi 352–3
Bendix, Reinhard 214
Beynon, Huw 431, 432
Biddle, Stephen 20, 38
Bidermann, Gottlob Herbert 462 n.164
Bird (Marshall) 58–9
Black, Jeremy 21
Black, Major 193
Black September 238
Blaise, Karine 405
Bloody Sunday 12
Blumentritt, Gunther 64
Blunden, Edmund 3, 10
Boer War 103, 130
Bolt, Henry 75
Bonnland, Germany 313, 334, *335*
Boudreau, Tyler 336
Bourke, Joanna 98
Boyer, James 92
Bradford, Alfred 184
Bradley, Omar 180, 205–6
Braim, Paul 179
Bramall, Edwin 84
Bramley, V. 347
Brazilian army 423
Brennan, Frank 44
British forces 267–71, 274–7
   attitudinal conditioning 325
   battle preparation 159–60
   bayonet charges 98–103, 108–9, 115, 130
   conscription 209
   cowardice 363, 366, 370

CQB techniques 240, 241, 242,
   243–52, *245*, *246*, *247*, *252*,
   253, 254–9, *255*, *257*, *259*
  marksmanship 242–4, 319–21
  mental preparation 325–6, 328
  pivoting *322*
  visualization 325–6
 drills 151–6
 ethno-political motivation 91, 94–5
 fire and movement tactics 139–44,
   146–7, 151
 gender integration 383, 384, 385–6,
   393, 395–9, 400–1
  discrimination 416–17
 homosexuals, integration of 377
 individual action 120–1
 mass formations 167–70
 mnemonics 226–7
 platoons 17
 political motivation 74–6, 83–5
 professional solidarity 353–4, 357–8
 professionalism of 346–7
 professionalism, failures of 332–3
 and racism 94–5, 96–7
 ROC drills 288–9
 SAS 238–9
 Task Force Helmand 298–9, 353–6, 370
 urban training facilities 309–11
 wives and families, role of 70
 see also Parachute Regiment, UK;
   Royal Marines, UK
Britten, Benjamin 447 n.35
Browne, Thomas 271
Browning Automatic Rifles (BARs) 148
Browning, Christopher 461 n.132
Buchan, John 101
Bugeaud, Thomas 269–70
Buick, Bob 199
Bundeswehr, Germany 210–11
 discipline: shame 372
 gender integration 410
 professional solidarity 351–2
 Quick Reaction Force (QRF): fight
   with Taliban 303–4
 training 277–9
Burke, Carole 411

C-Can Village, Combat Training Centre,
   Canada 313, *318*
CAAT (Counter-Insurgency Advisory,
   Assistance, and Training)
   teams 425–6
Cadorna, Luigi 82, 83, 106–7, 176
Calley, William 185
Campbell, John 43, 53
Campbell, Ronald 101
Canadian Expeditionary Force 145–6,
   160, 170–6
Canadian forces 209
 Advanced Reconnaissance Patrol
   Course 282–3
 Airborne Regiment 31–2
 battle preparation 160–1
 bayonet charges 103
 close-quarters battle (CQB)
   techniques 240–1, 249, *322*, 326,
   328–30, 487 n.117
 conscription 209
 discipline 370–1, 372
 drills 153
 fire and movement tactics 145–6, 171
 gender integration 376, 384–5, 386,
   387–93, 398, 399–400, 403–4,
   405, 409–10
  discrimination 411–12
 Gunfighter programme 240–1, 242,
   247, *248*, 321, *323*
 individual action 121
 Joint Task Force 2: 274, 330–1, 372
 mass formations 170–6
 platoons 17
 political motivation 76
 professional solidarity 353
 professionalization 209
 and Russia 264
 Somalia 209
 and sports psychology 330–2, 372
 training 174
  marksmanship 242–3, 244–5, 247, *248*
  pizza-slicing 249
  resilience training 328–9
 urban training facilities 313, *318*
 Warrior's Ethos 330, 500 n.25

Caputo, Philip 11–12
Carentan, Battle of 48–9
Carreiras, Helena 416, 417
Carter, Nick 344
casualty levels 204–6, 405–6
Catch-22 (Heller) 5–7
Catholicism: and nationalism 81
CBUs (cluster bomb units) 182
CENZUB (Centre d'Entraînement aux Actions en Zone Urbaine) 241, *314*, 315, *315*, *316–17*
  battle drills 254
  ROC drills 290–1, *292*
Ceva, Lucia 83
Chelwood Gate, UK 55–6, 153
China: People's Liberation Army (PLA) 422–3
Chirac, Jacques 210
Clausewitz, Karl von 13–14
close-quarters battle (CQB) techniques 237–65, 315–32
  Afghanistan 240
  attitudinal conditioning 325, 328
  Australian forces 241
  body armour to body armour 258–9, *259*
  British forces 240, 241, 242–52, *245*, *246*, *247*, *252*, *253*, 254–9, *255*, *257*, *259*, 319–21, *322*, 325–6, 328
  Canadian forces 240–1, 249, *322*, 326, 328–30, 487 n.117
  combat glide 246–7, *248*
  Cooper Colour Code 326–7
  and credentialism 215–16, 265
  five-step entry 251, 254–7, *255*, *256*
  French forces 241, 253–4
  house clearances 249–51
  Initiative-Based Tactics (IBT) 260–3
  Iraq war 234–6, 239–40
  marksmanship 242–4, 319–21
  mental preparation 325–6, 328
  pivoting 244–6, *322*, *322*
  popping 248, 249
  resilience training 328–30
  room clearances 249–58
  shooting and scanning 248–9
  stack formation 249, 251–4, *252*, *253*, 260
  US forces 239–40, 252–3, 319
  visualization 325–6
  see also urban training facilities
Close Quarters Battle Instructor: course manual (Royal Marines manual) 241, 243
cluster bomb units (CBUs) 182
Cockerham, William 72, 267
cohesion 345, 347–51
  bayonet charge and 98–100
  of citizen armies 60–1
  definitions 24–30
    re-definition of 30–9
  deviant cohesion 31–2
  and infantry tactics 13–23
  primary group cohesion 24–33, 345
  pure cohesion 347–50
  social cohesion 27–8, 31, 34, 35
  task cohesion 34–6, 267, 377–8
  training and 222–3, 273
Cole, Hugh 49
Cole, Robert 48–9, 109–10, 205
collective self-protection 33–4
Collins, Randall 10–12, 50, 51
  credentialism 215–16, 264–5
  training and social solidarity 345–6
combat
  as collective action 10–12
  and religion 16
  sociology of 7–12
combat performance
  and cohesion 26, 27, 30, 31, 32–8
  and small-group ties 25
  of Wehrmacht 24, 36–7
Commando Training Centre, UK 309, *309*, *310*, *311*
community
  decline of 438–45
  dissolution of 4–5, 6–7
comradeship 33–4
  in citizen armies 358–61
  professional 345–62
  see also friendship; mateship

conscription 21, 208
    Australia 209–210
    Canada 209
    France 166–7, 210
    Germany 134, 202, 210–11
    Italy 81–2, 210
    UK 209
    USA 209
Cooper, Lieutenant Colonel, John P. 53
Cooper Colour Code 326–7
Copp, Terry 172–3, 174, 187
corporateness 342–4
Cortes (2nd Platoon Battle Company member) 348, 349
Coss, Edward 267–72
Cota, Brigadier General 180
Counter-Insurgency Advisory, Assistance, and Training (CAAT) teams 425–6
counter-terrorist units 238–9
Cours de tactique générale d'après l'expérience de la Grande Guerre (Culmann) 139, 158
courts-martial 373–4
cowardice
    Afghanistan 370
    British forces 363, 366, 370
    execution for 363–4
    Falklands War 368, 369–70
    Iraq war 368
credentialism 215–16, 264–5
Crerar, Lieutenant General Harry 173
cricket: changes in 435
Crocker, John 173
Culmann, F. 139, 158
Currie, General 171

D-Day 43, 120–1, 180
    battle preparation 156, 161–2, 174, 180
    Severloh and 366–7, 420
Dahomey, Kingdom of: Amazon Corps 378
Dak To, Battle of 59
Danish forces: gender integration 386, 403

Darby, Bill 156
Davie, Grace 426
Davies, Christie 82
de Gaulle, Charles 78, 202
decimation 364
Defence Policy Guidelines (Germany, 2003) 210
Delbrück, Hans 131, 272, 443
Delta Force, USA 238–9
Depuy, Trevor 63
DePuy, William 183
desertion 363–4
D'Este, Carlo 120, 170
deviant cohesion 31–2
Dicks, Henry V. 65–6
Dienstunterricht (*Service Instruction*, Reibert) 64–5, 89–90, 150
Dinter, Elmar 66, 67, 69
discipline 362–75
    in citizen armies 363–8
    courts-martial 373–4
    execution for cowardice 363–4
    First World War 363–5
    in professional armies 368
    Second World War 363, 365–7
    shame 371–2
Dix, Otto 447 nn.35 & 38
Dobrofsky, L. 72
Dolchstoss (stab-in-the-back) myth 89
Don't ask, don't tell policy (USA) 377
Doughty, Robert 167
Drapier, Lieutenant 167
Draude, Thomas V. 401–2
Drill Regulations (Prussian manual) 130
drills 148–57
    Australian forces 156–7
    British forces 151–6
    Canadian forces 153
    First World War 150–1
    French forces 150
    German forces 150–1
    Second World War 151–4
    US forces 156
    Vietnam War 156–7
drug abuse 184, 185–6
du Picq, Ardant 11, 98, 99

Dundas, David 270
Dunn, John 3, 10, 167–8
Durkheim, Émile 15–16, 21, 291, 293–4, 442
  anomie 339–40
  solidarity 293, 339–41, 356–7
  status groups 340–2
Dürr, Emil 122

Eckhold, Robert 372–3
Edmonds, James 101, 168–9
Einzelausbildung am Leicht Maschinengewehr (ALMG, *Individual Training with the Light Machine-Gun*) 134–5
Eisenhart, Robert 71–2
Elshtain, Jean Bethke 381–2
Engall, Jack 74
English, John 174, 209
Enloe, Cynthia 380–1, 383
esprit de corps 18
ethno-political motivation
  British forces 91, 94–5
  German forces 87–91
  US forces 91–4
ethnomethodology 300–1

Falklands War 346–7, 423
  bayonet charges 115
  cowardice 368, 369–70
Fallujah, Second Battle of 234–6, 239–40
Fanon, Frantz 96
female soldiers 376–418
  acceptance of, in current operations 387–407
  and armed forces 376–82
  casualties 405–6
  discrimination against, in current operations 407–18
  gender integration 376, 383–6
  in guerrilla warfare 378–9
  harassment of 411–12
  in IDF 379–80
  in Kingdom of Dahomey 378
  lesbian gangs 499 n.155
  in Soviet Union 378

Festinger, Leon 28–9
Field Manual *The Infantry Rifle Platoon and Squad* (US manual) 148, 156, 225–6
Field Service Regulations (UK manual) 100, 130
fire and movement tactics 129–48, 151
  in battle drills 225
  British forces 139–44, 146–7, 151
  Canadian forces 145–6, 171
  First World War 132–4, 137–46, 147–8
  French forces 137–9
  Gatigal ambush 230–2
  German forces 132–7, 150, 194–5, 196–7
  US forces 147–8, 180
firing rates
  UK citizen army 55–6
  US citizen army 41–7, 48–51, 57
'1st Corps Tactical Notes' (Alexander) 152–3
First World War
  battle preparation 157–61
  bayonet charges 100–2, 103, 112, 114
  deviant cohesion in 32
  disciplinary measures 363–5
  drills 150–1
  fire and movement tactics 132–4, 137–46, 147–8
  gender integration 383
  individual action 117–18
  as industrial war 20
  literature 1–5
  mass formations 165–6, 167–9, 170–1, 176, 178–80, 186–7
  platoons 17
  tactics 163
  trench warfare 9–10
FOB cohesion 354–6, 397
Folgore Division, Italy 176, 177
football: changes in 434–5
Ford, Henry 430
Fordism 430–1, 432–3
Förster, Jürgen 90
forward panic 11–12, 50–1

Foulkes, Charles 172, 173
fraggings 94, 184, 185–6
Franco-Prussian War 130
Fraser, David 297
Freakley, Ben 297
Freikorps 64
French, David 170
French forces
  attaque à outrance 103, 108
  battle preparation 157–8
  bayonet charges 103–5, 112–13
  Boer War 103
  conscription 166–7, 210
  CQB techniques 241, 253–4
  deviant cohesion in First World War 32
  drills 150
  execution for cowardice 363–4
  fire and movement tactics 137–9
  First World War literature 2
  gender integration 383, 384, 410–11
  machine guns 138
  mass formations 165–7
  nationalism 76–7
  offense à outrance 103–4, 137
  platoons 17
  political motivation 76–9
  professionalization 210
  and racism 96–7
  ROC drills 290–1, *292*
  tactics in First World War 165–6
  tactics in Second World War 166–7
  urban training facilities 241, 254, 290–1, *292*, 311–13, *314*, 315, *315*, *316–17*
Freytag-Loringhoven, Freiherrn von 106
friendship 25–6, 28–30, 33–4: *see also* comradeship; mateship
Frieser, Karl-Heinz 189
Fritz, Stephen 87–8
Führung und Gefecht der verbundenen Waffen (FUG, *Combined Arms Command and Combat*), Germany 134–5
Fuller, J. G. 32
Fussell, Paul 8

Gamelin, Maurice 202
Garfinkel, Harold 300–1
Gatigal ambush, Afghanistan 230–2, 302–3
Gavin, James 43
gender integration 376, 383–6
  Australian forces 386
  British forces 383, 384, 385–6, 393, 395–9, 400–1, 416–17
  Canadian forces 376, 384–5, 386, 387–93, 398, 399–400, 403–4, 405, 409–10, 411–12
  Danish forces 386, 403
  discrimination 407–18
  German forces 410
  IDF 379–80
  Italian forces 385
  Soviet Union 378, 404–5
Gerhardt, Charles H. 148
German forces
  battle preparation 158–9
  bayonet charges 105–6, 112
  Boer War 103
  Bundeswehr 210–11, 277–9, 303–4, 351–2, 372, 410
  close-quarters battle (CQB) techniques 241
  conscription 134, 202, 210–11
  disciplinary measures 363, 366–8
  drills 150–1
  ethno-political motivation 87–91
  execution for cowardice 363
  fire and movement tactics 132–7, 150, 194–5, 196–7
  gender integration 385, 393, 410, 411
  Grenzeschutzgruppe (GSG) 9, Germany 238–9
  individual action 122–3
  Landsers 87–8
  masculinity 64–6
  mass formations 186–98
  platoons 17
  political motivation 86–91
  professionalism, failures of 333–5
  Reichswehr 88–9
  status of wives and families 66

German forces (*cont.*)
  stormtroops 132–3, 150, 158–9, 186–7, 195–6
  training 277–9
  urban training facilities 313, 334, *335*
  see also Wehrmacht
Germany
  Defence Policy Guidelines (2003) 210
  First World War literature 1–2, 4–5
  Second World War literature 5
Giezl, George 53
Girton, George 320
Giunta, Sal 231–2, 302–3
Glantz, David 188, 194
Glenn, Russell 57–8, 59–60
Goddard, Nichola 388–9
Godefroy, Andrew 171
Goldhagen, Daniel 461 n.132
Goldstein, Joshua 378–9, 386, 404
Goldstein, Lyle 423
Gooch, John 81–2
Goya, Michael 104, 105, 137
Granatstein, J. 209
Grandmaison, Louis de 104–5
Graves, Robert 10, 57
Graziani, General Rodolfo 177–8
Great Generation 60, 165, 360, 420, 439–40, 443
Great War, *see* First World War
Greek phalanxes 131
Grenzeschutzgruppe (GSG) 9, Germany 238–9
Griffin, Phillip 175
Griffith, James 27
Griffith, Paddy 101, 139–40, 167, 168, 186, 187
Griffiths, Wyn 3
Groningen urban combat facility, Combat Training Centre, Gagetown, Canada *318*
Gross, Neal 30
Grossman, David 51
group pride 35
Groupement de Sécurité et d'Intervention de la Gendarmerie Nationale, France 238–9
groupthink 31

Gudmundsson, Bruce 186
Gulf War (1991) 210, 307, 385
Gustavus Adolphus 149
Guttenberg, Karl-Theodore zu 211

Haditha killings 12
Haig, Alexander 182
Haig, Douglas 101
Hall, Henry 154–5
Hamilton, Laird 437, 438
Harding Brigadier, G.P. 108
Hart, Peter 101
Hartley, Jason 408–9
Hastie, William 92
Hastings, Max 120–1
hazing 31–2
Head, Lisa 406
Heller, Joseph 5–7, 8
Henderson, Darryl 26
Henessey, Patrick 413
Henriksen, Rune 213, 272
heroism 230
  and nationalism/masculinity 206
  as negative concept 68, 70
  US forces 234–6
Hetherington, Tim 288
Heydte, Friedrich, Freiherr von der 191
Hicks, Judith 408, 409, 411–12
Hickson, Major 153
Higate, Paul 407–8
Hitler, Adolf 89–90
Hobbes, Thomas 14–15
Hogg, Michael 37–8
Holl, Adelbert 90, 188
Hollis, Stan 121
homosexuals: integration of 377
Honig, Fritz 98
Horrocks, Brian 84
Horsfall, Robin 347
hostage rescues 238–9, 242
house clearances 239–40, 250
Human Rights Commission Tribunal, Canada 384
Humphries, M. 171
Huntington, Samuel 217–19, 220–1, 264–5, 427, 429, 444
  on corporateness 342–3

Index

IBT (Initiative-Based Tactics) 260–3
IEDs (improvised explosive devices) 275
Indian army: and racism 95
individual action 116–28, 206–7
  Australian forces 121
  British forces 120–1
  Canadian forces 121
  First World War 117–18
  German forces 122–3
  Korean War 123
  Second World War 118–21
  as solution to inertia problem 116, 127
  US forces 117–20, 123–6, 234–6
  Vietnam War 123–6
  see also heroism
individualism 6–7
  in British forces 116–17
  in US forces 26
industrial manufacturing 430–3
  flexible specialization 433–4
  Fordism 430–1, 432–3
  Stakhanovites 431–2
Infantry Immersion Trainer (IIT), Camp Pendleton 305–6, *306*
Infantry Rifle Platoon and Squad, The (US manual) 239
Infantry Section Leading (UK manual) 146–7
Infantry Tactics (UK manual) 146–7
Infantry Training (UK manual) 67, 100, 108, 130, 146
  battle preparations 159–60
  drills 151, 152
  fire and movement tactics 146–7
Initiative-Based Tactics (IBT) 260–3
Instruction sur le combat offensif des petites unités (French manual) 17, 138, 147, 158
Instructions for the Offensive Combat of Small Units (US translated manual) 147–8, 160
Instructions for the Training of Platoons for Offensive Action (UK manual) 17, 140–3, *144*, *145*
  open warfare attacks 142–3, *143*
  trench attacks *141*, 141–2, *142*

Instructor's Handbook on Fieldcraft and Battle Drill, The (UK manual) 116–17, 151–2, 160, 196
interpersonal relationships
  cohesion and 24–33
  and combat performance 32–3, 34–5
  deviant cohesion and 31–2
Iraq war 424
  bayonet charges 115
  close-quarters battle 234–6, 239–40
  cowardice 368
  Fallujah, Second Battle of 234–6, 239–40
  invasion of/ROC drill 286–9
  Al Kut 232–3, 328, 372
isolation
  in battle 42–3, 46–7, 53
  in literature 7, 8
Israeli Defence Force (IDF) 352–3, 379–80
Italian forces
  Alpini regiments 176, 177
  Arditi 176, 177
  bayonet charges 106–7
  conscription 81–2, 210
  disciplinary measures 364–5, 366
    decimation 364
    execution for desertion 364
  First World War 176, 177, 364–5
  Folgore Division 176, 177
  gender integration 385
  mass formations 176–8
  political motivation 81–3
  Second World War 176–7, 366

Janis, Irving 31
Janowitz, Morris 24–5, 36–7, 86, 219–21, 264–5, 344, 429
  primary groups 63, 345
Jary, Sidney 56, 192
Jeska, Andrea 410
Jim Crow laws 91
Johnson, Samuel: army quote: viii
Johnston, Mark 360
Joint Task Force 2, Canada 274, 330–1, 372

Jokelson, John 153
Jolly, Margaretta 76
Jones, Lieutenant Colonel H. 115
Jones, James 5
Jordan, Kelly 57
Jung, Franz-Josef 210
Jünger, Ernst 1-2, 4-5, 6-7, 8, 12, 16, 150, 159, 167, 186, 187
Junger, Sebastian 230-2, 288, 302, 393
  cohesion 347-51
Jürgens, Ulrich 433

Kamneva, Nina 404
Kant, Immanuel 291-3
Kanter, Rosabeth Moss 394
Kean, William B. 181
Keller, Major General Rodney 173
Kellett, Anthony 67
Kelly, D. 18
Kennedy, Paul 169
Kestnbaum, Meyer 28, 86
Khan Neshin, Afghanistan 276-7, 369
Kier, Elizabeth 34, 35, 345
Kiesling, Eugenia 166
Kiggell, Launcelot 99, 102
Killingray, David 94-5
Kindsvatter, Peter 80
Kissel, Hans 188-9
Knox, Macgregor 177-8
Korean War
  bayonet charges 112-13
  Imjin 206
  individual action 123
  mass formations 181-2
  political ideology and 80
  US forces 57, 181-2
Krulak, Charles 427
Krylova, Anna 404
Kuss, Hans 367
Al Kut, Iraq 232-3, 328, 372
Kwajalein Atoll, Battle of 45, 48, 454 n.58

Laffargue, André 17, 138, 157-8
Lafontaine, General 167
Landers, Daniel 33
Landsers, Germany 87-8
Lane, Christine 433

Langer, Philip 303-4
Lanz, Albrecht 189
Lasswell, Harold 218
Laure, Lieutenant 104
Lear, Ben 55
Lee, David 51
Lee, Ulysses 92-3
Leinbaugh, Harold 43, 44, 53
Lejeune, Major-General John 179
Lentz, John 52
Liddell Hart, Basil 9, 101-2, 139, 146, 159-60, 200
Lindsay, Martin 155, 161
Little, Roger 67, 70
Livre blanc 210
Long Tan, Battle of 198-200
Lopez, Gerry 437
Lord of the Rings, The (Tolkien) 3-4
Lott, A. & B. 29, 34
Lüschen, Gunter 33
Luttwak, Edward 204-5
Lydd, UK 310-11

McAulay, Alex 199
McChrystal, Stanley 344, 425-6
MacCoun, Robert 18, 34, 35, 345
McCoy, Bryan 219, 232-3, 273, 328, 371-2, 488 n.53
McDade, Robert 183
Machiavelli, Niccolò 13
machine guns
  French forces 138
  German 135, 136-7, 150, 192, 193-4, 196-7
McNair, Lesley 55, 92
McNaughton, General Andrew 174
McNeill, William 346
Magrin-Vernerey, Lieutenant General Raoul (Lieutenant Colonel Monclar) 112
Mailer, Norman 5
Makin Islands, Battle of 45-8, 47, 50
Malone, Bradley 303
Manuel du chef de section d'infanterie (French manual) 17, 138-9, 150, 158
marching fire 111-12

MARILYN Review, UK 385
Marine Corps Air Ground Combat Centre
    (MCAGCC), 29 Palms 307
  Range 220: 307, *308*, 315
Mark, Jason 196
marksmanship
  British forces 242–4, 319–21
  Canadian forces 174, 242–3, 244–5,
    247, *248*, 249
Marlantes, Karl 16, 125–6, 127, 156
Marlowe, D. H. 18
Marshall, Samuel Lyman Attwood
    40–50, 120
  on cohesion problem 62
  criticisms of 43–5
  on drills in Vietnam War 156
  First World War experience 44–5
  as historian of US Army 40–3
  Makin Islands fight 45–7
  US citizen army firing rates 41–7, 48–51
  on Vietnam War 58–9, 184
Marston, Daniel 95
Martens, Rainer 33
Martin, Michel 210
Martin, William 30
masculinity 62–73
  in Australian forces 68–9
  in British forces 66–7
  and combat motivation 62
  in First World War 66–7
  of Freikorps 64
  in German forces 64–6
  national identity and 62
  parachuting/airborne training and 72
  in Second World War 67
  in US forces 67–8, 70–2
  violence as measure of 71
mass formations 164–207
  Australian forces 198–200
  bayonets 98–115
  British forces 167–70
  Canadian forces 170–6
  First World War 165–6, 167–9, 170–1,
    176, 178–80, 186–7
  French forces 165–7
  German forces 186–98

  and individual action 116–28
  Italian forces 176–8
  Korean War 181–2
  Second World War 166–7, 169–78,
    180–1, 187–98
  and training 203–4
  US forces 178–86
  Vietnam War 182–6, 198–9
mass killings 12
Masters, Lieutenant 53–4
mateship 68–9, 198, 359–61: *see also*
    comradeship; friendship
Mattis, James 286–8, 305, 344, 427
Maurer, E. 190
Maxse, Ivor 139, 151, 159
Maxwell, Frank 168
Mead, Margaret 380
mechanical solidarity 339–40, 356
Meckel, Jakob 98
Megill, Brigadier William 175
Men against Fire (Marshall) 41–2, 46–7
Men of Company K, The (Leinbaugh) 44
Menard, Daniel 370–1
Mendes, Michelle 405
mental preparation
  Royal Marines 325–7
  US Marines 327–8
mercenaries 13
Merritt, Lieutenant-Colonel Charles,
    VC 121
military life
  cohesion and infantry tactics 13–23
  cultural representation of war 1–7
  sociology of combat 7–12
Miller, Steven 422
Millett, Allan 185
Milne, William 121
mnemonics 226–9
model building 282–91, 295, 296
  ROC drills 284–91, *285*, *286*, *287*,
    295, 296, 299
Monclar, Lieutenant Colonel
    (Lieutenant General Raoul
    Magrin-Vernerey) 112
Montgomery, Bernard 84, 174,
    205–6, 344

Moore, Harold 182–3
morality 291–5
  moral obligation: and battle preparation 295–6, 338
  status groups and 340–2
Moskos, Charles 33–4, 184, 424
  on black soldiers in Vietnam 93
  institutional/occupational armies 211–12
  political ideology, role of 79–80
  on primary groups 70–1
Mosse, George 62–3
motivation
  cohesion as 24–31
  deviant motivation 31–2
Mountain Thrust offensive 297
Mullaney, Craig 413
Munich Olympics 238
Murphy, Audie 118–19
Mussolini, Benito 83
My Lai massacre 185

Napoleonic War Journal 271
Nash, Paul 447 n.35
National Training Centre (NTC), Fort Irwin 307
nationalism
  Catholicism and 81
  French forces 76–7
  and masculinity 62
  Polish forces and 85–6
  and political motivation 74–7
  and sexuality 62–3
  and sport 74–5
NATO 494 n.27
  battle orders 281
  gender integration 385
  ISAF, Afghanistan 297
Naval Defence Act (1889) 203
Nesbitt, Kate 396, 398–9, 400–1, 406, 416–17
Nevill, Captain Wilfred P. 75
Nicholson, Brigadier Claude 67
Night Drop (Marshall) 43, 120
Nijmegen Bridge 50–1
Nivelle Offensive, Chemin des Dames 105

Noll, Greg 435–6
Notes from the Theatre of War No. 4 Destruction of a German Battery by No. 4 Commando (Lindsay) 161

Objective Rugby 297, 298
O'Byrne, Sergeant Brendan 349
offensive action/attacking 38–9
Olstad, Keith 298
On War (Clausewitz) 13–14
Operation Aabi Toorah 2B: 276–7
Operation Anaconda 297
Operation Atlantic 174
Operation Entirety 310–11
Operation Epsom 173
Operation Market Garden 475 n.188
Operation Medusa 297–8
Operation Rock Avalanche 230–1, 288
Operation Sond Chara 275, 298–9, 368–9
Operation Spring 173, 174–5
Operation Totalise 173
Operation Tractable 173
Operational Training and Advisory Group (OPTAG) 353–4
organic solidarity 340, 356–7
Overy, Richard 431
Owen, Wilfred 447 n.35

Pals battalions 66–7, 374
Parachute Regiment, UK
  and atrocities 393
  cohesion 346–7
  CQB techniques 240
  Falklands War 115, 346–7, 368, 369–70
  gender integration 394, 396, 415–16
  motivation 296, 362
  training 240, 278
Parsons, Talcott 14–15, 36, 37, 300
patriotism 62
  political motivation as 74–7
Patton, George S. 52, 53, 344
  on bayonets 109
  on marching fire 111–12

PAWPERSO (protection, ammunition, weapons, personal camouflage, equipment, radios, special equipment, and orders) 227
peace-keeping 209, 407–8
People's Liberation Army (PLA), China 422–3
Pereira, Major-General Cecil Edward 169
Pershing, John J. 178–9
Peterson, James 33
Petraeus, David 344
Peyton, Bernard 193, 197
Pililaau, Herbert 123
PLA (People's Liberation Army), China 422–3
Place, Timothy Harrison 108–9, 116, 152–3, 169–70
platoons 17–19
  French structure 138–9
  as modern maniples/as basic unit in tactical manoeuvres 131–2
  platoon attack in open warfare 249–50
  platoon cohesion 201
  platoon tactics 200–1
Polish forces: and nationalism 85–6
political motivation 73–97
  Canadian forces 76
  ethno-political motivation 87–94
  French forces 76–9
  German forces 86–91
  Italian forces 81–3
  nationalism and 74–7
  as patriotism 74–7
  Polish forces 85–6
  and primary group cohesion 73–4
  UK forces 74–6, 83–5
  US forces 79–80
  Wehrmacht 86–91
Pope, George 53–4
Porteus, Pat 109
Portuguese forces: gender integration 416
post-traumatic stress disorder 328
Pratten, Garth 69
Preece, Steven 347

PREWAR (preparation, reaction to effective enemy fire, enemy location, win the fire-fight, assault, reorganize) 226–7
primary group cohesion 24–33, 345
  and political motivation 73–4
primitive civilizations 15–16
Prince, The (Machiavelli) 13
Prior, Robin 101
professionalism 338–75, 419–45
  British forces 332–3, 346–7
  community, decline of 438–45
  concept of 211–22
  discipline 362–75
  failures of 332–7
  national mission 424–8
  other forms of 428–38
  professional comradeship 345–62
  professional ethos 338–45
  spread of 419–23
professionalization 208–22
  and size of armies 212–13
Prussia 130
pure cohesion 347–50
Putnam, Robert 438–40

Quick Reaction Force (QRF): fight with Taliban 303–4

racism 91–5, 96–7
  in British forces 94–5, 96–7
  in French forces 96–7
  Indian army and 95
  in US forces 91–4, 96–7
Raths, Ralf 105–6
Rawling, Bill 161, 170
reality television shows 442
Règlement de l'infanterie (1940) 77–8
Règlement sur la conduite des grandes unités (1913) 105
Règlement sur le service des armées en campagne (1913) 105
Rehearsal of Concept (ROC) drills 284–91, *285, 286, 287*
  British forces 288–9
  French forces 290–1, *292*

Rehearsal of Concept (ROC) drills (cont.)
    Iraq invasion 286–9
    US forces 284–8
Reibert, W. 64–5, 89–90, 150–1
Reichswehr, Germany 88–9
religion 15–16, 293–4, 427
    Catholicism and nationalism 81
    combat and 16
    in USA 426–7
Remarque, Erich Maria 2
resilience training 328–30
Restrepo (documentary film) 230, 348
Richards, David 297, 310, 489 n.82
RIGS (Reconnaissance, Isolation, Gain a foothold, Secure) 228
Rochat, Giorgio 365
Romain, Jules 2
Roman legions 131
Rommel, Erwin 112, 132, 159, 186, 202
room clearances 249–58, *257*
    five-step entry 251, 254–7, *255, 256*
    stack formation 249, 251–4, *252, 253*
Rosen, Leora 31
Rosolie, Walter W. 124–5
Royal Marines, UK 240, 298–9, 347
    42 Commando 274–7, 288–9, 368–9
    Commando Training Centre, Lympstone 241
    CQB techniques 241, 242, 243–52, *245, 246, 247, 252, 253*, 254–61, *255, 257, 259*, 319–20, 324–6
        Cooper Colour Code 326–7
        mental preparation 325–7
    Fleet Protection Group 240
    and gender integration 408
    in Helmand 298
    marksmanship 319–20
    masculine rituals 393
    Platoon Weapons Course 241
Royal Welch Fusiliers 3, 10
Rules and Regulations for the Movements of His Majesty's Infantry (18th-century British manual) 270
Rupprecht, Crown Prince of Bavaria 4
Russian army 422
Russo-Japanese War 99–100, 103

Sabben, David 156–7
Sadkovich, John 81, 176
Sajer, Guy 5, 66
Salisbury Plain Training Area, UK 309–10
Sandys Defence Review, UK 209
SAS, UK 238–9
Sassoon, Siegfried 10, 101
Schachter, Stanley 28–9
Scholes, Paul 425
Schreiber, Shane 171
Schulte, Thomas 91
Searle, John 300
Second World War
    battle preparation 160, 161–2
    bayonet charges 108–12, 114, 121
    cowardice 363, 365–6
    D-Day 43, 120–1, 156, 161–2, 174, 180, 366–7, 420
    disciplinary measures 363, 365–7
    drills 151–4
    forward panic 50–1
    gender integration 383
    individual action 118–21
    literature 5–6
    mass formations 166–7, 169–78, 180–1, 187–98
    US citizen army firing rates 41–7, 48–54, 57
    see also Stalingrad
secularization thesis 426
Sedan, Battle of 103, 167
Seeckt, Hans von 134, 197
Segal, David 28, 86
Segal, Mady 384
Sennett, Richard 442
Sennybridge Training Area, UK 309–10, 333
Severloh, Heinrich 366–7, 420
sexuality: and nationalism 62–3
Sharma, Guatum 95
Shaw, Marvin 29
Sheffield, Gary 101
Sherman, William Tecumseh 1
Shils, Edward 24–5, 36–7, 63, 86, 345
Siebold, Guy 18, 27–8
Simkins, Peter 74

Simonds, Guy 176
simunitions 249, 261, 306, 309, 327
Skaine, Rosemarie 400
Sledge, Eugene 110–11
SMEAC (situation, mission, execution, administration, and command and signals) 281
Smith, Harry 156
Smith, Philip 272
Smoler, Frederic 49, 453 n.30
social capital 438–40
social cohesion 27–8, 31, 34, 35
social reality 300
social solidarity 18–19, 23
 anomic individuality and 72
 Durkheim on 293, 339–41
 ethnic identity and 93
 and groupthink 31
 training and 345–6
Solaro, Erin 402, 499 n.155
solidarity
 mechanical 339–40, 356
 organic 340, 356–7
 see also social solidarity
Somalia 209
Somme, Battle of the 101
Sorin, Katia 410–11
South African army 94–5
Southby-Tailyour, Ewen 276–7
Soviet Union (USSR) 378, 404–5
Special Operations Forces 213, 238–42, 259–60
 Delta Force, USA 238–9
 gender integration 402
 Joint Task Force 2, Canada 274, 330–1, 372
 SAS, UK 238–9
 training of 272
Spiller, Roger 41, 43–4, 49, 60
sport
 changes in 434–8
 national identity and 74–5
sports psychology 330–2, 372
Stacey, Charles Perry 172
Stakhanovites 431–2
Stalingrad 66, 90, 188, 192, 194–6, 237–8

Stanford Training Area, Norfolk, UK 310, *312*
Stanton, Shelby 183, 185
status groups 213–17, 264–5, 338–9
 and morality 340–2
 status honour 215, 338–9
status honour 215, 338–9, 340
Steglich, Martin 122
Steiner, Ivan Dale 35
Stewart, Kinzer 25–6
Stockwell, Hugh 94–5
stormtroops 150, 186–7
 battle preparation 158–9
 First World War 132–3
 Stalingrad 195–6
Stouffer, Sam 67, 91–2
Strachan, Hew 10, 222–3, 233–4, 267, 471 n.65
Struck, Peter 210
Sturm, Hans 122–3
Sullivan, Brian 81
summary justice: and morale 364–5
Summers, Harrison 120
surfing 435–8

tactics, modern 129–63
 battle preparation 157–63
 drills 148–57
 fire and manoeuvre 129–48
 see also mass formations
tactics, techniques and procedures (TTP) 224
Taliban 275, 297, 298–9
Tannenberg, Battle of 106
Tarawa battle 48
task cohesion 34–6, 267, 377–8
Taylor, Charles: secularization thesis 426
Taylor, Eleanor 389, 403–4, 406
Taylor, Frederick Winslow 430
teamwork, *see* task cohesion
terrorism
 9/11 attacks 425
 Munich Olympics 238
Theweleit, Klaus 64
Thompson, Mark 83, 107, 364–5
Tillman, Pat 425

Tolkien, J. R. R. 3–4
training 266–337
  battle preparation 280–99
  and cohesion 222–3, 273
  importance of 266–7, 273–80
  live firing 276
  mass formations 203–4
  new urban training facilities 305–15
  professionalism, failures of 332–7
  recent scholarship 266–73
  shared professional definitions 299–305
  and social solidarity 345–6
  see also close-quarters battle (CQB) techniques
Training and Doctrine Command (TRADOC), USA 427
Training and Employment of Platoons, The (UK manual) 144–5, 147
Trask, David 179–80
trench warfare
  live and let live systems 9–10
  trench assaults 225–6, *141*, 141–2, *142*
  trench raiding 103
Trotta, Thilo von 106
Truppenführung 196–7
Truscott, Lucian 51–2, 161, 344, 365–6
TTP (tactics, techniques and procedures) 224
Tunnell, Harry 280
Turner, Victor 154
Tziner, Ahoron 27

United Kingdom (UK)
  citizen army firing rates in Second World War 55–6
  First World War literature 3–4
United States of America (USA)
  First World War literature 5
  Second World War literature 5–6
Upton, Emory 98
urban combat, *see* close-quarters battle (CQB) techniques
urban training facilities 305–15
  Bonnland, Germany 313, 334, *335*
  C-Can Village, Combat Training Centre, Canada 313, *318*
  Centre d'Entraînement aux Actions en Zone Urbaine (CENZUB) 241, 254, 290–1, *292*, *314*, 315, *315*, *316–17*
  Commando Training Centre, UK 309, *309*, *310*, *311*
  Groningen urban combat facility, Combat Training Centre, Gagetown, Canada *318*
  Infantry Immersion Trainer (IIT), Camp Pendleton 305–6, *306*
  Lydd, UK 310–11
  Marine Corps Air Ground Combat Centre (MCAGCC), 29 Palms 307, *308*, 315
  National Training Centre (NTC), Fort Irwin 307
  Salisbury Plain Training Area, UK 309–10
  Sennybridge Training Area, UK 309–10, 333
  Stanford Training Area, Norfolk, UK 310, *312*
US Army Field Service Regulations (US manual) 147
US forces
  101st Airborne Division 43
  battle preparation 160, 162
  bayonet charges 109–12, 113
  buddy system 67–8
  citizen army firing rates 41–7, 48–54, 57–8
  cohesion in 26–7, 32
  contemporary military doctrine 39
  CQB techniques 230–40, 252–3, 319
  conscription 209
  decline in community 438–40
  deviant cohesion 32
  drills 156
  drug abuse 184, 185–6
  82nd Airborne Division 43
  ethno-political motivation 91–4
  execution for cowardice 363
  fire and movement tactics 147–8, 180
  forward panic 11–12

fraggings 94, 184, 185-6
gender integration 383, 384, 385, 386, 393, 400, 401-3, 405-6, 408-9
  harassment of female soldiers 411
homosexuals, integration of 377
individual action 117-20, 123-6, 234-6
individualism in 26
Korean War 57, 181-2
masculinity 67-8, 70-2
mass formations 178-86
and national mission 424-6
platoons 17
political motivation 79-80
professional solidarity 353
racial integration 376-7
and racism 91-4, 96-7
and religion 426-8
ROC drills 284-8
Training and Doctrine Command (TRADOC) 427
urban training facilities 305-7, *306*, *308*, 315
in Vietnam 32, 57-8
Warrior's Creed 427-8
wives and families, role of 70
see also US Marine Corps
US Marine Corps
  CQB techniques 240, 261, 327-8
  discipline 371-2
  Fallujah, assault on 234-6, 239-40
  gender integration 401-3
  marksmanship 243-4
  phase line battle drill/RIGS process 227-9, *229*
  professionalism, failures of 335-6
USA, *see* United States of America (USA)
USSR (Soviet Union) 378, 404-5
utilitarianism 14-15
Utterson-Kelso, John 152-3

van Creveld, Martin 63, 379-80, 391, 417
Van Riper, P. 403
Vardi, Y. 27
Vaughan, Charles 154
Verney, Gerald 84

Verrières Ridge 174-5
Vietnam War 184, 424
  Australian Task Force 198-9, 209-10
  bayonet charges 113
  beehive rounds 58-9
  black soldiers in 93-4
  collective self-protection 33-4
  deviant cohesion in US troops 32
  drills 156-7
  forward panic 11-12
  fraggings 94
  individual action 123-6
  and masculinity in US forces 70-1
  mass formations 182-6, 198-9
  political motivation and 79-80
  US citizen army firing rates 57-8
Vimy Ridge 103, 161, 171, 172
violence 343
  as collective action 10-12
  ethno-political motivation 93
  and love 348-9
  management of 218
  as measure of masculinity 71
visualization 325-6, 328
Vitai Lampada (Bolt) 75
von der Heydte, Friedrich, Freiherr 191
Von Seeckt, Hans 134, 197

walking fire 168
war
  cultural representation of 1-7
  irrationality of 6
Warr, Nicholas 124, 228, 229
Waterloo, Battle of 149
Watson, Alexander 85-6
Webb, James 409
Weber, Max 213-15, 264-5, 300, 338-9
Webster, David Kenyon 148
Wehrmacht 24-5, 63
  combat performance 24, 36-7
  and machine guns (MG 34 and MG 42), dependence on 192-4
  mass tactics 188-90, 196-8
  massacres of black troops 96
  political motivation 86-91
Weingärtner, Uli 194

Wellington's army 267–72
Wesbrook, Stephen 73–4, 366
Wessely, Simon 86
Westmoreland, William 182, 184–5
Westover, John 43–4
Wette, Wolfram 88–9
Whiteclay Chambers II, John 44
Wigram, Lionel 55–6, 153, 160
Williams, Ivor Gwyn 76
Williams, Philip 368
Wilson, Robert 46
Wilson, Trevor 101
Winslow, Donna 31–2, 393
Winters, Richard 188, 205
women

wives and families, status of 64, 66, 69–70
see also female soldiers
Women's Armed Service Integration Act, USA 383
Women's Royal Army Corps, UK 385
Women's Royal Auxiliary Corps, UK 383
Wong, Leonard 26–7
Wray, Waverly 120

X Generation 439

York, Alvin 117–18

Zander, Alvin 29

The manufacturer's authorised representative in the EU for product safety is Oxford University Press España S.A. of El Parque Empresarial San Fernando de Henares, Avenida de Castilla, 2 - 28830 Madrid (www.oup.es/en or product.safety@oup.com). OUP España S.A. also acts as importer into Spain of products made by the manufacturer.
Printed and bound by CPI Group (UK) Ltd, Croydon, CR0 4YY

20/03/2026

02075336-0014